Birkhäuser

Michael Falk · Jürg Hüsler · Rolf-Dieter Reiss

Laws of Small Numbers: Extremes and Rare Events

Third, revised and extended edition

 Birkhäuser

Michael Falk
Institute of Mathematics
University of Würzburg
Am Hubland
97074 Würzburg
Germany
e-mail: falk@mathematik.uni-wuerzburg.de

Jürg Hüsler
Department of Mathematical Statistics and
Actuarial Science
University of Berne
Sidlerstrasse 5
3012 Bern
Switzerland
e-mail: juerg.huesler@stat.unibe.ch

Rolf-Dieter Reiss
Department of Mathematics
University of Siegen
Walter Flex Str. 3
57068 Siegen
Germany
e-mail: reiss@stat.math.uni-siegen.de

ISBN 978-3-0348-0008-2   e-ISBN 978-3-0348-0009-9
DOI 10.1007/978-3-0348-0009-9

Library of Congress Control Number: 2010934383

Mathematics Subject Classification (2010): 60-02, 60G70, 62-02, 62G32, 60G55, 60G15, 60G10, 62H05, 62P99

1st and 2nd edition: © Birkhäuser Verlag Basel – Boston – Berlin 1994, 2004

Cover design: deblik, Berlin

Printed on acid-free paper

Springer Basel AG is part of Springer Science+Business Media

www.birkhauser-science.com

Preface to the Third Edition

The main focus of extreme value theory has been undergoing a dramatic change. Instead of concentrating on maxima of observations, large observations are now in the focus, defined as exceedances over high thresholds. Since the pioneering papers by Balkema and de Haan (1974) and Pickands (1975) it is well known that exceedances over high thresholds can reasonably be modeled only by a generalized Pareto distribution. But only in recent years has this fact been widely spread outside the academic world as well.

Just as multivariate extreme value theory was developed roughly thirty years after its univariate basis was established, we presently see the theory of multivariate exceedances and, thus, the theory of multivariate generalized Pareto distributions under extensive investigation.

For that reason, one emphasis of the third edition of the present book is given to multivariate generalized Pareto distributions, their representations, properties such as their peaks-over-threshold stability, simulation, testing and estimation. Concerning this matter, the third edition in particular benefits from the recent PhD-theses of René Michel and Daniel Hofmann, who both made substantial contributions to the theory of multivariate generalized Pareto distributions, mainly concentrated in Section 4.4, Chapter 5 and 6. We are in particular grateful to René Michel, who coauthored these parts of the present edition with high diligence.

Exceedances of stochastic processes and random fields have been further considered in recent years, since the publication of the second edition. These new developments are discussed in additional sections or paragraphs. For instance, we deal with crossings of random processes in a random environment or with random variances, and crossings or level sets of smooth processes. Also maxima of a multi-fractional process, a recently introduced new class of random processes, are investigated.

The following contributions of co-authors are also gratefully acknowledged:

- Isabel Fraga Alves, Claudia Neves and Ulf Cormann: the modeling and testing of super-heavy tails in conjunction with log-Pareto distributions and a class of slowly-varying tails in Section 2.7

- Melanie Frick: testing against residual dependence in Section 6.5.

We are thankful to Holger Drees for pointing out a misarrangement of the text in the first chapter and to Laurens de Haan for correcting the erroneously assigned von Mises condition in the second chapter of the second edition.

Würzburg   Michael Falk
Bern   Jürg Hüsler
Siegen   Rolf-Dieter Reiss

Preface to the Second Edition

Since the publication of the first edition of this seminar book in 1994, the theory and applications of extremes and rare events have received an enormous, increasing interest. This is primarily due to its practical relevance which is well recognized in different fields such as insurance, finance, engineering, environmental sciences and hydrology. The application of extreme value theory in hydrology has a particularly long and fruitful tradition. Meanwhile there are various specialized books available which focus on selected applications.

Different to that, the intention of the present book is to give a mathematically oriented development of the theory of rare events, underlying all applications. In the second edition we strengthen this characteristic of the book. One of the consequences is that the section on the statistical software Xtremes and the pertaining CD are omitted, this software is updated and well documented in [389]. Various new results, which are scattered in the statistical literature, are incorporated in the new edition on about 130 new pages.

The new sections of this edition are written in such a way that the book is again accessible to graduate students and researchers with basic knowledge in probability theory and, partly, in point processes and Gaussian processes. The required statistical prerequisites are minimal.

The book is now divided into three parts, namely,

Part I: The IID Case: Functional Laws of Small Numbers;

Part II: The IID Case: Multivariate Extremes;

Part III: Non-IID Observations.

Part II, which is added to the second edition, discusses recent developments in multivariate extreme value theory based on the Pickands representation of extreme value distributions. A detailed comparison to other representations of such distributions is included. Notable is particularly a new spectral decomposition of multivariate distributions in univariate ones which makes multivariate questions more accessible in theory and practice.

One of the most innovative and fruitful topics during the last decades was the introduction of generalized Pareto distributions in the univariate extreme value theory (by J. Pickands and, concerning theoretical aspects, by A.A. Balkema and L. de Haan). Such a statistical modelling of extremes is now systematically developed in the multivariate framework. It is verified that generalized Pareto distributions play again an exceptional role. This, in conjunction with the aforementioned spectral decomposition, is a relatively novel but rapidly increasing field. Other new sections concern the power normalization of extremes and a LAN theory for thinned empirical processes related to rare events.

The development of rare events of non-iid observations, as outlined in Part III, has seen many new approaches, e.g. in the context of risk analysis, of telecommunication modelling or of finance investigations during the last ten years. Very often these problems can be seen as boundary crossing probabilities. Some of these new aspects of boundary crossing probabilities are dealt with in this edition. Also a subsection on the recent simulation investigations of Pickands constants, which were unknown up to a few values, is added. Another new section deals in detail with the relations between the maxima of a continuous process and the maxima of the process observed at some discrete time points only. This relates the theoretical results to results which are applied and needed in practice.

The present book has benefitted a lot from stimulating discussions and suggestions. We are in particular grateful to Sreenivasan Ravi for contributing the section on power normalization of extremes, to René Michel, who helped with extensive simulations of multivariate extremes, and to Michael Thomas for the administration of our version control system (cvs) providing us with the technical facilities to write this book online. We thank Johan Segers for pointing out an error in one of the earlier definitions of multivariate generalized Pareto distributions in dimensions higher than two, which, on the other hand, actually links them to the field of quasi-copulas.

We would also like to thank the German Mathematical Society (DMV) for the opportunity to organize the symposium *Laws of small numbers: Extremes and rare events* during its annual meeting 2003 at the University of Rostock, Germany. This turned out to be quite a stimulating meeting during the writing of the final drafts of this book. Last, but not least we are grateful to Thomas Hempfling, Editor, Mathematics Birkhauser Publishing, for his continuous support and patience during the preparation of the second edition.

Würzburg   Michael Falk
Bern   Jürg Hüsler
Siegen   Rolf-Dieter Reiss

Preface to the First Edition

In the first part of this book we will develop a theory of *rare events* for which a handy name is *functional laws of small numbers*. Whenever one is concerned with rare events, events with a small probability of occurrence, the *Poisson* distribution shows up in a natural way.

So the basic idea is simple, but its applications are nevertheless far-reaching and require a certain mathematical machinery. The related book by David Aldous entitled "Probability Approximations via the Poisson Clumping Heuristic" demonstrates this need in an impressive way. Yet this book focuses on examples, ranging over many fields of probability theory, and does not try to constitute a complete account of any field.

We will try to take another approach by developing a general theory first and then applying this theory to specific subfields. In prose: If we are interested only in those random elements among independent replicates of a random element Z, which fall into a given subset A of the sample space, a reasonable way to describe this *random* sample (with binomial sample size) is via the concept of *truncated empirical point processes*. If the probability for Z falling into A is small, then the Poisson approximation entails that we can approximate the truncated empirical point process by a Poisson point process, with the sample size now being a Poisson random variable. This is what we will call *first step Poisson process approximation*.

Often, those random elements falling into A follow closely an ideal or limiting distribution; replacing their actual distribution by this ideal one, we generate a *second step Poisson process approximation* to the initial truncated empirical process.

Within certain error bounds, we can therefore handle those observations among the original sample, which fall into the set A, like ideal observations, whose stochastic behavior depends solely upon a few (unknown) parameters. This approach permits the application of standard methods to statistical questions concerning the original and typically non-parametric sample.

If the subset A is located in the center of the distribution of Z, then *regression analysis* turns out to be within the scope of laws of small numbers. If the subset A is however located at the border, then *extreme value theory* is typically covered by our theory.

These specifications will lead to characteristic results in each case, and we will try in the following to convey the beauty of the laws of small numbers and several of their specific applications to the reader. In order to keep a more informal character, the proofs of several results are omitted, but references to detailed ones are given.

As the Hellinger distance provides a more accurate bound for the approximation of product measures in terms of their margins, as does the Kolmogorov-Smirnov or the variational distance, we will focus in the first part of this book on the formulation of laws of small numbers within the Hellinger distance.

The second part of the book concentrates on the theory of extremes and other rare events of non-iid random sequences. The rare events related to stationary sequences and independent sequences are considered as special cases of this general setup. The theory is presented in terms of extremes of random sequences as well as general triangular arrays of rare events.

Basic to the general theory is the restriction of the long range dependence. This enables the approximation of the point process of rare events by a Poisson process. The exact nature of this process depends also on the local behavior of the sequence of rare events. The local dependence among rare events can lead in the non-iid case to clustering, which is described by the compounding distribution of the Poisson process. Since non-stationarity causes the point process to be inhomogeneous, the occurrence of rare events is in general approximated by an extended compound Poisson process.

Part I of this book is organized as follows: In Chapter 1 the general idea of functional laws of small numbers is made rigorous. Chapter 2 provides basic elements from univariate extreme value theory, which enable particularly the investigation of the peaks over threshold method as an example of a functional law of small numbers. In Chapter 3 we demonstrate how our approach can be applied to regression analysis or, generally, to conditional problems. Chapter 4 contains basic results from multivariate extreme value theory including their extension to the continuous time setting. The multivariate peaks over threshold approach is studied in Chapter 5. Chapter 6 provides some elements of exploratory data analysis for univariate extremes.

Part II considers non-iid random sequences and rare events. Chapter 7 introduces the basic ideas to deal with the extremes and rare events in this case. These ideas are made rigorous in Chapter 8 presenting the general theory of extremes which is applied to the special cases of stationary and independent sequences. The extremes of non-stationary Gaussian processes are investigated in Chapter 9. Results for locally stationary Gaussian processes are applied to empirical characteristic functions. The theory of general triangular arrays of rare events is presented in Chapter 10, where we also treat general rare events of random sequences and the characterization of the point process of exceedances. This general approach provides a neat unification of the theory of extremes. Its application to multivariate non-stationary sequences is thus rather straightforward. Finally, Chapter 11 contains the statistical analysis of non-stationary ecological time series.

This book comes with the statistical software system X$_{\text{TR}}$E$_{\text{M}}$E$_{\text{S}}$, version 1.2, produced by Sylvia Haßmann, Rolf-Dieter Reiss and Michael Thomas. The disk runs on IBM-compatible personal computers under MS-DOS or compatible operating systems. We refer to the appendix (co-authored by Sylvia Haßmann and Michael Thomas) for a user's guide to X$_{\text{TR}}$E$_{\text{M}}$E$_{\text{S}}$. This software project was partially supported by the Deutsche Forschungsgemeinschaft by a grant.

This edition is based on lectures given at the DMV Seminar on "Laws of small numbers: Extremes and rare events", held at the Katholische Universität Eichstätt from October 20-27, 1991.

We are grateful to the Mathematisches Forschungsinstitut Oberwolfach and its director, Professor Dr. Martin Barner, and the Deutsche Mathematiker Vereinigung for their financial and organizational support. We are indebted to the participants for their engaged discussions and contributions, and to Birkhäuser Verlag for giving us the opportunity to publish these seminar notes.

It is a pleasure to thank Nese Catkan, Hans-Ueli Bräker, Frank Marohn and Sylvia Haßmann for their continuing support and Helma Höfter for her excellent typesetting of the manuscript using L^{A}T$_{\text{E}}$X, so that we could concentrate on the project.

Eichstätt   Michael Falk
Bern   Jürg Hüsler
Siegen   Rolf-Dieter Reiss

Contents

Part I

The IID Case:
Functional Laws
of Small Numbers

Chapter 1

Functional Laws
of Small Numbers

We will develop in the following a particular extension of the well-known Poisson approximation of binomial distributions with a small hitting probability, which is known as *the law of small numbers*. This extension, which one might call *functional laws of small numbers*, links such seemingly different topics like non-parametric regression analysis and extreme value theory.

1.1   Introduction

The economist Ladislaus von Bortkiewicz, born 1868 in St. Petersburg (that Russian town, whose name was changed several times during this century: 1703-1914 St. Petersburg, 1914-1924 Petrograd, 1924-1991 Leningrad, since 1991 St. Petersburg again), Professor in Berlin from 1901 until his death in 1931, was presumably one of the first to recognize the practical importance of the Poisson approximation of binomial distributions. His book *The law of small numbers* [51] popularized the Poisson distribution although - or perhaps because - his examples on the number of children suicides per year in certain parts of the population or of accidental deaths per year in certain professions are a bit macabre. His most popular example is on the number of Prussian cavalrymen killed by friendly horse-kick: The following table summarizes the frequencies n_k of the number k of cavalrymen killed in a regiment by horse-kicks within one year in ten particular regiments of the Royal Prussian army over a twenty years period

number of victims	k	0	1	2	3	4	≥ 5
frequency	n_k	109	65	22	3	1	0

.

M. Falk et al., *Laws of Small Numbers: Extremes and Rare Events*, 3rd ed.,
DOI 10.1007/978-3-0348-0009-9_1, © Springer Basel AG 2011

As observed by von Bortkiewicz [51], a Poisson distribution P_λ with parameter $\lambda = .61$ fits these data quite well:

number	k	0	1	2	3	4	≥ 5
relative frequency	$n_k/200$	.545	.325	.11	.015	.005	0
theoretical probability	$P_{.61}(\{k\})$	.543	.331	.101	.021	.003	.001.

THE LAW OF SMALL NUMBERS

If we model the event that an individual trooper is killed by a horse-kick within one year by a binary random variable (rv) R that is, $R \in \{0,1\}$ with $R = 1$ representing accidental death, then the total number $K(n)$ of victims in a regiment of size n is binomial distributed (supposing independence of individual lifetimes):

$$P(K(n) = k) = \binom{n}{k} p^k(1-p)^{n-k}$$

$$=: B(n,p)(\{k\}), \qquad k = 0, 1, \ldots, n,$$

where

$$p = P(R = 1)$$

is the mortality rate. Usually, p was small and n reasonably large, in which case $R = 1$ became a *rare event* and the binomial distribution $B(n,p)$ can be approximated within a reasonable remainder term by the Poisson distribution

$$P_\lambda(\{k\}) := e^{-\lambda}\frac{\lambda^k}{k!}, \qquad k = 0, 1 \ldots$$

with $\lambda = np$. This becomes obvious by writing, for $k \in \{1, \ldots, n\}$, $n \in \mathbb{N}$,

$$B(n,p)(\{k\}) = \binom{n}{k} p^k(1-p)^{n-k}$$

$$= \frac{n!}{k!(n-k)!} \frac{(np)^k}{n^k}\left(1 - \frac{np}{n}\right)^n \frac{1}{(1-p)^k}$$

$$= \left(\prod_{i\leq k}\left(1 - \frac{i-1}{n}\right)\right) \frac{1}{(1-\frac{\lambda}{n})^k} \frac{\lambda^k}{k!}\left(1 - \frac{\lambda}{n}\right)^n$$

$$= P_\lambda(\{k\})\left(\frac{(1-\frac{\lambda}{n})^n}{e^{-\lambda}} \frac{1}{(1-\frac{\lambda}{n})^k}\prod_{i\leq k}\left(1 - \frac{i-1}{n}\right)\right).$$

For this reason, the Poisson distribution is sometimes called the *distribution of rare events*, and the approximation

$$B(n,p) \sim P_{np} \tag{1.1}$$

is a *law of small numbers* (following von Bortkiewicz [51]). The quantification and application of the preceding approximation is still a hot topic, as the increasing number of publications shows. References are the books by Aldous [4] and Barbour et al. [29], and the survey article by Arratia et al. [17]. For remarks about the early history of Poisson distributions we refer to Haight [195].

POINT PROCESS APPROXIMATIONS

Consider a binomial $B(n,p)$-distributed rv $K(n)$ and a Poisson P_{np}-distributed rv $\tau(n)$. Then (1.1) can be rewritten as

$$K(n) \sim_D \tau(n), \tag{1.2}$$

where $\sim_D$ denotes approximate equality of the distributions $\mathcal{L}(K(n)) = B(n,p)$ and $\mathcal{L}(\tau(n)) = P_{np}$ of $K(n)$ and $\tau(n)$, respectively.

Let now $V_1, V_2, \ldots$ be a sequence of independent copies of a random element V with values in a sample space S equipped with a σ-field $\mathcal{B}$. We assume that the rv $V_1, V_2, \ldots$ are also independent of $K(n)$ and $\tau(n)$.

The approximation (1.2) suggests its following extension

$$(V_1, \ldots, V_{K(n)}) \sim_D (V_1, \ldots, V_{\tau(n)}). \tag{1.3}$$

If we choose $V \equiv 1$, the left- and right-hand sides of (1.3) are sequences $1, 1, \ldots, 1$ of ones, the left one being of length $K(n)$, the right one of length $\tau(n)$, and (1.3) is obviously an extension of (1.2). As the approximation of $K(n)$ by $\tau(n)$ is known as a law of small numbers, one might call the approximation (1.3) a *functional* law of small numbers.

But now we face the mathematical problem to make the approximation of $\mathcal{L}(V_1, \ldots, V_{K(n)})$ by $\mathcal{L}(V_1, \ldots, V_{\tau(n)})$ meaningful, as $(V_1, \ldots, V_{K(n)})$ and $(V_1, \ldots, V_{\tau(n)})$ are random vectors (rv) of usually different (random) length. An appealing way to overcome this dimensionality problem by dropping the (random) length is offered by the concept of point processes.

We identify any point $x \in S$ with the pertaining Dirac-measure

$$\varepsilon_x(B) = \begin{cases} 1 & x \in B \\ & \text{if} \\ 0 & x \notin B \end{cases}, \qquad B \in \mathcal{B}.$$

Thus we identify the random element V with the random measure ε_V, and the rv $(V_1, \ldots, V_{K(n)})$ and $(V_1, \ldots, V_{\tau(n)})$ turn into the random finite measures

$$N_n(B) := \sum_{i \leq K(n)} \varepsilon_{V_i}(B) \quad \text{and} \quad N_n^*(B) := \sum_{i \leq \tau(n)} \varepsilon_{V_i}(B), \qquad B \in \mathcal{B}.$$

We can equip the set $\mathbb{M}$ of finite point measures on $(S, \mathcal{B})$ with the smallest σ-field $\mathcal{M}$ such that for any $B \in \mathcal{B}$ the *projections* $\mathbb{M} \ni \mu \to \mu(B)$ are measurable. Then

$N_n(\cdot)$, $N_n^*(\cdot)$ are (measurable) random elements in $(\mathbb{M}, \mathcal{M})$ and as such they are called *point processes*. Since the counting variable $K(n)$ is a binomial rv and $\tau(n)$ is a Poisson one, N_n is called a *binomial process* and N_n^* a *Poisson process*.

A more precise formulation of (1.3) is then the

first-order Poisson process approximation

$$N_n(\cdot) = \sum_{i \leq K(n)} \varepsilon_{V_i}(\cdot) \sim_D \sum_{i \leq \tau(n)} \varepsilon_{V_i}(\cdot) = N_n^*(\cdot). \qquad (1.4)$$

Recall that with $V \equiv 1$, the preceding approximation yields in turn the Poisson approximation of a binomial distribution (1.1), (1.2)

$$N_n = K(n)\varepsilon_1 \sim_D N_n^* = \tau(n)\varepsilon_1.$$

Suppose now that the distribution of the random element V is close to that of some random element W, which suggests the approximation

$$(V_1, \ldots, V_{\tau(n)}) \sim_D (W_1, \ldots, W_{\tau(n)}), \qquad (1.5)$$

where $W_1, W_2 \ldots$ are independent copies of W and also independent of $\tau(n)$ and $K(n)$. Speaking in terms of point processes, we obtain the approximation

$$N_n^* = \sum_{i \leq \tau(n)} \varepsilon_{V_i} \sim_D N_n^{**} := \sum_{i \leq \tau(n)} \varepsilon_{W_i}. \qquad (1.6)$$

Our particular extension of the Poisson approximation of a binomial distribution (1.1), (1.2) then becomes the

second-order Poisson process approximation

$$N_n = \sum_{i \leq K(n)} \varepsilon_{V_i} \sim_D N_n^{**} = \sum_{i \leq \tau(n)} \varepsilon_{W_i}. \qquad (1.7)$$

There is obviously one further approximation left, namely

$$(V_1, \ldots, V_{K(n)}) \sim_D (W_1, \ldots, W_{K(n)}),$$

where we do *not* replace $K(n)$ by $\tau(n)$, but replace V_i by W_i. There is no Poisson approximation of binomial distributions involved in this approximation, but it is nevertheless typically a law of small numbers (see Section 1.3), valid only if p is small. Due to a particular application in extreme value theory, one might call this approximation the

peaks-over-threshold method (POT)

$$N_n = \sum_{i \leq K(n)} \varepsilon_{V_i} \sim_D M_n := \sum_{i \leq K(n)} \varepsilon_{W_i}. \tag{1.8}$$

By choosing particular sequences $V_1, V_2 \ldots$ and $W_1, W_2 \ldots$ of random elements within these extensions of (1.1), (1.2) it turns out, for example, that apparently completely different topics such as *regression analysis* and *extreme value theory* are within the scope of these *functional laws of small numbers* (see Section 1.3).

1.2 Bounds for the Functional Laws of Small Numbers

The error of the preceding approximations (1.4), (1.7) and (1.8) will be measured with respect to the *Hellinger* distance. The Hellinger distance (between the distributions) of random elements X and Y with values in some measurable space $(S, \mathcal{B})$ is defined by

$$H(X, Y) := \left(\int (f^{1/2} - g^{1/2})^2 \, d\mu \right)^{1/2},$$

where f, g are densities of X and Y with respect to some dominating measure μ.

The use of the Hellinger distance in our particular framework is motivated by the well-known fact that for *vectors* of independent replicates $\boldsymbol{X} = (X_1, \ldots, X_k)$ and $\boldsymbol{Y} = (Y_1, \ldots, Y_k)$ of X and Y we have the bound

$$H(\boldsymbol{X}, \boldsymbol{Y}) \leq k^{1/2} H(X, Y),$$

whereas for the variational distance $d(X, Y) := \sup_{B \in \mathcal{B}} |P(X \in B) - P(Y \in B)|$, we only have the bound

$$d(\boldsymbol{X}, \boldsymbol{Y}) \leq k \, d(X, Y).$$

Together with the fact that the variational distance is in general bounded by the Hellinger distance, we obtain therefore the bound

$$d(\boldsymbol{X}, \boldsymbol{Y}) \leq k^{1/2} H(X, Y).$$

If $d(X, Y)$ and $H(X, Y)$ are of the same order, which is typically the case, the Hellinger distance approach provides a more accurate bound for the comparison of *sequences* of iid observations than the variational distance that is, $k^{1/2}$ compared with k, roughly. Observe that our particular extension (1.4)-(1.7) of the Poisson approximation of binomial distributions actually involves the comparison of *sequences* of random elements. (For the technical background see, for example, Section 3.3 of Reiss [385] and Section 1.3 of Reiss [387].)

MARKOV KERNELS

Note that the distributions $\mathcal{L}(N_n), \mathcal{L}(N_n^*)$ of the processes $N_n = \sum_{i \leq K(n)} \varepsilon_{V_i}$ and $N_n^* = \sum_{i \leq \tau(n)} \varepsilon_{V_i}$ on $(\mathbb{M}, \mathcal{M})$ can be represented by means of the *Markov kernel* $Q(\cdot \mid \cdot) : \mathcal{M} \times \{0, 1, 2, \ldots, \} \to [0, 1]$, defined by

$$Q(M \mid m) := \mathcal{L}\left(\sum_{i \leq m} \varepsilon_{V_i}\right)(B) = P\left(\sum_{i \leq m} \varepsilon_{V_i} \in B\right), \qquad m = 0, 1, 2, \ldots, \quad M \in \mathcal{M},$$

by conditioning on $K(n)$ and $\tau(n)$:

$$\mathcal{L}(N_n)(\cdot) = P(N_n \in \cdot) = \int P\left(\sum_{i \leq m} \varepsilon_{V_i} \in \cdot\right) \mathcal{L}(K(n))\,(dm)$$

$$= \int Q(\cdot \mid m)\mathcal{L}(K(n)) = E(Q(\cdot \mid K(n))) =: Q\mathcal{L}(K(n))(\cdot)$$

and

$$\mathcal{L}(N_n^*)(\cdot) = P(N_n^* \in \cdot) = \int P\left(\sum_{i \leq m} \varepsilon_{V_i} \in \cdot\right) \mathcal{L}(\tau(n))\,(dm)$$

$$= \int Q(\cdot \mid m)\mathcal{L}(\tau(n))\,(dm) = E(Q(\cdot \mid \tau(n))) =: Q\mathcal{L}(\tau(n))(\cdot).$$

In case of $m = 0$, interpret $\sum_{i \leq 0} \varepsilon_{V_i}$ as the *null-measure* on $\mathcal{B}$ that is, $\sum_{i \leq 0} \varepsilon_{V_i}(B) = 0$, $B \in \mathcal{B}$.

THE MONOTONICITY THEOREM

It is intuitively clear from the preceding representation that the error in the first-order Poisson process approximation (1.4) is determined by the error of the approximation of $K(n)$ by $\tau(n)$.

Lemma 1.2.1 (Monotonicity theorem). *We have*

(i) $d(N_n, N_n^*) = d(Q\mathcal{L}(K(n)), Q\mathcal{L}(\tau(n))) \leq d(K(n), \tau(n))$,

(ii) $H(N_n, N_n^*) = H(Q\mathcal{L}(K(n)), Q\mathcal{L}(\tau(n))) \leq H(K(n), \tau(n))$.

While part (i) of the preceding result is obvious, the second bound is a simple consequence of the *monotonicity theorem for f-divergences* (see, for example Theorem 1.24 in Liese and Vajda [313] or, for a direct proof, Lemma 1.4.2 in Reiss [387]).

By choosing $V_i \equiv 1$ that is, $N_n = K(n)\varepsilon_1$ and $N_n^* = \tau(n)\varepsilon_1$ it is only an exercise to show that we can achieve equality in Lemma 1.2.1.

Lemma 1.2.1 entails that in order to establish bounds for the first-order Poisson process approximation (1.4), we can benefit from the vast literature on bounds for the distance between $K(n)$ and $\tau(n)$ (an up-to-date reference is the book by Barbour et al. [29]). The following bound is a consequence of Theorem 1 in Barbour and Hall [27] and of Lemma 3 in Falk and Reiss [148]. See also the section on the Stein-Chen method below that is, formulae (1.14)-(1.17).

Lemma 1.2.2. *We have, for $0 \le p \le 1$ and $n \in \mathbb{N}$,*

(i) $d(K(n), \tau(n)) \le p$,

(ii) $H(K(n), \tau(n)) \le 3^{1/2}p$.

Bounds for the first-order Poisson process approximation (1.4) are now immediate from Lemma 1.2.1 and 1.2.2:

$$d(N_n, N_n^*) \le p, \ H(N_n, N_n^*) \le 3^{1/2}p. \tag{1.9}$$

Notice that these bounds are valid under no further restrictions on $V_1, V_2, \ldots,$ $K(n)$ and $\tau(n)$; moreover, they do not depend on n but only on p.

THE CONVEXITY THEOREM

To establish bounds for the second-order Poisson process approximation (1.7) and the POT approach (1.8), observe that the distributions $\mathcal{L}(N_n^*), \mathcal{L}(N_n^{**}), \mathcal{L}(M_n)$ of the processes $N_n^* = \sum_{i \le \tau(n)} \varepsilon_{V_i}$, $N_n^{**} = \sum_{i \le \tau(n)} \varepsilon_{W_i}$, and $M_n = \sum_{i \le K(n)} \varepsilon_{W_i}$ can be represented by means of $\tau(n)$ and the *two* Markov kernels

$$Q_V(\cdot \mid m) := \mathcal{L}\Big(\sum_{i \le m} \varepsilon_{V_i}\Big)(\cdot), \qquad m = 0, 1, 2, \ldots,$$

$$Q_W(\cdot \mid m) := \mathcal{L}\Big(\sum_{i \le m} \varepsilon_{W_i}\Big)(\cdot), \qquad m = 0, 1, 2, \ldots,$$

as

$$\mathcal{L}(N_n^*)(\cdot) = \int Q_V(\cdot \mid m)\mathcal{L}(\tau(n))\,(dm) = Q_V\mathcal{L}(\tau(n))(\cdot),$$

$$\mathcal{L}(N_n^{**})(\cdot) = \int Q_W(\cdot \mid m)\mathcal{L}(\tau(n))\,(dm) = Q_W\mathcal{L}(\tau(n))(\cdot)$$

and

$$\mathcal{L}(M_n)(\cdot) = \int Q_W(\cdot \mid m)\mathcal{L}(K(n))\,(dm) = Q_W\mathcal{L}(K(n))(\cdot).$$

Lemma 1.2.3 (Convexity theorem). *We have*

(i) $d(N_n^*, N_n^{**}) \le \int d(Q_V(\cdot \mid m), Q_W(\cdot \mid m))\,\mathcal{L}(\tau(n))\,(dm),$

(ii) $H(N_n^*, N_n^{**}) \leq \left(\int H^2(Q_V(\cdot \mid m),\ Q_W(\cdot \mid m))\, \mathcal{L}(\tau(n))\, (dm) \right)^{1/2}$,

(iii) $d(N_n, M_n) \leq \int d(Q_V(\cdot \mid m),\ Q_W(\cdot \mid m))\, \mathcal{L}(K(n))\, (dm)$,

(iv) $H(N_n, M_n) \leq \left(\int H^2(Q_V(\cdot \mid m),\ Q_W(\cdot \mid m))\, \mathcal{L}(K(n))\, (dm) \right)^{1/2}$.

     While the bound for the variational distance is obvious, the bound for the Hellinger distance is an application of the *convexity theorem for f-divergences* (see formula (1.53) in Liese and Vajda [313]; a direct proof is given in Lemma 3.1.3 in Reiss [387]).

WHY THE HELLINGER DISTANCE?

Observe now that $Q_V(\cdot \mid m) = \mathcal{L}(\sum_{i \leq m} \varepsilon_{V_i})(\cdot)$ and $Q_W(\cdot \mid m) = \mathcal{L}(\sum_{i \leq m} \varepsilon_{W_i})(\cdot)$ can be viewed as the distribution of the same functional $T : S^m \to \mathbb{M}$, evaluated at the rv $(V_1, \ldots, V_m)$ and $(W_1, \ldots, W_m)$,

$$Q_V(\cdot \mid m) = \mathcal{L}(T(V_1, \ldots, V_m))(\cdot),\ Q_W(\cdot \mid m) = \mathcal{L}(T(W_1, \ldots, W_m))(\cdot),$$

with

$$T(x_1, \ldots, x_m) := \sum_{i \leq m} \varepsilon_{x_i}$$

for $x_1, \ldots, x_m \in S$.

     The following consequence of Lemma 1.2.3 is therefore immediate from the monotonicity theorem for the Hellinger distance (cf. Lemma 1.4.2 in Reiss [387]).

Corollary 1.2.4. *We have*

(i) $d(N_n^*, N_n^{**}) \leq \int d((V_1, \ldots, V_m), (W_1, \ldots, W_m))\, \mathcal{L}(\tau(n))\, (dm)$

$$\leq d(V, W) \int m\, \mathcal{L}(\tau(n))\, (dm) = d(V, W)\, E(\tau(n)),$$

(ii) $H(N_n^*, N_n^{**}) \leq \left(\int H^2((V_1, \ldots, V_m), (W_1, \ldots, W_m))\, \mathcal{L}(\tau(n))\, (dm) \right)^{1/2}$

$$\leq H(V, W) \left(\int m\, \mathcal{L}(\tau(n)) \right)^{1/2} = H(V, W)\, E(\tau(n))^{1/2}.$$

We obtain by the same arguments

(iii) $d(N_n, M_n) \leq d(V, W)\, E(K(n))$,

(iv) $H(N_n, M_n) \leq H(V, W)\, E(K(n))^{1/2}$.

If $d(V, W)$ and $H(V, W)$ are approximately equal, the use of the Hellinger distance reduces the bound between the distributions of N_n^* and N_n^{**} from $O(E(\tau(n)))$ to $O(E(\tau(n))^{1/2})$, which is actually an improvement only if $E(\tau(n)) > 1$ that is, if our set of data consists on the average of more than one observation of V and W. But this is obviously a minimum condition. The same arguments apply to the approximation of N_n by M_n.

Combining 1.2.1-1.2.4 and the fact that in general the variational distance is bounded by the Hellinger distance, we obtain the following bound for the second-order Poisson process approximation.

Theorem 1.2.5. *With* $\mathcal{L}(K(n)) = B(n, p)$ *and* $\mathcal{L}(\tau(n)) = P_{np}$ *we have*

$$
\begin{aligned}
H(N_n, N_n^{**}) &\leq H(K(n), \tau(n)) + H(N_n^*, N_n^{**}) \\
&\leq H(K(n), \tau(n)) + H(V, W)\,(E(\tau(n)))^{1/2} \\
&\leq 3^{1/2}p + H(V, W)(np)^{1/2}.
\end{aligned}
$$

Note that the first two inequalities in the preceding result are true for arbitrary binomial rv $K(n)$ and Poisson rv $\tau(n)$, being independent of the sequences $V_1, V_2, \ldots$ and $W_1, W_2, \ldots$ Only the final inequality arises from the particular choice $\mathcal{L}(K(n)) = B(n, p)$, $\mathcal{L}(\tau(n)) = P_{np}$.

Theorem 1.2.5 describes in which way the accuracy of the functional law of small numbers (1.7) is determined by the distance between $K(n)$ and $\tau(n)$ that is, by the values of p and np.

SPECIFYING $H(V, W)$

Suppose now that the random elements V and W have densities g, f with respect to some dominating measure μ, such that the representation

$$
g^{1/2} = f^{1/2}(1 + h)
$$

holds with some error function h. Then we obtain

$$
\begin{aligned}
H(V, W) &= \left(\int (g^{1/2} - f^{1/2})^2 \, d\mu \right)^{1/2} \\
&= \left(\int h^2 \, d\mathcal{L}(W) \right)^{1/2} = E(h^2(W))^{1/2}.
\end{aligned} \tag{1.10}
$$

The preceding considerations can then be summarized by the bounds

$$
H(N_n, N_n^*) \leq 3^{1/2}p \tag{1.11}
$$

for the first-order Poisson process approximation,

$$
H(N_n, N_n^{**}) \leq 3^{1/2}p + E(h^2(W))^{1/2}(np)^{1/2} \tag{1.12}
$$

for the second-order Poisson process approximation and

$$H(N_n, M_n) \leq E(h^2(W))^{1/2}(np)^{1/2} \tag{1.13}$$

for the POT approach if $\mathcal{L}(K(n)) = B(n,p)$, $\mathcal{L}(\tau(n)) = P_{np}$.

These general error bounds lead to specific error bounds in regression analysis and extreme value theory by computing $E(h^2(W))$ for particular W and h (see Theorem 2.3.2, Theorem 3.1.3, and Corollary 3.1.6).

THE STEIN-CHEN METHOD

While we focus on the Hellinger distance in the first part of this book for the reasons given above, the Stein-Chen method offers a powerful tool to measure laws of small numbers in variational distance; in particular it is very useful for dependent observations. In 1970, Stein developed his very special idea for the normal approximation in case of dependent rv, Chen [64] worked it out for the Poisson approximation. For the sake of completeness, we provide its basic idea, essentially taken from the book by Barbour et al. [29], to which we refer as an ocean of refinements, extensions to dependent observations and applications.

Let A be an arbitrary subset of $\{0, 1, 2, \ldots\}$ and choose $\lambda > 0$; then we can find a bounded function $g = g_{A,\lambda} : \{0, 1, 2, \ldots\} \to \mathbb{R}$ which satisfies the basic recursion

$$\lambda g(j+1) - jg(j) = \varepsilon_j(A) - P_\lambda(A), \qquad j \geq 0; \tag{1.14}$$

see below for the definition of this function g. As a consequence we obtain for an arbitrary rv Z with values in $\{0, 1, 2, \ldots\}$,

$$
\begin{aligned}
P(Z \in A) - P_\lambda(A) &= E(\varepsilon_Z(A) - P_\lambda(A)) \\
&= E(\lambda g(Z+1) - Zg(Z)). \tag{1.15}
\end{aligned}
$$

The preceding equality is the crucial tool for the derivation of bounds for the Poisson approximation in variational distance. Let $\xi_1, \ldots, \xi_n$ be independent Bernoulli rv that is, $\mathcal{L}(\xi_i) = B(1, p_i)$, $1 \leq i \leq n$; put $Z := \sum_{i \leq n} \xi_i$ and $Z_i := Z - \xi_i$, $1 \leq i \leq n$. Observe that Z_i and ξ_i are independent. Then we obtain with $\lambda := \sum_{i \leq n} p_i$ for $A \subset \{0, 1, \ldots\}$,

$$
\begin{aligned}
P(Z \in A) - P_\lambda(A) &= E(\lambda g(Z+1) - Zg(Z)) \\
&= \sum_{i \leq n} p_i E(g(Z+1)) - \sum_{i \leq n} E(\xi_i g(Z_i + \xi_i)) \\
&= \sum_{i \leq n} p_i E(g(Z+1) - g(Z_i + 1)) \tag{1.16}
\end{aligned}
$$

by the independence of ξ_i and Z_i. Since $Z + 1 = Z_i + 1 + \xi_i$ and $Z_i + 1$ coincide unless ξ_i equals 1, we have

$$|E(g(Z+1) - g(Z_i + 1))| = |E(g(Z_i + 1 + \xi_i) - g(Z_i + 1))|$$
$$\leq \sup_{j \geq 1} |g(j+1) - g(j)| \, p_i.$$

Equation (1.16) then implies

$$\left| P\left(\sum_{i \leq n} \xi_i \in A \right) - P_\lambda(A) \right| \leq \sup_{j \geq 1} |g(j+1) - g(j)| \sum_{i \leq n} p_i^2,$$

and all that remains to be done is to compute bounds for $\sup_{j \geq 1} |g(j+1) - g(j)|$.

It is readily seen that the function g, defined by

$$g(j+1) := \lambda^{-j-1} j! \, e^\lambda (P_\lambda(A \cap \{0, 1, \ldots, j\}) - P_\lambda(A) P_\lambda(\{0, 1, \ldots, j\})), \quad j \geq 0$$
$$g(0) := 0,$$

satisfies the recursion equation (1.14) for arbitrary $A \subset \{0, 1, \ldots\}$ and $\lambda > 0$. As shown in Lemma 1.1.1 in Barbour et al. [29], the function g is bounded and satisfies

$$\sup_{j \geq 1} |g(j+1) - g(j)| \leq \lambda^{-1}(1 - e^{-\lambda}).$$

As a consequence we obtain the bound

$$\sup_{A \subset \{0, 1, \ldots\}} \left| P\left(\sum_{i \leq n} \xi_i \in A \right) - P_\lambda(A) \right| \leq \lambda^{-1}(1 - e^{-\lambda}) \sum_{i \leq n} p_i^2, \qquad (1.17)$$

where $\lambda = \sum_{i \leq n} p_i$, which is Theorem 1 in Barbour and Hall [27]. With $p_i = p$ we derive for $K(n)$ and $\tau(n)$ with $\mathcal{L}(K(n)) = B(n, p)$, $\mathcal{L}(\tau(n)) = P_{np}$,

$$d(K(n), \tau(n)) \leq (1 - e^{-np})p \leq p,$$

which provides Lemma 1.2.2 (i).

As pointed out by Barbour et al. [29], page 8, a main advantage of the preceding approach is that only little needs to be changed if the independence assumption on $\xi_1, \ldots, \xi_n$ is dropped. Observe that we have used independence only once, in equation (1.16), and all that has to be done in the dependent case is to modify this step of the argument appropriately. We refer to page 9ff in Barbour et al. [29] for a detailed discussion.

As an example we show how the preceding Markov kernel technique together with the bound (1.17) can be applied to risk theory, to entail a bound for the compound Poisson approximation of a portfolio having identical claim size distributions.

Let $V_1, V_2, \ldots$ be iid claims on $(0, \infty)$. Then the total amount of claim sizes in a portfolio of size n with individual risk p_i, $i = 1, \ldots, n$, can be described by the rv $\sum_{i \leq L} V_i$, where L is the sum of n independent rv, each following a $B(1, p_i)$-distribution, $i = 1, \ldots, n$, and being independent of $V_1, V_2, \ldots$. Observe that in the case of equal probabilities $p_1 = p_2 = \cdots = p_n = p$, the counting variable L coincides with $K(n)$ following a $B(n, p)$-distribution.

Replacing L by a rv τ, which is Poisson distributed with parameter $\lambda = \sum_{i \leq n} p_i$ and being independent of $V_1, V_2, \ldots$ as well, we obtain the *compound Poisson process* approximation $\sum_{i \leq \tau} V_i$ of $\sum_{i \leq L} V_i$. The distribution of these two sums of random length can obviously be generated by the rv τ and L and the Markov kernel

$$Q_c(B \mid m) := P\left(\sum_{i \leq m} V_i \in B \right), \qquad m = 0, 1, 2, \ldots, \quad B \in \mathbb{B},$$

where $\mathbb{B}$ denotes the Borel-σ-field in $\mathbb{R}$. As a consequence we obtain from (1.17) the bounds

$$d\left(\sum_{i \leq L} V_i, \sum_{i \leq \tau} V_i \right) = d(Q_c \mathcal{L}(L), Q_c \mathcal{L}(\tau)) \leq d(L, \tau)$$

$$\leq \lambda^{-1}(1 - e^{-\lambda}) \sum_{i \leq n} p_i^2 \leq \sum_{i \leq n} p_i^2 / \sum_{i \leq n} p_i,$$

which improve the bounds established by Michel [329] and by Gerber [174] for the compound Poisson approximation in the case of a portfolio with identical claim size distributions.

1.3   Applications

In this section we describe in which situation our functional laws of small numbers typically apply.

Let Z be a random element in some sample space S bearing a σ-field $\mathcal{B}$ and suppose that we are only interested in those realizations of Z which fall into a fixed subset $A \in \mathcal{B}$ of the sample space. Let $Z_1, \ldots, Z_n$ be independent replicates of Z and consider therefore only those observations among $Z_1, \ldots, Z_n$ which fall into that subset A. Arranged in the order of their outcome, we can denote these $Z_i \in A$ by $V_1, \ldots, V_{K_A(n)}$, where the random number

$$K_A(n) := \sum_{i \leq n} \varepsilon_{Z_i}(A)$$

is binomial distributed $B(n, p)$ with probability $p = P(Z \in A)$.

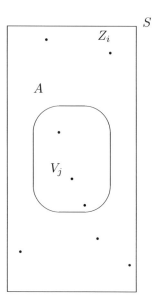

FIGURE 1.3.1. The setup for applications.

TRUNCATED EMPIRICAL PROCESSES

We can describe the set of data $V_1, \ldots, V_{K_A(n)}$ of those observations among $Z_1, \ldots,$ Z_n, which fall into A, by the *truncated empirical process*

$$N_{n,A}(\cdot) := \sum_{i \leq n} \varepsilon_{Z_i}(\cdot \cap A)$$

$$= \sum_{i \leq K_A(n)} \varepsilon_{V_i}(\cdot),$$

which is precisely that mathematical object which we have studied in the preceding sections. Note that the process $N_{n,A}$ does not only carry the information about the number $K_A(n) = N_{n,A}(S)$ of data in A, but it stores also their values.

It is intuitively clear that $V_1, V_2, \ldots$ are independent replicates of a random element V, whose range is the set A and whose distribution is the conditional distribution of Z given $Z \in A$:

$$P(V \in B) = P(Z \in B \mid Z \in A) = \frac{P(Z \in B \cap A)}{P(Z \in A)}, \qquad B \in \mathcal{B}.$$

It is probably less intuitively clear but nevertheless true that $K(n)$ and $V_1, V_2, \ldots$ are independent. The following crucial representation of $N_{n,A}$ is Theorem 1.4.1 in Reiss [387].

Theorem 1.3.1. *Let $X_1, X_2, \ldots$ be independent copies of the random element V, independent also from $K_A(n)$. Then,*

$$N_{n,A} = \sum_{i \leq K_A(n)} \varepsilon_{V_i} =_D \sum_{i \leq K_A(n)} \varepsilon_{X_i}.$$

We can therefore handle those data $V_1, \ldots, V_{K_A(n)}$ among $Z_1, \ldots, Z_n$, which fall into the set A, like independent copies of the random element V, whose distribution is the conditional distribution of Z given $Z \in A$, with their random number $K_A(n)$ being a $B(n, P(Z \in A))$ distributed rv and stochastically independent of $V_1, V_2, \ldots$

If the hitting probability $p = P(Z \in A)$ is small, then A is a rare event and the first-order Poisson process approximation applies to $N_{n,A}$,

$$N_{n,A} = \sum_{i \leq K_A(n)} \varepsilon_{V_i} \sim_D N_{n,A}^* := \sum_{i \leq \tau_A(n)} \varepsilon_{V_i}, \qquad (1.18)$$

where $\tau_A(n)$ is a Poisson rv with parameter np and also independent of the sequence $V_1, V_2, \ldots$

Note that in contrast to the *global Poissonization* technique, where the fixed sample size n is replaced by a Poisson rv $\tau(n)$ with parameter n that is,

$$\sum_{i \leq n} \varepsilon_{Z_i} \sim_D \sum_{i \leq \tau(n)} \varepsilon_{Z_i},$$

with $\tau(n)$ being independent of $Z_1, Z_2, \ldots$, the Poissonization described here is a *local* one in the set A. For further details on global Poissonization we refer to Section 8.3 of Reiss [387].

In the case, where the hitting probability $p = P(Z \in A)$ of A is small, the conditional distribution of Z, given $Z \in A$, can often be approximated by some ideal distribution

$$P(Z \in \cdot \mid Z \in A) = P(V \in \cdot) \sim P(W \in \cdot),$$

where W is a random element having this ideal distribution. We are therefore now precisely in a situation where we can apply the second-order Poisson approximation and the POT approach of $N_{n,A}$ from the preceding section. We expect therefore that the truncated empirical process $N_{n,A}$ behaves approximately like the Poisson process

$$N_{n,A}^{**} = \sum_{i \leq \tau_A(n)} \varepsilon_{W_i}$$

being its second-order Poisson process approximations or like

$$M_{n,A} = \sum_{i \leq K_A(n)} \varepsilon_{W_i},$$

which is the POT approach, with $W_1, W_2, \ldots$ being independent copies of W, and also independent of $\tau_A(n)$.

A GEOMETRIC CLASSIFICATION

The preceding model approximations entail that we can handle our actual data V_j that is, those Z_i among $Z_1, \ldots, Z_n$, which fall into the set A, within certain error bounds like some ideal W_j, with their counting random number being independent from their values. We will see in the following that *non-parametric regression analysis* as well as *extreme value theory* are within the scope of this approach: In the first case, the subset A is located in the center of the support of the distribution of Z, in the latter case it is located at the border.

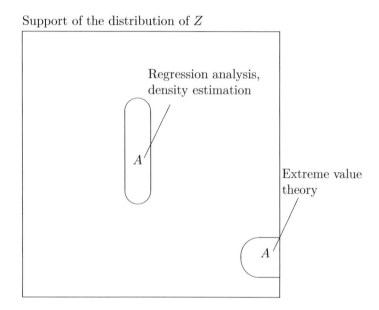

FIGURE 1.3.2. A geometric classification of typical applications.

EXAMPLES

The following examples highlight the wideranging applicability of our functional laws of small numbers.

Example 1.3.2 (Density estimation). Let Z be a rv with values in $\mathbb{R}$ and fix $x \in \mathbb{R}$. Suppose that Z has a density g, say, near x; our problem is to estimate $g(x)$ from a sample $Z_1, \ldots, Z_n$ of independent replicates of Z.

Choose to this end a window width $a_n > 0$ and consider only those observations Z_i which fall into the data window

$$A_n = [x - a_n/2, x + a_n/2].$$

In our previous notation these are $V_1, \ldots, V_{K(n)}$ with

$$K(n) := \sum_{i \leq n} \varepsilon_{Z_i}(A_n),$$

where $K(n)$ has distribution $B(n, p_n)$,

$$p_n = P(Z \in A_n) = \int_{-a_n/2}^{a_n/2} g(x + \varepsilon) \, d\varepsilon \sim g(x) a_n$$

for $a_n \to 0$ if g is continuous at x. If in addition $g(x) > 0$, we obtain for $t \in [0, 1]$ the approximation

$$\begin{aligned}
&P(V \leq x - a_n/2 + t a_n) \\
&= P(Z \leq x - a_n/2 + t a_n, \ Z \in A_n)/P(Z \in A_n) \\
&= P(x - a_n/2 \leq Z \leq x - a_n/2 + t a_n)/p_n \\
&= \frac{1}{p_n} \int_{-a_n/2}^{-a_n/2 + t a_n} g(x + \varepsilon) \, d\varepsilon \sim \frac{g(x) t a_n}{p_n} \xrightarrow[a_n \to 0]{} t.
\end{aligned}$$

Consequently, we obtain the approximation

$$N_n = \sum_{i \leq K(n)} \varepsilon_{V_i} \sim_D N_n^{**} = \sum_{i \leq \tau(n)} \varepsilon_{W_i}$$

where $W_1, W_2 \ldots$ are independent, on $[x - a_n/2, x + a_n/2]$ uniformly distributed, and independent from $\tau(n)$ which is Poisson distributed with parameter $n p_n \sim n a_n g(x)$.

Our approach entails therefore that the information we are interested in is essentially contained in the sample size $K(n)$ or $\tau(n)$, respectively.

Example 1.3.3 (Regression analysis). Let $Z = (X, Y)$ be a rv in $\mathbb{R}^2$ and fix $x \in \mathbb{R}$. Now we are interested in the conditional distribution function (df) of Y given $X = x$, denoted by $F(\cdot \mid x) := P(Y \leq \cdot \mid X = x)$.

In this case we choose the data window

$$A_n := [x - a_n/2, x + a_n/2] \times \mathbb{R}$$

with window width $a_n > 0$ for the data $Z_i = (X_i, Y_i)$, $i = 1, \ldots, n$:

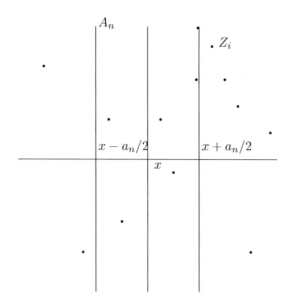

FIGURE 1.3.3. The setup for conditional problems.

Then,

$$K(n) := \sum_{i \leq n} \varepsilon_{Z_i}(A_n)$$

$$= \sum_{i \leq n} \varepsilon_{X_i}([x - a_n/2, x + a_n/2])$$

is $B(n, p)$-distributed with

$$p_n = P(Z \in A_n) = P(X \in [x - a_n/2, x + a_n/2]) \sim g(x)a_n,$$

where we assume that X has a density g, say, near x being continuous and positive at x.

If $Z = (X, Y)$ has a joint density f, say, on A_n, then we obtain for $t \in [0, 1]$ and $s \in \mathbb{R}$

$$P(V \leq (x - a_n/2 + ta_n, s))$$
$$= P(Z \leq (x - a_n/2 + ta_n, s), Z \in A_n)/P(Z \in A_n)$$
$$= P(x - a_n/2 \leq X \leq x - a_n/2 + ta_n, Y \leq s)/p_n$$
$$= \int_{x-a_n/2}^{x-a_n/2+ta_n} \int_{-\infty}^{s} f(u, w) \, dw \, du/p_n$$

$$= \int_{-\infty}^{s} a_n \int_{0}^{t} f(x + a_n u - a_n/2, w)\, du\, dw / p_n$$

$$\xrightarrow[a_n \to 0]{} \int_{-\infty}^{s} \int_{0}^{t} f(x, w)\, du\, dw / g(x)$$

$$= t\, \frac{\int_{-\infty}^{s} f(x, w)\, dw}{g(x)} = t\, F(s|x)$$

under suitable regularity conditions on f (near x).

Consequently, we obtain the approximation

$$N_n = \sum_{i \le K(n)} \varepsilon_{V_i} \sim_D N_n^{**} = \sum_{i \le \tau(n)} \varepsilon_{(U_i, W_i)},$$

where U is on $[x - a_n/2, x + a_n/2]$ uniformly distributed, W_i follows the conditional df $F(\cdot|x)$, $\tau(n)$ is Poisson $P_{np_n} \sim P_{na_n g(x)}$ and $\tau(n), W_1, W_2, \ldots, U_1, U_2, \ldots$ are all independent!

In this example our approach entails that the information we are interested in is essentially contained in the second component of V_i.

Example 1.3.4 (Extreme value theory). Let Z be $\mathbb{R}$-valued and suppose that we are only interested in large values of Z, where we call a realization of Z *large*, if it exceeds a given high *threshold* t. In this case we choose the data window $A = (t, \infty)$ or, better adapted to our purposes, we put $t \in \mathbb{R}$ on a linear scale and define

$$A_n = (a_n t + b_n, \infty)$$

for some norming constants $a_n > 0$, $b_n \in \mathbb{R}$.

Then $K(n) = \sum_{i \le n} \varepsilon_{Z_i}(A_n)$ is $B(n, p_n)$-distributed with $p_n = 1 - F(a_n t + b_n)$, where F denotes the df pertaining to Z. Then we obtain for $s \ge 0$,

$$P(V \le a_n(t + s) + b_n) = P(Z \le a_n(t + s) + b_n \mid Z > a_n t + b_n)$$

$$= P(a_n t + b_n < Z \le a_n(t + s) + b_n)/(1 - F(a_n t + b_n))$$

$$= 1 - \frac{1 - F(a_n(t + s) + b_n)}{1 - F(a_n t + b_n)},$$

thus facing the problem:

What is the limiting behavior of

$$(1 - F(a_n(t + s) + b_n))/(1 - F(a_n t + b_n)) \longrightarrow_{n \to \infty} \tag{1.19}$$

as n tends to infinity?

EXTREME VALUE DISTRIBUTIONS

Suppose that there exist $a_n > 0$, $b_n \in \mathbb{R}$ such that

$$F^n(a_n x + b_n) \longrightarrow_{n \to \infty} G(x), \qquad x \in \mathbb{R}, \tag{1.20}$$

for some (continuous) limiting distribution G. Then we say that G *belongs to the domain of attraction* of G, denoted by $F \in \mathcal{D}(G)$. In this case we deduce from the expansion $\log(1 + \varepsilon) = \varepsilon + O(\varepsilon^2)$ for $\varepsilon \to 0$ the equivalence

$$F^n(a_n x + b_n) \longrightarrow_{n \to \infty} G(x)$$
$$\iff n \log(1 - (1 - F(a_n x + b_n))) \longrightarrow_{n \to \infty} \log(G(x))$$
$$\iff n(1 - F(a_n x + b_n)) \longrightarrow_{n \to \infty} -\log(G(x))$$

if $0 < G(x) \le 1$, and hence,

$$\frac{1 - F(a_n(t + s) + b_n)}{1 - F(a_n t + b_n)} \longrightarrow_{n \to \infty} \frac{\log(G(t + s))}{\log(G(t))}$$

if $0 < G(t) < 1$.

From the now classical article by Gnedenko [176] (see also de Haan [184] and Galambos [167]) we know that $F \in \mathcal{D}(G)$ only if $G \in \{G_\beta : \beta \in \mathbb{R}\}$, where

$$G_\beta(t) := \exp(-(1 + \beta t)^{-1/\beta}), \qquad 1 + \beta t > 0,$$

is an *extreme value df*, abbreviated by EVD. Since $(1 + \beta t)^{-1/\beta} \longrightarrow_{\beta \to 0} \exp(-t)$, interpret $G_0(t)$ as $\exp(-e^{-t})$, $t \in \mathbb{R}$. We do not distinguish in our notation between distributions and their df.

GENERALIZED PARETO DISTRIBUTIONS

If we assume that $F \in \mathcal{D}(G_\beta)$, we obtain therefore

$$P(V \le a_n(t + s) + b_n) = 1 - \frac{n(1 - F(a_n(t + s) + b_n))}{n(1 - F(a_n t + b_n))}$$

$$\longrightarrow_{n \to \infty} 1 - \frac{\log(G_\beta(t + s))}{\log(G_\beta(t))} = 1 - \left(\frac{1 + \beta(t + s)}{1 + \beta t}\right)^{-1/\beta}$$

$$= 1 - \left(1 + \beta \frac{s}{1 + \beta t}\right)^{-1/\beta} = H_\beta\left(\frac{s}{1 + \beta t}\right), \qquad s \ge 0, \tag{1.21}$$

provided $1 + \beta t > 0$ and $1 + \beta(t + s) > 0$. The family

$$H_\beta(s) := 1 + \log(G_\beta(s)), \qquad s \ge 0$$

$$= 1 - (1 + \beta s)^{-1/\beta} \quad \text{for} \quad \begin{cases} s \ge 0 & \text{if} \quad \beta \ge 0 \\ 0 \le s \le -1/\beta & \text{if} \quad \beta < 0 \end{cases}$$

of df parameterized by $\beta \in \mathbb{R}$ is the class of *generalized Pareto df* (GPD) accompanying the family of EVD. Notice that H_β with $\beta > 0$ gives a Pareto distribution, H_{-1} is the uniform distribution on $(0,1)$ and H_0 has to be interpreted as the standard exponential distribution $H_0(s) = \lim_{\beta \to 0} H_\beta(s) = 1 - e^{-s}$, $s \geq 0$.

The Peaks-Over-Threshold Method

Formula (1.21) entails that under the condition $F \in \mathcal{D}(G_\beta)$, those observations in an iid sample generated independently according to F, which exceed the threshold $a_n t + b_n$, follow approximately a GPD. In this case our functional laws of small numbers specify to

$$N_n = \sum_{i \leq K(n)} \varepsilon_{V_i} \overset{\sim}{\mathcal{D}} M_n = \sum_{i \leq K(n)} \varepsilon_{W_i} \sim_D N_n^{**} = \sum_{i \leq \tau(n)} \varepsilon_{W_i}, \qquad (1.22)$$

where $W_1, W_2, \ldots$ follow a GPD and $\tau(n)$ is a Poisson rv with parameter $n(1 - F(a_n t + b_n)) \sim -\log(G_\beta(t)) = (1 + \beta t)^{-1/\beta}$.

The information on the probability of large values of Z that is, on $P(Z \leq a_n(t+s) + b_n)$ is now contained in the sample size $K(n)$ *and* the data V_i.

The approximation of N_n by M_n in (1.22) explains the idea behind the *peaks-over-threshold method* (POT), widely used for instance by hydrologists to model large floods having a GPD (cf. Todorovic and Zelenhasic [446], Hosking and Wallis [224]). Formula (1.21) suggests in particular that in case $F \in \mathcal{D}(G_\beta)$, the upper tail of F can be approximated by that of a GPD which was first observed and verified by Balkema and de Haan [22], see Theorem 2.7.1 below, and independently by Pickands [371].

It is worth mentioning that the class of GPD is still the only possible set of limiting distributions in (1.21), if we drop the assumption $F \in \mathcal{D}(G)$ and merely consider a sequence $a_n t + b_n$, $n \in \mathbb{N}$, of thresholds satisfying a certain regularity condition. The following theorem follows from results in Balkema and de Haan [22] or from Theorem 2.1 in Rychlik [403] .

Theorem 1.3.5. *Suppose that there exist* $a_n > 0$, $b_n \in \mathbb{R}$ *with* $1 - F(b_n) \longrightarrow_{n \to \infty} 0$ *and* $(1 - F(b_{n+1}))/(1 - F(b_n)) \longrightarrow_{n \to \infty} 1$ *such that for any* $s \geq 0$,

$$1 - \frac{1 - F(a_n s + b_n)}{1 - F(b_n)} \longrightarrow_{n \to \infty} L(s)$$

for some continuous df L. *Then* L *is a GPD that is, there exist* $\beta \in \mathbb{R}$ *and some* $a > 0$ *such that*

$$L(s) = 1 + \log(G_\beta(as)), \qquad s \geq 0.$$

If we drop the condition $(1 - F(b_{n+1}))/(1 - F(b_n)) \longrightarrow_{n \to \infty} 1$ in Theorem 1.3.5, then discrete limiting distributions promptly occur; consider for example the geometric df $F(k) = 1 - (1-p)^k$, $k = 0, 1, 2, \ldots$ for some $p \in (0,1)$. With

$a_n = 1$ and $b_n = n$, $n \in \mathbb{N}$ the ratio $(1 - F(a_n s + b_n))/(1 - F(b_n))$ then equals $1 - F(s)$, $s \geq 0$, which is well known. A complete characterization of possible limiting distributions is given in the paper by Balkema and de Haan [22].

Chapter 2

Extreme Value Theory

In this chapter we summarize results in extreme value theory, which are primarily based on the condition that the upper tail of the underlying df is in the δ-neighborhood of a generalized Pareto distribution (GPD). This condition, which looks a bit restrictive at first sight (see Section 2.2), is however essentially equivalent to the condition that rates of convergence in extreme value theory are at least of algebraic order (see Theorem 2.2.5). The δ-neighborhood is therefore a natural candidate to be considered, if one is interested in reasonable rates of convergence of the functional laws of small numbers in extreme value theory (Theorem 2.3.2) as well as of parameter estimators (Theorems 2.4.4, 2.4.5 and 2.5.4).

2.1 Domains of Attraction, von Mises Conditions

Recall from Example 1.3.4 that a df F belongs to the domain of attraction of an extreme value df (EVD) $G_\beta(x) = \exp(-(1 + \beta x)^{-1/\beta})$, $1 + \beta x > 0$, denoted by $F \in \mathcal{D}(G_\beta)$, iff there exist constants $a_n > 0$, $b_n \in \mathbb{R}$ such that

$$F^n(a_n x + b_n) \longrightarrow_{n \to \infty} G_\beta(x), \qquad x \in \mathbb{R}$$
$$\Longleftrightarrow \quad P((Z_{n:n} - b_n)/a_n \le x) \longrightarrow_{n \to \infty} G_\beta(x), \qquad x \in \mathbb{R},$$

where $Z_{n:n}$ is the sample maximum of an iid sample $Z_1, \ldots, Z_n$ with common df F. Moreover, $Z_{1:n} \le \cdots \le Z_{n:n}$ denote the pertaining order statistics.

THE GNEDENKO-DE HAAN THEOREM

The following famous result due to Gnedenko [176] (and partially due to de Haan [184]) provides necessary as well as sufficient conditions for $F \in \mathcal{D}(G_\beta)$.

M. Falk et al., *Laws of Small Numbers: Extremes and Rare Events*, 3rd ed.,
DOI 10.1007/978-3-0348-0009-9_2, © Springer Basel AG 2011

Theorem 2.1.1 (Gnedenko-de Haan). *Let G be a non-degenerate df. Suppose that F is a df with the property that for some constants $a_n > 0$, $b_n \in \mathbb{R}$,*

$$F^n(a_n x + b_n) \longrightarrow_{n \to \infty} G(x),$$

for any point of continuity x of G. Then G is up to a location and scale shift an EVD G_β, i.e., $F \in \mathcal{D}(G) = \mathcal{D}(G_\beta)$.

Put $\omega(F) := \sup\{x \in \mathbb{R} : F(x) < 1\}$. Then we have

(i) $F \in \mathcal{D}(G_\beta)$ *with* $\beta > 0 \iff \omega(F) = \infty$ *and*

$$\lim_{t \to \infty} \frac{1 - F(tx)}{1 - F(t)} = x^{-1/\beta}, \qquad x > 0.$$

The normalizing constants can be chosen as $b_n = 0$ and $a_n = F^{-1}(1 - n^{-1})$, $n \in \mathbb{N}$, where $F^{-1}(q) := \inf\{t \in \mathbb{R} : F(t) \geq q\}$, $q \in (0,1)$, denotes the quantile function or generalized inverse of F.

(ii) $F \in \mathcal{D}(G_\beta)$ *with* $\beta < 0 \iff \omega(F) < \infty$ *and*

$$\lim_{t \to \infty} \frac{1 - F(\omega(F) - \frac{1}{tx})}{1 - F(\omega(F) - \frac{1}{t})} = x^{1/\beta}, \qquad x > 0.$$

The normalizing constants can be chosen as $b_n = \omega(F)$ and $a_n = \omega(F) - F^{-1}(1 - n^{-1})$.

(iii) $F \in \mathcal{D}(G_0) \iff$ *there exists $t_0 < \omega(F)$ such that $\int_{t_0}^{\omega(F)} 1 - F(x) \, dx < \infty$ and*

$$\lim_{t \to \omega(F)} \frac{1 - F(t + xR(t))}{1 - F(t)} = \exp(-x), \qquad x \in \mathbb{R},$$

where $R(t) := \int_t^{\omega(F)} 1 - F(y) \, dy / (1 - F(t))$, $t < \omega(F)$. The norming constants can be chosen as $b_n = F^{-1}(1 - n^{-1})$ and $a_n = R(b_n)$.

It is actually sufficient to consider in part (iii) of the preceding Theorem 2.1.1 only $x \geq 0$, as shown by Worms [464]. In this case, the stated condition has a probability meaning in terms of conditional distributions, known as the *additive excess property*. We refer to Section 1.3 of Kotz and Nadarajah [293] for a further discussion.

Note that we have for any $\beta \in \mathbb{R}$,

$$F \in \mathcal{D}(G_\beta) \iff F(\cdot + \alpha) \in \mathcal{D}(G_\beta)$$

for any $\alpha \in \mathbb{R}$. Without loss of generality we will therefore assume in the following that $\omega(F) > 0$.

VON MISES CONDITIONS

The following sufficient condition for $F \in \mathcal{D}(G_\beta)$ goes essentially back to von Mises [336].

Theorem 2.1.2 (Von Mises conditions). *Let F have a positive derivative f on $[x_0, \omega(F))$ for some $0 < x_0 < \omega(F)$.*

(i) *If there exist $\beta \in \mathbb{R}$ and $c > 0$ such that $\omega(F) = \omega(H_\beta)$ and*

$$\lim_{x \uparrow \omega(F)} \frac{(1 + \beta x) f(x)}{1 - F(x)} = c, \qquad (VM)$$

then $F \in \mathcal{D}(G_{\beta/c})$.

(ii) *Suppose in addition that f is differentiable. If*

$$\lim_{x \uparrow \omega(F)} \frac{d}{dx} \left(\frac{1 - F(x)}{f(x)} \right) = 0, \qquad (VM_0)$$

then $F \in \mathcal{D}(G_0)$.

Condition (VM_0) is the original criterion due to von Mises [336, page 285] in case $\beta = 0$. Note that it is equivalent to the condition

$$\lim_{x \uparrow \omega(F)} \frac{1 - F(x)}{f(x)} \frac{f'(x)}{f(x)} = -1$$

and, thus, (VM) in case $\beta = 0$ and (VM_0) can be linked by l'Hôpital's rule. Condition (VM) will play a crucial role in what follows in connection with generalized Pareto distributions.

If F has ultimately a positive derivative, which is monotone in a left neighborhood of $\omega(F) = \omega(H_\beta)$ for some $\beta \neq 0$, and if $F \in \mathcal{D}(G_{\beta/c})$ for some $c > 0$, then F satisfies (VM) with β and c (see Theorems 2.7.1 (ii), 2.7.2 (ii) in de Haan [184]). Consequently, if F has ultimately a positive derivative f such that $\exp(-x)f(x)$ is non-increasing in a left neighborhood of $\omega(F) = \infty$, and if $F(\log(x))$, $x > 0$, is in $\mathcal{D}(G_{1/c})$ for some $c > 0$, then F satisfies (VM) with c and $\beta = 0$.

A df F is in $\mathcal{D}(G_0)$ iff there exists a df F^* with $\omega(F^*) = \omega(F)$, which satisfies (VM_0) and which is tail equivalent to F^*, i.e.,

$$\lim_{x \uparrow \omega(F)} \frac{1 - F(x)}{1 - F^*(x)} = 1,$$

see Balkema and de Haan [21].

Proof. We prove only the case $\beta = 0$ in condition (VM), the cases $\beta > 0$ and $\beta < 0$ can be shown in complete analogy (see Theorem 2.7.1 in Galambos [167] and Theorems 2.7.1 and 2.7.2 in de Haan [184]). Observe first that

$$\int_{t_0}^{\omega(F)} 1 - F(x) \, dx = \int_{t_0}^{\omega(F)} \frac{1 - F(x)}{f(x)} f(x) \, dx \leq 2/c$$

if t_0 is large. We have further by l'Hôpital's rule

$$\lim_{t \to \omega(F)} R(t) = \lim_{t \to \omega(F)} \frac{\int_t^{\omega(F)} 1 - F(x)\, dx}{1 - F(t)} = \lim_{t \to \omega(F)} \frac{1 - F(t)}{f(t)} = 1/c.$$

Put now $g(t) := -f(t)/(1 - F(t)) = (\log(1 - F(t)))'$, $t \geq t_0$. Then we have the representation

$$1 - F(t) = C \exp\Big(\int_{t_0}^t g(y)\, dy \Big), \qquad t \geq t_0,$$

with some constant $C > 0$ and thus,

$$\frac{1 - F(t + xR(t))}{1 - F(t)} = \exp\Big(\int_t^{t+xR(t)} g(y)\, dy \Big) \to_{t \to \omega(F)} \exp(-x), \qquad x \in \mathbb{R},$$

since $\lim_{y \to \omega(F)} g(y) = -c$ and $\lim_{t \to \omega(F)} R(t) = 1/c$. The assertion now follows from Theorem 2.1.1 (iii). $\qquad\qquad\qquad\qquad\qquad\qquad\qquad\qquad\qquad\qquad\qquad\qquad\square$

Distributions F with differentiable upper tail of Gamma type that is, $\lim_{x \to \infty}$ $F'(x)/((b^p/\Gamma(p))\, e^{-bx} x^{p-1}) = 1$ with $b, p > 0$ satisfy (VM) with $\beta = 0$. Condition (VM) with $\beta > 0$ is, for example, satisfied for F with differentiable upper tail of Cauchy-type, whereas triangular type distributions satisfy (VM) with $\beta < 0$. We have equality in (VM) with F being a GPD $H_\beta(x) = 1 - (1 + \beta x)^{-1/\beta}$, for all $x \geq 0$ such that $1 + \beta x > 0$.

The standard normal df Φ satisfies $\lim_{x \to \infty} x(1 - \Phi(x))/\Phi'(x) = 1$ and does not satisfy, therefore, condition (VM) but (VM$_0$).

The following result states that we have equality in (VM) *only* for a GPD. It indicates therefore a particular relationship between df with GPD-like upper tail and the von Mises condition (VM), which we will reveal later. Its proof can easily be established (see also Corollary 1.2 in Falk and Marohn [143]).

Proposition 2.1.3. *We have ultimately equality in (VM) for a df F iff F is ultimately a GPD. Precisely, we have equality in (VM) for $x \in [x_0, \omega(F))$ iff there exist $a > 0$, $b \in \mathbb{R}$ such that*

$$1 - F(x) = 1 - H_{\beta/c}(ax + b), \qquad x \in [x_0, \omega(F)),$$

where $b = (a - c)/\beta$ in case $\beta \neq 0$ and $a = c$ in case $\beta = 0$.

DIFFERENTIABLE TAIL EQUIVALENCE

Denote by h_β the density of the GPD H_β that is,

$$h_\beta(x) = \frac{G'_\beta(x)}{G_\beta(x)} = (1 + \beta x)^{-1/\beta - 1} \quad \text{for} \quad \left\{ \begin{array}{lll} x \geq 0 & \text{if} & \beta \geq 0 \\ 0 \leq x < -1/\beta & \text{if} & \beta < 0. \end{array} \right.$$

Note that with $b = (a - c)/\beta$ if $\beta \neq 0$ and $a = c$ if $\beta = 0$ we have

$$\frac{1 + \beta x}{c} = \frac{1 - H_{\beta/c}(ax + b)}{ah_{\beta/c}(ax + b)}$$

for all x in a left neighborhood of $\omega(H_\beta) = \omega(H_{\beta/c}(ax + b))$. If F satisfies (VM), we can write therefore for any $a > 0$ and $b \in \mathbb{R}$ such that $b = (a - c)/\beta$ if $\beta \neq 0$ and $a = c$ if $\beta = 0$,

$$
\begin{aligned}
1 &= \lim_{x \to \omega(F)} \frac{f(x)}{1 - F(x)} \frac{1 + \beta x}{c} \\
&= \lim_{x \to \omega(F)} \frac{f(x)}{ah_{\beta/c}(ax + b)} \frac{1 - H_{\beta/c}(ax + b)}{1 - F(x)}.
\end{aligned}
\qquad (2.1)
$$

As a consequence, we obtain that under (VM) a df F is *tail equivalent* to the GPD $H_{\beta/c}(ax + b)$, for some $a > 0$, $b \in \mathbb{R}$ with $b = (a - c)/\beta$ if $\beta \neq 0$ and $a = 1$ if $\beta = 0$, iff F and $H_{\beta/c}(ax + b)$ are *differentiable tail equivalent*. Precisely

$$\lim_{x \to \omega(F)} \frac{1 - F(x)}{1 - H_{\beta/c}(ax + b)} \quad \text{exists in } [0, \infty]$$

$$\Longleftrightarrow \quad \lim_{x \to \omega(F)} \frac{f(x)}{ah_{\beta/c}(ax + b)} \quad \text{exists in } [0, \infty]$$

and in this case these limits coincide. Note that the "if" part of this conclusion follows from l'Hôpital's rule anyway.

VON MISES CONDITION WITH REMAINDER

The preceding considerations indicate that the condition (VM) is closely related to the assumption that the upper tail of F is close to that of a GPD. This idea can be made rigorous if we consider the rate at which the limit in (VM) is attained.

Suppose that F satisfies (VM) with $\beta \in \mathbb{R}$ and $c > 0$ and define by

$$\eta(x) := \frac{(1 + \beta x)f(x)}{1 - F(x)} - c, \qquad x \in [x_0, \omega(F)),$$

the remainder function in condition (VM). Then we can write for any $a > 0$, $b \in \mathbb{R}$ with $b = (a - c)/\beta$ if $\beta \neq 0$ and $a = c$ if $\beta = 0$,

$$\frac{f(x)}{ah_{\beta/c}(ax + b)} = \frac{1 - F(x_1)}{1 - H_{\beta/c}(ax_1 + b)} \exp\left(-\int_{x_1}^{x} \frac{\eta(t)}{1 + \beta t} \, dt\right) \left(1 + \frac{\eta(x)}{c}\right), \qquad (2.2)$$

$x \in [x_1, \omega(F))$, where $x_1 \in [x_0, \omega(F))$ is chosen such that $ax_1 + b > 0$. Recall that for $\beta < 0$ we have $ax + b = ax + (a - c)/\beta \leq \omega(H_{\beta/c}) = -c/\beta \Longleftrightarrow x \leq -1/\beta = \omega(H_\beta) = \omega(F)$. The following result is now immediate from the preceding representation (2.2) and equation (2.1).

Proposition 2.1.4. *Suppose that F satisfies (VM) with $\beta \in \mathbb{R}$ and $c > 0$. Then we have for any $a > 0$, $b \in \mathbb{R}$, with $b = (a - c)/\beta$ if $\beta \neq 0$ and $a = c$ if $\beta = 0$,*

$$\lim_{x\uparrow\omega(F)} \frac{1 - F(x)}{1 - H_{\beta/c}(ax + b)} = \lim_{x\uparrow\omega(F)} \frac{f(x)}{ah_{\beta/c}(ax + b)} = \begin{cases} 0 \\ \alpha \in (0, \infty) \\ \infty \end{cases}$$

$$\Longleftrightarrow \int_{x_0}^{\omega(F)} \frac{\eta(t)}{1 + \beta t}\, dt = \begin{cases} \infty \\ d \in \mathbb{R} \\ -\infty. \end{cases}$$

Observe that, for any $a, c, \alpha > 0$,

$$\alpha\left(1 - H_{\beta/c}\left(ax + \frac{a - c}{\beta}\right)\right) = 1 - H_{\beta/c}\left(a\alpha^{-\beta/c}x + \frac{a\alpha^{-\beta/c} - c}{\beta}\right) \qquad (2.3)$$

if $\beta \neq 0$ and

$$\alpha(1 - H_0(ax + b)) = 1 - H_0(ax + b - \log(a)). \qquad (2.4)$$

Consequently, we can find by Proposition 2.1.4 constants $a > 0$, $b \in \mathbb{R}$, with $b = (a - c)/\beta$ if $\beta \neq 0$ and $a = c$ if $\beta = 0$, such that

$$\lim_{x\uparrow\omega(F)} \frac{1 - F(x)}{1 - H_{\beta/c}(ax + b)} = \lim_{x\uparrow\omega(F)} \frac{f(x)}{ah_{\beta/c}(ax + b)} = 1$$

iff

$$-\infty < \int_{x_0}^{\omega(F)} \frac{\eta(t)}{1 + \beta t}\, dt < \infty.$$

The preceding result reveals that a df F satisfying (VM) is tail equivalent (or, equivalently, differentiable tail equivalent) to a GPD iff the remainder function η converges to zero fast enough; precisely iff $\int_{x_0}^{\omega(F)} \eta(t)/(1 + \beta t)\, dt \in \mathbb{R}$.

Observe now that the condition

$$\eta(x) = O((1 - H_\beta(x))^\delta) \text{ as } x \to \omega(F) = \omega(H_\beta)$$

for some $\delta > 0$ implies that $\int_{x_0}^{\omega(F)} \eta(t)/(1 + \beta t)\, dt \in \mathbb{R}$ and

$$\int_{x}^{\omega(F)} \frac{\eta(t)}{1 + \beta t}\, dt = O((1 - H_\beta(x))^\delta) \text{ as } x \to \omega(F).$$

The following result is therefore immediate from equation (2.2) and Taylor expansion of exp at zero.

Proposition 2.1.5. *Suppose that F satisfies (VM) with $\beta \in \mathbb{R}$ and $c > 0$ such that $\eta(x) = O((1 - H_\beta(x))^\delta)$ as $x \to \omega(F)$ for some $\delta > 0$. Then there exist $a > 0$, $b \in \mathbb{R}$, with $b = (a - c)/\beta$ if $\beta \neq 0$ and $a = c$ if $\beta = 0$, such that*

$$f(x) = ah_{\beta/c}(ax + b)\left(1 + O((1 - H_\beta(x))^\delta)\right)$$

for any x in a left neighborhood of $\omega(F)$.

It is worth mentioning that under suitable conditions also the reverse implication in Proposition 2.1.5 holds. For the proof of this result, which is Proposition
2.1.7 below, we need the following auxiliary result.

Lemma 2.1.6. *Suppose that F and G are df having positive derivatives f and g
near $\omega(F) = \omega(G)$. If $\psi \geq 0$ is a decreasing function defined on a left neighborhood
of $\omega(F)$ with $\lim_{x \to \omega(F)} \psi(x) = 0$ such that*

$$|f(x)/g(x) - 1| = O(\psi(x)),$$

then

$$|(1 - G(x))/(1 - F(x)) - 1| = O(\psi(x))$$

as $x \to \omega(F) = \omega(G)$.

Proof. The assertion is immediate from the inequalities

$$\left| \frac{1 - G(x)}{1 - F(x)} - 1 \right| \leq \int_x^{\omega(F)} \left| \frac{f(t)}{g(t)} - 1 \right| dG(t)/(1 - F(x))$$

$$\leq C\psi(x)(1 - G(x))/(1 - F(x)),$$

where C is some positive constant. $\qquad\square$

Proposition 2.1.7. *Suppose that F satisfies condition (VM) with $\beta \in \mathbb{R}$ and
$c > 0$. We require further that, in a left neighborhood of $\omega(F)$,*

$$f(x) = ah_{\beta/c}(ax + b)\Big(1 + O((1 - H_\beta(x))^\delta)\Big)$$

*for some $\delta > 0$, $a > 0$, $b \in \mathbb{R}$, where $b = (a - c)/\beta$ if $\beta \neq 0$ and $a = c$ if $\beta = 0$.
Then the remainder function*

$$\eta(x) = \frac{f(x)(1 + \beta x)}{1 - F(x)} - c$$

is also of order $(1 - H_\beta(x))^\delta$ that is,

$$\eta(x) = O((1 - H_\beta(x))^\delta) \text{ as } x \to \omega(F).$$

Proof. Write for x, in a left neighborhood of $\omega(F) = \omega(H_\beta)$,

$$\eta(x) = c\frac{1 - H_{\beta/c}(ax + b)}{1 - F(x)} \frac{f(x)}{ah_{\beta/c}(ax + b)} - c$$

$$= c\left(\frac{1 - H_{\beta/c}(ax + b)}{1 - F(x)} - 1\right) \frac{f(x)}{ah_{\beta/c}(ax + b)} + c\left(\frac{f(x)}{ah_{\beta/c}(ax + b)} - 1\right)$$

$$= O((1 - H_\beta(x))^\delta)$$

by Lemma 2.1.6. $\qquad\square$

RATES OF CONVERGENCE OF EXTREMES

The growth condition $\eta(x) = O((1 - H_\beta(x))^\delta)$ is actually a fairly general one as revealed by the following result, which is taken from Falk and Marohn [143], Theorem 3.2. It roughly states that this growth condition is already satisfied, if $F^n(a_n x + b_n)$ approaches its limit G_β at a rate which is proportional to a power of n. For a multivariate version of this result we refer to Theorem 5.5.5.

Define the norming constants $c_n = c_n(\beta) > 0$ and $d_n = d_n(\beta) \in \mathbb{R}$ by

$$
c_n := \begin{cases} n^\beta & \beta \neq 0 \\ 1 & \beta = 0 \end{cases} \quad \text{if} \quad , \quad
d_n := \begin{cases} \frac{n^\beta - 1}{\beta} & \beta \neq 0 \\ \log(n) & \beta = 0. \end{cases} \quad \text{if}
$$

With these norming constants we have, for any $\beta \in \mathbb{R}$,

$$
H_\beta(c_n x + d_n) \longrightarrow_{n \to \infty} G_\beta(x), \qquad x \in \mathbb{R},
$$

as is seen immediately.

Theorem 2.1.8. *Suppose that F satisfies (VM) with $\beta \in \mathbb{R}$ and $c > 0$ such that $\int_{x_0}^{\omega(F)} \eta(t)/(1 + \beta t)\, dt \in \mathbb{R}$. Then we know from Proposition 2.1.4 and equations (2.3), (2.4) that*

$$
\lim_{x \uparrow \omega(F)} \frac{1 - F(x)}{1 - H_{\beta/c}(ax + b)} = 1
$$

for some $a > 0$, $b \in \mathbb{R}$, with $b = (a - c)/\beta$ if $\beta \neq 0$ and $a = c$ if $\beta = 0$. Consequently, we obtain with $a_n := c_n(\beta/c)/a$, $b_n := (d_n(\beta/c) - b)/a$ that

$$
\sup_{x \in \mathbb{R}} |F^n(a_n x + b_n) - G_{\beta/c}(x)| \longrightarrow_{n \to \infty} 0.
$$

If we require in addition that

$$
\lim_{x \uparrow \omega(F)} \frac{\eta(x)}{r(x)} = 1
$$

for some monotone function $r : (x_0, \omega(F)) \to \mathbb{R}$ and

$$
\sup_{x \in \mathbb{R}} |F^n(a_n x + b_n) - G_{\beta/c}(x)| = O(n^{-\delta})
$$

for some $\delta > 0$, then

$$
\eta(x) = O((1 - H_\beta(x))^{c\delta})
$$

as $x \to \omega(F) = \omega(H_\beta)$.

The following result is now immediate from Theorem 2.1.8 and Proposition 2.1.5.

Corollary 2.1.9. *Suppose that F satisfies (VM) with $\beta \in \mathbb{R}$, $c > 0$ such that $\int_{x_0}^{\omega(F)} \eta(t)/(1+\beta t)\,dt \in \mathbb{R}$ and*

$$\lim_{x \uparrow \omega(F)} \frac{\eta(x)}{r(x)} = 1$$

for some monotone function $r : (x_0, \omega(F)) \to \mathbb{R}$. If for some $\delta > 0$,

$$\sup_{x \in \mathbb{R}} |F^n(a_n x + b_n) - G_{\beta/c}(x)| = O(n^{-\delta}),$$

with $a_n > 0$, b_n as in Theorem 2.1.8, then there exist $a > 0$, $b \in \mathbb{R}$ with $b = (a-c)/\beta$ if $\beta \neq 0$ and $a = c$ if $\beta = 0$, such that

$$f(x) = ah_{\beta/c}(ax + b)\Big(1 + O((1 - H_\beta(x))^{c\delta})\Big)$$

for any x in a left neighborhood of $\omega(F) = \omega(H_\beta)$.

Our next result is a consequence of Corollary 5.5.5 in Reiss [385] and Proposition 2.1.5 (see also Theorems 2.2.4 and 2.2.5). By $\mathbb{B}^k$ we denote the Borel-σ-field in $\mathbb{R}^k$.

Theorem 2.1.10. *Suppose that F satisfies (VM) with $\beta \in \mathbb{R}$, $c > 0$ such that $\eta(x) = O((1 - H_\beta(x))^\delta)$ as $x \to \omega(F)$ for some $\delta > 0$. Then there exist $a_n > 0$, $b_n \in \mathbb{R}$ such that for $k \in \{1, \ldots, n\}$ and $n \in \mathbb{N}$,*

$$\sup_{B \in \mathbb{B}^k} |P(((Z_{n-i+1:n} - b_n)/a_n)_{i \leq k} \in B)$$

$$- \begin{cases} P((\beta(\sum_{j \leq i} \xi_j)^{-\beta})_{i \leq k} \in B)| & \beta \neq 0 \\ & \text{if} \\ P((-\log(\sum_{j \leq i} \xi_j))_{i \leq k} \in B)| & \beta = 0 \end{cases}$$

$$= O((k/n)^{\delta/c} k^{1/2} + k/n),$$

where $\xi_1, \xi_2, \ldots$ are independent and standard exponential rv.

BEST ATTAINABLE RATES OF CONVERGENCE

One of the significant properties of GPD is the fact that these distributions yield the best rate of joint convergence of the upper extremes, equally standardized, if the underlying df F is ultimately continuous and strictly increasing in its upper tail. This is captured in the following result. By $G_\beta^{(k)}$ we denote the distribution of $(\beta(\sum_{j \leq i} \xi_j)^{-\beta})_{i \leq k}$ if $\beta \neq 0$ and of $(-\log(\sum_{j \leq i} \xi_j))_{i \leq k}$ if $\beta = 0$, where $\xi_1, \xi_2, \ldots$ is again a sequence of independent and standard exponential rv and $k \in \mathbb{N}$. These distributions $G_\beta^{(k)}$ are the only possible classes of weak limits of the joint distribution of the k largest and equally standardized order statistics in an iid sample (see Theorem 2.2.2 and Remark 2.2.3).

Theorem 2.1.11. *Suppose that F is continuous and strictly increasing in a left neighborhood of $\omega(F)$. There exist norming constants $a_n > 0, b_n \in \mathbb{R}$ and a positive constant C such that, for any $k \in \{1, \ldots, n\}$, $n \in \mathbb{N}$,*

$$\sup_{B \in \mathbb{B}^k} \left| P\left(((Z_{n-i+1:n} - b_n)/a_n))_{i \le k} \in B \right) - G_\beta^{(k)}(B) \right| \le Ck/n$$

iff there exist $c > 0$, $d \in \mathbb{R}$ such that $F(x) = H_\beta(cx + d)$ for x near $\omega(F)$.

The *if*-part of this result is due to Reiss [383], Theorems 2.6 and 3.2, while the *only if*-part follows from Corollary 2.1.13 below.

The bound in Theorem 2.1.11 tends to zero as n tends to infinity for any sequence $k = k(n)$ such that $k/n \longrightarrow_{n \to \infty} 0$. The following result which is taken from Falk [129], reveals that this is a characteristic property of GPD that is, only df F, whose upper tails coincide with that of a GPD, entail approximation by $G_\beta^{(k)}$ for *any* such sequence k.

By $G_{\beta,(k)}$ we denote the k-th onedimensional marginal distribution of $G_\beta^{(k)}$ that is, $G_{\beta,(k)}$ is the distribution of $(\beta \sum_{j \le k} \xi_j)^{-\beta}$ if $\beta \ne 0$, and of $-\log(\sum_{j \le k} \xi_j)$ if $\beta = 0$. We suppose that F is ultimately continuous and strictly increasing in its upper tail.

Theorem 2.1.12. *If there exist $a_n > 0$, $b_n \in \mathbb{R}$ such that*

$$\sup_{t \in \mathbb{R}} \left| P((Z_{n-k+1:n} - b_n)/a_n \le t) - G_{\beta,(k)}(t) \right| \longrightarrow_{n \to \infty} 0$$

for any sequence $k = k(n) \in \{1, \ldots, n\}$, $n \in \mathbb{N}$, with $k/n \longrightarrow_{n \to \infty} 0$, then there exist $c > 0$, $d \in \mathbb{R}$ such that $F(x) = H_\beta(cx + d)$ for x near $\omega(F)$.

The following consequence is obvious.

Corollary 2.1.13. *If there exist constants $a_n > 0$, $b_n \in \mathbb{R}$ such that for any $k \in \{1, \ldots, n\}$, $n \in \mathbb{N}$,*

$$\sup_{t \in \mathbb{R}} \left| P((Z_{n-k+1:n} - b_n)/a_n \le t) - G_{\beta,(k)}(t) \right| \le g(k/n),$$

where $g : [0, 1] \to \mathbb{R}$ satisfies $\lim_{x \to 0} g(x) = 0$, then the conclusion of Theorem 2.1.12 holds.

With the particular choice $g(x) = Cx$, $x \in [0, 1]$, the preceding result obviously yields the *only if*-part of Theorem 2.1.11. A multivariate extension of Theorem 2.1.12 and Corollary 2.1.13 will be established in Theorem 5.4.7 and Corollary 5.4.8.

2.2 The δ-Neighborhood of a GPD

Distribution functions F, which satisfy the von Mises condition (VM) from Theorem 2.1.2 with rapidly vanishing remainder term η, are members of certain δ-neighborhoods $Q_i(\delta)$, $i = 1, 2, 3$, of GPD defined below. These classes $Q_i(\delta)$ will be our semiparametric models, underlying the upper tail of F, for statistical inference about extreme quantities such as extreme quantiles of F outside the range of given iid data from F (see Section 2.4).

The Standard Form of GPD

Define for $\alpha > 0$ the following df,

$$W_{1,\alpha}(x) := 1 - x^{-\alpha}, \qquad x \geq 1,$$

which is the usual class of Pareto distributions,

$$W_{2,\alpha}(x) := 1 - (-x)^{\alpha}, \qquad -1 \leq x \leq 0,$$

which consist of certain beta distributions as, e.g., the uniform distribution on $(-1, 0)$ for $\alpha = 1$, and

$$W_3(x) := 1 - \exp(-x), \qquad x \geq 0,$$

the standard exponential distribution.

Notice that W_i, $i = 1, 2, 3$, corresponds to H_β, $\beta > 0$, $\beta < 0$, $\beta = 0$, and we call a df $W \in \{W_{1,\alpha}, W_{2,\alpha}, W_3 : \alpha > 0\}$ a GPD as well. While $H_\beta(x) = 1 + \log(G_\beta(x))$, $x \geq 0$, was derived in Example 1.3.4 from the *von Mises representation*

$$G_\beta(x) = \exp(-(1 + \beta x)^{-1/\beta}), \qquad 1 + \beta x > 0, \quad \beta \in \mathbb{R},$$

of an EVD G_β, the df W_i can equally be derived from an EVD G_i given in its standard form. Put for $i = 1, 2, 3$ and $\alpha > 0$,

$$G_{1,\alpha}(x) := \begin{cases} 0, & x \leq 0 \\ \exp(-x^{-\alpha}), & x > 0, \end{cases}$$

$$G_{2,\alpha}(x) := \begin{cases} \exp(-(-x)^{\alpha}), & x \leq 0 \\ 1, & x > 0, \end{cases}$$

$$G_3(x) := \exp(-e^{-x}), \qquad x \in \mathbb{R},$$

being the Fréchet, (reversed) Weibull and Gumbel distribution. Notice that the Fréchet and Weibull df can be regained from G_β by the equations

$$G_{1,1/\beta}(x) = G_\beta((x - 1)/\beta) \qquad \qquad \beta > 0$$
$$\text{if}$$
$$G_{2,-1/\beta}(x) = G_\beta(-(x + 1)/\beta) \qquad \qquad \beta < 0.$$

Further we have for $G = G_{1,\alpha}$, $G_{2,\alpha}$, G_3 with $\alpha > 0$,

$$W(x) = 1 + \log(G(x)), \qquad \log(G(x)) > -1.$$

While we do explicitly distinguish in our notation between the classes of GPD H_β and W_i, we handle EVD G a bit laxly. But this should cause no confusion in the sequel.

δ-Neighborhoods

Suppose that the df F satisfies condition (VM) with $\beta \in \mathbb{R}$, $c > 0$ such that for some $\delta > 0$ the remainder term η satisfies $\eta(x) = O((1 - H_\beta(x))^\delta)$ as $x \to \omega(F)$. Then we know from Proposition 2.1.5 that for some $a > 0$, $b \in \mathbb{R}$, with $b = (a-c)/\beta$ if $\beta \neq 0$ and $a = c$ if $\beta = 0$,

$$f(x) = ah_{\beta/c}(ax + b)\Big(1 + O((1 - H_\beta(x))^\delta)\Big)$$

$$= \begin{cases} \tilde{a}w_{1,c/\beta}(\tilde{a}x)\Big(1 + O((1 - W_{1,c/\beta}(x))^{\tilde{\delta}})\Big), & \beta > 0 \\ \tilde{a}w_{2,-c/\beta}(\tilde{a}(x - \omega(F)))\Big(1 + O((1 - W_{2,-c/\beta}(x - \omega(F)))^{\tilde{\delta}})\Big), & \beta < 0 \\ \tilde{a}w_3(ax + b)\Big(1 + O((1 - W_3(ax))^{\delta/c})\Big), & \beta = 0, \end{cases}$$

for some $\tilde{a}, \tilde{\delta} > 0$, where we denote by w the density of W. As a consequence, F is a member of one of the following semiparametric classes $Q_i(\delta)$, $i = 1, 2, 3$ of df. In view of Corollary 2.1.9, theses classes $Q_i(\delta)$, which we call δ-*neighborhoods of GPD*, are therefore quite natural models for the upper tail of a df F. Such classes were first studied by Weiss [457]. Put for $\delta > 0$,

$Q_1(\delta) := \Big\{ F : \omega(F) = \infty$ and F has a density f on $[x_0, \infty)$ for some $x_0 > 0$ such that for some shape parameter $\alpha > 0$ and some scale parameter $a > 0$ on $[x_0, \infty)$,

$$f(x) = \frac{1}{a}w_{1,\alpha}\Big(\frac{x}{a}\Big)\Big(1 + O((1 - W_{1,\alpha}(x))^\delta)\Big)\Big\},$$

$Q_2(\delta) := \Big\{ F : \omega(F) < \infty$ and F has a density f on $[x_0, \omega(F))$ for some $x_0 < \omega(F)$ such that for some shape parameter $\alpha > 0$ and some scale parameter $a > 0$ on $[x_0, \omega(F))$,

$$f(x) = \frac{1}{a}w_{2,\alpha}\Big(\frac{x - \omega(F)}{a}\Big)\Big(1 + O\big((1 - W_{2,\alpha}(x - \omega(F)))^\delta\big)\Big)\Big\},$$

$Q_3(\delta) := \Big\{ F : \omega(F) = \infty$ and F has a density f on $[x_0, \infty)$ for some $x_0 > 0$ such that for some scale and location parameters $a > 0$, $b \in \mathbb{R}$ on $[x_0, \infty)$,

$$f(x) = \frac{1}{a}w_3\Big(\frac{x - b}{a}\Big)\Big(1 + O\big((1 - W_3\big(\frac{x}{a}\big))^\delta\big)\Big)\Big\}.$$

We will see that in case $F \in Q_i(\delta)$, $i = 1$ or 2, a suitable data transformation, which does not depend on the shape parameter $\alpha > 0$ and the scale parameter $a > 0$, transposes the underlying df F to $Q_3(\delta)$; this reduces for example the estimation of extreme quantiles of F to the estimation of the scale and location parameters a, b in the family $Q_3(\delta)$ see Section 2.4).

The EVD G_i lies in $Q_i(1)$, $i = 1, 2, 3$. The Cauchy distribution is in $Q_1(1)$, Student's t_n distribution with n degrees of freedom is in $Q_1(2/n)$, a triangular distribution lies in $Q_2(\delta)$ for any $\delta > 0$. Distributions F with upper Gamma tail that is, $f(x) = (c^p/\Gamma(p))e^{-cx}x^{p-1}$, $x \geq x_0 > 0$, with $c, p > 0$ and $p \neq 1$ do not belong to any class $Q_i(\delta)$.

A df F which belongs to one of the classes $Q_i(\delta)$ is obviously tail equivalent to the corresponding GPD $W_{i,\alpha}$ that is,

$$\lim_{x \to \omega(F)} \frac{1 - F(x)}{1 - W_{i,\alpha}((x-b)/a)} = 1 \qquad (2.5)$$

for some $a > 0$, $b \in \mathbb{R}$, with $b = 0$ in case $i = 1$ and $b = \omega(F)$ in case $i = 2$. Interpret $W_{3,\alpha}$ simply as W_3, as in the case $i = 3$ there is no shape parameter α. Consequently, we obtain from (2.5)

$$\lim_{q \to 0} \frac{F^{-1}(1-q)}{W_{i,\alpha}((\cdot - b)/a)^{-1}(1-q)} = \lim_{q \to 0} \frac{F^{-1}(1-q)}{aW_{i,\alpha}^{-1}(1-q) + b} = 1, \qquad (2.6)$$

and the estimation of large quantiles $F^{-1}(1-q)$ of F that is, for q near 0, then reduces within a certain error bound to the estimation of $aW_{i,\alpha}^{-1}(1-q) + b$.

The following result quantifies the error in (2.5) and (2.6) for a df F in a δ-neighborhood of a GPD.

Proposition 2.2.1. *Suppose that F lies in $Q_i(\delta)$ for some $\delta > 0$ that is, F is tail equivalent to some $W_{i,\alpha}((\cdot - b)/a)$, $i = 1, 2$ or 3, with $b = 0$ if $i = 1$ and $b = \omega(F)$ if $i = 2$. Then,*

(i) $$1 - F(x) = \left(1 - W_{i,\alpha}\left(\frac{x-b}{a}\right)\right)\left(1 + \psi_i(x)\right) \text{ as } x \to \omega(F),$$

where $\psi_i(x)$ decreases to zero at the order $O((1 - W_{i,\alpha}((x-b)/a))^\delta)$. We have in addition

(ii) $$F^{-1}(1-q) = \left(aW_{i,\alpha}^{-1}(1-q) + b\right)(1 + R_i(q)),$$

where

$$R_i(q) = \begin{cases} O(q^\delta) & i = 1 \text{ or } 2 \\ & \text{if} \\ O(q^\delta/\log(q)) & i = 3 \end{cases}$$

as $q \to 0$. Recall our convention $W_{3,\alpha} = W_3$.

Proof. Part (i) follows from elementary computations. The proof of part (ii) requires a bit more effort. From (i) we deduce the existence of a positive constant K such that, for q near zero with $W_{a,b}(t) := W_{i,\alpha}((t-b)/a)$,

$$F^{-1}(1-q) = \inf\{t \geq x_q : q \geq 1 - F(t)\}$$

$$= \inf\left\{t \geq x_q : q \geq \frac{1-F(t)}{1-W_{a,b}(t)}(1 - W_{a,b}(t))\right\}$$

$$\begin{cases} \leq \inf\{t \geq x_q : q \geq (1 + K \cdot r(t))(1 - W_{a,b}(t))\} \\ \geq \inf\{t \geq x_q : q \geq (1 - K \cdot r(t))(1 - W_{a,b}(t))\}, \end{cases}$$

where $r(x) = x^{-\alpha\delta}, |x - \omega(F)|^{\alpha\delta}, \exp(-(\delta/a)x)$ in case $i = 1, 2, 3$, and $x_q \to \omega(F)$ as $q \to 0$. Choose now

$$t_q^- := \begin{cases} aq^{-1/\alpha}(1 - K_1 q^\delta)^{-1/\alpha} & i = 1 \\ \omega(F) - aq^{1/\alpha}(1 - K_1 q^\delta)^{1/\alpha} & \text{in case } i = 2 \\ -a\log\{q(1 - K_1 q^\delta)\} + b & i = 3 \end{cases}$$

and

$$t_q^+ := \begin{cases} aq^{-1/\alpha}(1 + K_1 q^\delta)^{-1/\alpha} & i = 1 \\ \omega(F) - aq^{1/\alpha}(1 + K_1 q^\delta)^{1/\alpha} & \text{in case } i = 2 \\ -a\log\{q(1 + K_1 q^\delta)\} + b & i = 3, \end{cases}$$

for some large positive constant K_1. Then

$$(1 + Kr(t_q^-))(1 - W_{a,b}(t_q^-)) \leq q \quad \text{and} \quad (1 - Kr(t_q^+))(1 - W_{a,b}(t_q^+)) > q$$

for q near zero if K_1 is chosen large enough; recall that $b = 0$ in case $i = 1$. Consequently, we obtain for q near zero

$$t_q^+ \leq \inf\{t \geq x_q : q \geq (1 - Kr(t))(1 - W_{a,b}(t))\}$$

$$\leq F^{-1}(1-q)$$

$$\leq \inf\{t \geq x_q : q \geq (1 + Kr(t))(1 - W_{a,b}(t))\} \leq t_q^-.$$

The assertion now follows from the identity

$$W_{a,b}^{-1}(1-q) = aW_{i,\alpha}^{-1}(1-q) + b = \begin{cases} aq^{-1/\alpha} & i = 1 \\ \omega(F) - aq^{1/\alpha} & \text{in case } i = 2 \\ -a\log(q) + b & i = 3 \end{cases}$$

and elementary computations, which show that

$$t_q^+ = W_{a,b}^{-1}(1-q)(1 + O(R(q))), \quad t_q^- = W_{a,b}^{-1}(1-q)(1 + O(R(q))). \qquad \square$$

The approximation of the upper tail $1 - F(x)$ for large x by Pareto tails under von Mises conditions on F was discussed by Davis and Resnick [93]. New in the preceding result is the assumption that F lies in a δ-neighborhood of a GPD, which entails the handy error terms in the expansions of the tail and of large quantiles of F in terms of GPD ones. As we have explained above, this assumption $F \in Q_i(\delta)$ is actually a fairly general one.

Data Transformations

Suppose that F is in $Q_1(\delta)$. Then F has ultimately a density f such that, for some $\alpha, a > 0$,

$$f(x) = \frac{1}{a} w_{1,\alpha}\left(\frac{x}{a}\right)\left(1 + O((1 - W_{1,\alpha}(x))^\delta)\right)$$

as $x \to \infty$. In this case, the df with upper tail

$$F_1(x) := F(\exp(x)), \qquad x \ge x_0, \tag{2.7}$$

is in $Q_3(\delta)$. To be precise, F_1 has ultimately a density f_1 such that

$$f_1(x) = \alpha w_3\left(\frac{x - \log(a)}{1/\alpha}\right)\left(1 + O((1 - W_3(\alpha x))^\delta)\right)$$

$$= \frac{1}{a_0} w_3\left(\frac{x - b_0}{a_0}\right)\left(1 + O\left(\left(1 - W_3\left(\frac{x}{a_0}\right)\right)^\delta\right)\right), \qquad x \ge x_0,$$

with $a_0 = 1/\alpha$ and $b_0 = \log(a)$.

If we suppose that F is in $Q_2(\delta)$ that is,

$$f(x) = \frac{1}{a} w_{2,\alpha}\left(\frac{x - \omega(F)}{a}\right)\left(1 + O((1 - W_{2,\alpha}(x - \omega(F)))^\delta)\right)$$

as $x \to \omega(F) < \infty$ for some $\alpha, a > 0$, then

$$F_2(x) := F(\omega(F) - \exp(-x)), \qquad x \in \mathbb{R}, \tag{2.8}$$

is in $Q_3(\delta)$. The df F_2 has ultimately a density f_2 such that

$$f_2(x) = \alpha w_3\left(\frac{x + \log(a)}{1/\alpha}\right)\left(1 + O((1 - W_3(\alpha x))^\delta)\right)$$

$$= \frac{1}{a_0} w_3\left(\frac{x - b_0}{a_0}\right)\left(1 + O\left(\left(1 - W_3\left(\frac{x}{a_0}\right)\right)^\delta\right)\right), \qquad x \ge x_0,$$

with $a_0 = 1/\alpha$ and $b_0 = -\log(a)$.

The message of the preceding considerations can be summarized as follows. Suppose it is known that F is in $Q_1(\delta)$, $Q_2(\delta)$ or in $Q_3(\delta)$, but neither the particular shape parameter α nor the scale parameter a is known in case $F \in Q_i(\delta)$, $i = 1, 2$. Then a suitable data transformation which does not depend on α and a results in an underlying df F_i which is in $Q_3(\delta)$. And in $Q_3(\delta)$ the estimation of large quantiles reduces to the estimation of a scale and location parameter for the exponential distribution; this in turn allows the application of standard techniques. Details will be given in the next section. A brief discussion of that case, where F is in $Q_2(\delta)$ but $\omega(F)$ is unknown, is given after Lemma 2.4.3.

If it is assumed that F lies in a δ-neighborhood $Q_i(\delta)$ of a GPD for some $i \in \{1, 2, 3\}$, but the index i is unknown, then an initial estimation of the class index i is necessary. A suggestion based on Pickands [371] estimator of the extreme value index α is discussed in Section 2.5.

JOINT ASYMPTOTIC DISTRIBUTION OF EXTREMES

The following result describes the set of possible limiting distributions of the joint distribution of the k largest order statistics $Z_{n:n} \geq \cdots \geq Z_{n-k+1:n}$, equally standardized, in an iid sample $Z_1, \ldots, Z_n$. By $\longrightarrow_{\mathcal{D}}$ we denote the usual weak convergence.

Theorem 2.2.2 (Dwass [117]). *Let $Z_1, Z_2, \ldots$ be iid rv. Then we have for an EVD G and norming constants $a_n > 0$, $b_n \in \mathbb{R}$,*

$$\frac{Z_{n:n} - b_n}{a_n} \longrightarrow_D G$$

$$\Longleftrightarrow \quad \left(\frac{Z_{n-i+1:n} - b_n}{a_n} \right)_{i \leq k} \longrightarrow_D G^{(k)} \qquad for\ any \quad k \in \mathbb{N},$$

where the distribution $G^{(k)}/\mathbb{B}^k$ has Lebesgue density $g^{(k)}(x_1, \ldots, x_k) = G(x_k) \prod_{i \leq k} G'(x_i)/G(x_i)$ for $x_1 > \cdots > x_k$ and zero elsewhere.

Remark 2.2.3. Let $\xi_1, \xi_2, \ldots$ be a sequence of independent and standard exponential rv. Then $G_{1,\alpha}^{(k)}$ is the distribution of $((\sum_{j \leq i} \xi_j)^{-1/\alpha})_{i \leq k}$, $G_{2,\alpha}^{(k)}$ that of $(-(\sum_{j \leq i} \xi_j)^{1/\alpha})_{i \leq k}$ and $G_3^{(k)}$ that of $(-\log(\sum_{j \leq i} \xi_j))_{i \leq k}$. This representation was already utilized in Theorems 2.1.10-2.1.12.

Proof of Theorem 2.2.2. We have to show the only-if part of the assertion. Consider without loss of generality $Z_i = F^{-1}(U_i)$, where $U_1, U_2, \ldots$ are independent and uniformly on (0,1) distributed rv, and where F denotes the df of Z_i. Then we have the representation $(Z_{i:n})_{i \leq n} = (F^{-1}(U_{i:n}))_{i \leq n}$, and by the equivalence

$$F^{-1}(q) \leq t \iff q \leq F(t), \qquad q \in (0,1), \quad t \in \mathbb{R},$$

we can write

$$P\Big((Z_{n-i+1:n} - b_n)/a_n \leq x_i,\ 1 \leq i \leq k\Big)$$

$$= P\Big(F^{-1}(U_{n-i+1:n}) \leq a_n x_i + b_n,\ 1 \leq i \leq k\Big)$$

$$= P\Big(U_{n-i+1:n} \leq F(a_n x_i + b_n),\ 1 \leq i \leq k\Big)$$

$$= P\Big(n(U_{n-i+1:n} - 1) \leq n(F(a_n x_i + b_n) - 1),\ 1 \leq i \leq k\Big).$$

As the convergence $F^n(a_n x + b_n) \longrightarrow_{n \to \infty} G(x)$, $x \in \mathbb{R}$, is equivalent to $n(F(a_n x + b_n) - 1) \longrightarrow_{n \to \infty} \log(G(x))$, $0 < G(x) \leq 1$, and, as is easy to see, $(n(U_{n-i+1:n} - 1))_{i \leq k} \longrightarrow_D G_{2,1}^{(k)}$ with density $g_{2,1}^{(k)}(x_1, \ldots, x_k) = \exp(x_k)$ if $0 > x_1 > \cdots > x_k$ and 0 elsewhere, we obtain

$$P\Big((Z_{n-i+1:n} - b_n)/a_n \leq x_i,\ 1 \leq i \leq k\Big) \longrightarrow_{n \to \infty} G_{2,1}^{(k)}\Big((\log(G(x_i)))_{i \leq k}\Big).$$

This implies the assertion.   □

For a proof of the following result, which provides a rate of convergence in the preceding theorem if the upper tail of the underlying distribution is in a δ-neighborhood of a GPD, we refer to Corollary 5.5.5 of Reiss [385] (cf. also Theorem 2.1.10).

Theorem 2.2.4. *Suppose that the df F is in a δ-neighborhood $Q_i(\delta)$ of a GPD $W_i = W_{1,\alpha}$, $i = 1, 2$ or 3. Then there obviously exist constants $a > 0$, $b \in \mathbb{R}$, with $b = 0$ if $i = 1$, $b = \omega(F)$ if $i = 2$, such that*

$$af(ax + b) = w_i(x)\Big(1 + O((1 - W_i(x))^\delta)\Big) \tag{2.9}$$

for all x in a left neighborhood of $\omega(W_{i,\alpha})$. Consequently, we obtain from Corollary 5.5.5 in Reiss [385]

$$\sup_{B \in \mathbb{B}^k} \left| P\left(\left(\left(\frac{Z_{n-j+1:n} - b}{a} - d_n\right)/c_n\right)_{j \le k} \in B\right) - G^{(k)}(B) \right|$$

$$= O((k/n)^\delta k^{1/2} + k/n),$$

where $d_n = 0$ for $i = 1, 2$; $d_n = \log(n)$ for $i = 3$; $c_n = n^{1/\alpha}$, $n^{-1/\alpha}$, 1 for $i = 1, 2, 3$.

Notice that df F whose upper tails *coincide* with that of a GPD, are actually the only ones where the term $(k/n)^\delta k^{1/2}$ in the preceding bound can be dropped (cf. Theorem 2.1.11). This is indicated by Theorem 2.2.4, as δ can then and only then be chosen arbitrarily large.

SUMMARIZING THE RESULTS

The following list of equivalences now follows from Proposition 2.1.4, 2.1.5 and Theorem 2.1.8, 2.2.4. They summarize our considerations of this section and the preceding one.

Theorem 2.2.5. *Suppose that F satisfies condition (VM) from the preceding section with $\beta \in \mathbb{R}$ and $c > 0$, such that the remainder function $\eta(x)$ is proportional to some monotone function as $x \to \omega(F) = \omega(H_\beta)$ and $\int_{x_0}^{\omega(F)} \eta(t)/(1 + \beta t)\, dt \in \mathbb{R}$. Then there exist $a > 0$, $b \in \mathbb{R}$ with $b = -1/\beta$ if $\beta \ne 0$, such that*

$$\lim_{x \uparrow \omega(W_i)} \frac{1 - F(ax + b)}{1 - W_i(x)} = \lim_{x \uparrow \omega(W_i)} \frac{af(ax + b)}{w_i(x)} = 1,$$

where $i = 1, 2, 3$ if $\beta > 0, < 0, = 0$ and $W_i = W_{1,c/\beta}, W_{2,c/\beta}, W_3$. Consequently, with c_n, d_n as in the preceding result

$$\sup_{x \in \mathbb{R}} \left| P\left(\left(\frac{Z_{n:n} - b}{a} - d_n\right)/c_n \le x\right) - G_i(x) \right| \longrightarrow_{n \to \infty} 0,$$

where $Z_1, \ldots, Z_n$ are iid with common df F. Moreover, we have the following list of equivalences:

$$\sup_{x \in \mathbb{R}} \left| P\left(\left(\frac{Z_{n:n} - b}{a} - d_n \right) / c_n \leq x \right) - G_i(x) \right| = O(n^{-\delta}) \text{ for some } \delta > 0$$

$\iff$ *there exists $\delta > 0$ such that for $x \to \omega(F)$*

$$\eta(x) = O((1 - H_\beta(x))^\delta)$$

$\iff$ *F is in a δ-neighborhood $Q_i(\delta)$ of the GPD W_i*

$\iff$ *there exists $\delta > 0$ such that, for $k \in \{1, \ldots, n\}$, $n \in \mathbb{N}$,*

$$\sup_{B \in \mathbb{B}^k} \left| P\left(\left(\left(\frac{Z_{n-j+1:n} - b}{a} - d_n \right) / c_n \right)_{j \leq k} \in B \right) - G^{(k)}(B) \right|$$
$$= O\left((k/n)^\delta k^{1/2} + k/n \right).$$

2.3 The Peaks-Over-Threshold Method

The following example seems to represent one of the first applications of the POT approach (de Haan [189]).

Example 2.3.1. After the disastrous flood of February 1st, 1953, in which the sea-dikes broke in several parts of the Netherlands and nearly two thousand people were killed, the Dutch government appointed a committee (so-called Delta-committee) to recommend an appropriate level for the dikes (called Delta-level since) since no specific statistical study had been done to fix a safer level for the sea-dikes before 1953. The Dutch government set as the standard for the sea-dikes that at any time in a given year the sea level exceeds the level of the dikes with probability 1/10,000. A statistical group from the Mathematical Centre in Amsterdam headed by D. van Dantzig showed that high tides occurring during certain dangerous windstorms (to ensure independence) within the dangerous winter months December, January and February (for homogeneity) follow closely an exponential distribution if the smaller high tides are neglected.

If we model the annual maximum flood by a rv Z, the Dutch government wanted to determine therefore the $(1 - q)$-quantile

$$F^{-1}(1 - q) = \inf\{t \in \mathbb{R} : F(t) \geq 1 - q\}$$

of Z, where F denotes the df of Z and q has the value 10^{-4}.

THE POINT PROCESS OF EXCEEDANCES

From the past we have observations $Z_1, \ldots, Z_n$ (annual maximum floods), which we assume to be independent replicates of Z. With these rv we define the truncated empirical point process

$$N_n^{(t)}(\cdot) := \sum_{j \le n} \varepsilon_{Z_j}(\cdot \cap (t, \infty))$$

that is, we consider only those observations which exceed the *threshold* t. The process $N_n^{(t)}$ is therefore called the point process of the exceedances.

From Theorem 1.3.1 we know that we can write

$$N_n^{(t)}(\cdot) = \sum_{j \le K_t(n)} \varepsilon_{V_j^{(t)} + t}(\cdot),$$

where the *excesses* $V_1^{(t)}, V_2^{(t)}, \ldots$ are independent replicates of a rv $V^{(t)}$ with df $F^{(t)}(\cdot) := P(Z \le t + \cdot | Z \ge t)$, and these are independent of the sample size $K_t(n) := \sum_{i \le n} \varepsilon_{Z_i}((t, \infty))$.

Without specific assumptions, the problem to determine $F^{-1}(1 - q)$ is a non-parametric one. If we require however that the underlying df F is in a δ-neighborhood of a GPD, then this non-parametric problem can be approximated within a reasonable error bound by a parametric one.

Approximation of Excess Distributions

Suppose therefore that the df F of Z is in a δ-neighborhood $Q_i(\delta)$ of a GPD W_i that is, there exist δ, $a > 0$, $b \in \mathbb{R}$, with $b = 0$ if $i = 1$ and $b = \omega(F)$ if $i = 2$, such that, for $x \to \omega(F)$,

$$f(x) = \frac{1}{a} w_i \left(\frac{x - b}{a} \right) \left(1 + O \left(\left(1 - W_i \left(\frac{x - b}{a} \right) \right)^\delta \right) \right),$$

where F has density f in a left neighborhood of $\omega(F)$.

In this case, the df $F^{(t)}(s)$, $s \ge 0$, of the excess $V^{(t)}$ has density $f^{(t)}$ for all t in a left neighborhood of $\omega(F)$, with the representation

$$f^{(t)}(s) = \frac{f(t + s)}{1 - F(t)}$$

$$= \frac{\frac{1}{a} w_i \left(\frac{t+s-b}{a} \right)}{1 - W_i \left(\frac{t-b}{a} \right)} \cdot \frac{1 + O((1 - W_i(\frac{t+s-b}{a}))^\delta)}{1 + O((1 - W_i(\frac{t-b}{a}))^\delta)}$$

$$= \frac{\frac{1}{a} w_i \left(\frac{t+s-b}{a} \right)}{1 - W_i \left(\frac{t-b}{a} \right)} \left(1 + O \left(\left(1 - W_i \left(\frac{t-b}{a} \right) \right)^\delta \right) \right), \quad s \ge 0.$$

Note that $a^{-1} w_i((t + s - b)/a)/(1 - W_i((t - b)/a))$, $s \ge 0$, with $0 < W_i((t - b)/a) < 1$ and $b = 0$ if $i = 1$, $b = \omega(F)$ if $i = 2$, is again the density of a GPD $W_i^{(t)}$, precisely of

$$
W_i^{(t)}(s) = \begin{cases} W_1\left(1 + \frac{s}{t}\right) & i = 1 \\ W_2\left(-1 + \frac{s}{\omega(F)-t}\right) & \text{if} \quad i = 2, \quad s \geq 0, \\ W_3\left(\frac{s}{a}\right) & i = 3. \end{cases}
$$

We can consequently approximate the truncated empirical point process

$$
N_n^{(t)}(\cdot) = \sum_{j \leq n} \varepsilon_{Z_j}(\cdot \cap (t, \infty))
$$

$$
= \sum_{j \leq K_t(n)} \varepsilon_{V_j^{(t)}+t}(\cdot),
$$

pertaining to the $K_t(n)$ exceedances $V_1^{(t)} + t, \dots, V_{K_t(n)}^{(t)} + t$ over the threshold t, by the binomial point process

$$
M_n^{(t)} = \sum_{j \leq K_t(n)} \varepsilon_{c\xi_j+d+t},
$$

where $c = t$, $d = -t$ in case $i = 1$, $c = d = \omega(F) - t$ in case $i = 2$ and $c = a$, $d = 0$ in case $i = 3$, and $\xi_1, \xi_2 \dots$ are independent copies of a rv ξ having df W_i, and independent also from their random counting number $K_t(n)$.

BOUNDS FOR THE PROCESS APPROXIMATIONS

Choose the particular threshold

$$
t = aW_i^{-1}\left(1 - \frac{r}{n}\right) + b,
$$

with r/n less than a suitable positive constant c_0 such that t is in a proper left neighborhood of $\omega(F) = a\,\omega(W_i) + b$. By Corollary 1.2.4 (iv) we obtain for the Hellinger distance $H(N_n^{(t)}, M_n^{(t)})$ between $N_n^{(t)}$ and $M_n^{(t)}$ uniformly for $0 < r/n < c_0$ the bound

$$
H(N_n^{(t)}, M_n^{(t)}) \leq H(V^{(t)}, c\xi + d)\,(E(K_t(n)))^{1/2}
$$

$$
= O((1 - W_i((t-b)/a))^\delta)\,(E(K_t(n)))^{1/2}
$$

$$
= O((r/n)^\delta\,(n(1 - F(t)))^{1/2}) \;=\; O((r/n)^\delta\, r^{1/2}).
$$

As the Hellinger distance is in general bounded by $2^{1/2}$, we can drop the assumption $r/n \leq c_0$ and the preceding bound is therefore true uniformly for $0 < r < n$.

The preceding inequality explains the exponential fit of the high tides by van Dantzig in Example 2.3.1, if the smaller high tides are neglected. This peaks-over-threshold method is not only widely used by hydrologists to model large floods (Smith [420], Davison and Smith [94]), but also in insurance mathematics for modeling large claims (Teugels [441], [442], Kremer [297], Reiss [384]). For thorough discussions of the peaks-over-threshold approach in the investigation of extreme values and further references we refer to Section 6.5 of Embrechts et al. [122], Section 4 of Coles [71] and to Reiss and Thomas [389].

Replacing $M_n^{(t)}$ by the Poisson process

$$N_n^{(t)**}(\cdot) := \sum_{j \leq \tau_t(n)} \varepsilon_{c\xi_j + d + t}(\cdot),$$

with $\tau_t(n)$ being a Poisson rv with parameter $n(1 - F(t))$, we obtain therefore by Theorem 1.2.5 the following bound for the second-order Poisson process approximation of $N_n^{(t)}$ by $N_n^{(t)**}$,

$$H(N_n^{(t)}, N_n^{(t)**}) = O\Big(r^{1/2}(r/n)^\delta + (1 - F(t))\Big)$$
$$= O\Big(r^{1/2}(r/n)^\delta + r/n\Big),$$

uniformly for $0 < r < n$ and $n \in \mathbb{N}$.

The preceding considerations are summarized in the following result providing bounds for functional laws of small numbers in an EVD model.

Theorem 2.3.2. *Suppose that F is in the δ-neighborhood $Q_i(\delta)$ of some GPD W_i, $i = 1, 2$ or 3. Then there exist $a > 0$, $b \in \mathbb{R}$, with $b = 0$ in case $i = 1$ and $b = \omega(F)$ in case $i = 2$ such that*

$$\lim_{x \uparrow \omega(W_i)} \frac{1 - F(ax + b)}{1 - W_i(x)} = 1.$$

Define for $r \in (0, n)$ the threshold

$$t := t(n) := aW_i^{-1}\left(1 - \frac{r}{n}\right) + b$$

and denote by

$$N_n^{(t)} = \sum_{j \leq n} \varepsilon_{Z_j}(\cdot \cap (t, \infty)) = \sum_{j \leq K_t(n)} \varepsilon_{V_j^{(t)} + t}(\cdot)$$

the point process of the exceedances among $Z_1, \ldots, Z_n$ over t.

Define the binomial process

$$M_n^{(t)} := \sum_{j \leq K_t(n)} \varepsilon_{c\xi_j + d + t}$$

and the Poisson processes

$$N_n^{(t)*} := \sum_{j \leq \tau_t(n)} \varepsilon_{V_j^{(t)}+t}, \quad N_n^{(t)**} := \sum_{j \leq \tau_t(n)} \varepsilon_{c\xi_j+d+t},$$

where $c = t$, $d = -t$, if $i = 1$, $c = d = \omega(t) - t$ if $i = 1$, $c = a$, $d = 0$ if $i = 3$; $\xi_1, \xi_2, \ldots$ are iid rv with common df W_i and $\tau_t(n)$ is a Poisson rv with parameter $n(1 - F(t))$, independent of the sequences $\xi_1, \xi_2, \ldots$ and of $V_1^{(t)}, V_2^{(t)}, \ldots$ Then we have the following bounds, uniformly for $0 < r < n$ and $n \in \mathbb{N}$,

$$H(N_n^{(t)}, M_n^{(t)}) = O(r^{1/2}(r/n)^\delta)$$

for the POT method,

$$H(N_n^{(t)}, N_n^{(t)*}) = O(r/n)$$

for the first-order Poisson process approximation and

$$H(N_n^{(t)}, N_n^{(t)**}) = O(r/n + r^{1/2}(r/n)^\delta)$$

for the second-order Poisson process approximation.

A binomial process approximation with an error bound based on the remainder function of the von Mises condition (VM$_0$) in Theorem 2.1.2 was established by Kaufmann and Reiss [287] (cf. also [389], 2nd ed., Section 6.4).

2.4 Parameter Estimation
in δ-Neighborhoods of GPD

Suppose that we are given an iid sample of size n from a df F, which lies in a δ-neighborhood $Q_i(\delta)$ of a GPD W_i. Then there exist α, $a > 0$, $b \in \mathbb{R}$, with $b = 0$ if $i = 1$ and $b = \omega(F)$ if $i = 2$ such that $F(x)$ and $1 - W_{i,\alpha}((x - b)/a)$ are tail equivalent. Interpret again $W_{3,\alpha}$ as W_3. We assume that the class index $i = 1, 2, 3$ and $\omega(F)$ are known. As shown in (2.7) and (2.8) in Section 2.2, a suitable data transformation, which does not depend on α or a, transposes $F \in Q_i(\delta)$, $i = 1$ or 2, to a df F_i which is in $Q_3(\delta)$. And in $Q_3(\delta)$ the estimation of upper tails reduces to the estimation of a scale and location parameter $a_0 > 0$, $b_0 \in \mathbb{R}$ for the exponential distribution, which in turn allows the application of standard techniques. A brief discussion of that case, where F is in $Q_2(\delta)$ but $\omega(F)$ is unknown, is given after Lemma 2.4.3.

If it is assumed that F lies in a δ-neighborhood $Q_i(\delta)$ of a GPD, but the class index i is unknown, then an initial estimation of the class index i is necessary. A suggestion based on Pickands [371] estimator of the extremal index α is discussed in the next section.

Our considerations are close in spirit to Weissman [458], who considers n iid observations with common df F being the EVD $G_3((x - b_0)/a_0)$ with unknown

2.4. Parameter Estimation in δ-Neighborhoods of GPD

scale and location parameter $a_0 > 0$, $b_0 \in \mathbb{R}$. Based on the upper k order statistics in the sample, he defines maximum-likelihood and UMVU estimators of a_0 and b_0 and resulting estimators of extreme quantiles $F^{-1}(1 - c/n)$. Equally, he proposes the data transformations (2.7) and (2.8) in case $F = G_{1,\alpha}$ or $G_{2,\alpha}$ but considers no asymptotics.

Viewing F as an element of $Q_i(\delta)$, we can establish asymptotics for UMVU estimators of a, b and of resulting estimators of extreme quantiles $F^{-1}(1 - q_n)$ with $q_n \to 0$ and $k = k(n) \to \infty$ as $n \to \infty$. It follows in particular from Corollary 2.4.6 that the error of the resulting estimator of $F^{-1}(1 - q_n)$ is of the order $O_p(q_n^{-\gamma(i)} k^{-1/2} (\log^2(nq_n/k) + 1)^{1/2})$, where $\gamma(i) = 1/\alpha, -1/\alpha, 0$ if $F \in Q_i(\delta)$, $i = 1, 2, 3$.

This demonstrates the superiority of the estimators to the ones proposed by Dekkers and de Haan [108] and Dekkers et al. [107], if nq_n is of smaller order than k. Note however that our estimators are based on the assumption that the class index i of the condition $F \in Q_i(\delta)$ is known, whereas those estimators proposed by Dekkers et al. are uniformly consistent.

The Basic Approximation Lemma

The following consequence of Theorem 2.2.4 is crucial.

Lemma 2.4.1. *Suppose that F is in $Q_3(\delta)$ for some $\delta > 0$. Then there exist $a_0 > 0$, $b_0 \in \mathbb{R}$ such that*

$$\sup_{B \in \mathbb{B}^{k+1}} \left| P\left(((Z_{n-j+1:n} - Z_{n-k:n})_{j=k}^1, \, Z_{n-k:n}) \in B \right) \right.$$
$$\left. - P\left(((a_0 X_{j:k})_{j \le k}, \, (a_0/k^{1/2})Y + a_0 \log(n/k) + b_0) \in B \right) \right|$$
$$= O(k/n + (k/n)^\delta k^{1/2} + k^{-1/2}),$$

where $Y, X_1, \ldots, X_k$ are independent rv, Y is standard normal and $X_1, \ldots, X_k$ are standard exponential distributed.

Proof. By Theorem 2.2.4 there exist $a_0 > 0$ and $b_0 \in \mathbb{R}$ such that

$$\sup_{B \in \mathbb{B}^k} \left| P\left(((Z_{n-j+1:n} - b_0)/a_0 - \log(n))_{j \le k} \in B \right) - G_3^{(k+1)}(B) \right|$$
$$= O((k/n)^\delta k^{1/2} + k/n).$$

Recall that $G_3^{(k+1)}$ is the distribution of the vector $(-\log(\sum_{j \le r} \xi_j))_{r \le k+1}$, where $\xi_1, \xi_2, \ldots$ are independent and standard exponential distributed rv (Remark 2.2.3). Within the preceding bound, the rv $((Z_{n-j+1:n} - Z_{n-k:n})/a_0)_{j=k}^1$, $(Z_{n-k:n} - b_0)/a_0 - \log(n))$ behaves therefore like

$$\left(\left(-\log\left(\sum_{j\leq r}\xi_j\right)+\log\left(\sum_{j\leq k+1}\xi_j\right)\right)^1_{r=k}, -\log\left(\sum_{j\leq k+1}\xi_j\right)\right)$$

$$=\left(\left(-\log\left(\sum_{j\leq r}\xi_j\Big/\sum_{j\leq k+1}\xi_j\right)\right)^1_{r=k}, -\log\left(\sum_{j\leq k+1}\xi_j\right)\right)$$

$$=_D\left((X_{r:k})_{r\leq k}, -\log\left(\sum_{j\leq k+1}\xi_j\right)\right),$$

where $X_1, X_2, \ldots, \xi_1, \xi_2, \ldots$ are independent sets of independent standard exponential rv. By $=_D$ we denote equality of distributions. This follows from the facts that $(\sum_{j\leq r}\xi_j/\sum_{j\leq k+1}\xi_j)_{r\leq k}$ and $\sum_{j\leq k+1}\xi_j$ are independent (Lemma 1.6.6 in Reiss [385]), that $(\sum_{j\leq r}\xi_j/\sum_{j\leq k+1}\xi_j)_{r\leq k} =_D (U_{r:k})_{r\leq k}$, where $U_1, \ldots, U_k$ are independent and uniformly on $(0,1)$ distributed rv (Corollary 1.6.9 in Reiss [385]), and that $-\log(1-U)$ is a standard exponential rv if U is uniformly on $(0,1)$ distributed. Finally it is straightforward to show that $-\log(\sum_{j\leq k+1}\xi_j)$ is in variational distance within the bound $O(k^{-1/2})$ distributed like $Y/k^{1/2} - \log(k)$. $\square$

The preceding result shows that within a certain error bound depending on δ, the k excesses $(Z_{n-j+1:n} - Z_{n-k:n})^1_{j=k}$ over the *random threshold* $Z_{n-k:n}$ can be handled in case $F \in Q_3(\delta)$ like a complete set $(a_0X_{j:k})_{j\leq k}$ of order statistics from an exponential distribution with unknown scale parameter $a_0 > 0$, whereas the random threshold $Z_{n-k:n}$ behaves like $a_0 k^{-1/2}Y + a_0 \log(n/k) + b_0$, where Y is a standard normal rv being independent of $(X_{j:k})_{j\leq k}$. Notice that no information from $(Z_{n-j+1:n})_{j\leq k+1}$ is lost if we consider $((Z_{n-j+1:n} - Z_{n-k:n})^1_{j=k}, Z_{n-k:n})$ instead.

EFFICIENT ESTIMATORS OF a_0 AND b_0

After the transition to the model $((a_0X_{j:k})_{j\leq k}, (a_0/k^{1/2})Y + a_0\log(n/k) + b_0)$, we search for efficient estimators of a_0 and b_0 within this model.

Ad hoc estimators of the parameters $a_0 > 0$, $b_0 \in \mathbb{R}$ in the model

$$\{(V_{j:k})_{j\leq k}, \xi\}$$
$$= \left\{(a_0X_{j:k})_{j\leq k}, (a_0/k^{1/2})Y + a_0\log(n/k) + b_0 : a_0 > 0, b_0 \in \mathbb{R}\right\}$$

are

$$\hat{a}_k := k^{-1}\sum_{j\leq k} V_{j:k}$$

and

$$\hat{b}_{k,n} := \xi - \hat{a}_k \log(n/k).$$

The joint density f_{a_0,b_0} of $((V_{j:k})_{j\le k}, \xi)$ is

$$f_{a_0,b_0}(\boldsymbol{x}, y) = \frac{k!\, k^{1/2}}{a_0^{k+1}(2\pi)^{1/2}} \exp\left(-a_0^{-1} \sum_{j\le k} x_j\right)$$
$$\times \exp\left(-\frac{(y - a_0 \log(n/k) - b_0)^2}{2a_0^2/k}\right),$$

for $\boldsymbol{x} = (x_1, \ldots, x_k) \in \mathbb{R}^k$, if $0 < x_1 < \cdots < x_k$, $y \in \mathbb{R}$, and zero elsewhere (Example 1.4.2 (i) in Reiss [385]). This representation implies with respect to the family $\mathcal{P} := \{f_{a_0,b_0} : a_0 > 0,\ b_0 \in \mathbb{R}\}$. It is straightforward to show that $\mathcal{P}$ is an exponential family, and by using standard arguments from the theory of such families (see e.g. Chapter 3 of the book by Witting [463]), it is elementary to prove that $(\hat{a}_k, \hat{b}_{k,n})$ is a *complete* statistic as well. Altogether we have the following result.

Proposition 2.4.2. *The estimators $\hat{a}_k, \hat{b}_{k,n}$ are UMVU (uniformly minimum variance unbiased) estimators of a_0, b_0 for the family $\mathcal{P} = \{f_{a_0,b_0} : a_0 > 0,\ b_0 \in \mathbb{R}\}$.*

It is straightforward to show that $k^{-1/2} \sum_{i\le k}(X_i - 1)$ approaches the standard normal distribution $N(0,1)$ within the error bound $O(k^{-1/2})$ in variational distance. The following auxiliary result is therefore obvious.

Lemma 2.4.3. *We have uniformly in $\mathcal{P}$ the bound*

$$\sup_{B\in\mathbb{B}^2} \left| P\Big(((k^{1/2}(\hat{a}_k - a_0),\ (k^{1/2}/\log(n/k))(\hat{b}_{k,n} - b_0)) \in B\Big) \right.$$
$$\left. - P\Big((a_0\xi_1, (a_0/\log(n/k))\xi_2 - a_0\xi_1) \in B\Big) \right|$$
$$= O(k^{-1/2}),$$

where ξ_1, ξ_2 are independent standard normal rv.

HILL'S ESTIMATOR AND FRIENDS

If we plug our initial data $(Z_{n-j+1:n} - Z_{n-k:n})_{j=k}^1$, $Z_{n-k:n}$ into $\hat{a}_k$ and $\hat{b}_{k,n}$, we obtain the estimators

$$\hat{a}_{n,3} := \hat{a}_k((Z_{n-j+1:n} - Z_{n-k:n})_{j=k}^1)$$
$$= k^{-1} \sum_{j\le k} Z_{n-j+1:n} - Z_{n-k:n},$$

and

$$\hat{b}_{n,3} := Z_{n-k:n} - \log(n/k)\hat{a}_{n,3}$$

of a_0 and b_0 in case $F \in Q_3(\delta)$.

If we suppose that F is in $Q_1(\delta)$, then we know already that $F_1(x) = F(\exp(x))$ is in $Q_3(\delta)$, where the initial shape α parameter of F becomes the scale parameter $a_0 = 1/\alpha$ (cf. (2.7)).

We replace therefore $Z_{j:n}$ in this case by the log-transformed data $\log(Z_{j:n} \wedge 1) = \log(\max\{Z_{j:n}, 1\})$ and define the estimators

$$\hat{a}_{n,1} := \hat{a}_{n,3}\Big((\log(Z_{n-j+1:n} \wedge 1) - \log(Z_{n-k:n} \wedge 1))_{j=k}^1\Big)$$
$$= k^{-1} \sum_{j \le k} \log(Z_{n-j+1:n} \wedge 1) - \log(Z_{n-k:n} \wedge 1),$$

and

$$\hat{b}_{n,1} := \hat{b}_{n,3}\Big((\log(Z_{n-j+1:n} \wedge 1) - \log(Z_{n-k:n} \wedge 1))_{j=k}^1, \ \log(Z_{n-k:n} \wedge 1)\Big)$$
$$= \log(Z_{n-k:n} \wedge 1) - \log(n/k)\hat{a}_{n,1}$$

of a_0 and b_0. The estimator $\hat{a}_{n,1}$ is known as the *Hill estimator* (Hill [217]). It actually estimates $1/\alpha$, the reciprocal of the initial shape parameter α of F. Note that the upper tail of the df of $Z \wedge 1$ and of Z coincide as $\omega(F) = \infty$. Asymptotic normality of $k^{1/2}(\hat{a}_{n,1} - 1/\alpha)$ with mean 0 and variance $1/\alpha^2$ under suitable conditions on F and the sequence $k = k(n)$ is well known (see, for example, Hall [202], Csörgő and Mason [84], Hall and Welsh [204], Häusler and Teugels [212]). For a thorough discussion of the Hill estimator we refer to Section 6.4 of Embrechts et al. [122].

If the underlying F is in $Q_2(\delta)$, the transformation $-\log(\omega(F) - Z_j)$ of our initial data Z_j leads us back to $Q_3(\delta)$ with particular scale and location parameters $a_0 > 0$, $b_0 \in \mathbb{R}$ (cf. (2.8)). The pertaining estimators are now

$$\hat{a}_{n,2} := \hat{a}_{n,3}\Big((-\log(\omega(F) - Z_{n-j+1:n}) + \log(\omega(F) - Z_{n-k:n}))_{j=k}^1\Big)$$
$$= \log(\omega(F) - Z_{n-k:n}) - k^{-1} \sum_{j \le k} \log(\omega(F) - Z_{n-j+1:n})$$

and

$$\hat{b}_{n,2} := -\log(\omega(F) - Z_{n-k:n}) - \log(n/k)\hat{a}_{n,2}.$$

If the endpoint $\omega(F)$ of F is finite but unknown, then we can replace the transformation $-\log(\omega(F) - Z_{j:n})$ of our initial data Z_j by the data-driven transformation $-\log(Z_{n:n} - Z_{j:n})$ and j running from 1 to $n-1$. This yields the modified versions

$$\hat{a}'_{n,2} := \log\Big(Z_{n:n} - Z_{n-k:n}\Big) - (k-1)^{-1} \sum_{2 \le j \le k} \log(Z_{n:n} - Z_{n-j+1:n})$$

and

$$\hat{b}'_{n,2} := -\log\Big(Z_{n:n} - Z_{n-k:n}\Big) - \log(n/k)\hat{a}'_{n,2}.$$

of the estimators $\hat{a}_{n,2}$ and $\hat{b}_{n,2}$.

One can show (cf. Falk [134]) that in case $0 < \alpha < 2$, the data-driven estimators $\hat{a}'_{n,2}, \hat{b}'_{n,2}$ perform asymptotically as good as their counterparts $\hat{a}_{n,2}, \hat{b}_{n,2}$ with known $\omega(F)$. Precisely, if $k = k(n)$ satisfies $k/n \to 0$, $\log(n)/k^{1/2} \to 0$ as n tends to infinity, we have

$$k^{1/2}|\hat{a}_{n,2} - \hat{a}'_{n,2}| = o_P(1)$$

and

$$(k^{1/2}/\log(n/k))|\hat{b}_{n,2} - \hat{b}'_{n,2}| = o_P(1).$$

As a consequence, the asymptotic normality of $(\hat{a}_{n,2}, \hat{b}_{n,2})$, which follows from the next result if in addition $(k/n)^\delta k^{1/2} \to 0$ as n increases, carries over to $(\hat{a}'_{n,2}, \hat{b}'_{n,2})$. If $\alpha \geq 2$, then maximum likelihood estimators of $\omega(F)$, a and $1/\alpha$ can be obtained, based on an increasing number of upper-order statistics. We refer to Hall [201] and, in case α known, to Csörgő and Mason [85]. For a discussion of maximum likelihood estimation of general EVD we refer to Section 6.3.1 of Embrechts et al. [122], Section 1.7.5 of Kotz and Nadarajah [293] and to Section 4.1 of Reiss and Thomas [389]. The following result summarizes the preceding considerations and Proposition 2.2.1.

Theorem 2.4.4. *Suppose that F is in $Q_i(\delta)$, $i = 1, 2$ or 3 for some $\delta > 0$ that is, F is in particular tail equivalent to a GPD $W_{i,\alpha}((x - b)/a)$, where $b = 0$ if $i = 1$ and $b = \omega(F)$ if $i = 2$. Then we have in case*

$$\left. \begin{array}{llll} i = 1: & 1 - F_1(x) & = 1 - F(\exp(x)) \\ i = 2: & 1 - F_2(x) & = 1 - F(\omega(F) - \exp(-x)) \\ i = 3: & 1 - F_3(x) & := 1 - F(x) \end{array} \right\}$$

$$= (1 - W_3((x - b_0)/a_0))\Big(1 + O(\exp(-(\delta/a_0)x))\Big)$$

with $a_0 = 1/\alpha$, $b_0 = \log(a)$ if $i = 1$; $a_0 = 1/\alpha$, $b_0 = -\log(a)$ if $i = 2$ and $a_0 = a$, $b_0 = b$ if $i = 3$. Furthermore,

$$f_3(x) = a_0^{-1} w_3((x - b_0)/a_0)\Big(1 + O(\exp(-(\delta/a_0)x))\Big)$$

for $x \to \infty$ and

$$F_i^{-1}(1 - q) = \Big((a_0 W_3^{-1}(1 - q) + b_0)\Big)\Big(1 + O(q^\delta/\log(q))\Big), \quad i = 1, 2, 3$$

as $q \to 0$. Finally, we have for $i = 1, 2, 3$ the representations

$$\sup_{B \in \mathbb{B}^2} \left| P\Big(\Big(k^{1/2}(\hat{a}_{n,i} - a_0), (k^{1/2}/\log(n/k))(\hat{b}_{n,i} - b_0)\Big) \in B\Big) \right.$$

$$\left. - P\Big(\Big(a_0\xi_1, (a_0/\log(n/k))\xi_2 - a_0\xi_1\Big) \in B\Big) \right|$$

$$= O(k/n + (k/n)^\delta k^{1/2} + k^{-1/2}),$$

where ξ_1, ξ_2 are independent standard normal rv.

Note that in cases $i = 1$ and 2, estimators of the initial scale parameter a in the model $F \in Q_i(\delta)$ are given by $\exp(\hat{b}_{n,1}) \sim \exp(b_0) = a$ and $\exp(-\hat{b}_{n,2}) \sim \exp(-b_0) = a$, respectively. Their asymptotic behavior can easily be deduced from the preceding theorem and Taylor expansion of the exponential function.

THE PARETO MODEL WITH KNOWN SCALE FACTOR

Suppose that the df F underlying the iid sample $Z_1, \ldots, Z_n$ is in $Q_1(\delta)$. Then F has a density f on (x_0, ∞) such that

$$f(x) = \frac{1}{a} w_{1,1/\alpha}\left(\frac{x}{a}\right)\left(1 + O((1 - W_{1,1/\alpha}(x))^\delta)\right), \quad x > x_0, \qquad (2.10)$$

for some $\alpha, \delta, a > 0$. Note that we have replaced α by $1/\alpha$. The preceding result states that for the Hill estimator

$$\hat{a}_{n,1} = k^{-1} \sum_{j \leq k} \log(Z_{n-j+1:n} \wedge 1) - \log(Z_{n-k:n} \wedge 1)$$

of $a_0 = \alpha$ we have

$$\sup_{B \in \mathbb{B}} \left| P\left(\frac{k^{1/2}}{\alpha}(\hat{a}_{n,1} - \alpha) \in B\right) - N(0,1)(B) \right|$$
$$= O(k/n + (k/n)^\delta k^{1/2} + k^{-1/2}),$$

where $N(0,1)$ denotes the standard normal distribution on $\mathbb{R}$. If the scale parameter a in (2.10) is however *known*, then the Hill estimator is outperformed by

$$\hat{\alpha}_n := \frac{\log(Z_{n-k:n} \wedge 1) - \log(a)}{\log(n/k)},$$

as by Lemma 2.4.1 and the transformation (2.7)

$$\sup_{B \in \mathbb{B}} \left| P\left(\frac{k^{1/2} \log(n/k)}{\alpha}(\hat{\alpha}_n - \alpha) \in B\right) - N(0,1)(B) \right|$$
$$= O(k/n + (k/n)^\delta k^{1/2} + k^{-1/2}),$$

showing that $\hat{\alpha}_n$ is more concentrated around α than $\hat{a}_{n,1}$.

This result, which looks strange at first sight, is closely related to the fact that $Z_{n-k:n}$ is the central sequence generating *local asymptotic normality* (LAN) of the loglikelihood processes of $(Z_{n-j+1:n})_{j \leq k+1}$, indexed by α. In this sense, $Z_{n-k:n}$ carries asymptotically the complete information about the underlying shape parameter α that is contained in $(Z_{n-j+1:n})_{j \leq k}$ (see Theorems 1.3, 2.2 and 2.3 in Falk [133]).

THE EXTREME QUANTILE ESTIMATION

Since $W_3^{-1}(1-q) = -\log(q)$, we obtain from the preceding theorem the following result on the estimation of extreme quantiles outside the range of our actual data. We adopt the notation of the preceding result.

Theorem 2.4.5. *Suppose that F is in $Q_i(\delta)$, $i = 1, 2$ or 3. Then we have, for $i = 1, 2, 3$,*

$$\sup_{B \in \mathbb{B}} \left| P\left(F_i^{-1}(1-q) - (\hat{a}_{n,i} W_3^{-1}(1-q) + \hat{b}_{n,i}) \in B \right) \right.$$

$$\left. - P\left(a_0 \xi_1 \log(qn/k)/k^{1/2} + a_0 \xi_2/k^{1/2} + O(q^\delta) \in B \right) \right|$$

$$= O(k/n + (k/n)^\delta k^{1/2} + k^{-1/2})$$

uniformly for $q \to 0$, where ξ_1, ξ_2 are independent and standard normal rv.

The preceding result entails in particular that $\hat{a}_{n,i} W_3^{-1}(1-q) + \hat{b}_{n,i} = \hat{b}_{n,i} - \hat{a}_{n,i} \log(q)$ is a consistent estimator of $F_i^{-1}(1-q)$ for any sequence $q = q_n \longrightarrow_{n\to\infty} 0$ such that $\log(qn)/k^{1/2} \to 0$ with $k = k(n) \to \infty$ satisfying $(k/n)^\delta k^{1/2} \to 0$.

The bound $O(k/n + (k/n)^\delta k^{1/2} + k^{-1/2})$ for the normal approximation in Theorem 2.4.5 suggests that an optimal choice of $k = k(n) \to \infty$ is of order $n^{2\delta/(2\delta+1)}$, in which case the error bound $(k/n)^\delta k^{1/2}$ does not vanish asymptotically.

Note that $F_1^{-1}(1-q) = \log(F^{-1}(1-q))$ and $F_2^{-1}(1-q) = -\log(\omega(F) - F^{-1}(1-q))$. We can apply therefore the transformation $T_1(x) = \exp(x)$ and $T_2(x) = \omega(F) - \exp(-x)$ in case $i = 1, 2$ to the estimators of $F_i^{-1}(1-q)$ in Theorem 2.4.5, and we can deduce from this theorem the asymptotic behavior of the resulting estimators of the initial extreme quantile $F^{-1}(1-q)$.

Theorem 2.4.5 implies the following result.

Corollary 2.4.6. *Suppose that F is in $Q_i(\delta)$, $i = 1, 2$ or 3. Then we have*

(i) $$q_n^{\gamma(i)}(F^{-1}(1-q_n) - T_i(\hat{b}_{n,i} - \hat{a}_{n,i}\log(q_n))) \longrightarrow_{n\to\infty} 0$$

in probability, with $\gamma(i) = 1/\alpha, -1/\alpha, 0$ if $i = 1, 2, 3$, and $T_1(x) = \exp(x)$, $T_2(x) = \omega(F) - \exp(-x)$, $T_3(x) = x$, $x \in \mathbb{R}$, for any sequence $q_n \longrightarrow_{n\to\infty} 0$ such that $\log(q_n n)/k^{1/2} \longrightarrow_{n\to\infty} 0$, where $k = k(n)$ satisfies $(k/n)^\delta k^{1/2} \longrightarrow_{n\to\infty} 0$.

(ii) $$\sup_{t \in \mathbb{R}} \left| P\left(\frac{q_n^{\gamma(i)} k^{1/2}}{a(i)(\log^2(q_n n/k) + 1)^{1/2}} (F^{-1}(1-q_n) - T_i(\hat{b}_{n,i} - \hat{a}_{n,i}\log(q_n))) \le t \right) \right.$$

$$\left. - \Phi(t) \right| \longrightarrow_{n\to\infty} 0,$$

for any sequence $q_n \longrightarrow_{n\to\infty} 0$ such that $k^{1/2} q_n^\delta$ is bounded and $\log(q_n n)/k^{1/2} \longrightarrow_{n\to\infty} 0$, where $k \to \infty$ satisfies $(k/n)^\delta k^{1/2} \longrightarrow_{n\to\infty} 0$, $a(i) = a/\alpha, a/\alpha, a$ if $i = 1, 2, 3$ and Φ denotes the standard normal df.

Proof. Theorem 2.4.5 implies

(i) $F_i^{-1}(1 - q_n) - (\hat{b}_{n,i} - \hat{a}_{n,i}W_3^{-1}(1 - q_n)) \longrightarrow_{n\to\infty} 0$ in probability for any sequence $q_n \longrightarrow_{n\to\infty} 0$ such that $\log(q_n n)/k^{1/2} \longrightarrow_{n\to\infty} 0$, where $k = k(n) \longrightarrow_{n\to\infty} \infty$ satisfies $(k/n)^\delta k^{1/2} \longrightarrow_{n\to\infty} 0$.

(ii) $\displaystyle \sup_{t\in\mathbb{R}} \left| P\left(\frac{k^{1/2}}{a(\log^2(q_n n/k) + 1)^{1/2}} \left(F_i^{-1}(1 - q_n) - (\hat{b}_{n,i} - \hat{a}_{n,i}W_3^{-1}(1 - q_n)) \right) \right.\right.$

$$\left.\left. \le t \right) - \Phi(t) \right| \longrightarrow_{n\to\infty} 0$$

for any sequence $q_n \longrightarrow_{n\to\infty} 0$ such that $k^{1/2}q_n^\delta$ is bounded, where $k \to \infty$ satisfies $(k/n)^\delta k^{1/2} \longrightarrow_{n\to\infty} 0$.

The assertion of Corollary 2.4.6 now follows from the equation $F^{-1}(1 - q_n) = T_i(F_i^{-1}(1-q_n))$, $i = 1, 2, 3$, Taylor expansion of T_i and the equation $F^{-1}(1-q_n) = aq_n^{-1/\alpha}(1 + o(1))$ if $i = 1$; $\omega(F) - F^{-1}(1 - q_n) = aq_n^{1/\alpha}(1 + o(1))$ if $i = 2$ (see Proposition 2.2.1). $\qquad\qquad\square$

CONFIDENCE INTERVALS

Theorem 2.4.5 can immediately be utilized to define confidence intervals for the extreme quantile $F^{-1}(1 - q_n)$. Put for $q_n \in (0,1)$,

$$\widehat{F}_n^{-1}(1 - q_n) := \hat{a}_{n,i}W_3^{-1}(1 - q_n) + \hat{b}_{n,i}$$
$$= \hat{b}_{n,i} - \hat{a}_{n,i}\log(q_n),$$

and define the interval

$$I_n := \Big[\hat{F}_n^{-1}(1 - q_n) - \hat{a}_{n,i}(\log^2(q_n n/k) + 1)^{1/2}k^{-1/2}\Phi^{-1}(1 - \beta_1),$$
$$\hat{F}_n^{-1}(1 - q_n) + \hat{a}_{n,i}(\log^2(q_n n/k) + 1)^{1/2}k^{-1/2}\Phi^{-1}(1 - \beta_2) \Big],$$

where $0 < \beta_1, \beta_2 < 1/2$. For $F \in Q_i(\delta)$ we obtain that I_n is a confidence interval for $F_i^{-1}(1 - q_n)$ of asymptotic level $1 - (\beta_1 + \beta_2)$ that is, $\lim_{n\to\infty} P(F_i^{-1}(1 - q_n) \in I_n) = 1 - (\beta_1 + \beta_2)$. Consequently, we obtain from the equation $T_i(F_i^{-1}(1 - q_n)) = F^{-1}(1 - q_n)$,

$$\lim_{n\to\infty} P(F^{-1}(1 - q_n) \in T_i(I_n)) = 1 - (\beta_1 + \beta_2)$$

with $T_1(x) = \exp(x)$, $T_2(x) = \omega(F) - \exp(-x)$ and $T_3(x) = x$, $x \in \mathbb{R}$, for any sequence $q_n \longrightarrow_{n\to\infty} 0$ such that $k^{1/2}q_n^\delta$ is bounded, where $k \to \infty$ satisfies $(k/n)^\delta k^{1/2} \longrightarrow_{n\to\infty} 0$. Note that T_i are strictly monotone and continuous functions. The confidence interval $T_i(I_n)$ can therefore be deduced from I_n immediately by just transforming its endpoints. Note further that the length of I_n is in probability proportional to $(\log^2(q_n n/k) + 1)^{1/2}k^{-1/2}$, which is a convex function in q_n with the minimum value $k^{-1/2}$ at $q_n = k/n$.

2.5 Initial Estimation of the Class Index

If it is assumed that F is in $Q_i(\delta)$ but the index i is unknown, an initial estimation of the index $i \in \{1, 2, 3\}$ is necessary. Such a decision can be based on graphical methods as described in Castillo et al. [61] or on numerical estimates like the following Pickands [371] estimator (for a discussion we refer to Dekkers and de Haan [108]).

THE PICKANDS ESTIMATOR

Choose $m \in \{1, \ldots, n/4\}$ and define

$$\hat{\beta}_n(m) := (\log(2))^{-1} \log\left(\frac{Z_{n-m+1:n} - Z_{n-2m+1:n}}{Z_{n-2m+1:n} - Z_{n-4m+1:n}} \right).$$

This estimator is an asymptotically consistent estimator of $\beta := 1/\alpha, -1/\alpha, 0$ in case $F \in Q_i(\delta)$ with pertaining shape parameter α. A stable positive or negative value of $\hat{\beta}_n(m)$ indicates therefore that F is in $Q_1(\delta)$ or $Q_2(\delta)$, while $\hat{\beta}_n(m)$ near zero indicates that $F \in Q_3(\delta)$. By $N(\mu, \sigma^2)$ we denote the normal distribution on $\mathbb{R}$ with mean μ and variance σ^2.

Proposition 2.5.1. *Suppose that F is in $Q_i(\delta)$, $i = 1, 2, 3$. Then we have*

$$\sup_{t \in \mathbb{R}} \left| P\left(m^{1/2}(\hat{\beta}_n(m) - \beta) \leq t \right) - N(0, \sigma_\beta^2)((-\infty, t]) \right|$$
$$= O((m/n)^\delta m^{1/2} + m/n + m^{-1/2}),$$

where

$$\sigma_\beta^2 := \frac{1 + 2^{-2\beta-1}}{2 \log^2(2)} \left(\frac{\beta}{1 - 2^{-\beta}} \right)^2, \qquad \beta \in \mathbb{R}.$$

Interpret $\sigma_0^2 = \lim_{\beta \to 0} \sigma_\beta^2 = 3/(4 \log(2)^4)$.

The estimator $\hat{\beta}_n(m)$ of β can easily be motivated as follows. One expects by Proposition 2.2.1,

$$\frac{Z_{n-m+1:n} - Z_{n-2m+1:n}}{Z_{n-2m+1:n} - Z_{n-4m+1:n}} \sim \frac{F^{-1}(1 - \frac{m}{n+1}) - F^{-1}(1 - \frac{2m}{n+1})}{F^{-1}(1 - \frac{2m}{n+1}) - F^{-1}(1 - \frac{4m}{n+1})}$$

$$\sim \frac{W_i^{-1}(1 - \frac{m}{n+1}) - W_i^{-1}(1 - \frac{2m}{n+1})}{W_i^{-1}(1 - \frac{2m}{n+1}) - W_i^{-1}(1 - \frac{4m}{n+1})},$$

with $W_i \in \{W_1, W_2, W_3\}$ being the GPD pertaining to F. Since location and scale shifts are canceled by the definition of the estimator $\hat{\beta}_n(m)$, we can assume without loss of generality that W_i has standard form. Now

$$W_i^{-1}(1 - q) = \begin{cases} q^{-1/\alpha} & i = 1, \\ -q^{1/\alpha} & \text{in case} \quad i = 2, \\ -\log(q) & i = 3, \end{cases}$$

$q \in (0,1)$ and, thus,

$$\frac{W_i^{-1}(1 - \frac{m}{n+1}) - W_i^{-1}(1 - \frac{2m}{n+1})}{W_i^{-1}(1 - \frac{2m}{n+1}) - W_i^{-1}(1 - \frac{4m}{n+1})} = \begin{cases} 2^{1/\alpha}, & i = 1, \\ 2^{-1/\alpha}, & i = 2, \\ 1, & i = 3, \end{cases}$$

which implies the approximation

$$\hat{\beta}_n(m) \sim (\log(2))^{-1} \log \left(\frac{W_i^{-1}(1 - \frac{m}{n+1}) - W_i^{-1}(1 - \frac{2m}{n+1})}{W_i^{-1}(1 - \frac{2m}{n+1}) - W_i^{-1}(1 - \frac{4m}{n+1})} \right)$$

$$= \begin{cases} 1/\alpha, & i = 1, \\ -1/\alpha, & i = 2, \\ 0, & i = 3, \end{cases} = \beta.$$

Weak consistency of $\hat{\beta}_n(m)$ actually holds under the sole condition that F is in the domain of attraction of an EVD and $m = m(n)$ satisfies $m \to \infty$, $m/n \to 0$ as $n \to \infty$ (see Theorem 2.1 of Dekkers and de Haan [108]). Asymptotic normality of $\hat{\beta}_n(m)$ however, requires additional conditions on F (see also Theorem 2.3 of Dekkers and de Haan [108]).

CONVEX COMBINATIONS OF PICKANDS ESTIMATORS

The limiting variance of Pickands estimator $\hat{\beta}_n(m)$ can considerably be reduced by a simple convex combination. Choose $p \in [0,1]$ and define for $m \in \{1, \ldots, n/4\}$,

$$\hat{\beta}_n(m,p) := p \cdot \hat{\beta}_n([m/2]) + (1-p) \cdot \hat{\beta}_n(m)$$

$$= (\log(2))^{-1} \log \left\{ \left(\frac{Z_{n-[m/2]+1:n} - Z_{n-2[m/2]+1:n}}{Z_{n-2[m/2]+1:n} - Z_{n-4[m/2]+1:n}} \right)^p \right.$$

$$\left. \times \left(\frac{Z_{n-m+1:n} - Z_{n-2m+1:n}}{Z_{n-2m+1:n} - Z_{n-4m+1:n}} \right)^{1-p} \right\},$$

where $[x]$ denotes the integral part of $x \in \mathbb{R}$. If m is even, $[m/2]$ equals $m/2$, and the preceding notation simplifies.

We consider the particular convex combination $\hat{\beta}_n(m,p)$ to be a natural extension of Pickands estimator $\hat{\beta}_n(m)$: As $\hat{\beta}_n(m)$ is built upon powers of 2 that is, of $2^0 m, 2^1 m, 2^2 m$, it is only natural (and makes the computations a bit easier) to involve the next smaller integer powers $2^{-1}m, 2^0 m, 2^1 m$ of 2 and to combine $\hat{\beta}_n(m)$ with $\hat{\beta}_n([m/2])$. As a next step one could consider linear combinations $\sum_{i \leq k} p_i \hat{\beta}_n([m/2^{i-1}])$, $\sum_{i \leq k} p_i = 1$, of length k. But as one uses the $4m$ largest observations in a sample of size n in the definition of $\hat{\beta}_n(m)$, with $4m$ having to be relatively small to n anyway, it is clear that $m/2$ is already a rather small number for making asymptotics ($m \to \infty$). For moderate sample sizes n, the length m will therefore be limited to 2 in a natural way. Higher linear combinations nevertheless perform still better in an asymptotic model (cf. Drees [114]).

In the following result we establish asymptotic normality of $\hat{\beta}_n(m, p)$. With $p = 0$ it complements results on the asymptotic normality of $\hat{\beta}_n(m) = \hat{\beta}_n(m, 0)$ (cf. also Dekkers and de Haan ([108], Theorems 2.3, 2.5)). Its proof is outlined at the end of this section. A careful inspection of this proof also implies Proposition 2.5.1.

Lemma 2.5.2. *Suppose that F is in $Q_i(\delta)$, $i = 1, 2, 3$ for some $\delta > 0$. Then we have, for $m \in \{1, \ldots, n/4\}$ and $p \in [0, 1]$,*

$$\sup_{B \in \mathbb{B}} \left| P(m^{1/2}(\hat{\beta}_n(m, p) - \beta) \in B) - P(\sigma_\beta \nu_\beta(p) X + O_p(m^{-1/2}) \in B) \right|$$
$$= O\left((m/n)^\delta m^{1/2} + m/n + m^{-1/2} \right),$$

where X is a standard normal rv and

$$\nu_\beta(p) := 1 + p^2 \left(3 + \frac{4 \cdot 2^{-\beta}}{2^{-2\beta} + 2} \right) - p \left(2 + \frac{4 \cdot 2^{-\beta}}{2^{-2\beta} + 2} \right).$$

THE ASYMPTOTIC RELATIVE EFFICIENCY

The following result is an immediate consequence of Lemma 2.5.2.

Corollary 2.5.3. *Under the conditions of the preceding lemma we have, for $m = m(n) \longrightarrow_{n \to \infty} \infty$ such that $(m/n)^\delta m^{1/2} \to 0$ as $n \to \infty$,*

$$m^{1/2}(\hat{\beta}_n(m, p) - \beta) \to_D N(0, \sigma_\beta^2 \nu_\beta^2(p)).$$

By $\to_D$ we denote again convergence in distribution. Recall that σ_β^2 is the variance of the limiting central normal distribution of the standardized Pickands estimator $\sqrt{m} \, (\hat{\beta}_n(m) - \beta)$. The factor $\nu_\beta^2(p)$ is now the *asymptotic relative efficiency (ARE)* of $\hat{\beta}_n(m)$ with respect to $\hat{\beta}_n(m, p)$, which we define by the ratio of the variances of the limiting central normal distributions of $m^{1/2}(\hat{\beta}_n(m, p) - \beta)$ and $m^{1/2}(\hat{\beta}_n(m) - \beta)$:

$$\text{ARE} \, (\hat{\beta}_n(m) | \hat{\beta}_n(m, p)) := \nu_\beta^2(p).$$

THE OPTIMAL CHOICE OF p

The optimal choice of p minimizing $\nu_\beta^2(p)$ is

$$p_{\text{opt}}(\beta) := \frac{(2^{-2\beta} + 2) + 2 \cdot 2^{-\beta}}{3(2^{-2\beta} + 2) + 4 \cdot 2^{-\beta}},$$

in which case $\nu_\beta^2(p)$ becomes

$$\nu_\beta^2(p_{\mathrm{opt}}(\beta)) = 1 - p_{\mathrm{opt}}(\beta) \cdot \left(1 + 2\frac{2^{-\beta}}{2^{-2\beta}+2}\right).$$

As $p_{\mathrm{opt}}(\beta)$ is strictly between 0 and 1, we have $\nu_\beta^2(p_{\mathrm{opt}}(\beta)) < 1$ and the convex combination $\hat{\beta}_n(m, p_{\mathrm{opt}}(\beta))$ is clearly superior to the Pickands estimator $\hat{\beta}_n(m)$.

The following figure displays the ARE function $\beta \mapsto \nu_\beta^2(p_{\mathrm{opt}}(\beta))$. As $\nu_\beta^2(p_{\mathrm{opt}}(\beta)) =: g(2^{-\beta})$ depends upon β through the transformation $2^{-\beta}$, we have plotted the function $g(x)$, $x \in \mathbb{R}$, with $x = 2^{-\beta}$. Notice that both for $x \to 0$ (that is, $\beta \to \infty$) and $x \to \infty$ (that is, $\beta \to -\infty$) the pertaining ARE function converges to $2/3$ being the least upper bound, while .34 is an approximate infimum.

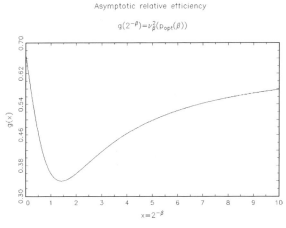

FIGURE 2.5.1. $g(x) = 1 - \frac{(x^2+2)+2x}{3(x^2+2)+4x}\left(1 + \frac{2x}{x^2+2}\right).$

DATA-DRIVEN OPTIMAL ESTIMATORS

The optimal p depends however on the unknown β and it is therefore reasonable to utilize the data-driven estimator

$$p_{\mathrm{opt}}(\tilde{\beta}_n) := \frac{(2^{-2\tilde{\beta}_n}+2) + 2 \cdot 2^{-\tilde{\beta}_n}}{3(2^{-2\tilde{\beta}_n}+2) + 4 \cdot 2^{-\tilde{\beta}_n}},$$

where $\tilde{\beta}_n$ is an initial estimator of β. If $\tilde{\beta}_n$ is asymptotically consistent, then, using Lemma 2.5.2, it is easy to see that the corresponding data-driven convex combination $\hat{\beta}_n(m, p_{\mathrm{opt}}(\tilde{\beta}_n))$ is asymptotically equivalent to the optimal convex combination $\hat{\beta}_n(m, p_{\mathrm{opt}}(\beta))$ with underlying β that is,

$$m^{1/2}\left(\hat{\beta}_n(m, p_{\mathrm{opt}}(\tilde{\beta}_n)) - \hat{\beta}_n(m, p_{\mathrm{opt}}(\beta))\right) = o_P(1),$$

so that their ARE is one.

A reasonable initial estimator of β is suggested by the fact that the particular parameter $\beta = 0$ is crucial as it is some kind of change point: If $\beta < 0$, then the right endpoint $\omega(F)$ of F is finite, while in case $\beta > 0$ the right endpoint of F is infinity. The question $\omega(F) = \infty$ or $\omega(F) < \infty$ is in case $\beta = 0$ numerically hard to decide, as an estimated value of β near 0 can always be misleading. In this case, graphical methods such as the one described in Castillo et al. [61] can be helpful. To safeguard oneself against this kind of a least favorable value β, it is therefore reasonable to utilize as an initial estimator $\tilde{\beta}_n$ that convex combination $\hat{\beta}_n(m, p)$, where p is chosen optimal for $\beta = 0$ that is, $p_{\text{opt}}(0) = 5/13$. We propose as an initial estimator therefore $\tilde{\beta}_n = \hat{\beta}_n(m, 5/13)$.

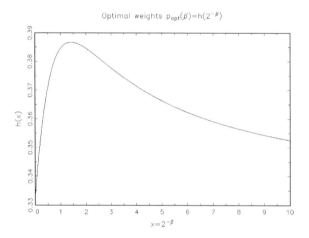

FIGURE 2.5.2. $h(x) = ((x^2 + 2) + 2x)/(3(x^2 + 2) + 4x)$.

Figure 2.5.2 displays the function of optimal weights $p_{\text{opt}}(\beta)$, again after the transformation $x = 2^{-\beta}$ that is, $p_{\text{opt}}(\beta) =: h(2^{-\beta})$. These weights do not widely spread out, they range between .33 and .39, roughly.

Note that the ARE of the Pickands estimator $\hat{\beta}_n(m)$ with respect to the optimal convex combination $\hat{\beta}_n(m, 5/13)$ in case $\beta = 0$ is 14/39. We summarize the preceding considerations in the following result.

Theorem 2.5.4. *Suppose that F is in $Q_i(\delta)$, $i = 1, 2$ or 3 for some $\delta > 0$. Then we have, for $m \to \infty$ such that $(m/n)^\delta m^{1/2} \to 0$ as $n \to \infty$,*

$$m^{1/2}\left(\hat{\beta}_n(m, p_{opt}(\tilde{\beta}_n)) - \beta\right) \to_D N\left(0, \sigma_\beta^2\left(1 - p_{opt}(\beta)\left(1 + 2\frac{2^{-\beta}}{2^{-2\beta} + 2}\right)\right)\right)$$

for any initial estimator sequence $\tilde{\beta}_n$ of β which is asymptotically consistent.

Dropping the δ-Neighborhood

The crucial step in the proof of Lemma 2.5.2 is the approximation

$$
\Delta_{n,m} := \sup_{B \in \mathbb{B}^m} \left| P(((Z_{n-j+1:n} - b_n)/a_n)_{j \leq m} \in B) \right.
$$

$$
- \begin{cases} P\left(\left(\beta(\textstyle\sum_{r \leq j} \xi_r)^{-\beta} \right)_{j \leq m} \in B \right) \Big| & \text{if } \beta \neq 0, \\ P\left(\left(-\log(\textstyle\sum_{r \leq j} \xi_r) \right)_{j \leq m} \in B \right) \Big| & \text{if } \beta = 0, \end{cases}
$$

$$
= O((m/n)^\delta m^{1/2} + m/n)
$$

of Theorem 2.2.4, where $\xi_1, \xi_2, \ldots$ are independent and standard exponential rv and $a_n > 0$, $b_n \in \mathbb{R}$ are suitable constants (see also Remark 2.2.3).

Lemma 2.5.2, Corollary 2.5.3 and Theorem 2.5.4 remain however true with $(m/n)^\delta m^{1/2} + m/n$ replaced by $\Delta_{n,m}$, if we drop the condition that F is in $Q_i(\delta)$ and require instead $\Delta_{n,m} \to 0$ for some sequence $m = m(n) \leq n/4$, $m \to \infty$ as $n \to \infty$.

Then we can consider for example the case, where F is the standard normal df, which is not in any $Q_i(\delta)$; but in this case we have $\Delta_{n,m} = O(m^{1/2} \log^2(m+1)/\log(n))$ (cf. Example 2.33 in Falk [126]), allowing however only rather small sizes of $m = m(n)$ to ensure asymptotic normality.

Simulation Results

In this section we briefly report the results of extensive simulations which we have done for the comparison between $\hat{\beta}_n(m, \hat{p}_{\mathrm{opt}}) = \hat{\beta}_n(m, p_{\mathrm{opt}}(\tilde{\beta}_n))$, based on the initial estimate $\tilde{\beta}_n = \hat{\beta}_n(m, 5/13)$ and the Pickands estimator $\hat{\beta}_n(m)$.

These simulations with various choices of n, m and underlying df F support the theoretical advantage of using $\hat{\beta}_n(m, \hat{p}_{\mathrm{opt}})$ in those cases, where F is in a δ-neighborhood of a GPD. Figures 2.5.3-2.5.6 exemplify the gain of relative performance which is typically obtained by using $\hat{\beta}_n(m, \hat{p}_{\mathrm{opt}})$. In these cases we generated $n = 50/100/200/400$ replicates $Z_1, \ldots, Z_n$ of a (pseudo-) rv Z with different distribution F in each case. The estimators $\hat{\beta}_n(m, \hat{p}_{\mathrm{opt}})$ and $\hat{\beta}_n(m)$ of the pertaining values of β with $m = 6/8/12/40$ were computed, and we stored by $B := |\hat{\beta}_n(m) - \beta|$, $C := |\hat{\beta}_n(m, \hat{p}_{\mathrm{opt}}) - \beta|$ their corresponding absolute deviations. We generated $k = 1000$ independent replicates $B_1, \ldots, B_k$ and $C_1, \ldots, C_k$ of B and C, with their sample quantile functions now visualizing the concentration of $\hat{\beta}_n(m)$ and $\hat{\beta}_n(m, \hat{p}_{\mathrm{opt}})$ around β.

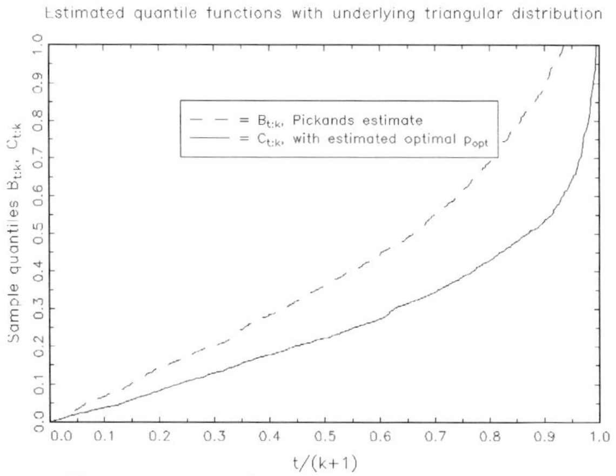

FIGURE 2.5.3. $\beta = -.5$, $n = 50$, $m = 6$.

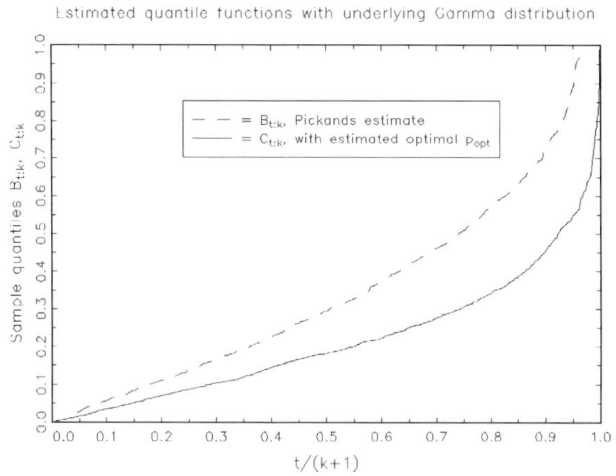

FIGURE 2.5.4. $\beta = 0$, $n = 100$, $m = 8$.

Figures 2.5.3-2.5.6 display the pertaining sample quantile functions $(t/(k+1),\ B_{t:k})$ and $(t/(k+1),\ C_{t:k})$, $t = 1, \dots, k = 1000$ for Z. By $B_{1:k} \le \cdots \le B_{k:k}$ and $C_{1:k} \le \cdots \le C_{k:k}$ we denote the ordered values of $B_1, \dots, B_k$ and $C_1, \dots, C_k$. In Figure 2.5.3, Z equals the sum of two independent and uniformly on (0,1) distributed rv ($\beta = -.5$); in Figure 2.5.4 the df F is a Gamma distribution ($Z = X_1 + X_2 + X_3$, X_1, X_2, X_3 independent and standard exponential, $\beta = 0$) and in Figure 2.5.5, F is a Cauchy distribution ($\beta = 1$). Elementary computations show that these distributions satisfy (VM) with rapidly decreasing remainder. The triangular distribution is in particular a GPD. In Figure 2.5.6, F is the normal distribution.

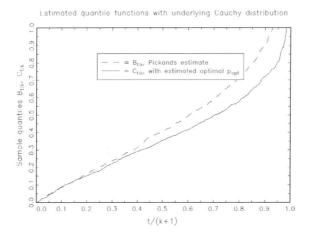

Estimated quantile functions with underlying Cauchy distribution

FIGURE 2.5.5. $\beta = 1$, $n = 200$, $m = 12$.

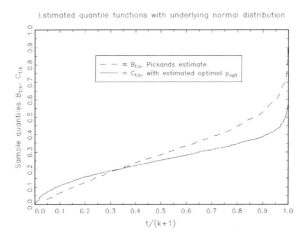

Estimated quantile functions with underlying normal distribution

FIGURE 2.5.6. $\beta = 0$, $n = 400$, $m = 40$.

Except the normal distribution underlying Figure 2.5.6, these simulations are clearly in favor of the convex combination $\hat{\beta}_n(m, \hat{p}_{\text{opt}})$ as an estimator of β, even for moderate sample sizes n. In particular Figure 2.5.3 shows in this case with $\beta = -.5$, that $\hat{\beta}_n(m)$ would actually give a negative value between -1 and 0 with approximate probability .67, whereas the corresponding probability is approximately .87 for $\hat{\beta}_n(m, \hat{p}_{\text{opt}})$. Recall that a negative value of β implies that the underlying df has finite right endpoint. In Figure 2.5.4 with $\beta = 0$ the sample medians $B_{k/2:k}$ and $C_{k/2:k}$ are in particular interesting: While with approximate probability $1/2$

the Pickands estimate $\hat{\beta}_n(m)$ has an absolute value less than .3, the combination $\hat{\beta}_n(m, \hat{p}_{\text{opt}})$ is less than .18 apart from $\beta = 0$ with approximate probability $1/2$. Figure 2.5.6 is not in favor of $\hat{\beta}_n(m, \hat{p}_{\text{opt}})$. But this can be explained by the fact that the normal distribution does not belong to a δ-neighborhood of a GPD and the choice $m = 40$ is too large. This observation underlines the significance of δ-neighborhoods of GPD.

Our simulations showed that the relative performance of $\hat{\beta}_n(m, \hat{p}_{\text{opt}})$ is quite sensitive to the choice of m which corresponds to under- and oversmoothing in bandwidth selection problems in non-parametric curve estimation (see Marron [322] for a survey). Our simulations suggest as a rule of thumb the choice $m = (2/25)n$ for not too large sample size n that is, $n \leq 200$, roughly.

NOTES ON COMPETING ESTIMATORS

If one knows the sign of β, then one can use the estimators $\hat{a}_{n,i}$ of the shape parameter $1/\alpha = |\beta|$, which we have discussed in Section 2.4. Based on the $4m$ largest order statistics, $m^{1/2}(\hat{a}_{n,i} - |\beta|)$ is asymptotically normal distributed under appropriate conditions with mean 0 and variance $\beta^2/4$ in case $i = 1, 2$ and $\beta \neq 0$ (Theorem 2.4.4), and therefore outperforms $\hat{\beta}_n(m, \hat{p}_{\text{opt}})$ (see Theorem 2.5.4).

A competitor of $\hat{\beta}_n(m, p)$ is the moment estimator investigated by Dekkers et al. [107], which is outperformed by $\hat{\beta}_n(m, \hat{p}_{\text{opt}})$ if $\beta < 0$ is small enough. Alternatives such as the maximum-likelihood estimator or the method of probability-weighted moment (PWM) considered by Hosking and Wallis [224] require restrictions on β such as $\beta > -1$ (for the PWM method) and are therefore not universally applicable. A higher linear combination of Pickands estimators with estimated optimal scores was studied by Drees [114]. For thorough discussions of different approaches we refer to Section 9.6 of the monograph by Reiss [385], Sections 6.3 and 6.4 of Embrechts et al. [122], Section 5.1 of Reiss and Thomas [389], Section 1.7 of Kotz and Nadarajah [293] as well as to Beirlant et al. [32] and de Haan and Ferreira [190].

OUTLINE OF THE PROOF OF LEMMA 2.5.2: Put

$$A_n := \frac{Z_{n-[m/2]+1:n} - Z_{n-2[m/2]+1:n}}{Z_{n-2[m/2]+1:n} - Z_{n-4[m/2]+1:n}} - 2^\beta$$

and

$$B_n := \frac{Z_{n-m+1:n} - Z_{n-2m+1:n}}{Z_{n-2m+1:n} - Z_{n-4m+1:n}} - 2^\beta.$$

We will see below that A_n and B_n are both of order $O_P(m^{-1/2})$, so that we obtain, by the expansion $\log(1 + \varepsilon) = \varepsilon + O(\varepsilon^2)$ as $\varepsilon \to 0$,

$$\hat{\beta}_n(m,p) - \beta = \frac{1}{\log(2)} \left\{ p\Big(\log(2^\beta + A_n) - \log(2^\beta) \Big) \right.$$

$$\left. + (1-p)\Big(\log(2^\beta + B_n) - \log(2^\beta) \Big) \right\}$$

$$= \frac{1}{\log(2)} \left\{ p\log\Big(1 + \frac{A_n}{2^\beta}\Big) + (1-p)\log\Big(1 + \frac{B_n}{2^\beta}\Big) \right\}$$

$$= \frac{1}{2^\beta \log(2)} (pA_n + (1-p)B_n) + O_P(m^{-1}).$$

From Theorem 2.2.4 we obtain within the error bound $O((m/n)^\delta m^{1/2} + m/n)$ in variational distance the representation

$$pA_n + (1-p)B_n$$

$$= \begin{cases} p\left(\dfrac{\Big(1 + [m/2]^{-1}\sum_{j\le[m/2]}\eta_j\Big)^{-\beta} - \Big(2 + [m/2]^{-1}\sum_{j\le2[m/2]}\eta_j\Big)^{-\beta}}{\Big(2 + [m/2]^{-1}\sum_{j\le2[m/2]}\eta_j\Big)^{-\beta} - \Big(4 + [m/2]^{-1}\sum_{j\le4[m/2]}\eta_j\Big)^{-\beta}} - 2^\beta \right) \\[6pt]
\quad + (1-p)\left(\dfrac{\Big(1 + m^{-1}\sum_{j\le m}\eta_j\Big)^{-\beta} - \Big(2 + m^{-1}\sum_{j\le2m}\eta_j\Big)^{-\beta}}{\Big(2 + m^{-1}\sum_{j\le2m}\eta_j\Big)^{-\beta} - \Big(4 + m^{-1}\sum_{j\le4m}\eta_j\Big)^{-\beta}} - 2^\beta \right), \\[6pt]
\qquad\qquad\qquad \text{if } \beta \ne 0 \\[10pt]
p\left(\dfrac{\log\Big(\frac{2 + [m/2]^{-1}\sum_{j\le2[m/2]}\eta_j}{1 + [m/2]^{-1}\sum_{j\le[m/2]}\eta_j} \Big)}{\log\Big(\frac{4 + [m/2]^{-1}\sum_{j\le4[m/2]}\eta_j}{2 + [m/2]^{-1}\sum_{j\le2[m/2]}\eta_j} \Big)} - 1 \right) \\[6pt]
\quad + (1-p)\left(\dfrac{\log\Big(\frac{2 + m^{-1}\sum_{j\le2m}\eta_j}{1 + m^{-1}\sum_{j\le m}\eta_j} \Big)}{\log\Big(\frac{4 + m^{-1}\sum_{j\le4m}\eta_j}{2 + m^{-1}\sum_{j\le2m}\eta_j} \Big)} - 1 \right), \qquad \text{if } \beta = 0, \end{cases}$$

where $\eta_1 + 1, \eta_2 + 1, \ldots$ are independent and standard exponential rv. Now elementary computations show that the distribution of $k^{-1/2}\sum_{j\le k}\eta_j$ approaches the standard normal distribution uniformly over all Borel sets at the rate $O(k^{-1/2})$ and thus, within the bound $O(m^{-1/2})$, we can replace the distribution of the right-hand side of the preceding equation by that of

$$p\left(\frac{\Big(1 + \frac{X}{\sqrt{m/2}}\Big)^{-\beta} - \Big(2 + \frac{X+Y}{\sqrt{m/2}}\Big)^{-\beta}}{\Big(2 + \frac{X+Y}{\sqrt{m/2}}\Big)^{-\beta} - \Big(4 + \frac{X+Y+\sqrt{2}Z}{\sqrt{m/2}}\Big)^{-\beta}} - 2^\beta \right)$$

$$+(1-p)\left(\frac{\left(1+\frac{X+Y}{\sqrt{2m}}\right)^{-\beta}-\left(2+\frac{X+Y+\sqrt{2}Z}{\sqrt{2m}}\right)^{-\beta}}{\left(2+\frac{X+Y+\sqrt{2}Z}{\sqrt{2m}}\right)^{-\beta}-\left(4+\frac{X+Y+\sqrt{2}Z+2W}{\sqrt{2m}}\right)^{-\beta}}-2^{\beta}\right),\quad\text{if }\beta\neq0,$$

and by

$$p\left(\frac{\log\left(\frac{2+\frac{X+Y}{\sqrt{m/2}}}{1+\frac{X}{\sqrt{m/2}}}\right)}{\log\left(\frac{4+\frac{X+Y+\sqrt{2}Z}{\sqrt{m/2}}}{2+\frac{X+Y}{\sqrt{m/2}}}\right)}-1\right)+(1-p)\left(\frac{\log\left(\frac{2+\frac{X+Y+\sqrt{2}Z}{\sqrt{2m}}}{1+\frac{X+Y}{\sqrt{2m}}}\right)}{\log\left(\frac{4+\frac{X+Y+\sqrt{2}Z+2W}{\sqrt{2m}}}{2+\frac{X+Y+\sqrt{2}Z}{\sqrt{2m}}}\right)}-1\right),\quad\text{if }\beta=0,$$

where X,Y,W,Z are independent standard normal rv. We have replaced $[m/2]$ in the preceding formula by $m/2$, which results in an error of order $O_P(1/m)$.

By Taylor's formula we have $(1+\varepsilon)^{-\beta}=1-\beta\varepsilon+O(\varepsilon^2)$ for $\beta\neq0$ and $\log(1+\varepsilon)=\varepsilon+O(\varepsilon^2)$ as $\varepsilon\to0$. The assertion of Lemma 2.5.2 now follows from Taylor's formula and elementary computations. $\qquad\square$

2.6 Power Normalization and p-Max Stable Laws

Let $Z_1,\dots Z_n$ be iid rv with common df F. In order to derive a more accurate approximation of the df of $Z_{n:n}$ by means of limiting df, Weinstein [456] and Pancheva [360] used a nonlinear normalization for $Z_{n:n}$. In particular, F is said to belong to the domain of attraction of a df H *under power normalization*, denoted by $F\in\mathcal{D}_p(H)$ if, for some $\alpha_n>0$, $\beta_n>0$,

$$F^n\big(\alpha_n|x|^{\beta_n}\operatorname{sign}(x)\big)\longrightarrow_\omega H(x),\qquad n\to\infty,\qquad(2.11)$$

or in terms of rv,

$$(|Z_{n:n}|/\alpha_n)^{1/\beta_n}\operatorname{sign}(Z_{n:n})\longrightarrow_D Z,\qquad n\to\infty,$$

where Z is a rv with df H and $\operatorname{sign}(x)=-1,0$ or 1 according as $x<,=$ or >0, respectively.

The Power-Max Stable Distributions

Pancheva [360] (see also Mohan and Ravi [339]) showed that a non-degenerate limit df H in (2.11) can up to a possible power transformation $H(\alpha|x|^\beta\operatorname{sign}(x))$, $\alpha,\beta>0$, only be one of the following six different df H_i, $i=1,\dots,6$, where with $\gamma>0$,

$$H_1(x) = H_{1,\gamma}(x) = \begin{cases} 0 & \text{if } x \leq 1, \\ \exp\big(-(\log(x))^{-\gamma}\big) & \text{if } x > 1, \end{cases}$$

$$H_2(x) = H_{2,\gamma}(x) = \begin{cases} 0 & \text{if } x \leq 0, \\ \exp\big(-(-\log(x))^{\gamma}\big) & \text{if } 0 < x < 1, \\ 1 & \text{if } x \geq 1, \end{cases}$$

$$H_3(x) = \begin{cases} 0 & \text{if } x \leq 0, \\ \exp(-1/x) & \text{if } x > 0, \end{cases}$$

$$H_4(x) = H_{4,\gamma}(x) = \begin{cases} 0 & \text{if } x \leq -1, \\ \exp\big(-(-\log(-x))^{-\gamma}\big) & \text{if } -1 < x < 0, \\ 1 & \text{if } x \geq 0, \end{cases}$$

$$H_5(x) = H_{5,\gamma}(x) = \begin{cases} \exp\big(-(\log(-x))^{\gamma}\big) & \text{if } x < -1, \\ 1 & \text{if } x \geq -1, \end{cases}$$

$$H_6(x) = \begin{cases} \exp(x) & \text{if } x < 0, \\ 1 & \text{if } x \geq 0. \end{cases}$$

A df H is called *power-max stable* or *p-max stable* for short by Mohan and Ravi [339] if it satisfies the stability relation

$$H^n\big(\alpha_n |x|^{\beta_n} \text{sign}(x)\big) = H(x), \qquad x \in \mathbb{R}, \, n \in \mathbb{N},$$

for some $\alpha_n > 0$, $\beta_n > 0$. The df H_i, $i = 1, \ldots, 6$, are p-max stable and, from a result of Pancheva [360], these are the only non-degenerate ones. Necessary and sufficient conditions for F to satisfy (2.11) were given by Mohan and Ravi [339], Mohan and Subramanya [340], and Subramanya [433]. In view of these considerations one might label the max stable EVD more precisely l-max stable, because they are max stable with respect to a linear normalization.

MAX AND MIN STABLE DISTRIBUTIONS

We denote by $F \in \mathcal{D}_{\max}(G)$ the property that F is in the max-domain of attraction of an EVD G if

$$\frac{Z_{n:n} - b_n}{a_n} \longrightarrow_D G$$

for some norming constants $a_n > 0$, $b_n \in \mathbb{R}$. We denote by $F \in \mathcal{D}_{\min}(L)$ the property that F is in the min-domain of attraction of an EVD L if

$$\frac{Z_{1:n} - b_n}{a_n} \longrightarrow_D L.$$

From Theorem 2.1.1 we know that there are only three different types of possible non-degenerate limiting df G and L: G equals up to a possible linear transformation (i.e., location and scale shift) G_1, G_2 or G_3, where

$$G_1(x) = G_{1,\gamma}(x) = \begin{cases} 0 & \text{if } x \leq 0, \\ \exp(-x^{-\gamma}) & \text{if } x > 0, \end{cases}$$

$$G_2(x) = G_{2,\gamma}(x) = \begin{cases} \exp(-(-x)^\gamma) & \text{if } x < 0, \\ 1 & \text{if } x \geq 0, \end{cases}$$

for some $\gamma > 0$, and $G_3(x) = \exp(-e^{-x})$, $x \in \mathbb{R}$. Note that $G_{1,1} = H_3$, $G_{2,1} = H_6$ and that G_3 is not a p-max stable law. The df L is up to a possible linear transformation equal to L_1, L_2 or L_3, where

$$L_1(x) = L_{1,\gamma}(x) = \begin{cases} 1 - \exp(-(-x)^{-\gamma}) & \text{if } x < 0, \\ 1 & \text{if } x \geq 0, \end{cases}$$

$$L_2(x) = L_{2,\gamma}(x) = \begin{cases} 0 & \text{if } x < 0, \\ 1 - \exp(-x^\gamma) & \text{if } x \geq 0, \end{cases}$$

for some $\gamma > 0$, and $L_3(x) = 1 - \exp(-e^x)$, $x \in \mathbb{R}$.

THE CHARACTERIZATION THEOREM

The right endpoint $\omega(F) := \sup\{x : F(x) < 1\} \in (-\infty, \infty]$ of the df F plays a crucial role in the sequel.

The following result by Christoph and Falk [69] reveals that the upper *as well as* the lower tail behavior of F determine whether $F \in \mathcal{D}_p(H)$: The right endpoint $\omega(F) > 0$ yields the max stable distributions G, and $\omega(F) \leq 0$ results in the min stable distributions L; this explains the number of six power types of p-max stable df.

Moreover, if $\omega(F) < \infty$ is *not* a point of continuity of F, i.e., if $P(Z_1 = \omega(F)) =: \rho > 0$, then $F \notin \mathcal{D}_p(H)$ for any non-degenerate df H. In this case we have

$$P(Z_{n:n} = \omega(F)) = 1 - P(Z_{n:n} < \omega(F)) = 1 - (1 - \rho)^n \longrightarrow 1 \quad \text{as} \quad n \to \infty,$$

and

$$F^n\big(\alpha_n |x|^{\beta_n} \operatorname{sign}(x)\big) \begin{cases} \leq (1 - \rho)^n & \text{if } \alpha_n |x|^{\beta_n} \operatorname{sign}(x) < \omega(F), \\ = 1 & \text{if } \alpha_n |x|^{\beta_n} \operatorname{sign}(x) \geq \omega(F). \end{cases}$$

Hence, a limiting df H is necessarily degenerate in this case.

Theorem 2.6.1. *We have the following characterizations of the domains of attraction.*

(i) *Suppose that $\omega(F) > 0$. Put $F^*(x) = 0$ if $x \leq \min\{\log(\omega(F)/2), 0\}$ and $F^*(x) = F(\exp(x))$ elsewhere. Then F^* is a df and*

$$F \in \mathcal{D}_p(H) \text{ for some non-degenerate } H \iff F^* \in \mathcal{D}_{\max}(G). \quad (2.12)$$

In this case we have $H(x) = G((\log(x) - a)/b)$, $x > 0$, for some $b > 0$, $a \in \mathbb{R}$.

(ii) *Suppose that $\omega(F) \leq 0$ and put $F_*(x) := 1 - F(-\exp(x))$, $x \in \mathbb{R}$. Then,*

$$F \in \mathcal{D}_p(H) \text{ for some non-degenerate } H \iff F_* \in \mathcal{D}_{\min}(L). \quad (2.13)$$

In this case we have $H(x) = 1 - L((\log(-x) - a)/b)$, $x < 0$, for some $b > 0$, $a \in \mathbb{R}$.

As the domains of attraction of G and L as well as sequences of possible norming constants are precisely known (see Theorem 2.1.1), the preceding result together with its following proof characterizes the p-max stable distributions, their domains of attraction and the class of possible norming constants α_n, β_n in (2.11). In particular, we have $H_i(x) = G_i(\log(x))$, $x > 0$, and $H_{i+3}(x) = 1 - L_i(\log(-x))$, $x < 0$, $i = 1, 2, 3$. Subramanya [433], Remark 2.1, proved the special case $F \in \mathcal{D}_p(H_3)$ iff $F^* \in \mathcal{D}_{\max}(G_3)$.

Proof. (i) Suppose that $\omega(F) > 0$. In this case we have, for $x \leq 0$ and any sequence $\alpha_n > 0$, $\beta_n > 0$,

$$F^n(\alpha_n|x|^{\beta_n}\operatorname{sign}(x)) = P(Z_{n:n} \leq \operatorname{sign}(x)\,\alpha_n|x|^{\beta_n})$$
$$\leq P(Z_{n:n} \leq 0) \longrightarrow 0 \quad \text{as} \quad n \to \infty.$$

Let now $x > 0$ be a point of continuity of the limiting df H in (2.11) and put $c := \min\{\omega(F)/2, 1\}$. Then, as $n \to \infty$,

$$P(Z_{n:n} \leq \operatorname{sign}(x)\,\alpha_n|x|^{\beta_n}) \longrightarrow H(x)$$
$$\iff P(\log(Z_{n:n}) \leq \beta_n\log(x) + \log(\alpha_n), Z_{n:n} \geq c) \longrightarrow H(x)$$
$$\iff P((Y_{n:n} - a_n)/b_n \leq \log(x)) \longrightarrow H(x),$$

where $b_n := \beta_n$, $a_n := \log(\alpha_n)$ and $Y_i := \log(Z_i)1_{[c,\infty)}(Z_i) + \log(c)1_{(-\infty,c)}(Z_i)$, $i = 1, \ldots, n$ and $1_A(\cdot)$ is the indicator function of the set A. The rv $Y_1, \ldots, Y_n$ are iid with common df satisfying

$$1 - P(Y_i \leq t) = P(Y_i > t)$$
$$= P(Z_i > \exp(t)) = 1 - F(\exp(t)), \qquad \exp(t) \geq c.$$

Therefore, by the Gnedenko-de Haan Theorem 2.1.1 we obtain the equivalence (2.12) for some $G \in \{G_1, G_2, G_3\}$ with $H(x) = G_i(\log(x) - a)/b)$, $x > 0$, for some $b > 0$, $a \in \mathbb{R}$ and some $i \in \{1, 2, 3\}$. This completes the proof of part (i).

(ii) Suppose that $\omega(F) \leq 0$. Then we have for $x \geq 0$ and any $\alpha_n > 0$, $\beta_n > 0$

$$F^n\big(\alpha_n|x|^{\beta_n}\mathrm{sign}(x)\big) = P(Z_{n:n} \leq \alpha_n x^{\beta_n}) = 1.$$

Let now $x < 0$ be a point of continuity of a non-degenerate df H. Then, as $n \to \infty$,

$$P\big(Z_{n:n} \leq \mathrm{sign}(x)\alpha_n|x|^{\beta_n}\big) \longrightarrow H(x)$$
$$\Longleftrightarrow\ P(-Z_{n:n} \geq \alpha_n|x|^{\beta_n}) \longrightarrow H(x)$$
$$\Longleftrightarrow\ P\big(\log(-Z_{n:n}) \geq \beta_n \log(|x|) + \log(\alpha_n)\big) \longrightarrow H(x)$$
$$\Longleftrightarrow\ P\big((X_{1:n} - A_n)/B_n \geq \log(-x)\big) \longrightarrow H(x),$$

where $B_n := \beta_n$, $A_n := \log(\alpha_n)$ and $X_i := \log(-Z_i)$, $i = 1,\ldots,n$ with df $F_*(x) = 1 - F(-\exp(x))$. Notice that the rv X_i is well defined, since $P(Z_i \geq 0) = 0$. In case $\omega(F) = 0$ and $\rho = P(Z_i = 0) > 0$, the limit H would necessarily be degenerate from (2.12). Hence, with $H_*(x) = 1 - H(-\exp(x))$ we find

$$F \in \mathcal{D}_p(H) \Longleftrightarrow F_*(x) = 1 - F\big(\exp(x)\big) \in \mathcal{D}_{\min}(H_*),$$

and Theorem 2.1.1 leads to the representation

$$H(x) = 1 - L_i\big((\log(-x) - a)/b\big), \qquad x < 0,$$

for some $b > 0$, $a \in \mathbb{R}$ and some $i \in \{1,2,3\}$. This completes the proof. $\qquad\square$

The characterization of p-max domains of attraction by the tail behavior of F and the sign of $\omega(F)$ as given in Theorems 3.1 - 3.3 by Mohan and Subramanya [340] (they are reproduced in Subramanya [433] as Theorems A, B, and C) now follows immediately from Theorem 2.6.1 and Theorem 2.1.1 for max domains under linear transformation. On the other hand, Theorems 2.2, 3.1, and 3.2 of Subramanya [433] are now a consequence of the above Theorem 2.6.1 and Theorems 10, 11 and 12 of de Haan [185] using only some substitutions.

COMPARISON OF MAX DOMAINS OF ATTRACTION UNDER LINEAR AND POWER NORMALIZATIONS

Mohan and Ravi [339] compared the max domains of attraction under linear and power normalizations and proved the following result which shows that the class of max domains of attraction under linear normalization is a proper subset of the class of max domains of attraction under power normalization. This means that any df belonging to the max domain of attraction of some EVD limit law under linear normalization definitely belongs to the max domain of attraction of some p-max stable law under power normalization. Also, one can show that there are infinitely many df which belong to the max domain of attraction of a p-max stable law but do not belong to the max domain of attraction of any EVD limit law.

Theorem 2.6.2. *Let F be any df. Then*

(a) (i) $F \in \mathcal{D}_{\max}(G_{1,\gamma})$
    (ii) $F \in \mathcal{D}_{\max}(G_3), \omega(F) = \infty$ $\left.\right\} \Longrightarrow F \in \mathcal{D}_p(H_3),$

(b) $F \in \mathcal{D}_{\max}(G_3), 0 < \omega(F) < \infty \Longleftrightarrow F \in \mathcal{D}_p(H_3), \omega(F) < \infty,$

(c) $F \in \mathcal{D}_{\max}(G_3), \omega(F) < 0 \Longleftrightarrow F \in \mathcal{D}_p(H_6), \omega(F) < 0,$

(d) (i) $F \in \mathcal{D}_{\max}(G_3), \omega(F) = 0$
    (ii) $F \in \mathcal{D}_{\max}(G_{2,\gamma}), \omega(F) = 0$ $\left.\right\} \Longrightarrow F \in \mathcal{D}_p(H_6),$

(e) $F \in \mathcal{D}_{\max}(G_{2,\gamma}), \omega(F) > 0 \Longleftrightarrow F \in \mathcal{D}_p(H_{2,\gamma}),$

(f) $F \in \mathcal{D}_{\max}(G_{2,\gamma}), \omega(F) < 0 \Longleftrightarrow F \in \mathcal{D}_p(H_{4,\gamma}).$

Proof. Let $F \in \mathcal{D}_{\max}(G)$ for some $G \in \{G_1, G_2, G_3\}$. Then for some $a_n > 0, b_n \in \mathbb{R}$,

$$\lim_{n \to \infty} F^n (a_n x + b_n) = G(x), \ x \in \mathbb{R}.$$

(a) (i) If $F \in \mathcal{D}_{\max}(G_{1,\gamma})$, then one can take $b_n = 0$ and hence setting $\alpha_n = a_n, \beta_n = 1/\gamma,$

$$\lambda^{(1)}(x) = \lambda_n^{(1)}(x) = \begin{cases} 0 & \text{if } x < 0, \\ x^{1/\alpha} & \text{if } 0 \le x, \end{cases}$$

we get

$$\lim_{n \to \infty} F^n \left(\alpha_n |x|^{\beta_n} \operatorname{sign}(x)\right) = \lim_{n \to \infty} F^n \left(a_n \lambda_n^{(1)}(x) + b_n\right) = G_{1,\gamma}(\lambda^{(1)}(x)) = H_3(x).$$

(ii) If $F \in \mathcal{D}_{\max}(G_3), \omega(F) = \infty$, then $b_n > 0$ for n large and $\lim_{n \to \infty} a_n/b_n = 0$. Setting $\alpha_n = b_n, \beta_n = a_n/b_n,$

$$\lambda_n^{(2)}(x) = \begin{cases} -1/\beta_n & \text{if } x \le 0, \\ \left(x^{\beta_n} - 1\right)/\beta_n & \text{if } 0 < x, \end{cases}$$

and

$$\lambda^{(2)}(x) = \begin{cases} -\infty & \text{if } x \le 0, \\ \log(x) & \text{if } 0 < x; \end{cases}$$

and proceeding as in the proof of (a) (i), we get $F \in \mathcal{D}_p(H_3)$ since $G_3(\lambda^{(2)}(x)) = H_3(x)$.

(b) If $F \in \mathcal{D}_{\max}(G_3), 0 < \omega(F) < \infty$, then the proof that $F \in \mathcal{D}_p(H_3)$ is the same as that in the case $\omega(F) = \infty$ above. So let $F \in \mathcal{D}_p(H_3)$ with $\omega(F) < \infty$. Then $\lim_{n \to \infty} \alpha_n = \omega(F), \lim_{n \to \infty} \beta_n = 0$. Setting $a_n = \alpha_n \beta_n, b_n = \alpha_n,$

$$u_n^{(1)}(x) = \begin{cases} 0 & \text{if } x \le -1/\beta_n, \\ (1 + \beta_n x)^{1/\beta_n} & \text{if } -1/\beta_n < x, \end{cases}$$

$$u^{(1)}(x) = \exp(x),$$

we get

$$\lim_{n\to\infty} F^n(a_n x + b_n) = \lim_{n\to\infty} F^n\left(\alpha_n |u_n^{(1)}(x)|^{\beta_n} \operatorname{sign}(x)\right) = H_3(u^{(1)}(x)) = G_3(x).$$

(c) If $F \in \mathcal{D}_{\max}(G_3), \omega(F) < 0$, then since $b_n < 0$ and $\lim_{n\to\infty} a_n = 0$, setting $\alpha_n = -b_n, \beta_n = -a_n/b_n$,

$$\lambda_n^{(3)}(x) = \begin{cases} (1 - |x|^{\beta_n})/\beta_n & \text{if } x < 0, \\ 1/\beta_n & \text{if } 0 \le x; \end{cases}$$

and

$$\lambda^{(3)}(x) = \begin{cases} -\log(-x) & \text{if } x < 0, \\ \infty & \text{if } 0 \le x; \end{cases}$$

we get $G_3(\lambda^{(3)}x) = H_6(x)$ and the claim follows as in the proof of (a) (i).
Now if $F \in \mathcal{D}_p(H_6), \omega(F) < 0$, then $\lim_{n\to\infty} \alpha_n = -\omega(F), \lim_{n\to\infty} \beta_n = 0$.
Setting $a_n = \alpha_n \beta_n, b_n = -\alpha_n$,

$$u_n^{(2)}(x) = \begin{cases} -(1 - \beta_n x)^{1/\beta_n} & \text{if } x < 1/\beta_n, \\ 0 & \text{if } 1/\beta_n \le x, \end{cases}$$

$$u^{(2)}(x) = -\exp(-x),$$

and proceeding as in the proof of (b), we get the result since $H_3(u^{(2)}(x)) = G_3(x)$.
(d) (i) Suppose $F \in \mathcal{D}_{\max}(G_3), \omega(F) = 0$. Then $\lim_{n\to\infty} b_n = 0$ and $\lim_{n\to\infty} a_n/b_n = 0$. Proceeding as in the proof of (c) we show that $F \in \mathcal{D}_p(H_6)$.
(ii) Now if $F \in \mathcal{D}_{\max}(G_{2,\gamma}), \omega(F) = 0$, then $b_n = 0$, and setting $\alpha_n = a_n, \beta_n = 1/\gamma$,

$$\lambda^{(4)}(x) = \lambda_n^{(4)}(x) = \begin{cases} -|x|^{1/\alpha} & \text{if } x < 0, \\ 0 & \text{if } 0 \le x, \end{cases}$$

we prove the claim as in (a)(i) using the fact that $G_{2,\gamma}(\lambda^{(4)})(x) = H_6(x)$.
(e) If $F \in \mathcal{D}_{\max}(G_{2,\gamma}), \omega(F) > 0$, then since $b_n = \omega(F)$ and $\lim_{n\to\infty} a_n = 0$, setting $\alpha_n = b_n, \beta_n = a_n/b_n$,

$$\lambda_n^{(5)}(x) = \begin{cases} -1/\beta_n & \text{if } x \le 0, \\ (x^{\beta_n} - 1)/(\beta_n) & \text{if } 0 < x \le 1, \\ 0 & \text{if } 1 < x, \end{cases}$$

and

$$\lambda^{(5)}(x) = \begin{cases} -\infty & \text{if } x \le 0, \\ \log(x) & \text{if } 0 < x \le 1, \\ 0 & \text{if } 1 < x; \end{cases}$$

we get $G_{2,\gamma}(\lambda^{(5)}(x)) = H_{2,\gamma}(x)$ and the claim follows as in the proof of (a) (i).
Now if $F \in \mathcal{D}_p(H_{2,\gamma})$, then $\lim_{n\to\infty} \alpha_n = \omega(F), \lim_{n\to\infty} \beta_n = 0$. Setting $a_n = \alpha_n \beta_n, b_n = \alpha_n$,

$$u_n^{(3)}(x) = \begin{cases} 0 & \text{if } x \le -1/\beta_n, \\ (1 + \beta_n x)^{1/\beta_n} & \text{if } 0 < x \le 1, \\ 1 & \text{if } 0 < x, \end{cases}$$

$$u^{(3)}(x) = \begin{cases} \exp(x) & \text{if } x < 0, \\ 1 & \text{if } 0 \leq x, \end{cases}$$

and proceeding as in the proof of (b), we get the result since $H_{2,\gamma}(u^{(3)}(x)) = G_{2,\gamma}(x)$.

(f) If $F \in \mathcal{D}_{\max}(G_{2,\gamma}), \omega(F) < 0$, then since $b_n = \omega(F)$ and $\lim_{n\to\infty} a_n = 0$, setting $\alpha_n = -b_n, \beta_n = -a_n/b_n$,

$$\lambda_n^{(6)}(x) = \begin{cases} (1 - |x|^{\beta_n})/\beta_n & \text{if } x < -1, \\ 0 & \text{if } -1 \leq x, \end{cases}$$

and

$$\lambda^{(6)}(x) = \begin{cases} -\log(-x) & \text{if } x < -1, \\ 0 & \text{if } -1 \leq x; \end{cases}$$

we get $G_{2,\gamma}(\lambda^{(6)}(x)) = H_{4,\gamma}(x)$ and the claim follows as in the proof of (a) (i). Now if $F \in \mathcal{D}_p(H_{4,\gamma})$, then $\omega(F) < 0, \alpha_n = -\omega(F), \lim_{n\to\infty} \beta_n = 0$. Setting $a_n = \alpha_n\beta_n, b_n = -\alpha_n$,

$$u_n^{(4)}(x) = \begin{cases} -(1 - \beta_n x)^{\beta_n} & \text{if } x < 0, \\ -1 & \text{if } 0 \leq x, \end{cases}$$

$$u^{(4)}(x) = \begin{cases} -\exp(-x) & \text{if } x < 0, \\ -1 & \text{if } 0 \leq x, \end{cases}$$

and proceeding as in the proof of (b), we get the result since $H_{4,\gamma}(u^{(4)}(x)) = G_{2,\gamma}(x)$. The proof of the theorem is complete. $\square$

COMPARISON OF MAX DOMAINS OF ATTRACTION UNDER LINEAR AND POWER NORMALIZATIONS - THE MULTIVARIATE CASE

In this section we generalize Theorem 2.6.2 to the multivariate case. If $F \in \mathcal{D}_{\max}(G)$ for some max stable law G on $\mathbb{R}^d$ then we denote the normalizing constants by $a_n(i) > 0$ and $b_n(i), i \leq d$, so that

$$\lim_{n\to\infty} F^n(a_n(i)x_i + b_n(i), 1 \leq i \leq d) = G(\mathbf{x}), \qquad \mathbf{x} = (x_1, \ldots, x_d) \in \mathbb{R}^d.$$

Similarly, if $F \in \mathcal{D}_p(H)$ for some p-max stable law H on $\mathbb{R}^d$ then we denote the normalizing constants by $\alpha_n(i) > 0$ and $\beta_n(i), 1 \leq i \leq d$, so that

$$\lim_{n\to\infty} F^n\left(\alpha_n(i)|x_i|^{\beta_n(i)}, 1 \leq i \leq d\right) = H(\mathbf{x}), \qquad \mathbf{x} \in \mathbb{R}^d.$$

For a df F on $\mathbb{R}^d$, let $F_{i(1)\ldots i(k)}$ denote the $(i(1)\ldots i(k))$-th k-variate marginal df, $1 \leq i(1) < \cdots < i(k) \leq d, 2 \leq k \leq d$.

Theorem 2.6.3. *Let $F \in \mathcal{D}_{\max}(G)$ for some max stable law G under linear normalization. Then there exists a p-max stable law H on $\mathbb{R}^d$ such that $F \in \mathcal{D}_p(H)$.*

Proof. Let $F \in \mathcal{D}_{\max}(G)$. Then for all $i \leq d, F_i \in \mathcal{D}_{\max}(G_i)$. Hence by Theorem 2.6.2, $F_i \in D_p(H_i)$, for some p-max stable law H_i which must be necessarily a p-type of one of the four p-max stable laws $H_{2,\gamma}, H_{4,\gamma}, H_3, H_6$. The normalization constants $\alpha_n(i), \beta_n(i)$ are determined by $a_n(i), b_n(i)$ as in the proof of Theorem 2.6.2. Further, it follows from the proof of Theorem 2.6.2 that there exists $\theta_n^{(i)}(x_i)$ such that

$$\lim_{n\to\infty} F_i^n\left(\alpha_n(i)|x_i|^{\beta_n(i)}\text{sign}(x_i)\right) = \lim_{n\to\infty} F_i^n\left(a_n(i)\theta_n^{(i)}(x_i) + b_n(i)\right)$$
$$= G_i\left(\theta^{(i)}(x_i)\right),$$

where $\theta_n^{(i)}$ is one of the $\lambda_n^{(j)}$, $j \leq 6$, defined in the proof of Theorem 2.6.2 depending upon which one of the conditions is satisfied by F_i and $\theta^{(i)} = \lim_{n\to\infty} \theta_n^{(i)}$. So, $H_i(x_i) = G_i\left(\theta^{(i)}(x_i)\right)$, $i \leq d$. Now fix $\mathbf{x} = (x_1,\ldots,x_d) \in \mathbb{R}^d$. If, for some $j \leq d, H_j(x_j) = 0$, then by Theorem 2.6.2, we have

$$F_i^n\left(\alpha_n(i)|x_i|^{\beta_n(i)}\text{sign}(x_i), i \leq d\right) \leq F_j^n\left(\alpha_n(j)|x_j|^{\beta_n(j)}\text{sign}(x_j)\right) \to 0,$$

as $n \to \infty$. Suppose now that for some integers $k, i(1),\ldots,i(k)$, we have $0 < H_{i(j)}\left(x_{i(j)}\right) < 1, j \leq k$, and $H_i(x_i) = 1, i \neq i(1),\ldots,i(k)$. Using uniform convergence, we have

$$\liminf_{n\to\infty} F^n\left(\alpha_n(i)|x_i|^{\beta_n(i)}\text{sign}(x_i), i \leq d\right)$$
$$\geq \lim_{n\to\infty} F^n\left(a_n(i)\theta_n^{(i)}(x_i) + b_n(i), i \leq d\right)$$
$$= G\left(\theta^{(i)}x_i, i \leq d\right)$$
$$= G_{i(1)\ldots i(k)}\left(\theta^{(i(j))}(x_{i(j)}), j \leq k\right),$$

since $H_i(x_i) = G_i\left(\theta^{(i)}(x_i)\right) = 1, i \neq i(1),\ldots,i(k)$. Again

$$\limsup_{n\to\infty} F^n\left(\alpha_n(i)|x_i|^{\beta_n(i)}\text{sign}(x_i), i \leq d\right)$$
$$\leq \lim_{n\to\infty} F^n\left(a_n(i(j))\theta_n^{(i(j))}(x_{i(j)}) + b_n(i(j)), j \leq k\right)$$
$$= G_{i(1)\ldots i(k)}\left(\theta^{(i(j))}(x_{i(j)}), j \leq k\right).$$

The claim now follows with

$$H(\mathbf{x}) = G\left(\theta^{(1)}(x_1),\ldots,\theta^{(d)}(x_d)\right). \qquad \square$$

In view of Theorems 2.6.2 and 2.6.3, it is clear that p-max stable laws collectively attract more distributions than do max stable laws under linear normalization collectively.

EXAMPLES

If F is the uniform distribution on $(-1, 0)$, then $\omega(F) = 0$ and (2.11) holds with $H = H_6$, $\alpha_n = 1/n$ and $\beta_n = 1$. Since $H_{2,1}$ is the uniform distribution on $(0, 1)$, it is p-max stable and (2.11) holds with $F = H = H_{2,1}$, $\alpha_n = 1$ and $\beta_n = 1$. For the uniform distribution F on $(-2, -1)$ we find $\omega(F) = -1$ and (2.11) holds with $H = H_{5,1}$, $\alpha_n = 1$ and $\beta_n = 1/n$.

If $F = F_\varepsilon$ is the uniform distribution on $(-1 + \varepsilon, \varepsilon)$ with $0 \le \varepsilon < 1$, then $\omega(F_\varepsilon) = \varepsilon$. Here $F_\varepsilon \in \mathcal{D}_p(H_{2,1})$, and (2.11) holds with $\alpha_n = \varepsilon$ and $\beta_n = 1/(\varepsilon n)$ if $\varepsilon > 0$, whereas for $\varepsilon = 0$ we find (as mentioned above) $F_0 \in \mathcal{D}_p(H_6)$ with power-norming constants $\alpha_n = 1/n$ and $\beta_n = 1$. On the other hand for any fixed $n \ge 1$ we find $F_\varepsilon^n(\text{sign}(x)\varepsilon|x|^{1/(\varepsilon n)}) \to 1_{(-\infty, -1]}(x)$ as $\varepsilon \to 0$. Here the limit distribution is degenerate.

The min stable df $L_{2,1}$ is an exponential law. On the other hand, $L_{2,1} \in \mathcal{D}_{\max}(G_3)$. If $F = L_{2,1}$, then $F(\exp(x)) = L_3(x) \in \mathcal{D}_{\max}(G_3)$. It follows from Theorem 2.6.1 that (2.11) holds with $H = H_3$, $\beta_n = 1/\log(n)$ and $\alpha_n = \log(n)$.

Let $F(x) = 1 - x^{-k}$ for $x \ge 1$, where $k > 0$. Then $F^n(n^{1/k} x) \to G_{1,k}(x)$ as $n \to \infty$, whereas by power normalization $F^n(n^{1/k} x^{1/k}) \to H_3(x)$. Note that $G_{1,1} = H_3$.

The df $G_{1,\gamma}$ for $\gamma > 0$ are under power transformation of type H_3, whereas the df $L_{1,\gamma}$ for $\gamma > 0$ are under power transformation of type H_6.

The max stable and min stable df are connected by the equation

$$L_i(x) = 1 - G_i(-x), \qquad x \in \mathbb{R}, \, i = 1, 2, 3.$$

Under the six p-max stable df $H_1, \ldots, H_6$ there are again three pairs. If the rv U has the df H_i, then the rv $V = -1/U$ has df H_{i+3}, $i = 1, 2$ or 3. The set of possible limit distributions of $Z_{1:n}$ under power normalization

$$(|Z_{1:n}|/\alpha_n^*)^{1/\beta_n^*} \, \text{sign}\,(Z_{1:n}) \longrightarrow_D Z, \qquad n \to \infty,$$

for some suitable constants $\alpha_n^* > 0$ and $\beta_n^* > 0$ can be obtained from Theorem 2.6.1: The limit df equal up to a possible power transformation $1 - H_i(-x)$, $i = 1, \ldots, 6$.

Put $F_1(x) = 1 - (\log(x))^{-1}$ for $x \ge e$. Then F_1 does not belong to any of $\mathcal{D}_{\max}$, but $F_1 \in \mathcal{D}_p(H_{1,1})$ with $\alpha_n = 1$ and $\beta_n = n$, see Galambos [167], Example 2.6.1, and Subramanya [433], Example 1. Taking now $F_2(x) = 1 - (\log\log(x))^{-1}$ for $x \ge \exp(e)$, then without any calculations it follows that F_2 does not belong to any of $\mathcal{D}_p$ since $F_2(\exp(x)) = F_1(x)$ does not belong to any of $\mathcal{D}_{\max}$.

If

$$F(x) = \begin{cases} 0 & \text{if } x < 1, \\ 1 - \exp(-(\log(x))^2) & \text{if } 1 \le x, \end{cases}$$

then $F \in \mathcal{D}_p(H_3)$ with $\alpha_n = \exp(\sqrt{\log(n)})$, $\beta_n = 1/(2\sqrt{\log(n)})$. However, F does not belong to $\mathcal{D}_{\max}(G_{1,\gamma})$ or to $\mathcal{D}_{\max}(G_3)$.

If

$$F(x) = \begin{cases} 0 & \text{if} \quad x < -1, \\ 1 - \exp\left(-(\sqrt{-\log(-x)})\right) & \text{if} \quad 1 \le x < 0, \\ 1 & \text{if} \quad 0 \le x, \end{cases}$$

then $F \in \mathcal{D}_p(H_6)$ with $\alpha_n = \exp(-(\log(n))^2)$, $\beta_n = 2\log(n)$. Note that F does not belong to any $\mathcal{D}_{\max}(G_i)$ for any $i = 1, 2, 3$.

Note that df belonging to $\mathcal{D}_p(H_{1,\gamma})$ or $\mathcal{D}_p(H_{3,\gamma})$ do not belong by Theorem 2.6.2 to the max domain of attraction of any EVD limit law under linear normalization.

Applications of power-normalizations to the analysis of super-heavy tails are included in the subsequent section.

2.7   Heavy and Super-Heavy Tail Analysis

Distributions with light tails like the normal or the exponential distributions have been of central interest in classical statistics. Yet, to cover in particular risks in fields like flood frequency analysis, insurance and finance in an appropriate manner, it became necessary to include distributions which possess heavier tails. An early reference is the modeling of incomes by means of Pareto distributions.

One standard method to get distributions with heavier tails is the construction of log-distributions. Prominent examples are provided by log-normal and log-exponential distributions whereby the latter belongs to the above mentioned class of Pareto distributions. Normal and log-normal distributions possess an exponential decreasing upper tail and, as a consequence all moments of these distributions are finite. However, a log-normal distribution exhibits a higher kurtosis than the normal and, in this sense, its upper tail can be also considered heavier than that of the normal one. It is also a general rule that the mixture of distributions, as a model for heterogeneous populations, leads to heavier tails. For instance, log-Pareto df can be deduced as mixtures of Pareto df with respect to gamma df.

In contrast to normal, log-normal and exponential distributions, one can say that Pareto distributions are prototypes of distributions with heavy, upper tails. An important characteristic of this property is that not all moments are finite. Other prominent distributions of this type are, e.g., Student and sum-stable distributions with the exception of the normal one. All these distributions have power decreasing upper tails, a property shared by Pareto distributions.

Distributions with heavy tails have been systematically explored within the framework of extreme value theory with special emphasis laid on max-stable Fréchet and Pareto distributions where the latter ones possess a peaks-over-threshold (POT) stability. More precisely, one may speak of linearly max-stable or linearly POT-stable (l-max or l-POT) distributions in view of our explanations on page 66 and the remainder of this section. Related to this is the property that a distribution belongs to the l-max or l-POT domain of attraction of a Fréchet or

Pareto distribution if, and only if, the upper tail is regularly varying with negative index. In contrast to this, slowly varying upper tails will be of central importance in the subsequent context.

This means that we are out of the *"power-law-world"*, as Taleb's book, *"The Black Swan: the Impact of the Highly Improbable"* [439] entitles the class of distributions possessing a regularly varying upper tail or, equivalently, with polynomially decreasing upper tails. The designation of super-heavy concerns right tails decreasing to zero at a slower rate, as logarithmic, for instance. This also means that the classical *bible* for inferring about rare events, the Extreme Value Theory, is no longer applicable, since we are in presence of distributions with slowly varying tails.

We give a short overview of the peaks-over-threshold approach which is the recent common tool for statistical inference of heavy tailed distributions. Later on, we present extensions to the super-heavy tailed case.

Heavy Tail Analysis

We shortly address the peaks-over-threshold approach as already described at the end of Chapter 1 but take a slightly different point of view. We do not start with the assumption that a df F is in the max domain of attraction of some EVD G but we consider limiting distributions of exceedances in their own right. Recently, this has been the most commonly used statistical approach for heavy tailed distributions.

Recall that (1.22) indicates that, if $F \in \mathcal{D}(G)$ then GPD are the only possible limiting distributions of the linear normalized df $F^{[u_n]}$ of exceedances over thresholds u_n tending to $\omega(F)$. Hereby, $\omega(F) = \sup\{x : F(x) < 1\}$ is the right endpoint of the support of the df F and

$$F^{[u]}(x) = \frac{F(x) - F(u)}{1 - F(u)}, \quad x \geq u,$$

is the common df of exceedances above the threshold u. If X is a rv with df F then the exceedance df may be written as $F^{[u]}(x) = P(X \leq x \mid X > u)$ for $x \geq u$.

In what follows we assume that there exist real-valued functions $a(\cdot)$ and $b(\cdot) > 0$ such that

$$F^{[u]}(a(u) + b(u)x) \longrightarrow_{u \to \omega(F)} L(x) \tag{2.14}$$

for some non-degenerate df L. If (2.14) holds for df F and L we say that F is in the POT domain of attraction of L denoted by $F \in \mathcal{D}_{\mathrm{POT}}(L)$.

Notice that (2.14) can be formulated in terms of the survivor function $1 - F$ as

$$\frac{1 - F(a(u) + b(u)x)}{1 - F(u)} \longrightarrow_{u \to \omega(F)} 1 - L(x),$$

which corresponds to formula (1.21).

Due to results in Balkema and de Haan [22], stated as Theorem 2.7.1 below, we know that the limiting df L is POT-stable. Hereby, a df F is POT-stable if there exists constants $a(u)$ and $b(u) > 0$ such that

$$F^{[u]}(a(u) + b(u)x) = F(x) \tag{2.15}$$

for all points of continuity u in the interior of the support of F and $F(x) > 0$. The class of continuous POT-stable df, and, therefore, that of continuous limiting df of exceedances in (2.14) is provided by the family of generalized Pareto df (GPD). This result is stated in the following theorem which can be regarded as an extension of Theorem 1.3.5.

Theorem 2.7.1. *Let F be an df and L be a non-degenerate df. Suppose there exists real-valued functions $a(\cdot)$ and $b(\cdot) > 0$ such that*

$$F^{[u]}(a(u) + b(u)x) \longrightarrow_{u \to \omega(F)} L(x)$$

for all points of continuity x of F in the interior of its support. Then,

(i) *the limiting df L is POT-stable,*

(ii) *if L is continuous, then L is up to a location and scale shift a GPD W_γ,*

(iii) *the POT domain of attraction of a GPD W_γ coincides with the max domain of attraction of the corresponding EVD G_γ, thus $\mathcal{D}_{POT}(W_\gamma) = \mathcal{D}(G_\gamma)$.*

It is evident that all POT-stable df L appear as limiting df in Theorem 2.7.1 by choosing $F = L$. Therefore, GPD are the only continuous, POT-stable df.

For statistical applications, e.g., high quantile estimation, these results are of particular importance. Note that high q-quantiles $F^{-1}(q)$ of a df F with $q > F(u)$ for some threshold u only depend on the tail of F, thus $F(x)$ for $x > u$. Notice that for a df F and $x > u$,

$$\begin{aligned}F(x) &= F(u) + (1 - F(u))F^{[u]}(x)\\ &\approx F(u) + (1 - F(u))W_{\gamma,u,\sigma}(x)\end{aligned} \tag{2.16}$$

where the approximation is valid if $F \in \mathcal{D}_{\mathrm{POT}}(W_\gamma)$ and u is sufficiently large. Now (2.16) provides a certain parametric model for the tail of F where a non-parametric part $F(u)$ can be replaced by an empirical counterpart. A similar "piecing-together approach" can also be utilized in the multivariate framework, cf. Chapter 5.

In what follows we concentrate on the heavy tail analysis, that is, on df F belonging to $\mathcal{D}_{\mathrm{POT}}(W_\gamma)$ for some $\gamma > 0$, for which case $\omega(F) = \infty$. The model (2.16) offers the possibility to carry out statistical inference for such df. These df have the special property that their pertaining survivor function $1 - F$ is of *regular variation at infinity*. We include some remarks concerning the theory of regular varying functions and point out relations to the concept of POT-stability.

Regular and Slow Variation

Consider the Pareto distribution $W_{1,\alpha,0,\sigma}(x) = 1 - (x/\sigma)^{-\alpha}$, $x > \sigma > 0$, in the α-parametrization, with $\alpha > 0$. Recall that for any df F the survivor function of the pertaining exceedance df $F^{[u]}$ satisfies

$$\overline{F^{[u]}}(x) = \bar{F}(x)/\bar{F}(u), \quad x > u.$$

For $F = W_{1,\alpha,0,\sigma}$ and replacing x by ux one gets

$$\bar{F}(ux)/\bar{F}(u) = \overline{F^{[u]}}(ux) = x^{-\alpha}$$

which is the POT-stability of $W_{1,\alpha,0,\sigma}$. If F is an arbitrary Pareto df $W_{1,\gamma,\mu,\sigma}$ with additional location parameter μ this relation holds in the limit. We have

$$\bar{F}(ux)/\bar{F}(u) = \overline{F^{[u]}}(ux) = \left(\frac{x - \mu/u}{1 - \mu/u}\right)^{-\alpha} \longrightarrow_{u\to\infty} x^{-\alpha}, \qquad x \geq 1.$$

This implies that $\bar{F}$ is regularly varying at infinity according to the following definition: A measurable function $R : (0,\infty) \to (0,\infty)$ is called regularly varying at infinity with index (exponent of variation) ρ, denoted by $R \in RV_\rho$, if

$$R(tx)/R(t) \longrightarrow_{t\to\infty} x^\rho, \quad x > 0. \tag{2.17}$$

A comprehensive treatment of the theory of regular variation may, e.g., be found in Bingham et al. [46]. If $\rho = 0$ we have

$$R(tx)/R(t) \longrightarrow_{t\to\infty} 1;$$

in this case, R is said to be of slow variation at infinity ($R \in RV_0$). Recall from Theorem 2.1.1 together with Theorem 2.7.1, part (iii), that a df F is in the POT domain of attraction of some GPD W_γ, $\gamma > 0$, if $\bar{F} \in RV_{-1/\gamma}$. For any $R \in RV_\rho$ we have the representation

$$R(x) = x^\rho U(x), \quad x > a,$$

for some $a > 0$ sufficiently large and $U \in RV_0$. If a df F is in the POT domain of attraction of some GPD W_γ for $\gamma > 0$ (thus, $\bar{F} \in RV_{-1/\gamma}$) we call F heavy tailed.

The existence of finite β-power moments is restricted to the range $\beta < 1/\gamma$. Although there is no unified agreement on terminology, in literature the term *very heavy tailed* case has been attached to a degree of tail heaviness given by $0 < 1/\gamma < 1$.

Super-Heavy Tails and Slow Variation

The use of heavy tailed distributions constitutes a fundamental tool in the study of rare events and have been extensively used to model phenomena for which extreme values occur with a relatively high probability. Here, emphasis lies on the modelling of extreme events, i.e., events with a low frequency, but mostly with a high and often disastrous impact. For such situations it has became reasonable to consider an underlying distribution function F with polynomially decaying right tail, i.e., with tail distribution function

$$\bar{F} := 1 - F \in RV_{-\alpha}, \quad \alpha > 0. \tag{2.18}$$

Generalizing this heavy tail framework, it is also possible to consider the so-called *super-heavy tailed* case, for which $\alpha = 0$, i.e., $1 - F$ is a slowly varying function, decaying to zero at a logarithmic rate, for instance. We will consider two special classes of such super-heavy tailed df.

Class A. Notice that if X has a df F such that $\bar{F} \in RV_{-\alpha}$, for some positive α, then $Z := e^X$ has the df H with

$$\bar{H}(x) \sim (\log(x))^{-\alpha} U(\log(x)) \tag{2.19}$$

as $x \to \infty$, with $U \in RV_0$, meaning that the tail decays to zero at a logarithmic rate raised to some power. Although this transformation leads to a super-heavy tailed df it does not exhaust all possible slowly varying tail types. On the other hand, for the super-heavy tailed case there is no possible linear normalization of the maxima such that F belongs to any max-domain of attraction. Consider the case $U \equiv 1$ in (2.19). This gives the super-heavy tailed df $F(x) = 1 - \log(x)^{-\alpha}$, $x \geq e$. The pertaining survivor function satisfies

$$\bar{F}\left(x^{\log(u)}\right) / \bar{F}(u) = \bar{F}(x).$$

Subsequently, this property will be called the power-POT (p-POT) stability of F, it characterizes the class of limiting df of exceedances under power-normalization, cf. Theorem 2.7.2. Corresponding to the case of heavy tailed df an asymptotic version of this relation will be identified as an necessary and sufficient condition of a df to belong to certain p-POT domains of attraction, cf. Theorem 2.7.5.

Class B. The df F satisfies (2.18) if, and only if, there exists a positive function a such that

$$\lim_{t \to \infty} \frac{F(tx) - F(t)}{a(t)} = \frac{1 - x^{-\alpha}}{\alpha}, \quad x > 0. \tag{2.20}$$

For the latter it is enough to consider the auxiliary function $a = \alpha \bar{F}$ and thus $a \in RV_{-\alpha}$, $\alpha > 0$. A sub-class of slowly varying df is deduced from (2.20) by extension to the case of $\alpha = 0$, through the limit of the right-hand side of (2.20), as $\alpha \to 0$:

$$\lim_{t \to \infty} \frac{F(tx) - F(t)}{a(t)} = \log(x). \tag{2.21}$$

The above relation identifies the well-known class Π (cf., e.g., de Haan and Ferreira [190]). The class of super-heavy tailed distributions is characterized by (2.21).

More details about all distributions satisfying (2.20) with $\alpha \geq 0$ will be provided at the end of this section, together with testing procedures for super-heavy tails (see Theorems 2.7.12 and 2.7.13).

For the time being notice that the df given by $1 - 1/\log(x)$, $x > e$ belongs to both Classes A and B. Moreover, according to Proposition 2.7.10, any df H in Class A and resulting from composition with a df F such that the density $F' =: f$ exists, also belongs to the Class B. However, the reverse is not always true: for instance, the df $H(x) = 1 - 1/\log(\log(x))$, for $x > e^e$, belongs to B but not to A. Note that H is obtained by iterated log-transforms upon a Pareto df. Distributions of this type are investigated in Cormann [77] and Cormann and Reiss [76].

In the remainder of this section we study two special aspects of the statistical analysis of super-heavy tailed df. First we deal with asymptotic models for certain super-heavy tailed df related to df given by Class A. The second part concerns testing the presence of a certain form of super-heavy tails, namely Π-varying tailed distributions given by Class B.

SUPER-HEAVY TAILS IN THE LITERATURE

We first give a short outline of the statistical literature dealing with super-heavy tails. Although there is no uniform agreement on terminology, the term *super-heavy tailed* has been attached, in the literature, to a degree of tail heaviness associated with slow variation. Examples of models with slowly varying tail are the log-Pareto, log-Fréchet and log-Cauchy distributions. We say that the rv X is a log-Pareto rv if $\log(X)$ is a Pareto rv.

In Galambos [166], Examples 1.3.3 and 2.6.1, the log-Pareto df

$$F(x) = 1 - 1/\log(x), \quad x > e, \tag{2.22}$$

serves as a df under which maxima possess "shocking" large values, not attained under the usual linear normalization pertaining to the domain of attraction of an EVD. Some theoretical results for super-heavy tailed distributions can be found in Resnick [392], Section 5, which is devoted to fill some "interesting gaps in classical limit theorems, which require the assumption that tails are even fatter than regularly varying tails". Two cases are considered in some probabilistic descriptions of *"fat"* tails, under the context of point process convergence results: slowly varying tails and its subclass of Π-varying distribution functions.

Another early reference to df with slowly varying tails, in conjunction with extreme value analysis, can be found in the book by Reiss and Thomas [388], Section 5.4, where log-Pareto distributions are regarded as mixtures of Pareto df with respect to gamma df. The authors have coined all log-Pareto df with the term "super-heavy" because the log-transformation leads to a df with a heavier tail than the tail heaviness of Pareto type.

Log-Pareto df within a generalized exponential power model are studied by Desgagné and Angers [100]. In a technical report, see Diebolt et al. [110], associated to Diebolt et al. [111], the authors mention another mixture distribution, different from the log-Pareto one, with super-heavy tails.

Moreover, Zeevi and Glynn [470] have studied properties of autoregressive processes with super-heavy tailed innovations, specifically, the case where the innovations are log-Pareto distributed. Their main objective was to illustrate the range of behavior that AR processes can exhibit in this super-heavy tailed setting. That paper studies recurrence properties of autoregressive (AR) processes with "super-heavy tailed" innovations. Specifically, they study the case where the innovations are distributed, roughly speaking, as log-Pareto rvs (i.e., the tail decay is essentially a logarithm raised to some power).

In Neves and Fraga Alves [352] and in Fraga Alves et al. [159] the tail index α is allowed to be 0, so as to embrace the class of super-heavy tailed distributions. Statistical tests then are developed in order to distinguish between heavy and super-heavy tailed probability distributions. This is done in a semi-parametric way, i.e., without specifying the exact df underlying the data in the sense of composite hypothesis testing. Therein, the authors present some simulation results concerning estimated power and type I error of the test. Application to data sets in teletraffic and seismology fields is also given.

Cormann and Reiss [76] introduced a full-fledged statistical model of log-Pareto distributions parametrized with two shape parameters and a scale parameter and show that these distributions constitute an appropriate model for super-heavy tailed phenomena. Log-Pareto distributions appear as limiting distributions of exceedances under power-normalization. Therein it is shown, that the well-known Pareto model is included in the proposed log-Pareto model for varying shape-parameters whence the log-Pareto model can be regarded as an extension of the Pareto model. This article also explores an hybrid maximum likelihood estimator for the log-Pareto model. The testing of the Pareto model against the log-Pareto model is considered in Villaseñor-Alva et al. [450].

The need of distributions with heavier tails than the Pareto type has also been claimed in Earth Sciences research. A usual statistical data analysis in seismology is done through the scalar seismic moment M, which is related to the earthquake moment magnitude m according to: $M = 10^{3(m+6)/2}$ (notice the power transformation with consequences on the distribution tail weight). Zaliapin et al. [469] presents an illustration of the distribution of seismic moment for Californian seismicity ($m \geq 5.5$), during the last two centuries, using an earthquake catalog and converting its magnitudes into seismic moments. They observed that

> *... with such a data set one does not observe fewer earthquakes of large seismic moment than expected according to the Pareto law. Indeed, ... may even suggest that the Pareto distribution underestimates the frequency of earthquakes in this seismic moment range.*

In fact, these authors called the attention to the practitioners that:

> *Statistical data analysis, a significant part of modern Earth Sciences research, is led by the intuition of researchers traditionally trained to think in terms of "averages", "means", and "standard deviations". Curiously, an essential part of relevant natural processes does not allow such an interpretation, and appropriate statistical models do not have finite values of these characteristics.*

The same data set has been analyzed by Neves and Fraga Alves [352] in the context of detecting super-heavy tails.

The *P*-Pot Stable Distributions

Recall the log-Pareto df $F(x) = 1 - 1/\log(x)$, $x \geq e$ mentioned above as an important example of a super-heavy tailed df. Such distributions cannot be studied within the POT-framework outlined in Sections 2.1 to 2.4 because they possess slowly varying tails. Nevertheless, p-max domains of attraction in Section 2.6 contain certain super-heavy tailed distributions.

We have noted in the previous section that the distribution of the largest order statistic $Z_{n:n}$ out of an iid sample $Z_1, \ldots, Z_n$ can be approximated by certain p-max stables laws even if the common df belongs to a certain subclass of distributions with slowly varying distribution tails. In the present section we derive an asymptotic model for the upper tail of such a df F. Similarly to the linear normalization we consider the asymptotic relation

$$F^{[u]}\left(\mathrm{sign}(x)\alpha(u)|x|^{\beta(u)}\right) \longrightarrow_{u \to \omega(F)} L(x) \qquad (2.23)$$

for all points of continuity x of L, where L is a non-degenerate df and $\alpha(\cdot)$, $\beta(\cdot) > 0$. Notice that (2.23) is equivalent to

$$\frac{1 - F\left(\mathrm{sign}(x)\alpha(u)|x|^{\beta(u)}\right)}{1 - F(u)} \longrightarrow_{u \to \omega(F)} 1 - L(x). \qquad (2.24)$$

Recall that limiting distributions of exceedances under linear normalization are POT-stable. A similar results holds for limiting distributions under power-normalization. These distributions satisfy the p-POT stability property. A df F is p-POT stable if there are positive constants $\beta(u)$ and $\alpha(u)$ such that

$$F^{[u]}(\mathrm{sign}(x)\alpha(u)|x|^{\beta(u)}) = F(x) \qquad (2.25)$$

for all x with $F(x) > 0$ and all continuity points u of F with $0 < F(u) < 1$.

Due to Theorem 2.7.1 we know that GPD are the only continuous POT-stable distributions. According to Theorem 1 in Cormann and Reiss [76], stated below as Theorem 2.7.2, we know that for every p-POT stable df L there is a POT-stable df W such that $L(x) = W(\log(x))$ if $\omega(L) > 0$, or $L(x) = W(-\log(-x))$ if $\omega(L) \leq 0$. As in Mohan and Ravi [339] we call a df F_1 a p-type of F_2, if $F_1(x) = F_2\left(\mathrm{sign}(x)\alpha|x|^{\beta}\right)$ for positive constants α and β.

Given a df F we define auxiliary df F^{**} and F_{**} by

$$F^{**}(x) = \frac{F(\exp(x)) - F(0)}{1 - F(0)}, \quad x \in \mathbb{R}, \tag{2.26}$$

if $\omega(F) > 0$, and

$$F_{**}(x) = F(-\exp(-x)), \quad x \in \mathbb{R}, \tag{2.27}$$

if $\omega(F) \le 0$. These auxiliary df play a similar role for limiting df of exceedances as do the df F^* and F_* in Theorem 2.6.1 in the context of limiting df of maxima.

Theorem 2.7.2. *Let F be a df which is p-POT stable, cf. (2.25). Then,*

$$F(x) = W\left(\log(x)\right),$$

or

$$F(x) = W\left(-\log(-x)\right),$$

where W denotes a POT-stable df.

Proof. Let $0 < F(u) < 1$. First note that (2.25) is equivalent to

$$\frac{1 - F(\mathrm{sign}(x)\alpha(u)|x|^{\beta(u)})}{1 - F(u)} = 1 - F(x).$$

Let $F(0) > 0$. Then

$$\frac{1 - F(0)}{1 - F(u)} = 1 - F(0)$$

and $F(0) = 1$ because $0 < F(u) < 1$. Thus, we have $F(0) = 0$ or $F(0) = 1$ and, consequently, F has all the mass either on the positive or negative half-line.

(a) Let $F(0) = 0$ and, therefore, $F(x) = 0$ for all $x < 0$. It suffices to consider $x, u > 0$. Let $x > 0$, $F(x) > 0$ and $0 < F(u) < 1$. Then, (2.25) yields

$$\frac{1 - F\left(\alpha(u)x^{\beta(u)}\right)}{1 - F(u)} = 1 - F(x).$$

It follows that

$$\frac{1 - F\left(\alpha(\exp(u))\exp(x)^{\beta(\exp(u))}\right)}{1 - F(\exp(u))} = 1 - F(\exp(x))$$

for all x and continuity points u of $F(\exp(\cdot))$ with $F\left(\exp(x)\right) > 0$ and $0 < F\left(\exp(u)\right) < 1$. Furthermore,

$$\frac{1 - F\left(\alpha(\exp(u))\exp(x)^{\beta(\exp(u))}\right)}{1 - F(\exp(u))} = 1 - F(\exp(x))$$

$$\Longleftrightarrow \frac{1 - F\left(\exp\left(\log(\alpha(\exp(u))) + \beta(\exp(u))x\right)\right)}{1 - F\left(\exp(u)\right)} = 1 - F\left(\exp(x)\right).$$

Observe that $F^{**} := F(\exp(\cdot))$ since $F(0) = 0$. The above computations yield

$$\frac{1 - F^{**}\left(\tilde{\alpha}(u) + \tilde{\beta}(u)x\right)}{1 - F^{**}(u)} = 1 - F^{**}(x)$$

with $\tilde{\alpha}(u) = \log\left(\alpha(\exp(u))\right)$ and $\tilde{\beta}(u) = \beta(\exp(u))$. Consequently, $F^{**} = W$ for some POT-stable df W and $F(\cdot) = W(\log(\cdot))$.

(b) Next assume that $F(0) = 1$. Let (2.25) hold for $x < 0$, $F(x) > 0$ and all continuity points u of F with $0 < F(u) < 1$. Then, similar arguments as in part (a) yield that (2.25) is equivalent to

$$\frac{1 - F_{**}(\tilde{\alpha}(u) + \tilde{\beta}(u)x)}{1 - F_{**}(u)} = 1 - F_{**}(x)$$

with $F_{**}(x) := F(-\exp(-x))$ where $\tilde{\alpha}(u)$ can be chosen as $\alpha(-\exp(-u))$ and $\tilde{\beta}(u) = \beta(-\exp(-u))$. Thus, F_{**} is a POT-stable df W and $F(x) = W\left(-\log(-x)\right)$, $x \leq 0$.   □

Due to the foregoing theorem all continuous p-POT stable df are p-types of the df

$$L_\gamma(x) = 1 - (1 + \gamma \log(x))^{-1/\gamma}, \quad x > 0,\ \gamma \in \mathbb{R} \tag{2.28}$$

which is a generalized log-Pareto distribution (GLPD), or

$$V_\gamma(x) := 1 - (1 - \gamma \log(-x))^{-1/\gamma}, \quad x < 0,\ \gamma \in \mathbb{R} \tag{2.29}$$

which may be addressed as negative generalized log-Pareto df. The case $\gamma = 0$ is again taken as the limit $\gamma \to 0$. Notice that only the p-POT stable law L_γ, $\gamma > 0$ is super-heavy tailed, while L_γ, $\gamma < 0$ and V_γ possess finite right endpoints. The df L_0 is a Pareto df and, therefore, heavy tailed.

RELATIONS TO P-MAX STABLE LAWS

We start with a representation of log-Pareto df by means of p-max stable df. Recall that a df F is p-max stable if there exist sequences α_n, $\beta_n > 0$ such that

$$F^n(\text{sign}(x)\alpha_n|x|^{\beta_n}) = F(x), \quad x \in \mathbb{R}$$

and all positive integers n, cf. Section 2.6.

For the special p-max stable df

$$H_{1,\gamma}(x) = \exp(-(\log(x))^{-\gamma}), \quad x \geq 1,$$

with $\gamma > 0$, define

$$\begin{aligned} F_\gamma(x) &= 1 + \log\left(H_{1,\gamma}(x)\right) \\ &= 1 - (\log(x))^{-\gamma}, \quad x \geq \exp(1), \end{aligned} \tag{2.30}$$

which is a log-Pareto df with shape parameter $1/\gamma$.

In analogy to (2.30), the whole family of GLPDs in (2.28) can be deduced from p-max stable laws $H_{i,\gamma,\beta,\sigma}(x) = H_{i,\gamma}((x/\sigma)^\beta)$, $i = 1, 2, 3$. This relationship makes the theory of p-max df applicable to log-Pareto df to some extent.

LIMITING DISTRIBUTIONS OF EXCEEDANCES

In the subsequent lines we present some unpublished material. We identify the limiting distributions of exceedances under power normalization in (2.23) as the class of p-POT stable df. We start with a technical lemma concerning F^{**} and F_{**}, cf. (2.26) and (2.27).

Lemma 2.7.3. *Let L be a non-degenerate limiting df in (2.23) for some df F. Then, for each point of continuity x in the interior of the support of L^{**} or L_{**}, respectively,*

(i) *there are functions $a(\cdot)$ and $b(\cdot) > 0$ such that*

$$\bar{F}^{**}(a(u) + b(u)x)/\bar{F}^{**}(u) \longrightarrow_{u\to\omega(F^{**})} \bar{L}^{**}(x)$$

if $\omega(F) > 0$, and

$$\bar{F}_{**}(a(u) + b(u)x)/\bar{F}_{**}(u) \longrightarrow_{u\to\omega(F_{**})} \bar{L}_{**}(x)$$

if $\omega(F) \leq 0$.

(ii) *L^{**} and, respectively, L_{**} are POT-stable df.*

Proof. We outline the proof for both assertions merely for $\omega(F) > 0$. The case of $\omega(F) \leq 0$ follows by similar arguments. Under (2.23) we first prove that the total mass of L is concentrated on the positive half-line and, therefore,

$$L(\exp(x)) = L^{**}(x), \quad x \in \mathbb{R}, \tag{2.31}$$

if $\omega(L) > 0$.

If $x < 0$, we have

$$F^{[u]}\left(\text{sign}(x)\alpha(u)|x|^{\beta(u)}\right) \leq F^{[u]}(0) \longrightarrow_{u\to\omega(F)} 0 \tag{2.32}$$

because $\omega(F) > 0$. This implies $L(x) = 0$ for all $x \leq 0$.

Next consider $x > 0$. From (2.24) one gets

$$\bar{F}\left(\alpha(u)x^{\beta(u)}\right)/\bar{F}(u) \longrightarrow_{u\to\omega(F)} \bar{L}(x).$$

By straightforward computations,

$$\frac{\bar{F}\left(\exp\left(a(u) + b(u)x\right)\right)}{\bar{F}(\exp(u))} \longrightarrow_{\exp(u)\to\omega(F)} \bar{L}(\exp(x)) \tag{2.33}$$

for all $x \in \mathbb{R}$ with $a(u) = \log(\alpha(\exp(u))) \in \mathbb{R}$ and $b(u) = \beta(\exp(u)) > 0$. Therefore,

$$\frac{\bar{F}^{**}(a(u) + b(u)x)}{\bar{F}^{**}(u)} \longrightarrow_{u \to \omega(F^{**})} \bar{L}(\exp(x)) = \bar{L}^{**}(x), \qquad (2.34)$$

and assertion (i) is verified. This also implies (ii) because limiting df under the linear normalization are necessarily POT-stable, cf. Theorem 2.7.1. □

Lemma 2.7.3 now offers the prerequisites to prove the the announced result concerning limiting df of exceedances under power-normalizations.

Theorem 2.7.4. *Every non-degenerate limiting df L in (2.23) is p-POT stable.*

Proof. Again, we merely prove the case $\omega(F) > 0$. From Lemma 2.7.3 (ii) we know that L^{**} is POT-stable. Thus, there are $a(u) \in \mathbb{R}$ and $b(u) > 0$ such that

$$\bar{L}^{**}(a(u) + b(u)x)/\bar{L}^{**}(u) = \bar{L}^{**}(x),$$

for each point of continuity u of L^{**} with $0 < L^{**}(u) < 1$ and $L^{**}(x) > 0$. This yields for $x, u > 0$,

$$\bar{L}^{**}(a(u) + b(u)\log(x))/\bar{L}^{**}(\log(u)) = \bar{L}^{**}(\log(x)).$$

Choosing $\alpha(u)$ and $\beta(u)$ as in the proof of Lemma 2.7.3 one gets from the equation $\bar{L}^{**}(a(u) + b(u)\log(x)) = \bar{L}^{**}\left(\log\left(\alpha(u)x^{\beta(u)}\right)\right)$ that

$$\bar{L}\left(\alpha(u)x^{\beta(u)}\right)/\bar{L}(u) = \bar{L}(x)$$

for all $x, u > 0$ with $0 < L^{**}(\log(u)) < 1$ and $L^{**}(\log(x)) > 0$. Notice that $L(x) = L^{**}(\log(x))$ if $x > 0$, and $L(x) = 0$ if $x \leq 0$. This yields the p-POT stability of L according to the preceding equation. □

It is evident that the converse implication is also true, that is, every p-POT stable df L is a limiting df in (2.23) by choosing $F = L$. Summarizing the previous results we get that L is a limiting df of exceedances pertaining to a df F under power-normalization, if and only if, L^{**} (if $\omega(F) > 0$) or L_{**} (if $\omega(F) \leq 0$) are POT-stable.

DOMAINS OF ATTRACTION

Recall that within the linear framework, a df F belongs to the POT domain of attraction of a df W, denoted by $F \in \mathcal{D}_{\mathrm{POT}}(W)$, if there are functions $a(\cdot)$ and $b(\cdot) > 0$ such that

$$F^{[u]}(a(u) + b(u)x) \longrightarrow_{u \to \omega(F)} W(x). \qquad (2.35)$$

Correspondingly, if relation (2.23) holds for df F and L, then F belongs to the p-POT domain of attraction of L, denoted by $F \in \mathcal{D}_{p\text{-POT}}(L)$.

We characterize p-POT domains of attraction of a p-POT stable df L by means of POT domains of attraction of L^{**} or L_{**} which are POT-stable according to Theorem 2.7.4. As a direct consequence of Lemma 2.7.3(i) one gets Theorem 2.7.5.

Theorem 2.7.5. *For the p-POT domain of attraction $\mathcal{D}_{p\text{-}POT}(L)$ of a p-POT stable law L we have*

$$\mathcal{D}_{p\text{-}POT}(L) = \{F : F^{**} \in \mathcal{D}_{POT}(L^{**})\},$$

if $\omega(L) > 0$, and

$$\mathcal{D}_{p\text{-}POT}(L) = \{F : F_{**} \in \mathcal{D}_{POT}(L_{**})\}$$

if $\omega(L) \leq 0$.

P-POT domains of attraction of continuous p-POT stable laws can be deduced from p-max domains of attractions due to the identity of POT- and max-domains of attraction in the linear setup. The domains of attraction of the discrete p-POT stable laws have no counterpart in the framework of max-stable df. Their domains of attraction can be derived from the above theorem and Section 3 of Balkema and de Haan [22].

In the framework of super-heavy tail analysis we are merely interested in the super-heavy tailed p-POT stable laws, thus log-Pareto df. We also make use of a parametrization of log-Pareto df which is different from that in (2.28). Let

$$\tilde{L}_\gamma(x) = 1 - (\log(x))^{-1/\gamma}, \quad \gamma > 0, \, x \geq e. \tag{2.36}$$

It is apparent that $\tilde{L}_\gamma$ is a p-type of L_γ. Such df can be regarded as prototypes of p-POT stable df with slowly varying tails.

Corollary 2.7.6. *We have $F \in \mathcal{D}_{p\text{-}POT}(\tilde{L}_\gamma)$ if, and only if, there is a slowly varying function U and some $c > 0$ such that*

$$F(x) = 1 - (\log(x))^{-1/\gamma} U(\log(x)), \quad x > c. \tag{2.37}$$

Proof. This is a direct consequence of Theorem 2.7.5. We have for $x > 0$ that $\bar{F}(x) = \bar{F}(0) F^{**}(\log(x))$ for the df F^{**} which is in the POT domain of attraction of a Pareto df and, therefore, $\bar{F}^{**}$ is regularly varying at infinity. □

The p-POT domain of attraction of a log-Pareto df $\tilde{L}_\gamma$ can as well be characterized by a property with is deduced from regular variation, which characterizes the POT domain of attraction of Pareto df under linear transformation. Observe that

$$\bar{\tilde{L}}_\gamma\left(x^{\log(u)}\right) / \bar{\tilde{L}}_\gamma(u) = (\log(x))^{-1/\gamma},$$

which is the p-POT stability of $\tilde{L}_\gamma$. For the domain of attraction this relations holds in the limit and, furthermore, this yields a characterization of the domain attraction.

Corollary 2.7.7. *We have* $F \in \mathcal{D}_{p\text{-POT}}(\tilde{L}_\gamma)$ *if, and only if,*

$$\bar{F}\left(x^{\log(u)}\right)/\bar{F}(u) \longrightarrow_{u \to \infty} (\log(x))^{-1/\gamma}, \quad x > 1. \tag{2.38}$$

Proof. If $F \in \mathcal{D}_{p\text{-POT}}(\tilde{L}_\gamma)$ we have

$$\bar{F}(x) = (\log(x))^{-1/\gamma} U(\log(x)), \quad x > c$$

for some slowly varying function U and some $c > 0$. Therefore,

$$\begin{aligned}
\frac{\bar{F}\left(x^{\log(u)}\right)}{\bar{F}(u)} &= \frac{\left(\log\left(x^{\log(u)}\right)\right)^{-1/\gamma} U\left(\log\left(x^{\log(u)}\right)\right)}{(\log(u))^{-1/\gamma} U(\log(u))} \\
&= (\log(x))^{-1/\gamma} \frac{U(\log(u)\log(x))}{U(\log(u))} \\
&\to (\log(x))^{-1/\gamma} \qquad \text{for } u \to \infty.
\end{aligned}$$

Conversely, let

$$\lim_{u \to \infty} \bar{F}(x^{\log(u)})/\bar{F}(u) = (\log(x))^{-1/\gamma}$$

for $x > 1$. It follows that

$$\lim_{u \to \infty} \bar{F}\left(\exp(uy)\right)/\bar{F}\left(\exp(u)\right) = y^{-1/\gamma}.$$

for all $y > 0$. Thus, $F^{**} \in \mathcal{D}_{\text{POT}}(W_\gamma)$ and, consequently, $F \in \mathcal{D}_{p\text{-POT}}\left(\tilde{L}_\gamma\right)$. □

We include a result about the invariance of $\mathcal{D}_{p\text{-POT}}\left(\tilde{L}_\gamma\right)$ under shift and power transformations.

Corollary 2.7.8. *The following equivalences hold true for* $\mu \in \mathbb{R}$ *and* $\gamma, \beta, \sigma > 0$:

$$F(\cdot) \in \mathcal{D}_{p\text{-POT}}\left(\tilde{L}_\gamma\right) \Leftrightarrow F\left((\cdot - \mu)\right) \in \mathcal{D}_{p\text{-POT}}\left(\tilde{L}_\gamma\right),$$

and

$$F(\cdot) \in \mathcal{D}_{p\text{-POT}}\left(\tilde{L}_\gamma\right) \Leftrightarrow F\left(\sigma(\cdot)^\beta\right) \in \mathcal{D}_{p\text{-POT}}\left(\tilde{L}_\gamma\right).$$

Proof. We only prove the first assertion because the second one is straightforward. Putting $F_\mu(x) = F(x - \mu)$ for $F \in \mathcal{D}_{p\text{-POT}}\left(\tilde{L}_\gamma\right)$, we get

$$\begin{aligned}
\frac{1 - (F_\mu)^*(tx)}{1 - (F_\mu)^*(t)} &= \frac{\bar{F}(\exp(tx) - \mu)}{\bar{F}(\exp(t) - \mu)} \\
&= \frac{\bar{F}\left(\frac{\exp(tx) - \mu}{\exp(tx)} \exp(tx)\right)}{\bar{F}\left(\frac{\exp(t) - \mu}{\exp(t)} \exp(t)\right)}
\end{aligned}$$

$$= \frac{\bar{F}\left(\exp\left(tx + \log\left(\frac{\exp(tx)-\mu}{\exp(tx)}\right)\right)\right)}{\bar{F}\left(\exp\left(t + \log\left(\frac{\exp(t)-\mu}{\exp(t)}\right)\right)\right)}$$

$$= \frac{F^{**}(tx + a_t)}{F^{**}(t + b_t)}$$

with

$$a_t = \log\left(\frac{\exp(tx)-\mu}{\exp(tx)}\right) \quad \text{and} \quad b_t = \log\left(\frac{\exp(t)-\mu}{\exp(t)}\right).$$

Obviously $a_t \to 0$ and $b_t \to 0$, hence using uniform convergence

$$\frac{\bar{F}^{**}(tx + a_t)}{\bar{F}^{**}(t + b_t)} \longrightarrow_{t\to\infty} x^{-1/\gamma}$$

and, thus, $F(\cdot - \mu) \in \mathcal{D}_{p\text{-POT}}$. $\qquad\qquad\square$

The previous result yields that

$$\mathcal{D}_{p\text{-POT}}(L) = \mathcal{D}_{p\text{-POT}}(\tilde{L}_\gamma) \qquad (2.39)$$

for all p-types L of $\tilde{L}_\gamma$. It is easily seen that this result is valid for a *p*-POT domain of attraction of an arbitrary *p*-POT stable law. The result concerning location shifts cannot be extended to *p*-POT stable laws with finite right endpoints.

MIXTURES OF REGULARLY VARYING DFs

We also deal with super-heavy tailed df given as mixtures of regularly varying df. We start with a result in Reiss and Thomas [389] concerning a relation of log-Pareto df,

$$\tilde{L}_\gamma(x) = 1 - (\log(x))^{-1/\gamma}, \quad x > e, \gamma > 0$$

and Pareto df,

$$\widetilde{W}_{\gamma,\sigma}(x) = 1 - (x/\sigma)^{-1/\gamma}, \quad x > \sigma, \gamma > 0.$$

Log-Pareto df can be represented as mixtures of certain Pareto df with respect to gamma densities. We have

$$\tilde{L}_\gamma(x) = \int_0^\infty \widetilde{W}_{1/z,e}(x) h_{1/\gamma}(z) dz \qquad (2.40)$$

where h_α is the gamma density

$$h_\alpha(x) = \frac{1}{\Gamma(\alpha)} \exp(-x) x^{\alpha-1}. \qquad (2.41)$$

We prove that this result can be extended to df in the domains of attraction of log-Pareto and Pareto df under power and, respectively, linear normalization. Assertion (ii) of the subsequent theorem is a modification and extension of Lemma 1 in Meerschaert and Scheffler [326], cf. also Cormann [77].

Theorem 2.7.9. *The following properties hold for the p-POT domain of attraction of a log-Pareto df $\tilde{L}_\gamma$:*

(i) *Let $F \in \mathcal{D}_{p\text{-}POT}(\tilde{L}_\gamma)$ for some $\gamma > 0$. Then there is a family of df G_z, with $G_z \in \mathcal{D}_{POT}(\widetilde{W}_{1/z})$, such that*

$$F(x) = \int_0^\infty G_z(x)p(z)dz,$$

where p is a density which is ultimately monotone (monotone on $[x_0, \infty)$ for some $x_0 > 0$) and regularly varying at zero with index $1/\gamma - 1$.

(ii) *Let G_z be a family of df with $G_z \in \mathcal{D}(W_{1/z})$ with representation*

$$G_z(x) = 1 - x^{-z}U\left(\log(x)\right), \quad x > a_1,$$

for some slowly varying function U and some $a_1 > 0$. Then the mixture

$$F(x) := \int_0^\infty G_z(x)p(z)dz,$$

where p is a density as in (i), has the representation

$$F(x) = 1 - \left(\log(x)\right)^{-1/\gamma}V\left(\log(x)\right), \quad x > a_2,$$

for some slowly varying function V and some $a_2 > 0$ and, thus, $F \in \mathcal{D}_{p\text{-}POT}(\tilde{L}_\gamma)$.

Proof. To prove (i) observe that the gamma density $h_{1/\gamma}$ in (2.41) satisfies the conditions imposed on p. Therefore, (i) is a direct consequence of (2.40) and Corollary 2.7.6. Therefore the statement is still true with p replaced by $h_{1/\gamma}$.

To prove (ii) notice that

$$1 - F(x) = \int_0^\infty e^{-z\log(x)}p(z)dz\, U\left(\log(x)\right).$$

The integral is now a function $\hat{p}(\log(\cdot))$ where $\hat{p}$ denotes the Laplace transform of p. Since p is assumed to be ultimately monotone and regularly varying at zero with index $1/\gamma - 1$ one can apply Theorem 4 on page 446 of Feller [156] getting

$$\int_0^\infty e^{-z\log(x)}p(z)dz = \log(x)^{-1/\gamma}\widetilde{V}(\log(x)), \quad x > a_3,$$

for some slowly varying function $\widetilde{V}$ and $a_3 > 0$. Now $V(x) := U(x)\widetilde{V}(x)$ is again slowly varying which completes the proof. $\qquad\square$

TESTING FOR SUPER-HEAVY TAILS

In the subsequent lines the focus will be on statistical inference for distributions in Class B, namely on testing procedures for detecting the presence of a df F with Π-varying tail (2.21) underlying the sampled data. The main concern is to discriminate between a super-heavy tailed distribution and a distribution with a regularly varying tail. Since the non-negative parameter α is regarded as a gauge of tail heaviness, it can well serve the purpose of providing a straightforward distinction between super-heavy ($\alpha = 0$) and heavy tails ($\alpha > 0$). Moreover, note that if X is a rv with absolutely continuous df F in the Fréchet domain of attraction, i.e., satisfying (2.20), then e^X has a df H such that (2.21) holds. This is verified by the following proposition.

Proposition 2.7.10. *Let X be a rv with df F such that (2.20) holds and denote by $f := F'$ the corresponding density function. Define $Z := e^X$ with the df H. Then (2.21) holds with auxiliary function $a(t) := f(\log t)$, i.e., $H \in \Pi(a)$.*

Proof. The df of rv Z is related to the df of rv X through

$$H(x) = F(\log(x)) = (F \circ \log)(x).$$

Now notice that f is regularly varying with index $-\alpha - 1 > -1$. Following the steps in the proof of Proposition B.2.15 (1) of de Haan and Ferreira [190], the following statements hold for the composition $F \circ \log$, since $\log \in \Pi$ and $F \in RV_{-\alpha}$: for some $\theta = \theta(x, t) \in (0, 1)$,

$$
\begin{aligned}
\frac{H(tx) - H(t)}{f(\log(t))} &= \frac{F(\log(tx)) - F(\log(t))}{f(\log(t))} \\
&= (\log(tx) - \log(t)) \frac{f(\log(t) + \theta\{\log(tx) - \log(t)\})}{f(\log(t))} \\
&= (\log(x)) \frac{f(\log(t) + \theta \log(x))}{f(\log(t))} \\
&= (\log(x)) \frac{f\left(\log(t)\{1 + \theta\frac{\log(x)}{\log(t)}\}\right)}{f(\log(t))} \\
&\to_{t \to \infty} \log(x)
\end{aligned}
$$

by uniform convergence. $\qquad\square$

Although the transformation via exponentiation projects a Pareto tailed distribution (2.20) into (2.21) as stated in Proposition 2.7.10, it is also possible to obtain a super-heavy tailed distribution in the sense of (2.21) via a similar transformation upon exponentially tailed distributions, i.e., departing from a df in the Gumbel max-domain of attraction. This is illustrated in Example 2.7.11 where a log-Weibull(β), $\beta \in (0, 1)$, distribution is considered.

For the purpose of statistical inference, let $X_1, X_2, \ldots, X_n, \ldots$ be a sequence of rvs with common df F and let $X_{1,n} \leq X_{2,n} \leq \cdots \leq X_{n,n}$ be their ascending order statistics. Furthermore, assume that F is a continuous and strictly increasing function.

In this context, in Fraga Alves et al. [159] and Neves and Fraga Alves [352] two test statistics have been developed to distinguish between heavy and super-heavy tailed probability distributions, i.e., for testing

$$H_0 : \alpha = 0 \text{ [super-heavy]} \qquad vs. \qquad H_1 : \alpha > 0 \text{ [heavy]} \qquad (2.42)$$

in the framework carried out by the Class B of distribution functions (see equations (2.21) and (2.20)).

TEST 1. In Fraga Alves et al. [159], the asymptotic normality of the proposed statistic for testing (2.42) is proven under suitable and reasonable conditions. In particular, we need to require second-order refinements of (2.20) and (2.21) in order to specify the inherent rate of convergence. Hence, suppose there exists a positive or negative function A with $A(t) \to_{t \to \infty} 0$ and a second-order parameter $\rho \leq 0$ such that

$$\lim_{t \to \infty} \frac{\frac{F(tx) - F(t)}{a(t)} - \frac{1 - x^{-\alpha}}{\alpha}}{A(t)} = \frac{1}{\rho} \left(\frac{x^{-\alpha+\rho} - 1}{-\alpha + \rho} - \frac{1 - x^{-\alpha}}{\alpha} \right) =: H_{\alpha,\rho}(x), \qquad (2.43)$$

for all $x > 0$ and some $\alpha \geq 0$. Appendix B of de Haan and Ferreira [190] offers a thorough catalog of second-order conditions, where it is also shown that necessarily $|A(t)| \in RV_\rho$.

Example 2.7.11 (log-Weibull distribution). *Let W be a random variable with min-stable Weibull(β) df, for $0 < \beta < 1$,*

$$F_W(x) = 1 - \exp(-x^\beta), \qquad x \geq 0.$$

Then the rv $X := e^W$ is log-Weibull distributed with df

$$F(x) = 1 - \exp(-(\log(x))^\beta), \qquad x \geq 1, \ 0 < \beta < 1.$$

From Taylor expansion of $F(tx) - F(t)$ one concludes that condition (2.43) holds with $\alpha = \rho = 0$, auxiliary functions

$$a(t) = \beta(\log(t))^{\beta-1} \exp(-(\log(t))^\beta)$$

and $A(t) = (\beta - 1)/\log t$, $0 < \beta < 1$. Hence F belongs to the subclass defined by condition (2.43) is, consequently, in Class B. However, F is not in Class A since $\log X$ has df F_W with an exponentially decaying tail.

The test statistic introduced in Fraga Alves et al. [159] for discerning between super-heavy and heavy tailed distributions, as postulated in (2.42), is closely related to a new estimator for $\alpha \geq 0$. Both estimator and testing procedure evolve from the limiting relation below (with $j > 0$) which follows in turn from condition (2.20):

$$\lim_{t \to \infty} \int_1^\infty \frac{F(tx) - F(t)}{a(t)} \frac{dx}{x^{j+1}} = \int_1^\infty \frac{1 - x^{-\alpha}}{\alpha} \frac{dx}{x^{j+1}} = \frac{1}{j(j+\alpha)}.$$

The above equation entails that

$$\lim_{t \to \infty} \frac{\int_1^\infty \left(F(tx) - F(t)\right) dx/x^3}{\int_1^\infty \left(F(tx) - F(t)\right) dx/x^2} = \frac{1+\alpha}{2(2+\alpha)} \tag{2.44}$$

for $0 \leq \alpha < \infty$. Equation (2.44) can, furthermore, be rephrased as

$$\frac{\int_t^\infty (t/u)^2\, dF(u)}{\int_t^\infty (t/u)\, dF(u)} \to_{t \to \infty} \frac{1+\alpha}{2+\alpha} =: \psi(\alpha). \tag{2.45}$$

Replacing F by the empirical df F_n and t by the intermediate order statistic $X_{n-k,n}$, with $k = k_n$ a sequence of intermediate integers such that

$$k = k_n \to \infty, \quad k/n \to 0 \quad \text{as } n \to \infty, \tag{2.46}$$

the left-hand side of (2.45) becomes $\hat{\psi}_n(k)$, defined as

$$\hat{\psi}_n(k) := \frac{\sum_{i=0}^{k-1} \left(X_{n-k,n}/X_{n-i,n}\right)^2}{\sum_{i=0}^{k-1} X_{n-k,n}/X_{n-i,n}}. \tag{2.47}$$

On the other hand, the limiting function $\psi(\alpha)$ in (2.45) is a monotone continuous function. Hence, by simple inversion, we obtain the following estimator of $\alpha \geq 0$:

$$\hat{\alpha}_n(k) := \frac{2\sum_{i=0}^{k-1} \left(X_{n-k,n}/X_{n-i,n}\right)^2 - \sum_{i=0}^{k-1} \left(X_{n-k,n}/X_{n-i,n}\right)}{\sum_{i=0}^{k-1} \left(X_{n-k,n}/X_{n-i,n}\right) - \sum_{i=0}^{k-1} \left(X_{n-k,n}/X_{n-i,n}\right)^2}. \tag{2.48}$$

In the next theorem we establish without proof the asymptotic normality of the statistic $\hat{\psi}_n(k)$ introduced in (2.45), albeit under a mild second-order condition involving the intermediate sequence $k = k_n$. The result is akin to Theorem 2.4 in Fraga Alves et al. [159].

Theorem 2.7.12. *Let $k = k_n$ be a sequence of intermediate integers as in (2.46) and such that*

$$\left(n/\sqrt{k}\right) a(U(n/k)) \to \infty \qquad (2.49)$$

as $n \to \infty$, where the function a is given in (2.43) and U denotes the generalized inverse $U(t) := \left(\frac{1}{1-F}\right)^{\leftarrow}(t) = \inf\left\{x : F(x) \geq 1 - \frac{1}{t}\right\}$, for $t > 1$. If the second-order condition (2.43) holds with $\alpha \geq 0$ and

$$(na(U(n/k)))^{1/2} A\left(U(n/k)\right) \to_{n\to\infty} \lambda \in \mathbb{R}, \qquad (2.50)$$

then

$$\left(\sum_{i=0}^{k-1} \frac{X_{n-k:n}}{X_{n-i:n}}\right)^{1/2} \left(\hat{\psi}_n(k) - \frac{1+\alpha}{2+\alpha}\right) \to_D N(b, \sigma^2) \qquad (2.51)$$

as $n \to \infty$, where

$$b := \frac{-\lambda\sqrt{1+\alpha}}{(2+\alpha)(1+\alpha-\rho)(2+\alpha-\rho)},$$

$$\sigma^2 := \frac{(1+\alpha)(4+3\alpha+\alpha^2)}{(2+\alpha)^3(3+\alpha)(4+\alpha)}.$$

An alternative formulation of (2.51) is

$$\left(n\, a\big(U(n/k)\big)\right)^{1/2} \left(\hat{\psi}_n(k) - \frac{1+\alpha}{2+\alpha}\right) \to_D N(b^\star, \sigma^{\star 2}),$$

as $n \to \infty$, where

$$b^\star := \frac{-\lambda(1+\alpha)}{(2+\alpha)(1+\alpha-\rho)(2+\alpha-\rho)},$$

$$\sigma^{\star 2} := \frac{(1+\alpha)^2(4+3\alpha+\alpha^2)}{(2+\alpha)^3(3+\alpha)(4+\alpha)}.$$

Theorem 2.7.12 has just provided a way to assess the presence of an underlying super-heavy tailed distribution. Taking k upper-order statistics from a sample of size n such that k accounts only for a small top sample fraction, in order to attain (2.46), we now define the test statistic

$$S_n(k) := \sqrt{24} \left(\sum_{i=0}^{k-1} \frac{X_{n-k,n}}{X_{n-i,n}}\right)^{1/2} \left(\hat{\psi}_n(k) - \frac{1}{2}\right). \qquad (2.52)$$

The critical region for the one-sided test (2.42) at the nominal size $\bar{\alpha}$ is given by

$$\mathcal{R} := \{S_n(k) > z_{1-\bar{\alpha}}\},$$

where z_ε denotes the ε-quantile of the standard normal distribution.

It is worthwhile to mention that our null hypothesis is not only that the distribution F is in Class B defined in (2.21), but also F satisfies the second-order condition (2.43). Moreover, we should perform the test with a sequence k_n such that (2.50) holds with $\lambda = 0$. Condition (2.50) imposes an upper bound on the sequence k_n. For $\alpha = \rho = 0$, it seems difficult to prove that conditions (2.49) and (2.50) are never contradictory. However, if we replace (2.50) by the somewhat stricter condition $\sqrt{k_n}\,A(U(n/k_n)) \to_{n\to\infty} \lambda_1 \in \mathbb{R}$, we never hinder (2.49) from being valid. So, for any $\alpha \geq 0$, if $\sqrt{k_n}\,A(U(n/k_n)) \to_{n\to\infty} \lambda_1$ holds, then (2.50) holds with $\lambda = \sqrt{\alpha}\,\lambda_1$. The estimator of $\alpha \geq 0$ introduced in (2.48) is regarded as a way of testing for super-heavy tails. As an estimator for $\alpha > 0$ only, the present one is not really competitive.

TEST 2. The second proposal for testing (2.42) comes from Neves and Fraga Alves [352]; therein the test statistic $T_n(k)$, consisting of the ratio of maximum to the sum of log-excesses:

$$T_n(k) := \frac{\log(X_{n,n}) - \log(X_{n-k,n})}{\frac{1}{\log(k)} \sum_{i=0}^{k-1} \left(\log(X_{n-i,n}) - \log(X_{n-k,n})\right)} \tag{2.53}$$

proves to attain a standard Fréchet limit, as long as $k = k_n$ remains an intermediate sequence, under the simple null-hypothesis of condition (2.21) being fulfilled.

Theorem 2.7.13 below encloses a general result for heavy and super-heavy distributions belonging to the Class B (see (2.20) and (2.21)) thus suggesting a possible normalization for the test statistic $T_n(k)$ to attain a non-degenerate limit as n goes to infinity. Furthermore, results (i) and (ii) of Corollary 2.7.14 expound eventual differences in the stochastic behavior between super-heavy and heavy tailed distributions, accounting for power and consistency of the test, respectively.

First note that an equivalent characterization of the Class B can be formulated in terms of the tail quantile-type function U:

$$\lim_{t\to\infty} \frac{U\left(t + x\,q(t)\right)}{U(t)} = (1 + \alpha\,x)^{1/\alpha} \tag{2.54}$$

for all $1 + \alpha x > 0$, $\alpha \geq 0$, with a positive measurable function q such that

$$\lim_{t\to\infty} \frac{q\left(t + x\,q(t)\right)}{q(t)} = 1 + \alpha x \tag{2.55}$$

(cf. Lemma 2.7.15 below). This function q is called an auxiliary function for U.

If $\alpha = 0$, the right-hand side of (2.54) should be understood in the limiting sense as e^x while q becomes a self-neglecting function. This corresponds to an equivalent characterization of Class B as defined by (2.21). According to de Haan [184], Definition 1.5.1, we then say that the tail quantile function U belongs to the class Γ of functions of rapid variation (notation: $U \in \Gamma$).

Theorem 2.7.13. *Suppose the function U is such that condition (2.54) holds for some $\alpha \geq 0$. Let $k = k_n$ be a sequence of intermediate integers as in (2.46). Then*

$$T_n(k) = O_p\left(\frac{1}{\log(k)}\right),$$

with $T_n(k)$ as defined in (2.53).

Corollary 2.7.14. *Under the conditions of Theorem 2.7.13,*

   (i) *if $\alpha = 0$,*

$$\log(k)T_n(k) \to_D T^*, \tag{2.56}$$

     where the limiting random variable T^* has a Fréchet df $\Phi(x) = \exp(-x^{-1})$, $x \geq 0$;

  (ii) *if $\alpha > 0$,*

$$\log(k)T_n(k) \to_P 0. \tag{2.57}$$

Corollary 2.7.14 suffices to determine the critical region for assessing an underlying super-heavy tailed distribution. Considering the k upper-order statistics from a sample of size n such that k satisfies (2.46), we obtain the critical region for the one-sided test (2.42) at a nominal size $\bar{\alpha}$:

$$\mathcal{R} := \left\{\log(k)T_n(k) < \Phi^{-1}(\bar{\alpha})\right\},$$

where Φ^{-1} denotes the inverse of the standard Fréchet df Φ.

For the proof of Theorem 2.7.13 two auxiliary results are needed.

Lemma 2.7.15. *Suppose the function U is such that relation (2.54) holds with some $\alpha \geq 0$. Then, the auxiliary function q satisfies*

$$\lim_{t \to \infty} \frac{q(t)}{t} = \alpha \tag{2.58}$$

and

- *if $\alpha > 0$, then $U(\infty) := \lim_{t \to \infty} U(t) = \infty$ and U is of regular variation near infinity with index $1/\alpha$, i.e., $U \in RV_{1/\alpha}$;*

- *if $\alpha = 0$, then $U(\infty) = \infty$ and U is ∞-varying at infinity.*

Furthermore, for $\alpha = 0$,

$$\lim_{t \to \infty} \left(\log(U(t + xq(t))) - \log(U(t))\right) = x \quad \text{for every } x \in \mathbb{R}. \tag{2.59}$$

Lemma 2.7.15 coupled with condition (2.54) imposes the limit (2.55) on the auxiliary function $q(t)$.

Proof. In case $\alpha > 0$, the first part of the lemma follows directly from (2.54), whereas in case $\alpha = 0$ it is ensured by Lemma 1.5.1 and Theorem 1.5.1 of de Haan [184]. Relation (2.59) follows immediately from (2.54) with respect to $\alpha = 0$. $\qquad\square$

Proposition 2.7.16. *Suppose condition (2.54) holds for some $\alpha \geq 0$.*

(i) *If $\alpha > 0$, then for any $\varepsilon > 0$ there exists $t_0 = t_0(\varepsilon)$ such that for $t \geq t_0$, $x \geq 0$,*

$$(1 - \varepsilon)(1 + \alpha x)^{\frac{1}{\alpha} - \varepsilon} \leq \frac{U(t + x\, q(t))}{U(t)} \leq (1 + \varepsilon)(1 + \alpha x)^{\frac{1}{\alpha} + \varepsilon}. \quad (2.60)$$

(ii) *If (2.54) holds with $\alpha = 0$ then, for any $\varepsilon > 0$, there exists $t_0 = t_0(\varepsilon)$ such that for $t \geq t_0$, for all $x \in \mathbb{R}$,*

$$\frac{U(t + x\, q(t))}{U(t)} \leq (1 + \varepsilon)\exp\big(x(1 + \varepsilon)\big). \quad (2.61)$$

Proof. Inequalities in (2.60) follow immediately from Proposition 1.7 in Geluk and de Haan [170] when we settle $q(t) = \alpha t$ (see also (2.58) in Lemma 2.7.15) while (2.61) was extracted from Beirlant and Teugels [34], p.153. $\qquad\square$

Lemma 2.7.17.

(i) *If $U \in RV_{1/\alpha}$, $\alpha > 0$, then, for any $\varepsilon > 0$, there exists $t_0 = t_0(\varepsilon)$ such that for $t \geq t_0$ and $x \geq 1$,*

$$(1 - \varepsilon)\frac{1}{\alpha}\log(x) \leq \log(U(tx)) - \log(U(t)) \leq (1 + \varepsilon)\frac{1}{\alpha}\log(x). \quad (2.62)$$

(ii) *If $U \in \Gamma$ then, for any $\varepsilon > 0$, there exists $t_0 = t_0(\varepsilon)$ such that for $t \geq t_0$ and for all $x \in \mathbb{R}$,*

$$\log(U(t + xq(t))) - \log(U(t)) \leq \varepsilon + x(1 + \varepsilon). \quad (2.63)$$

Proof. Notice that once we apply the logarithmic transformation to relation (2.60) for large enough t, it becomes

$$(1 - \varepsilon)\log\left((1 + \alpha x)^{1/\alpha}\right) \leq \log(U(t + xq(t))) - \log(U(t))$$
$$\leq (1 + \varepsilon)\log\left((1 + \alpha x)^{1/\alpha}\right).$$

As before, the precise result is obtained by taking $q(t) = \alpha t$ with the concomitant translation of (2.54) for $\alpha > 0$ into the regularly varying property of U (cf. Lemma 2.7.15 again). The proof for (2.63) is similar and therefore omitted. $\qquad\square$

Proof of Theorem 2.7.13. Let $(Y_{i,n})_{i=1}^n$ be the order statistics corresponding to the iid rv $(Y_i)_{i=1}^n$ with standard Pareto df $1 - y^{-1}$, for all $y \geq 1$. Taking into account the equality in distribution

$$(X_{i,n})_{i=1}^n =_D (U(Y_{i,n}))_{i=1}^n, \tag{2.64}$$

and defining

$$Q_n^{(i)} := \frac{Y_{n-i,n} - Y_{n-k,n}}{q(Y_{n-k,n})}, \quad i = 0, 1, \ldots, k-1, \tag{2.65}$$

as well as

$$M_n^{(1)} := \frac{1}{k} \sum_{i=0}^{k-1} \log(U(Y_{n-i,n})) - \log(U(Y_{n-k,n})), \tag{2.66}$$

we get in turn

$$T_n(k) =_D \frac{\log(U(Y_{n,n})) - \log(U(Y_{n-k,n}))}{k \, M_n^{(1)}} \tag{2.67}$$

$$= \frac{\log(U(Y_{n,n})) - \log(U(Y_{n-k,n}))}{\sum_{i=0}^{k-1} \left(\log(U(Y_{n-i,n})) - \log(U(Y_{n-k,n})) \right)}$$

$$= \frac{\log\left(U\left(Y_{n-k,n} + Q_{k,n}^{(0)} \, q(Y_{n-k,n})\right)\right) - \log(U(Y_{n-k,n}))}{\sum_{i=0}^{k-1} \left(\log\left(U\left(Y_{n-k,n} + Q_{k,n}^{(i)} \, q(Y_{n-k,n})\right)\right) - \log(U(Y_{n-k,n})) \right)}. \tag{2.68}$$

Bearing on the fact that the almost sure convergence $Y_{n-k,n} \to \infty$ holds with an intermediate sequence $k = k_n$ (cf. Embrechts *et al.* [122], Proposition 4.1.14), we can henceforth make use of condition (2.54). For ease of exposition, we shall consider the cases $\alpha > 0$ and $\alpha = 0$ separately.

Case $\alpha > 0$: As announced, the core of this part of the proof lies at relation (2.54). Added (2.62) from Lemma 2.7.17, we obtain the following inequality for any $\varepsilon > 0$ and sufficiently large n:

$$M_n^{(1)} = \frac{1}{k} \sum_{i=0}^{k-1} \log \left(U\left(\frac{Y_{n-i,n}}{Y_{n-k,n}} \, Y_{n-k,n} \right) \right) - \log(U(Y_{n-k,n}))$$

$$\leq (1+\varepsilon) \frac{1}{k} \sum_{i=0}^{k-1} \frac{1}{\alpha} \left(\log(Y_{n-i,n}) - \log(Y_{n-k,n}) \right).$$

Owing to Rényi's important representation for exponential spacings,

$$E_{k-i,k} =_D E_{n-i,n} - E_{n-k,n} = \log(Y_{n-i,n}) - \log(Y_{n-k,n}), \tag{2.69}$$

where $E_{n-i,n}$, $i = 0, 1, \ldots, k-1$, are the order statistics pertaining to independent standard exponential rv $E_i = \log(Y_i)$, we thus obtain

$$M_n^{(1)} = \frac{1}{k} \sum_{i=0}^{k-1} \log(U(Y_{n-i,n})) - \log(U(Y_{n-k,n}))$$

$$\leq \frac{1}{\alpha}(1+\varepsilon)\frac{1}{k}\sum_{i=0}^{k-1}\log(Y_{k-i,k}). \tag{2.70}$$

We can also establish a similar lower bound. The law of large numbers ensures the convergence in probability of the term on the right-hand side of (2.70) since, for an intermediate sequence $k = k_n$, as $n \to \infty$,

$$\frac{1}{k}\sum_{i=0}^{k-1}\log(Y_i) \to_P \int_1^\infty \frac{\log(y)}{y^2}\,dy = 1.$$

In conjunction with (2.58), the latter entails

$$L_n(k) := \frac{q(Y_{n-k,n})}{Y_{n-k,n}}M_n^{(1)} = 1 + o_p(1) \tag{2.71}$$

as $n \to \infty$. Hence, using (2.62) followed by (2.69) upon (2.67), we obtain, as $n \to \infty$,

$$T_n(k) =_D \frac{1}{k}\frac{q(Y_{n-k,n})}{Y_{n-k,n}}\frac{\log(U(Y_{n,n})) - \log(U(Y_{n-k,n}))}{L_n(k)}$$

$$= \frac{1}{k}(E_{k,k} - \log(k))(1 + o_p(1)) + \frac{\log(k)}{k}(1 + o_p(1)). \tag{2.72}$$

Finally, by noting that $E_{k,k} - \log(k) \to_D \Lambda$, as $k \to \infty$, where Λ is denoting a Gumbel rv, we obtain a slightly stronger result than the one stated in the present theorem. More specifically, we get from (2.72) that $T_n(k) = o_p(k^{-1/2})$, for any intermediate sequence $k = k_n$.

Case $\alpha = 0$: The proof concerning this case of super-heavy tailed distributions, virtually mimics the steps followed in the heavy tailed case ($\alpha > 0$). We get from (2.68) that $M_n^{(1)}$ as defined in (2.66) can be written as

$$M_n^{(1)} = \frac{1}{k}\sum_{i=0}^{k-1}\log\left(U\left(Y_{n-k,n} + Q_{k,n}^{(i)}\,q(Y_{n-k,n})\right)\right) - \log(U(Y_{n-k,n})).$$

Giving heed to the fact that, for each $i = 0, 1, \ldots, k-1$,

$$Q_{k,n}^{(i)} = \frac{Y_{n-k,n}}{q(Y_{n-k,n})}\left(\frac{Y_{n-i,n}}{Y_{n-k,n}} - 1\right)$$

is stochastically bounded away from zero (see Lemma 2.7.15), we can thus apply relation (2.63) from Lemma 2.7.17 in order to obtain, for any intermediate sequence $k = k_n$,

$$\frac{1}{k}\sum_{i=0}^{k-1}\log\left(U\left(Y_{n-k,n}+Q_{k,n}^{(i)}q(Y_{n-k,n})\right)\right)-\log(U(Y_{n-k,n})) \le (1+\varepsilon)\frac{1}{k}\sum_{i=0}^{k-1}Q_{k,n}^{(i)},$$

as $n \to \infty$. Using Rényi's representation (2.69), we get

$$\frac{q(Y_{n-k,n})}{Y_{n-k,n}}M_n^{(1)} \le (1+\varepsilon)\frac{1}{k}\sum_{i=0}^{k-1}\left(Y_{k-i,k}-1\right). \tag{2.73}$$

It is worth noticing at this point that with constants $a_k^* > 0$, $b_k^* \in \mathbb{R}$ such that $a_k^* \sim k\pi/2$ and $b_k^*a_k^*/k \sim \log(k)$ as $k \to \infty$, this new random variable S_k^* defined by

$$S_k^* := \frac{1}{a_k^*}\sum_{i=0}^{k-1}\left(Y_i-1\right)-b_k^*, \tag{2.74}$$

converges in distribution to a sum-stable law (cf. Geluk and de Haan [171]). Embedding S_k^* defined above in the right-hand side of (2.73), we ensure that $L_n(k)$ as introduced in (2.71) satisfies $L_n(k) = O_p(\log(k))$. Therefore, in view of (2.68), the proof is concluded by showing that it is possible to normalize the maximum of the log-spacings in such a way as to exhibit a non-degenerate behavior eventually. Since $U \in \Gamma$ we get in a similar way as before, for large enough n,

$$\frac{q(Y_{n-k,n})}{kY_{n-k,n}}\left(\log\left(U\left(Y_{n-k,n}+Q_{k,n}^{(0)}q(Y_{n-k,n})\right)\right)-\log(U(Y_{n-k,n}))\right)$$
$$= k^{-1}\left(\frac{Y_{n,n}}{Y_{n-k,n}}-1\right)\left(1+o_p(1)\right)$$
$$= k^{-1}\left(Y_{k,k}-1\right)\left(1+o_p(1)\right) = O_p(1). \qquad \square$$

Proof of Corollary 2.7.14. *(i)* For $\alpha = 0$, the last part of the proof of Theorem 2.7.13 emphasizes that, as $n \to \infty$,

$$\log(k)T_n(k) =_D \frac{k^{-1}\left(\log(U(Y_{n,n}))-\log(U(Y_{n-k,n}))\right)}{L_n(k)/\log(k)}\frac{q(Y_{n-k,n})}{Y_{n-k,n}}$$
$$= \left(T^*+o_p(1)\right)/\left(1+o_p(1)\right)$$
$$= T^*\left(1+o_p(1)\right)$$

because, after suitable normalization by $a_k = k^{-1}$, the maximum of a sample of size k with standard Pareto parent distribution is attracted to a Fréchet law. *(ii)* The precise result follows from (2.72) by straightforward calculations. $\square$

Neves and Fraga Alves [352] present a finite (large) sample simulation study which seems to indicate that the conservative extent of Test 2 opens a window of opportunity for its applicability as a complement to Test 1. This is particularly true for the less heavy distributions lying in the class of super-heavy tails since in this case the number of wrong rejections is likely to rise high above the nominal level of the test based on (2.52). Moreover, the asymptotics pertaining to the test statistic $S_n(k)$ in (2.52) (cf. Theorem 2.7.12) require a second-order refinement of (2.21) (as in (2.43)), while the asymptotic behavior of the test statistic $T_n(k)$ only relies on the first-order conditions on the tail of F, meaning that we are actually testing $F \in$ Class B.

Chapter 3

Estimation of Conditional Curves

In this chapter we will pick up Example 1.3.3 again, and we will show how the Poisson approximation of truncated empirical point processes enables us to reduce conditional statistical problems to unconditional ones.

A nearest neighbor alternative to this applications of our functional laws of small numbers is given in Sections 3.5 and 3.6.

3.1 Poisson Process Approach

Let $Z = (X, Y)$ be a rv, where X is $\mathbb{R}^d$-valued any Y is $\mathbb{R}^m$-valued, and denote by

$$F(\cdot \mid x) := P(Y \leq \cdot \mid X = x)$$

the conditional df of Y given $X = x$, $x \in \mathbb{R}^d$. Applying the approach developed in Chapter 1, one may study the fairly general problem of evaluating a functional parameter $T(F(\cdot \mid x))$ based on independent replicates $Z_i = (X_i, Y_i)$, $i = 1, \ldots, n$, of Z. This will be exemplified in the particular cases of non-parametric estimation of regression means and quantiles that is, for the functionals $T_1(F) = \int t\, F(dt)$ and $T_2(F) = F^{-1}(q)$, $q \in (0, 1)$.

Example 3.1.1. When a child is born in Germany, the parents are handed out a booklet showing on its back cover a *somatogram*, where the average height (in cm) of a child is plotted against its weight (in kg) together with a .95 per cent confidence bound. If, for example, a child has a height of about 80cm but a weight of less than 9 kg (more than 13.25 kg) it is rated significantly light (heavy).

If we model the (height, weight) of a randomly chosen child by a rv $Z = (X, Y) \in \mathbb{R}^2$, then the confidence curves in this somatogram represent the curves of the 2.5 per cent and 97.5 per cent quantiles of the conditional df $F(\cdot \mid x)$ of

M. Falk et al., *Laws of Small Numbers: Extremes and Rare Events*, 3rd ed., DOI 10.1007/978-3-0348-0009-9_3, © Springer Basel AG 2011

Y given $X = x$ with $50 \leq x \leq 120$ that is, of $T_2(F(\cdot \mid x)) = F(\cdot \mid x)^{-1}(q)$ with $q_1 = 25/1000$ and $q_2 = 975/1000$.

While classical non-parametric regression analysis focuses on the problem of estimating the conditional mean $T_1(F(\cdot \mid x)) = \int t \, F(dt|x)$ (see, for example, Eubank [124]), the estimation of general regression functionals $T(F(\cdot \mid x))$ has been receiving increasing interest only some years ago (see, for example, Stute [424], Härdle et al. [206], Samanta [404], Truong [447], Manteiga [317], Hendricks and Koenker [216], Goldstein and Messer [177]).

TRUNCATED EMPIRICAL PROCESS

Statistical inference based on $(X_1, Y_1), \ldots, (X_n, Y_n)$ of a functional $T(F(\cdot \mid x))$ has obviously to be based on those Y_i among $Y_1, \ldots, Y_n$, whose corresponding X_i-values are *close* to x. Choose therefore as in Example 1.3.3 a window width $a_n = (a_{n1}, \ldots, a_{nd}) \in (0, \infty)^d$ and define as the data-window for X_i

$$S_n := \mathsf{X}_{j \leq d}[x_j - a_{nj}^{1/d}/2, x_j + a_{nj}^{1/d}/2]$$
$$=: [x - a_n^{1/d}/2, x + a_n^{1/d}/2],$$

where the operations $a_n^{1/d}/2$ are meant componentwise. The data set Y_i with $X_i \in S_n$ is described in a mathematically precise way by the truncated empirical point process

$$N_n(B) = \sum_{i \leq n} \varepsilon_{Y_i}(B) \varepsilon_{X_i}(S_n)$$

$$= \sum_{i \leq K(n)} \varepsilon_{V_i}(B), \qquad B \in \mathbb{B}^m,$$

where

$$K(n) := \sum_{i \leq n} \varepsilon_{X_i}(S_n)$$

is the number of those Y_i with $X_i \in S_n$ which we denote by $V_1, V_2, \ldots$. From Theorem 1.3.1 we know that $K(n)$ and $V_1, V_2, \ldots$ are independent, where V_i are independent copies of a rv V with distribution

$$P(V \in \cdot) = P(Y \in \cdot \mid X \in S_n),$$

and $K(n)$ is $B(n, p_n)$-distributed with $p_n = P(X \in S_n) \sim vol(S_n) =$ volume of S_n if $|a_n|$ is small (under suitable regularity conditions). By $|\cdot|$ we denote the common Euclidean distance in $\mathbb{R}^d$.

THE FIRST-ORDER POISSON APPROXIMATION

If we replace in N_n the sample size $K(n)$ by a Poisson rv $\tau(n)$ with parameter $E(K(n)) = np_n$, which is stochastically independent of $V_1, V_2, \ldots$, then we obtain the first-order Poisson process approximation N_n^* of N_n from Section 1.1, given by

$$N_n^*(B) = \sum_{i \leq \tau(n)} \varepsilon_{V_i}(B), \qquad B \in \mathbb{B}^m.$$

The error of this approximation is determined only by the error of the approximation of $K(n)$ by $\tau(n)$ (see Lemma 1.2.1).

Theorem 3.1.2. *We have for the Hellinger distance*

$$H(N_n, N_n^*) \leq 3^{1/2} P(X \in S_n).$$

THE SECOND-ORDER POISSON APPROXIMATION

It is intuitively clear and was already shown in Example 1.3.3 that for $|a_n| \to 0$ under suitable regularity conditions

$$P(V \in \cdot) = P(Y \in \cdot \mid X \in S_n) \to_{|a_n| \to 0} P(Y \in \cdot \mid X = x).$$

This implies the approximation of N_n^* by the Poisson process

$$N_n^{**}(B) = \sum_{i \leq \tau(n)} \varepsilon_{W_i}(B), \qquad B \in \mathbb{B}^m,$$

where $W_1, W_2, \ldots$ are independent replicates of a rv W with target df $F(\cdot|x)$, and $\tau(n)$ is a Poisson rv with parameter $np_n = P(X \in S_n)$. The rv $W_1, W_2, \ldots$ are again chosen independent of $\tau(n)$ and $K(n)$.

From Theorem 1.2.5 we obtain the following bound for the second-order Poisson approximation

$$H(N_n, N_n^{**}) \leq H(N_n, N_n^*) + H(N_n^*, N_n^{**})$$

$$\leq 3^{1/2} P(X \in S_n) + H(V, W)(nP(X \in S_n))^{1/2}$$

$$= 3^{1/2} p_n + H(V, W)(np_n)^{1/2}.$$

THE POT-APPROACH

If we denote again by

$$M_n = \sum_{i \leq K(n)} \varepsilon_{W_i}$$

the binomial process pertaining to the POT approach, then we obtain from Corollary 1.2.4 (iv) the bound

$$H(N_n, M_n) \leq H(V, W) E(K(n))^{1/2} = H(V, W)(np_n)^{1/2}.$$

It is therefore obvious that we have to seek conditions on the joint distribution of (X, Y) for X near x, such that we obtain reasonable bounds for $H(V, W)$.

BASIC SMOOTHNESS CONDITIONS

An obvious condition is to require that the conditional distribution $P(Y \in \cdot \mid X \in [x - \varepsilon, x + \varepsilon])$ of Y given $X \in [x - \varepsilon, x + \varepsilon]$ has a density $f(y \mid [x - \varepsilon, x + \varepsilon])$ for $\varepsilon \in (0, \varepsilon_0)^d$ such that

$$f(y \mid [x - \varepsilon, x + \varepsilon])^{1/2} = f(y \mid x)^{1/2}(1 + R(y \mid [x - \varepsilon, x + \varepsilon])) \qquad (3.1)$$

for $y \in \mathbb{R}^m$, where $f(y \mid x)$ denotes the density of $F(\cdot \mid x)$, which we assume to exist as well. If we require that

$$\int R^2(y \mid [x - \varepsilon, x + \varepsilon])f(y \mid x)\, dy = O(|\varepsilon|^4) \qquad (3.2)$$

as $|\varepsilon| \to 0$, then we obtain from equation (1.10) in Chapter 1 the bound

$$H(V, W) = \left(\int R^2(y \mid S_n) f(y \mid x)\, dy \right)^{1/2} = O(|a_n^{1/d}|^2)$$

as $|a_n| \to 0$.

If we require further that X has a density g near x which is continuous at x and $g(x) > 0$, then we obtain

$$p_n = P(X \in S_n) = \int_{S_n} g(y)\, dy = \int_{[-a_n^{1/d}/2, a_n^{1/d}/2]} g(x + \varepsilon)\, d\varepsilon$$

$$= g(x) \int_{[-a_n^{1/d}/2, a_n^{1/d}/2]} 1 + \frac{g(x + \varepsilon) - g(x)}{g(x)}\, d\varepsilon$$

$$= g(x)\, vol(S_n)(1 + o(1)) = g(x) \left(\prod_{j \leq d} a_{nj}^{1/d} \right)(1 + o(1)) \qquad (3.3)$$

as $|a_n| \to 0$.

BOUNDS FOR THE APPROXIMATIONS

The following result is now obvious.

Theorem 3.1.3. *Suppose that the rv (X, Y) satisfies conditions (3.1), (3.2) and that X has a density g near x which is continuous at x with $g(x) > 0$. Then we have for $|a_n| \to 0$,*

$$H(N_n, N_n^*) = O(vol(S_n)) = O\left(\prod_{j \leq d} a_{nj}^{1/d}\right)$$

for the first-order Poisson approximation,

$$H(N_n, N_n^{**}) = O\left(vol(S_n) + (n\, vol(S_n))^{1/2} |a_n^{1/d}|^2\right)$$

$$= O\left(\prod_{j \leq d} a_{nj}^{1/d} + \left(n \prod_{j \leq d} a_{nj}^{1/d}\right)^{1/2} |a_n^{1/d}|^2\right)$$

for the second-order Poisson approximation and

$$H(N_n, M_n) = O((n\, vol(S_n))^{1/2} |a_n^{1/d}|^2)$$

$$= O\left(\left(n \prod_{j \leq d} a_{nj}^{1/d}\right)^{1/2} |a_n^{1/d}|^2\right)$$

for the POT approach.

With equal bin widths $a_{nj}^{1/d} = c^{1/d}$, $j = 1, \ldots, d$, we have $vol(S_n) = c$ and the preceding bounds simplify to

$$H(N_n, N_n^*) = O(c),$$
$$H(N_n, N_n^{**}) = O(c + (nc^{(d+4)/d})^{1/2}),$$
$$H(N_n, M_n) = O((nc^{(d+4)/d})^{1/2}), \tag{3.4}$$

uniformly for $c > 0$ and $n \in \mathbb{N}$. These results give a precise description of how $c = c(n)$ can be chosen depending on the sample size n, in order to ensure convergence of the Hellinger distances to zero.

THE THIRD-ORDER POISSON APPROXIMATION

Once we know that

$$p_n = P(X \in S_n) = g(x)\, vol(S_n)(1 + o(1)),$$

a further approximation of N_n suggests itself namely, its approximation by

$$M_n^* := \sum_{i \leq \tau^*(n)} \varepsilon_{W_i},$$

where $W_1, W_2, \ldots$ are as before, but now $\tau^*(n)$ is an independent Poisson rv with parameter $ng(x)\,vol(S_n)$. This is the *third-order* Poisson process approximation of N_n. From the arguments in Lemma 1.2.1 (ii) we obtain the bound

$$H(N_n^{**}, M_n^*) \le H(\tau(n), \tau^*(n)).$$

If we require now in addition that g has bounded partial derivatives of second order near x, then the arguments in (3.3) together with Taylor expansion of $g(x + \varepsilon)$ imply the expansion

$$p_n = g(x)\,vol(S_n)(1 + O(|a_n^{1/d}|^2))$$

for $|a_n| \to 0$. The following lemma entails that

$$
\begin{aligned}
H(\tau(n), \tau^*(n)) &= O\!\left(\frac{np_n - ng(x)\,vol(S_n)}{(n\,vol(S_n))^{1/2}}\right) \\
&= O\!\left((n\,vol(S_n))^{1/2}|a_n^{1/d}|^2\right) \\
&= O\!\left(\Big(n\prod_{j\le d} a_{nj}^{1/d}\Big)^{1/2}|a_n^{1/d}|^2\right).
\end{aligned}
$$

We, therefore, obtain under this additional smoothness condition on g the bound

$$
\begin{aligned}
H(N_n, M_n^*) &\le H(N_n, N_n^{**}) + H(N_n^{**}, M_n^*) \\
&= O\!\left(\prod_{j\le d} a_{nj}^{1/d} + \Big(n\prod_{j\le d} a_{nj}^{1/d}\Big)^{1/2}|a_n^{1/d}|^2\right),
\end{aligned}
\tag{3.5}
$$

which coincides with the bound for $H(N_n, N_n^{**})$ established in Theorem 3.1.3.

Lemma 3.1.4. *Let τ_1, τ_2 be Poisson rv with corresponding parameters $0 < \lambda_1 \le \lambda_2$. Then we have for the Hellinger distance*

$$H(\tau_1, \tau_2) \le \frac{\lambda_2 - \lambda_1}{\sqrt{2\lambda_1}}.$$

Proof. As the squared Hellinger distance is bounded by the Kullback-Leibler distance (see, for example, Lemma A.3.5 in Reiss [385]), we obtain

$$
\begin{aligned}
H^2(\tau_1, \tau_2) &\le -\int \log(P(\tau_2 = k)/P(\tau_1 = k))\,\mathcal{L}(\tau_1)(dk) \\
&= -\int \log\Big(e^{\lambda_1 - \lambda_2}(\lambda_2/\lambda_1)^k\Big)\,\mathcal{L}(\tau_1)(dk) \\
&= \lambda_2 - \lambda_1 - \log(\lambda_2/\lambda_1)\,E(\tau_1)
\end{aligned}
$$

$$= \lambda_2 - \lambda_1 - \log\left(1 + \frac{\lambda_2 - \lambda_1}{\lambda_1}\right)\lambda_1$$

$$\leq \lambda_2 - \lambda_1 - \left(\frac{\lambda_2 - \lambda_1}{\lambda_1} - \frac{(\lambda_2 - \lambda_1)^2}{2\lambda_1^2}\right)\lambda_1 = \frac{(\lambda_2 - \lambda_1)^2}{2\lambda_1},$$

where the second to last line follows from the inequality $\log(1 + \varepsilon) \geq \varepsilon - \varepsilon^2/2$, which is true for any $\varepsilon > 0$. $\qquad\square$

A UNIFIED SMOOTHNESS CONDITION

Conditions (3.1) and (3.2) together with the assumption that the marginal density g of X has bounded partial derivatives of second order near x, can be replaced by the following handy condition on the joint density of (X, Y), which is suggested by Taylor's formula.

Suppose that the rv (X, Y) has a joint density f on the strip $[x - \varepsilon_0, x + \varepsilon_0] \times \mathbb{R}^m$ for some $\varepsilon_0 \in (0, \infty)^d$, which satisfies uniformly for $\varepsilon \in [-\varepsilon_0, \varepsilon_0] (\subset \mathbb{R}^d)$ and $y \in \mathbb{R}^m$ the expansion

$$f(x + \varepsilon, y) = f(x, y)\Big\{1 + \langle \varepsilon, h(y)\rangle + O(|\varepsilon|^2 r(y))\Big\}, \tag{3.6}$$

where $h : \mathbb{R}^m \to \mathbb{R}^d, r : \mathbb{R}^m \to \mathbb{R}$ satisfy $\int (|h(y)|^2 + |r(y)|^2) f(x, y)\, dy < \infty$ and $\langle \cdot, \cdot \rangle$ denotes the usual inner product on $\mathbb{R}^d$. Then

$$f(\cdot \mid [x - \varepsilon, x + \varepsilon]) = \frac{\int_{[x-\varepsilon, x+\varepsilon]} f(u, \cdot)\, du}{\int_\mathbb{R} \int_{[x-\varepsilon, x+\varepsilon]} f(u, y)\, du\, dy}$$

is the conditional density of Y given $X \in [x - \varepsilon, x + \varepsilon]$ and

$$f(\cdot \mid x) := \frac{f(x, \cdot)}{g(x)}$$

the conditional density of Y given $X = x$, where

$$g(\cdot) := \int_\mathbb{R} f(\cdot, y)\, dy$$

is the marginal density of X which we assume again to be positive at x. Elementary computations then show that conditions (3.1) and (3.2) are satisfied that is,

$$f(y \mid [x - \varepsilon, x + \varepsilon])^{1/2} = f(y \mid x)^{1/2}\Big\{1 + R(y \mid [x - \varepsilon, x + \varepsilon])\Big\}$$

for $y \in \mathbb{R}^m$ and $\varepsilon \in (-\varepsilon_0, \varepsilon_0)$, where

$$\int R^2(y \mid [x - \varepsilon, x + \varepsilon]) f(y \mid x)\, dy = O(|\varepsilon|^4)$$

as $|\varepsilon| \to 0$. We, furthermore, have in this case again

$$p_n = P(X \in S_n) = g(x)\,vol(S_n)(1 + O(|a_n^{1/d}|^2))$$

as $|a_n| \to 0$. Then the bounds on the approximation of the process N_n by the processes N_n^*, N_n^{**}, M_n and M_n^* as given in Theorem 3.1.3 and in (3.4) and (3.5) remain valid.

Example 3.1.5. Suppose that (X, Y) is bivariate normally distributed that is, (X, Y) has joint density

$$f(z, y) = \frac{1}{2\pi\sigma_1\sigma_2(1-\rho^2)} \exp\left\{ -\frac{1}{2(1-\rho^2)}\left(\left(\frac{z-\mu_1}{\sigma_1}\right)^2 \right.\right.$$

$$\left.\left. -2\rho\left(\frac{z-\mu_1}{\sigma_1}\right)\left(\frac{y-\mu_2}{\sigma_2}\right) + \left(\frac{y-\mu_2}{\sigma_2}\right)^2\right)\right\}, \qquad z, y \in \mathbb{R},$$

where $\mu_1, \mu_2 \in \mathbb{R}, \sigma_1, \sigma_2 > 0$ and $\rho \in (-1, 1)$. Taylor expansion of exp at 0 entails the expansion

$$\frac{f(x+\varepsilon, y)}{f(x, y)} = \exp\left\{ -\frac{1}{2(1-\rho^2)}\left(\frac{2(x-\mu_1)\varepsilon + \varepsilon^2}{\sigma_1^2} - \frac{2\rho\varepsilon}{\sigma_1\sigma_2}(y-\mu_2)\right)\right\}$$

$$= 1 + \varepsilon\frac{1}{1-\rho^2}\left(\frac{\rho}{\sigma_1\sigma_2}(y-\mu_2) - \frac{x-\mu_1}{\sigma_1^2}\right)$$

$$+ O\left(\varepsilon^2 \exp(c|y|)(1 + y^2)\right)$$

$$=: 1 + \varepsilon h(y) + O(\varepsilon^2 r(y))$$

with some appropriate positive constants c. We, obviously, have $\int (h^2(y) + r^2(y))\,f(x, y)\,dy < \infty$.

BOUNDS FOR EQUAL BIN WIDTHS

The preceding considerations are summarized in the following result.

Corollary 3.1.6. *Suppose that the rv* (X, Y) *satisfies condition* (3.6) *at the point* $x \in \mathbb{R}^d$ *and that the marginal density of* X *is positive at* x. *With equal bin widths* $a_{nj}^{1/d} = c^{1/d}$, $j = 1, \ldots, m$, *we obtain uniformly for* $c > 0$ *and* $n \in \mathbb{N}$ *the bound*

$$H(N_n, N_n^*) = O(c)$$

for the first-order Poisson approximation,

$$H(N_n, N_n^{**}) = O\left(c + (nc^{(d+4)/d})^{1/2}\right)$$

for the second-order Poisson approximation,

$$H(N_n, M_n) = O\left((nc^{(d+4)/d})^{1/2}\right)$$

for the POT approach and

$$H(N_n, M_n^*) = O\left(c + (nc^{(d+4)/d})^{1/2}\right)$$

for the third-order Poisson approximation.

The preceding approach will be extended to several points $x_1, \ldots, x_r$ in Section 3.3 with the corresponding bounds summing up.

3.2 Applications: The Non-parametric Case

In this section we assume for the sake of a clear presentation that the covariate Y of X is a one-dimensional rv.

LOCAL EMPIRICAL DISTRIBUTION FUNCTION

The usual non-parametric estimate of a functional $T(F)$, based on an iid sample $Y_1, \ldots, Y_n$ with common df F, is $T(F_n)$, where $F_n(t) := n^{-1} \sum_{i \leq n} \varepsilon_{Y_i}((-\infty, t])$ denotes the pertaining empirical df. Within our framework, the *local empirical* df

$$\hat{F}_n(t \mid S_n) := K(n)^{-1} \sum_{i \leq n} \varepsilon_{Y_i}((-\infty, t]) \varepsilon_{X_i}(S_n)$$
$$= N_n(\mathbb{R}^m)^{-1} N_n((-\infty, t]), \qquad t \in \mathbb{R},$$

pertaining to those Y_i among $(X_1, Y_1), \ldots, (X_n, Y_n)$ with $X_i \in S_n = [x - a_n^{1/d}/2, x + a_n^{1/d}/2]$, suggests itself as a non-parametric estimate of $F(\cdot \mid x)$. The resulting estimate of $T(F(\cdot \mid x))$ is $T(\hat{F}_n(\cdot \mid S_n))$. Observe that $\hat{F}_n$ is the df pertaining to the standardized random measure N_n.

KERNEL ESTIMATOR OF A REGRESSION FUNCTIONAL

For the mean value functional T_1 we obtain for example

$$T_1(\hat{F}_n(\cdot \mid S_n)) = \int t\, \hat{F}_n(dt \mid S_n)$$
$$= \frac{\sum_{i \leq n} Y_i\, \varepsilon_{X_i}(S_n)}{\sum_{i \leq n} \varepsilon_{X_i}(S_n)}$$

which is the Nadaraya-Watson estimator of $T_1(F(\cdot \mid x)) = \int t\, F(dt \mid x)$. Following Stone [428], [429] and Truong [447], we call $T(\hat{F}_n(\cdot \mid S_n))$ the *kernel estimator* of a *general regression functional* $T(F(\cdot \mid x))$.

The Basic Reduction Theorem

The following result is crucial. It shows how conditional estimation problems reduce to unconditional ones by the approach developed in the preceding section.

Theorem 3.2.1. *Suppose that for some $\sigma > 0$, $\delta \in (0, 1/2]$ and $C > 0$,*

$$\sup_{t \in \mathbb{R}} \left| P\left(\frac{k^{1/2}}{\sigma} (T(F_k(\cdot \mid x)) - T(F(\cdot \mid x))) \leq t \right) - \Phi(t) \right| \leq C k^{-\delta}, \quad k \in \mathbb{N}, \quad (3.7)$$

where $F_k(\cdot \mid x)$ denotes the empirical df pertaining to k independent rv with common df $F(\cdot \mid x)$, and Φ is the standard normal df. If the vector (X, Y) satisfies condition (3.6) and the marginal density of X is positive at x, then we obtain for the kernel estimator $T(\hat{F}_n(\cdot \mid S_n))$ with equal bin widths $a_{n1}^{1/d} = \cdots = a_{nd}^{1/d} = c^{1/d}$

$$\sup_{t \in \mathbb{R}} \left| P\left(\frac{(nc\,g(x))^{1/2}}{\sigma} (T(\hat{F}_n(\cdot \mid S_n)) - T(F(\cdot \mid x))) \leq t \right) - \Phi(t) \right|$$
$$= O\left((nc)^{-\delta} + c + (nc^{(d+4)/d})^{1/2} \right)$$

uniformly for $c > 0$ and $n \in \mathbb{N}$.

   With the particular choice $c = c_n = O(n^{-d/(d+4)})$, we, roughly, obtain the rate $O_P(n^{-2/(d+4)})$ for $T(\hat{F}_n(\cdot \mid S_n)) - T(F(\cdot \mid x))$, which is known to be the optimal attainable accuracy under suitable regularity conditions in case of the mean value functional (Stone [428], [429]), and quantile functional (Chaudhuri [63]) (for a related result for the quantile functional we refer to Truong [447], and for a discussion of a general functional T to Falk [130]; a version of this result based on the nearest neighbor approach is established in Section 3.6).
   The proof of Theorem 3.2.1 is based on the following elementary result (see Lemma 3 in Falk and Reiss [149]).

Lemma 3.2.2. *Let $V_1, V_2, \ldots$ be a sequence of rv such that for some $\sigma > 0$, $\mu \in \mathbb{R}$ and $\delta \in (0, 1/2]$,*

$$\sup_{t \in \mathbb{R}} \left| P\left(\frac{k^{1/2}}{\sigma} (V_k - \mu) \leq t \right) - \Phi(t) \right| \leq Ck^{-\delta}, \qquad k \in \mathbb{N}.$$

Then we have with τ being a Poisson rv with parameter $\lambda > 0$ and independent of each V_i, $i = 1, 2, \ldots$,

$$\sup_{t \in \mathbb{R}} \left| P\left(\frac{\lambda^{1/2}}{\sigma} (V_\tau - \mu) \leq t \right) - \Phi(t) \right| \leq D\lambda^{-\delta},$$

where D depends only on C (with the convention $V_\tau = 0$ if $\tau = 0$).

Proof of Theorem 3.2.1. Put $V_k := T(F_k(\cdot \mid x))$, $k = 1, 2, \ldots$, and $\mu := T(F(\cdot \mid x))$. Observe that $T(\hat{F}_n(\cdot \mid S_n))$ is a functional of the empirical point process N_n. If we replace therefore N_n by the Poisson process $M_n^* = \sum_{i \leq \tau^*(n)} \varepsilon_{W_i}$, where $W_i, W_2, \ldots$ are independent rv with common df $F(\cdot \mid x)$ and independent of $\tau^*(n)$, we obtain

$$\sup_{t \in \mathbb{R}} \left| P\left(\frac{(nc\,g(x))^{1/2}}{\sigma} (T(\hat{F}_n(\cdot \mid S_n)) - T(F(\cdot \mid x))) \leq t \right) - \Phi(t) \right|$$

$$= \sup_{t \in \mathbb{R}} \left| P\left(\frac{(nc\,g(x))^{1/2}}{\sigma} (V_{\tau^*(n)} - \mu) \leq t \right) - \Phi(t) \right|$$

$$+ O(H(N_n, M_n^*)),$$

where $\tau^*(n) = M_n^*(\mathbb{R}^d)$ is Poisson with parameter $\lambda = nc\,g(x)$ and independent of each $V_1, V_2, \ldots$. The assertion is now immediate from Lemma 3.2.2 and Corollary 3.1.6. $\qquad\square$

EXAMPLES

The following examples on regression quantiles and the regression mean highlight the wide-ranging applicability of the reduction theorem that is, of the approximation of N_n by M_n^*.

Example 3.2.3 (Regression quantiles). Put $T(F) = F^{-1}(q)$, $q \in (0, 1)$ fixed and assume that $F(\cdot \mid x)$ is continuously differentiable in a neighborhood of the conditional quantile $F(\cdot \mid x)^{-1}(q)$ with $f_x(F(\cdot \mid x)^{-1}(q)) > 0$, where $f_x = F(\cdot \mid x)'$ is the conditional density of $F(\cdot \mid x)$. Then, condition (3.7) is satisfied with $\sigma^2 = q(1 - q)/f_x^2(F(\cdot \mid x)^{-1}(q))$ and $\delta = 1/2$ (see Section 4.2 of Reiss [385]).

Consequently, we obtain with equal bin widths $a_{n1}^{1/d} = \cdots = a_{nd}^{1/d} = c^{1/d}$ uniformly for $c > 0$,

$$\sup_{t \in \mathbb{R}} \left| P\left(\frac{(ncg(x))^{1/2}}{\sigma} (\hat{F}_n(\cdot \mid S_n)^{-1}(q) - F(\cdot \mid x)^{-1}(q)) \leq t \right) - \Phi(t) \right|$$

$$= O((nc_n)^{-1/2}) + H(N_n, M_n^*).$$

Example 3.2.4 (Regression mean). *Assume that condition (3.7) holds for the mean value functional $T(F) = \int t\,F(dt)$ with $\sigma^2 := \int (t - \mu)^2\,F(dt \mid x)$, $\mu = \int t\,F(dt \mid x)$, (think of the usual Berry-Esseen theorem for sums of iid rv). Then, with $a_{n1}^{1/d} = \cdots = a_{nd}^{1/d} = c^{1/d}$,*

$$\sup_{t \in \mathbb{R}} \left| P\left(\frac{(nc\,g(x))^{1/2}}{\sigma} \left(\int t\hat{F}_n(dt \mid x) - \int t\,F(dt \mid x) \right) \leq t \right) - \Phi(t) \right|$$

$$= O((nc)^{-1/2}) + H(N_n, M_n^*)$$

uniformly for $c > 0$.

Condition (3.7) is satisfied for a large class of functionals T, for which a Berry-Esseen result is available that is, U- and V-statistics, M, L and R estimators. See, for example, the monograph by Serfling [408].

3.3 Applications: The Semiparametric Case

Assume that the conditional distribution $P(Y \in \cdot \mid X = x) = P_\vartheta(Y \in \cdot \mid X = x) = Q_\vartheta(\cdot)$ of $Y(\in \mathbb{R}^m)$, given $X = x \in \mathbb{R}^d$, is a member of a parametric family, where the parameter space Θ is an open subset of $\mathbb{R}^k$. Under suitable regularity conditions we establish asymptotically optimal estimates based on N_n of the true underlying parameter ϑ_0. Since the estimation problem involves the joint density of (X, Y) as an infinite dimensional nuisance parameter, we actually have to deal with a special semiparametric problem: Since we observe data Y_i whose X_i-values are only *close* to x, our set of data $V_1, \ldots, V_{K(n)}$, on which we will base statistical inference, is usually *not* generated according to our target conditional distribution $Q_{\vartheta_0}(\cdot)$ but to some distribution being close to $Q_{\vartheta_0}(\cdot)$. This error is determined by the joint density f of (X, Y), which is therefore an infinite dimensional nuisance parameter. As a main tool we utilize *local asymptotic normality* (LAN) of the Poisson process M_n^* (cf. the books by Strasser [430], LeCam [308], LeCam and Yang [309] and Pfanzagl [367]. For a general approach to semiparametric problems we refer to the books by Pfanzagl [366] and Bickel et al. [44]).

A Semiparametric Model

Suppose that for $\vartheta \in \Theta$ the probability measure $Q_\vartheta(\cdot)$ has Lebesgue-density q_ϑ. We suppose that the density f of the rv (X, Y) exists on a strip $[x - \varepsilon_0, x + \varepsilon_0] \times \mathbb{R}^m (\subset \mathbb{R}^d \times \mathbb{R}^m)$ and that it is a member of the following class of functions:

$$
\begin{aligned}
\mathcal{F}(C_1, C_2) \\
:= \Big\{ f : [x - \varepsilon_0, x + \varepsilon_0] \times \mathbb{R}^m \to [0, \infty) \text{ such that } 0 < g_f(x) \\
:= \int f(x, y) \, dy \le C_1, \text{ and for any } \varepsilon \in (0, \varepsilon_0] \\
\big| f(x + \varepsilon, y) - f(x, y)(1 + \langle \varepsilon, h_f(y) \rangle) \big| \le |\varepsilon|^2 r_f(y) f(x, y) \\
\text{for some functions } h_f : \mathbb{R}^m \to \mathbb{R}^d, r_f : \mathbb{R}^m \to [0, \infty) \text{ satisfying} \\
\int (|h_f(y)|^2 + r_f^2(y)) f(x, y) \, dy \le C_2 \Big\},
\end{aligned}
$$

where C_1, C_2 are fixed positive constants. Observe that the densities $f \in \mathcal{F}(C_1, C_2)$ uniformly satisfy condition (3.6) with the sum of the second moments bounded by C_2.

The class of possible distributions Q of (X, Y), which we consider, is then defined by

$$\mathcal{P} := \mathcal{P}(\mathcal{F}(C_1, C_2), \Theta)$$

$$:= \Big\{ P | \mathbb{R}^{d+m} : P \text{ has density } f \in \mathcal{F}(C_1, C_2) \text{ on } [x - \varepsilon_0, x + \varepsilon_0]$$

$$\text{such that the conditional density } f(\cdot \mid x) := f(x, \cdot) \Big/ \int f(x, y)\, dy$$

$$\text{is an element of } \{q_\vartheta : \vartheta \in \Theta\} \Big\}.$$

Note that $\mathcal{P}(\mathcal{F}(C_1, C_2), \Theta)$ is a semiparametric family of distributions, where the densities $f \in \mathcal{F}(C_1, C_2)$ form the non-parametric part, and where the k-dimensional parametric part (we are primarily interested in) is given by Θ. As a consequence, we index expectations, distributions etc. by $E_{f,\vartheta}, \mathcal{L}_{f,\vartheta}$ etc.

THE BASIC APPROXIMATION LEMMA

A main tool for the solution of our estimation problem is the following extension of Corollary 3.1.6 which follows by repeating the arguments of its derivation. By this result, we can handle our data $V_1, \ldots, V_{K(n)}$ within a certain error bound as being independently generated according to Q_ϑ, where the independent sample size is a Poisson rv $\tau^*(n)$ with parameter $n\, vol(S_n) g_f(x)$; in other words, we can handle the empirical point process N_n (which we observe) within this error bound as the ideal Poisson process $M_n^* = \sum_{i \le \tau^*(n)} \varepsilon_{W_i}$, where $W_1, W_2, \ldots$ are iid with common distribution Q_ϑ and independent of $\tau^*(n)$, uniformly in f and ϑ.

Lemma 3.3.1. *We have, for* $|a_n| \to 0$,

$$\sup_{\mathcal{P}(\mathcal{F}(C_1, C_2), \Theta)} H(N_n, M_n^*) = O\Big(vol(S_n) + (n\, vol(S_n))^{1/2} |a_n^{1/d}|^2 \Big).$$

Notice that in the preceding result the distribution of the Poisson process $M_n^*(\cdot) = \sum_{i \le \tau^*(n)} \varepsilon_{W_i}(\cdot)$ depends only on ϑ and the real parameter $g_f(x) = \int f(x, y)\, dy$, with $n\, vol(S_n)\, g_f(x)$ being the expectation of the Poisson rv $\tau^*(n)$. We index the distribution $\mathcal{L}_{g_f(x), \vartheta}(M_n^*)$ of M_n^* therefore only by $g_f(x)$ and ϑ.

By the preceding model approximation we can reduce the semiparametric problem $\mathcal{L}_{f,\vartheta}(N_n)$ with unknown $f \in \mathcal{F}(C_1, C_2)$ and $\vartheta \in \Theta$ to the $(k+1)$-dimensional parametric problem

$$\mathcal{L}_{b,\vartheta}(M_n^*) = \mathcal{L}_{b,\vartheta}\Big(\sum_{i \le \tau^*(n)} \varepsilon_{W_i} \Big),$$

where $\tau^*(n)$ is a Poisson rv with expectation $n\, vol(S_n)b$, $b \in (0, C_1]$, $W_1, W_2, \ldots$ are iid rv with distribution Q_ϑ and $\tau^*(n)$ and $W_1, W_2, \ldots$ are independent.

THE HELLINGER DIFFERENTIABILITY

We require *Hellinger differentiability* (cf. Section 1.3 of Groeneboom and Wellner [181]) of the family $\{q_\vartheta : \vartheta \in \Theta\}$ of densities at any point $\vartheta_0 \in \Theta$ that is, we require the expansion

$$q_\vartheta^{1/2}(\cdot) = q_{\vartheta_0}^{1/2}(\cdot)\Big(1 + \langle \vartheta - \vartheta_0, v_{\vartheta_0}(\cdot)\rangle/2 + |\vartheta - \vartheta_0|r_{\vartheta,\vartheta_0}(\cdot)\Big), \qquad (3.8)$$

for some measurable function $v_{\vartheta_0} = (v_{01}, \ldots, v_{0k})^t$, v_{0i} being square integrable with respect to the measure Q_{ϑ_0}, denoted by $v_{0i} \in L_2(Q_{\vartheta_0}), i = 1, \ldots, k$, and some remainder term $r_{\vartheta,\vartheta_0}$ satisfying

$$|r_{\vartheta,\vartheta_0}|_{L_2(Q_{\vartheta_0})} := \Big(\int r_{\vartheta,\vartheta_0}^2(y)\, Q_{\vartheta_0}(dy)\Big)^{1/2} \xrightarrow{\quad|\vartheta - \vartheta_0|\to 0\quad} 0.$$

Hellinger differentiability is also named L_2-*differentiability* (Witting ([463], Section 1.8.3)) or *differentiability in quadratic mean* (LeCam and Yang ([309], Section 5.2)).

LOCAL ASYMPTOTIC NORMALITY

Denote by $\mathbb{M}(\mathbb{R}^m)$ the space of all finite point measures on $\mathbb{R}^m$, endowed with the smallest σ-field $\mathcal{M}(\mathbb{R}^m)$ such that all projections $\mathbb{M}(\mathbb{R}^m) \ni \mu \mapsto \mu(B)$, $B \in \mathbb{B}^m$, are measurable. Define the statistical experiment $E_n = (\mathbb{M}(\mathbb{R}^m), \mathcal{M}(\mathbb{R}^m), \{\mathcal{L}_{\vartheta_0 + t\delta_n}(M_n^*) : t \in \Theta_n\})$, where $\delta_n = (n\, vol(S_n))^{-1/2}$ and $\Theta_n = \{t \in \mathbb{R}^k : \vartheta_0 + t\delta_n \in \Theta\}$. Throughout the rest we suppose that $n\, vol(S_n) \to \infty$ as $n \to \infty$.

It is well known that condition (3.8) implies local asymptotic normality (LAN) of the statistical experiments $(\mathbb{R}^m, \mathbb{B}^m, \{Q_{\vartheta_0 + tn^{-1/2}} : t \in \Theta_n\})$ (cf. Chapter 5 and Section 6.2 of LeCam and Yang [309]). The following result is adopted from Falk and Marohn [142].

Theorem 3.3.2 (LAN of E_n). *Fix $b > 0$. Under condition (3.8) we have with $b_n = b + o(\delta_n)$ and $\vartheta_n = \vartheta_0 + t\delta_n$,*

$$\frac{d\mathcal{L}_{b_n,\vartheta_n}(M_n^*)}{d\mathcal{L}_{b,\vartheta_0}(M_n^*)}(\cdot) = \exp\Big(\langle t, Z_{n,\vartheta_0}(\cdot)\rangle_{b,\vartheta_0} - \frac{1}{2}|t|_{b,\vartheta_0}^2 + R_{n,\vartheta_0,t}(\cdot)\Big)$$

with central sequence $Z_{n,\vartheta_0} : M(\mathbb{R}) \to \mathbb{R}^k$ given by

$$Z_{n,\vartheta_0}(\mu) = (\delta_n \mu(\mathbb{R}^m))^{-1}\Gamma^{-1}(\vartheta_0) \int v_{\vartheta_0}\, d\mu$$

and $R_{n,\vartheta_0,t} \to 0$ in $\mathcal{L}_{b,\vartheta_0}(M_n^*)$-probability, where $\langle s,t\rangle_{b,\vartheta_0} := s'b\,\Gamma(\vartheta_0)t$, $s,t \in \mathbb{R}^k$, and the $k \times k$-matrix $\Gamma(\vartheta_0) := (\int v_{0i}v_{0j}\, dQ_{\vartheta_0})_{i,j\in\{1,\ldots,k\}}$ is assumed to be positive definite.

The preceding result shows in particular that under alternatives of the form $b_n = b + o(\delta_n)$, $\vartheta_n = \vartheta_0 + t\,\delta_n$, the central sequence Z_{n,ϑ_0} *does not depend on* the nuisance parameter b, which was the value of the marginal density of X at x. If we allow the rate $b_n = b + O(\delta_n)$ instead, then LAN of $(E_n)_n$ still holds, but the central sequence depends on the nuisance parameter b, which cannot be estimated without affecting the asymptotics (see Falk and Marohn [142] for details).

THE HÁJEK-LECAM CONVOLUTION THEOREM

We recall the famous *convolution theorem* of Hájek-LeCam (see, for example, Section 8.4 in Pfanzagl [367]). Suppose that condition (3.8) holds for $\vartheta_0 \in \Theta$ and that $T_n(M_n^*)$ is an *asymptotically δ_n-regular* sequence of estimators in ϑ_0 based on M_n^* that is,

$$\delta_n^{-1}(T_n(M_n^*) - \vartheta_0 - t\delta_n) \to_D P \quad \text{for all } t \in \mathbb{R}^k$$

under $\vartheta_0 + t\delta_n$ for some probability measure P on $\mathbb{R}^k$, where $\to_D$ denotes convergence in distribution. Then there exists a probability measure H on $\mathbb{R}^k$ such that

$$P = H * N\left(0, b^{-1}\Gamma^{-1}(\vartheta_0)\right),$$

where $N(0, b^{-1}\Gamma^{-1}(\vartheta_0))$ with mean vector 0 and covariance matrix $b^{-1}\Gamma^{-1}(\vartheta_0)$ is the standard normal distribution on $(\mathbb{R}^k, \langle \cdot, \cdot \rangle_{b,\vartheta_0})$, and $*$ denotes convolution.

ASYMPTOTICALLY EFFICIENT ESTIMATION

In view of this convolution theorem, a δ_n-regular sequence of estimators $T_n(M_n^*)$ is called *asymptotically efficient* in ϑ_0 if

$$\delta_n^{-1}(T_n(M_n^*) - \vartheta_0) \to_D N\left(0, b^{-1}\Gamma^{-1}(\vartheta_0)\right)$$

under ϑ_0.

By Theorem 3.3.2 we know that Z_{n,ϑ_0} is central and hence,

$$\delta_n Z_{n,\vartheta_0}(M_n^*) + \vartheta_0 = \tau^*(n)^{-1}\Gamma^{-1}(\vartheta_0) \sum_{i \leq \tau^*(n)} v_{\vartheta_0}(W_i) + \vartheta_0$$

is asymptotically efficient in ϑ_0 for each $b > 0$. Note that this is true however only under the condition $b_n = b + o(\delta_n)$ in which case Z_{n,ϑ_0} is central. If we replace now the unknown underlying parameter ϑ_0 by any δ_n^{-1}-consistent estimator $\hat{\vartheta}_n(M_n^*)$ of ϑ_0 that is, $\delta_n^{-1}(\hat{\vartheta}_n(M_n^*) - \vartheta_0)$ is stochastically bounded under ϑ_0, we obtain that

$$\hat{\kappa}_n(M_n^*) := \delta_n Z_{n,\hat{\vartheta}_n(M_n^*)}(M_n^*) + \hat{\vartheta}_n(M_n^*)$$

is asymptotically efficient in ϑ_0, whenever

$$\sup_{|\vartheta_0-\vartheta|\leq K\delta_n} \left|\delta_n Z_{n,\vartheta_0}(M_n^*) + \vartheta_0 - \delta_n Z_{n,\vartheta}(M_n^*) - \vartheta\right| = o_P(\delta_n) \qquad (3.9)$$

under ϑ_0 (and b) for any $K > 0$.

Denote by $F = F_{\vartheta_0}$ the df of Q_{ϑ_0} and by $F_l(t) := l^{-1}\sum_{i\leq l}\varepsilon_{W_i}((-\infty,t])$, $t \in \mathbb{R}^m$, the empirical df pertaining to an iid sample $W_1,\ldots,W_l$ with common distribution Q_{ϑ_0}. Using conditioning techniques, elementary calculations show that condition (3.9) is satisfied, if the function $\vartheta \to \Gamma(\vartheta)$ is continuous at ϑ_0 and the following two conditions hold:

$$\sup_{|\vartheta_0-\vartheta|\leq Kl^{-1/2}} \left|l^{1/2}\int (v_{\vartheta_0}(s) - v_{\vartheta}(s))(F_l - F)(ds)\right| = o_P(1) \qquad (3.10)$$

as $l \to \infty$ for any $K > 0$ and

$$\left(\int v_{\vartheta}(s)\,F(ds) + \Gamma(\vartheta_0)(\vartheta - \vartheta_0)\right)\Big/|\vartheta - \vartheta_0| \longrightarrow_{|\vartheta-\vartheta_0|\to 0} 0. \qquad (3.11)$$

Note that $\sqrt{n}$-consistency of an estimator sequence $\vartheta_n(W_1,\ldots,W_n)$ of ϑ_0 implies δ_n^{-1}-consistency of $\hat{\vartheta}_n(M_n^*) = \hat{\vartheta}_{\tau^*(n)}(W_1,\ldots,W_{\tau^*(n)})$. We remark that under the present assumptions $\sqrt{n}$-consistent estimators actually exist (cf. LeCam [308], Proposition 1, p. 608).

EXPONENTIAL FAMILIES

In the following we discuss one standard family $\{Q_\vartheta : \vartheta \in \Theta\}$ (of possible conditional distributions) which satisfies conditions (3.8) and (3.9). Further examples can easily be constructed as well.

Example 3.3.3. Let $\{Q_\vartheta : \vartheta \in \Theta\}$, $\Theta \subset \Theta^*$ open, be a k-parametric exponential family of probability measures on $\mathbb{R}$ with natural parameter space $\Theta^* \subset \mathbb{R}^k$, i.e.,

$$q_\vartheta(x) = \frac{dQ_\vartheta}{d\nu}(x) = \exp(\langle\vartheta, T(x)\rangle - K(\vartheta)), \qquad x \in \mathbb{R},$$

for some σ-finite measure ν on $\mathbb{R}$ and some measurable map $T = (T_1,\ldots,T_k) :$ $\mathbb{R} \to \mathbb{R}^k$. The functions $\{1, T_1,\ldots,T_k\}$ are supposed to be linear independent on the complement of each ν-null set and $K(\vartheta) := \log\{\int \exp(\langle\vartheta, T(x)\rangle)\,\nu(dx)\}$. It is well known that the function $\vartheta \to E_\vartheta T$ is analytic in the interior of Θ^*. From Theorem 1.194 in Witting [463] we conclude that for $\vartheta_0 \in \Theta^*$ the family $\{Q_\vartheta\}$ is Hellinger-differentiable at ϑ_0 with derivative

$$v_{\vartheta_0}(x) = \nabla \log q_{\vartheta_0}(x) = T(x) - E_{\vartheta_0}T$$

where $\nabla = (\frac{\partial}{\partial \vartheta_i})_{i=1,\dots,k}$ denotes the nabla-operator. In this case we get $\Gamma(\vartheta_0) = Cov_{\vartheta_0}T$ and condition (3.11) is implied by

$$\frac{E_\vartheta T - E_{\vartheta_0}T - \nabla E_{\vartheta_0}T(\vartheta - \vartheta_0)}{|\vartheta - \vartheta_0|} \to 0$$

for $\vartheta \to \vartheta_0$ and $\nabla E_{\vartheta_0}T = Cov_{\vartheta_0}T$. Note that $Cov_{\vartheta_0}T$ is positive definite by the linear independence of $\{1, T_1, \dots, T_k\}$ (Witting [463, Theorem 1.153]). Condition (3.10) trivially holds since the integrand is independent of s.

EFFICIENT ESTIMATION BASED ON M_n^*

We can rewrite $\hat{\kappa}_n(M_n^*)$ in the form

$$\hat{\kappa}(M_n^*) = (M_n^*(\mathbb{R}))^{-1}\Gamma^{-1}(\hat{T}(M_n^*)) \int v_{\hat{T}(M_n^*)} dM_n^* + \hat{T}(M_n^*),$$

with $\hat{T} : M(\mathbb{R}^m) \to \mathbb{R}^k$ given by

$$\hat{T}(\mu) = \hat{\vartheta}_{\mu(\mathbb{R}^m)}(w_1, \dots, w_{\mu(\mathbb{R}^m)})$$

if $\mu = \sum_{i \leq \mu(\mathbb{R}^m)} \varepsilon_{w_i}$ is an atomization of μ.

The preceding considerations are summarized in the following result with Poisson process $M_n^* = \sum_{i \leq \tau^*(n)} \varepsilon_{W_i}$.

Theorem 3.3.4. *Fix $b > 0$ and suppose that the family $\{Q_\vartheta : \vartheta \in \Theta\}$ satisfies conditions (3.8) and (3.9) for any $\vartheta_0 \in \Theta(\subset \mathbb{R}^k)$. Let $\hat{\vartheta}_n = \hat{\vartheta}_n(W_1, \dots, W_n)$ be a $\sqrt{n}$-consistent estimator of each ϑ_0 and put $\hat{T}(M_n^*) := \hat{\vartheta}_{\tau^*(n)}(W_1, \dots, W_{\tau^*(n)})$. If $b_n = b + o(\delta_n)$, then*

$$\hat{\kappa}(M_n^*) = (M_n^*(\mathbb{R}^m))^{-1}\Gamma^{-1}(\hat{T}(M_n^*)) \int v_{\hat{T}(M_n^*)} dM_n^* + \hat{T}(M_n^*)$$

$$= \tau^*(n)^{-1}\Gamma^{-1}(\hat{T}(M_n^*)) \sum_{i \leq \tau^*(n)} v_{\hat{T}(M_n^*)}(W_i) + \hat{T}(M_n^*)$$

is an asymptotically efficient estimator that is, asymptotically efficient in ϑ_0 for all $\vartheta_0 \in \Theta$.

REGULAR PATHS

By means of Lemma 3.3.1 and the preceding result we can now establish asymptotic efficiency of an estimator $\hat{\kappa}(N_n)$ of ϑ_0 along *regular paths* in $\mathcal{P}(\mathcal{F}(C_1, C_2), \Theta)$.

Definition 3.3.5. *A path $\lambda \to P_{\vartheta_0 + \lambda t} \in \mathcal{P}(\mathcal{F}(C_1, C_2), \Theta)$, $t \in \mathbb{R}^k$, $\lambda \in (-\varepsilon, \varepsilon)$ for some $\varepsilon > 0$, is regular in ϑ_0, if the corresponding marginal densities of X satisfy $|g_{\vartheta_0 + \lambda t}(x) - g_{\vartheta_0}(x)| = o(\lambda)$ for $\lambda \to 0$.*

EFFICIENT ESTIMATION BASED ON N_n

Now we can state our main result.

Theorem 3.3.6. *Suppose that the family $\{Q_\vartheta : \vartheta \in \Theta\}$ satisfies conditions (3.8) and (3.9) for any $\vartheta_0 \in \Theta$. Let $vol(S_n) \to 0$, $|a_n| \to 0$, $n\,vol(S_n)|a_n|^{4/d} \to 0$ and $n\,vol(S_n) \to \infty$ as $n \to \infty$. Then*

$$\hat{\kappa}(N_n) := (N_n(\mathbb{R}^m))^{-1}\Gamma^{-1}(\hat{T}(N_n)) \int v_{\hat{T}(N_n)}\,dN_n + \hat{T}(N_n)$$

is asymptotically efficient in the sense that

$$\delta_n^{-1}(\hat{\kappa}(N_n) - \vartheta_0 - t\delta_n) \to_D N\left(0, \Gamma^{-1}(\vartheta_0)/g_{\vartheta_0}(x)\right)$$

under regular paths $P_{\vartheta_0 + t\delta_n}$ in $\mathcal{P}$, whereas for any other estimator sequence $T_n(N_n)$ of ϑ_0 based on N_n, which is asymptotically δ_n-regular along regular paths $P_{\vartheta_0 + t\delta_n}$, we have

$$\delta_n^{-1}(T_n(N_n) - \vartheta_0 - t\delta_n) \to_D H * N\left(0, \Gamma^{-1}(\vartheta_0)/g_{\vartheta_0}(x)\right)$$

for some probability measure H on $\mathbb{R}^k$.

Proof. By Lemma 3.3.1 we can replace N_n by M_n^* and hence, the assertion follows from the asymptotic efficiency of $\hat{\kappa}(M_n^*)$ established in Theorem 3.3.4 together with elementary computations. $\qquad\square$

Remark. If we choose equal bin widths $a_{n1} = \cdots = a_{nd} = c > 0$ for the data window $S_n = [x - a_n^{1/d}/2, x + a_n^{1/d}/2]$, then we obtain $vol(S_n) = c$, $n\,vol(S_n)|a_n|^{4/d} = O(nc^{(d+4)/d})$ and $\delta_n = (nc)^{-1/2}$. The choice $c = c_n = l^2(n)n^{-d/(d+4)}$ with $l(n) \to 0$, as $n \to \infty$, results in δ_n of minimum order $O(l(n)^{-1}\,n^{-2/(d+4)})$. The factor $l(n)^{-1}$, which may converge to infinity at an arbitrarily slow rate, actually ensures that the approximation of N_n by M_n^* is close enough, so that asymptotically the non-parametric part of the problem of the estimation of ϑ_0 that is, the joint density of (X, Y), is suppressed. In particular, it ensures the asymptotically unbiasedness of the optimal estimator sequence $\hat{\kappa}(N_n)$.

3.4   Extension to Several Points

In this section we will generalize the Poisson process approach, which we developed in Section 3.1 for a single point $x \in \mathbb{R}^d$, to a set $\{x_1, \ldots, x_r\}$ of several points, where $r = r(n)$ may increase as n increases.

Consider now only those observations $Y_i \in \mathbb{R}^m$ among $(X_1, Y_1), \ldots, (X_n, Y_n)$, with X_i falling into one of the cubes $S_\nu \subset \mathbb{R}^d$ with center x_ν, $\nu = 1, \ldots, r$ that is,

$$X_i \in \bigcup_{\nu \leq r} S_\nu,$$

where
$$S_\nu := S_{\nu n} = [x_\nu - a_{\nu n}^{1/d}/2, x_\nu + a_{\nu n}^{1/d}/2],$$
$x_\nu = (x_{\nu 1}, \ldots, x_{\nu d}) \in \mathbb{R}^d$, $a_{\nu n} = (a_{\nu n 1}, \ldots, a_{\nu n d}) \in (0, \infty)^d$, $\nu = 1, \ldots, r$.

We suppose in the sequel that the cubes $S_\nu, 1 \leq \nu \leq r$, are pairwise disjoint and that the marginal density of X, say g, is continuous at each x_ν with $g(x_\nu) > 0$.

VECTORS OF PROCESSES

Our data Y_i with $X_i \in \bigcup_{\nu \leq r} S_\nu$ can be described by the vector $(N_{n1}, \ldots, N_{nr})$ of truncated empirical point processes on $\mathbb{R}^m$, where

$$N_{n\nu}(B) := \sum_{i \leq n} \varepsilon_{Y_i}(B) \varepsilon_{X_i}(S_\nu), \qquad B \in \mathbb{B}^m, \quad \nu = 1, \ldots, r.$$

The rv $(N_{n1}, \ldots, N_{nr})$ will be approximated with respect to Hellinger distance by the vector $(M_{n1}^*, \ldots, M_{nr}^*)$ of independent Poisson processes, where

$$M_{n\nu}^* := \sum_{i \leq \tau_\nu^*(n)} \varepsilon_{W_{\nu i}};$$

$W_{\nu 1}, W_{\nu 2}, \ldots$ are independent rv on $\mathbb{R}^m$ with common df $F(\cdot \mid x_\nu)$, $\tau_\nu^*(n)$ is a Poisson rv with parameter $n \, vol(S_\nu) g(x_\nu)$ and $\tau_\nu^*(n), W_{\nu 1}, W_{\nu 2}, \ldots$ are mutually independent.

THE THIRD-ORDER POISSON APPROXIMATION

The following result extends Corollary 3.1.6 for the third-order Poisson process approximation at a single point, to the simultaneous approximation at several points.

Theorem 3.4.1. *We suppose that the rv (X, Y) has a joint density f on the strips $[x_\nu - \varepsilon_0, x_\nu + \varepsilon_0] \times \mathbb{R}^m$, $\nu = 1, \ldots, r$, for some $\varepsilon_0 \in (0, \infty)^d$, which satisfies uniformly for $\varepsilon \in (-\varepsilon_0, \varepsilon_0)(\subset \mathbb{R}^d)$, $y \in \mathbb{R}^m$ and $\nu = 1, \ldots, r$ the expansion*

$$f(x_\nu + \varepsilon, y) = f(x_\nu, y)\Big(1 + \langle \varepsilon, h_\nu(y) \rangle + O(|\varepsilon|^2 r_\nu(y))\Big), \tag{3.12}$$

where $\max_{1 \leq \nu \leq r} \int (|h_\nu(y)|^2 + |r_\nu(y)|^2) f(x_\nu, y) \, dy < \infty$. If $\max_{1 \leq \nu \leq r} |a_{\nu n}| \to 0$ as $n \to \infty$, we have

$$H\Big((N_{n\nu})_{\nu \leq r}, (M_{n\nu}^*)_{\nu \leq r}\Big) = O\Big(\sum_{\nu \leq r} vol(S_\nu) + \Big(\sum_{\nu \leq r} n \, vol(S_\nu) |a_{\nu n}^{1/d}|^4\Big)^{1/2}\Big).$$

If we chose equal bin widths $a_{\nu n 1}^{1/d} = \cdots = a_{\nu n d}^{1/d} = c^{1/d}$ for $\nu = 1, \ldots, r$, then the preceding bound is $O\left(rc + (rnc^{(d+4)/d})^{1/2}\right)$, uniformly for $c > 0$ and any $n \in \mathbb{N}$ as the Hellinger distance is in general bounded by $\sqrt{2}$.

Proof. Put for $B \in \mathbb{B}^{d+m}$,

$$\tilde{N}_n(B) := \sum_{i \le n} \varepsilon_{X_i \times Y_i} \left(B \cap \left(\bigcup_{\nu \le r} S_\nu \times \mathbb{R}^m \right) \right)$$

$$= \sum_{\nu \le r} \sum_{i \le n} \varepsilon_{X_i \times Y_i} (B \cap (S_\nu \times \mathbb{R}^m))$$

$$= \sum_{\nu \le r} \tilde{N}_n(B \cap (S_\nu \times \mathbb{R}^m)) =: \sum_{\nu \le r} \tilde{N}_{n\nu}(B).$$

Observe that the processes $N_{n\nu}$ can be derived from $\tilde{N}_{n\nu}$ by the projection

$$N_{n\nu}(\cdot) = \tilde{N}_{n\nu}(\mathbb{R}^d \times \cdot), \qquad \nu = 1, \dots, r.$$

By Theorem 1.3.1 we can write $\tilde{N}_n = \sum_{i \le K(n)} \varepsilon_{V_i}$, where $V_1, V_2, \dots$ are iid rv with common distribution $P(X \times Y \in \cdot | X \in \bigcup_{\nu \le r} S_\nu)$, and $K(n)$ is a $B(n, p_n)$ distributed rv, which is independent of $V_1, V_2, \dots$, with

$$p_n = P\left(X \in \bigcup_{\nu \le r} S_\nu \right) = \sum_{\nu \le r} P(X \in S_\nu) = O\left(\sum_{\nu \le r} vol(S_\nu) \right).$$

Define now by

$$\tilde{N}_n^* := \sum_{i \le \tau(n)} \varepsilon_{V_i} = \sum_{\nu \le r} \sum_{i \le \tau(n)} \varepsilon_{V_i} (\cdot \cap (S_\nu \times \mathbb{R}^m))$$

$$= \sum_{\nu \le r} \tilde{N}_n^* (\cdot \cap (S_\nu \times \mathbb{R}^m)) =: \sum_{\nu \le r} \tilde{N}_{n\nu}^*$$

the first-order Poisson approximation of $\tilde{N}_n$, where $\tau(n)$ is a Poisson rv with parameter np_n and also independent of $V_1, V_2, \dots$

Observe that by Lemma 1.2.1 and 1.2.2,

$$H\left((\tilde{N}_{n\nu})_{\nu \le r}, \ (\tilde{N}_{n\nu}^*)_{\nu \le r} \right) = H(\tilde{N}_n, \tilde{N}_n^*) \le 3^{1/2} p_n.$$

Put further for $A \in \mathbb{B}^m$,

$$N_n^*(A) := \tilde{N}_n^*(\mathbb{R}^d \times A) = \sum_{\nu \le r} \tilde{N}_{n\nu}^*(\mathbb{R}^d \times A) =: \sum_{\nu \le r} N_{n\nu}^*(A).$$

Then $N_{n\nu}^*$ is the first-order Poisson approximation of $N_{n\nu}$, $\nu = 1, \dots, r$. Note that $N_{n1}^*, \dots, N_{nr}^*$ as well as $\tilde{N}_{n1}, \dots, \tilde{N}_{nr}$ are sequences of independent Poisson processes, since $S_1, \dots, S_r$ are disjoint (see Section 1.1.2 in Reiss [387]).

We, consequently, have

$$H((N_{n\nu})_{\nu \le r}, \ (M_{n\nu}^*)_{\nu \le r})$$

$$\le H((N_{n\nu})_{\nu \le r}, \ (N_{n\nu}^*)_{\nu \le r}) + H((N_{n\nu}^*)_{\nu \le r}, \ (M_{n\nu}^*)_{\nu \le r})$$

$$\leq H((\tilde{N}_{n\nu})_{\nu\leq r},\ (\tilde{N}^*_{n\nu})_{\nu\leq r}) + H((N^*_{n\nu})_{\nu\leq r},\ (M^*_{n\nu})_{\nu\leq r})$$

$$= O\Big(\sum_{\nu\leq r} vol(S_\nu)\Big) + \Big(\sum_{\nu\leq r} H^2(N^*_{n\nu}, M^*_{n\nu})\Big)^{1/2}$$

$$= O\Big(\sum_{\nu\leq r} vol(S_\nu) + \Big(\sum_{\nu\leq r} n\, vol(S_\nu)|a_n^{1/d}|^4\Big)^{1/2}\Big)$$

by the arguments in the proof of Theorem 3.1.3 and (3.5). $\qquad\square$

THE FIRST-ORDER POISSON APPROXIMATION

The preceding proof entails the following extension of the first-order Poisson approximation in Theorem 3.1.2 to several points; we do not need the regularity condition (3.12).

Theorem 3.4.2. *We have, for disjoint S_ν, $\nu = 1,\dots,r$,*

$$H((N_{n\nu})_{\nu\leq r}, (N^*_{n\nu})_{\nu\leq r}) \leq 3^{1/2}\sum_{\nu\leq r} P(X\in S_\nu),$$

where $N^*_{n\nu}$, $\nu = 1,\dots,r$ are the (independent) first-order Poisson approximations of $N_{n\nu}$, $\nu = 1,\dots,r$.

ESTIMATION OVER COMPACT INTERVALS

Suppose now for the sake of simplicity that the rv (X,Y) is $\mathbb{R}^2$-valued. The preceding result can be utilized to derive the limiting distribution of the maximum error of an interpolated version of the kernel estimator $\hat{\vartheta}_n(x_\nu) := T(\hat{F}_n(\cdot\mid S_{\nu n}))$, $\nu = 1,\dots,r$, of $\vartheta(x) := T(F(\cdot\mid x))$ for x ranging over a compact interval $[a,b]$ in $\mathbb{R}$. For the definition of the local conditional empirical df $\hat{F}_n(\cdot\mid S_{\nu n})$ we refer to Section 3.2.

Choose a grid of $r+1$ equidistant points $x_0 = a < x_1 < \dots < x_r = b$ with $r = r_n \to \infty$ as $n\to\infty$ and define by *interpolation* the polygons

$$\hat{\vartheta}_n^i(x) := \hat{\vartheta}_n(x_j) + \frac{x - x_j}{x_{j+1} - x_j}(\hat{\vartheta}_n(x_{j+1}) - \hat{\vartheta}_n(x_j)), \qquad x_j \leq x \leq x_{j+1},$$

and

$$\vartheta^i(x) := \vartheta(x_j) + \frac{x - x_j}{x_{j+1} - x_j}(\vartheta(x_{j+1}) - \vartheta(x_j)), \qquad x_j \leq x \leq x_{j+1},$$

where we suppose that $\hat{\vartheta}_n(x_\nu) = T(\hat{F}_n(\cdot\mid x_\nu))$, $\nu = 0,\dots,r$, is defined with equal bin width $a_{\nu n} = c_n$ of order ε_n/r, $\varepsilon_n \to 0$ as $n\to\infty$.

If we suppose that $\vartheta(x) = T(F(\cdot \mid x))$ is twice differentiable on $[a, b]$ with bounded second derivative, then a Taylor expansion implies

$$\sup_{x \in [a,b]} |\vartheta(x) - \vartheta_n^i(x)| = O(1/r^2).$$

As a consequence, we obtain

$$\sup_{x \in [a,b]} |\hat{\vartheta}_n^i(x) - \vartheta(x)| = \sup_{x \in [a,b]} |\hat{\vartheta}_n^i(x) - \vartheta_n^i(x)| + O(1/r_n^2)$$

$$= \max_{0 \leq \nu \leq r} |\hat{\vartheta}_n(x_\nu) - \vartheta(x_\nu)| + O(1/r_n^2),$$

since $\hat{\vartheta}_n^i - \vartheta_n^i$ is again a polygon, which therefore attains its maximum and minimum on $[a, b]$ at the set $\{x_0, \ldots, x_r\}$ of grid points.

If we suppose that X has a continuous density on $[a, b]$, then Theorem 3.4.2 implies that within the error bound $\sum_{0 \leq \nu \leq r} P(X \in S_\nu) = O(rc_n) = O(\varepsilon_n) = o(1)$, the rv $\hat{\vartheta}_n(x_\nu) - \vartheta(x_\nu)$, $\nu = 0, \ldots, r$, may be replaced by *independent* rv $\xi_{\nu n}$, $\nu = 0, \ldots, r$, say.

The problem of computing the limiting distribution of the maximum error $\sup_{x \in [a,b]} |\hat{\vartheta}_n^i(x) - \vartheta(x)|$ therefore reduces to the problem of computing the limiting distribution of the maximum in a set of independent rv, which links the present problem to extreme value theory. With the particular choices of $r_n = O((n/\log(n))^{1/5})$ and $c_n = \varepsilon_n(\log(n)/n)^{1/5}$, $\varepsilon_n \log(n) \to \infty$ and $\varepsilon_n \log(n)^{2/5} \to 0$ as $n \to \infty$, it turns out that $\sup_{x \in [a,b]} |\hat{\vartheta}_n^i(x) - \vartheta(x)| = O_P((\log(n)/n)^{2/5} \varepsilon_n^{-1/2})$, (see Theorem 4.2 in Falk [130]), which is up to the factor $\varepsilon_n^{-1/2}$ the optimal global achievable rate of convergence in case of the mean value functional (Stone [428], [429] being twice differentiable with bounded second derivative (for related results we refer to Nussbaum [356] and, for the quantile functional, to Truong [447] and Chaudhuri [63]. Notice however that in Theorem 4.2 in Falk [130] actually the limiting distribution of $\sup_{x \in [a,b]} |\hat{\vartheta}_n^i - \vartheta(x)|$ is computed.

3.5   A Nearest Neighbor Alternative

Let again $(X_1, Y_1), \ldots, (X_n, Y_n)$ be independent replicates of the rv (X, Y) with values in $\mathbb{R}^{d+m}$. As pointed out at the beginning of this chapter, non-parametric estimators of regression functionals $T(F(\cdot \mid x))$ have to be based on those values among $Y_1, \ldots, Y_n$, whose first coordinate is within a small distance (in other words, bin width) of x. There are essentially two different ways of selecting the Y-values.

(a) The bin width is non-random and, hence, the number of selected Y-values is random. This has been our approach in the preceding sections.

(b) Take those Y, whose X-values are the k closest to x. This is the *nearest neighbor method*. Now the number of selected Y-values is non-random whereas the bin width is random.

Nearest neighbor estimators in regression analysis were first studied by Royall [402] and Cover [78]. Their consistency properties under very weak conditions were established by Stone [426], [427], Devroye [109] and Cheng [66]; weak convergence results were proved by Mack [316], Stute [431], Bhattacharya and Mack [42] and Bhattacharya and Gangopadhyay [41]. For a discussion of nearest neighbor estimators we refer also to Section 7.4 of the book by Eubank [124] and to Section 3.2 of the one by Härdle [205].

In the present section we will focus on the second point (b) by considering a fixed sample size $k = k(n)$ of observations among $Y_1, \ldots, Y_n$ with X-values close to x. In analogy to the POT process approximation in Theorem 3.1.3, we will establish in Theorem 3.5.2 a bound for the Hellinger distance between these $k(n)$ observations among $Y_1, \ldots, Y_n$ coming from the nearest neighbor approach and $k(n)$ independent rv from the ideal df $F(\cdot \mid x)$. The expansion (3.6) of the joint density f of (X, Y) will again be a crucial condition.

Denote by
$$R_i := |X_i - x|, \ 1 \le i \le n,$$
the (Euclidean) distances of X_i from x. Obviously, $R_1, \ldots, R_n$ are iid replicates of the rv $R := |X - x|$ with values in $[0, \infty)$. The corresponding order statistics are $R_{1:n}, \ldots, R_{n:n}$. By $B(x, r) := \{z \in \mathbb{R}^d : |z - x| < r\}$ we denote the open ball in $\mathbb{R}^d$ with center x.

We assume that R has a continuous df; this condition is indispensable in the following lemma. Then, there will be exactly k nearest neighbors among $X_1, \ldots, X_n$ with probability 1 that fall into $B(x, R_{k+1:n})$, with corresponding Y-values $V_1, \ldots, V_k$, say, in the original order of their outcome. Denote, for $r > 0$, by
$$P_r = P(Y \in \cdot \mid X \in B(x, r))$$
the conditional distribution of Y given $X \in B(x, r)$.

The Basic Representation Lemma

The following lemma is statistical folklore (see, for example, Lemma 1 in Bhattacharya [40]). For a rigorous and appealing proof of that result we refer to Kaufmann and Reiss [284], where the general case of conditioning on *g-order statistics* is dealt with.

Lemma 3.5.1. *Assume that R has a continuous df. Then the rv $V_1, \ldots, V_k$ are iid, conditional on $R_{k+1:n}$. Precisely, we have for $r > 0$ and $k \in \{1, \ldots, n\}$, $n \in \mathbb{N}$:*
$$P\big((V_1, \ldots, V_k) \in \cdot \mid R_{k+1:n} = r\big) = P_r^k,$$
where P_r^k denotes the k-fold product of P_r.

According to Lemma 3.5.1, the unconditional joint distribution of $V_1, \ldots, V_k$ may be represented by P_r^k and the distribution $\mathcal{L}(R_{k+1:n})$ of $R_{k+1:n}$ as
$$P\big((V_1, \ldots, V_k) \in \cdot\big) = \int P_r^k(\cdot) \, \mathcal{L}(R_{k+1:n})(dr).$$

An Approximation Result

The probability measure $P_r = P\big(Y \in \cdot \mid X \in B(x,r)\big)$ approximates for small $r > 0$ the conditional distribution of Y given $X = x$ that is,

$$P(\cdot \mid x) := P(Y \in \cdot \mid X = x),$$

with df $F(\cdot \mid x)$. We, therefore, expect the approximation

$$P((V_1, \ldots, V_k) \in \cdot) = \int P_r^k(\cdot)\, \mathcal{L}(R_{k+1:n})(dr)$$

$$\sim \int P(\cdot \mid x)^k(\cdot)\, \mathcal{L}(R_{k+1:n})(dr) = P(\cdot \mid x)^k = P((W_1, \ldots, W_k) \in \cdot),$$

where $W_1, \ldots, W_k$ are iid rv with common df $F(\cdot \mid x)$.

In other words, we expect that the Y_i-values $V_1, \ldots, V_k$ pertaining to the k nearest neighbors of x among $X_1, \ldots, X_n$, can approximately be handled like independent rv $W_1, \ldots, W_k$, equally distributed according to the target df $F(\cdot \mid x)$. This observation corresponds to the POT process approximation of the empirical truncated point process established in Theorem 3.1.3. We want to quantify the error of this approximation in the following by establishing a bound for the Hellinger distance H between the distributions of $(V_1, \ldots, V_k)$ and $(W_1, \ldots, W_k)$.

Within this error bound, the estimation of conditional parameters $\vartheta(x) = T(F(\cdot \mid x))$, based on $V_1, \ldots, V_k$, can therefore again be carried out within the classical statistical framework of sequences of iid observations, but this time with a *non-random* sample size k. We will exemplify this consequence in the next section, where we establish asymptotic optimal accuracy in a certain sense of an estimator sequence $\hat\vartheta_n(x)$ of $\vartheta(x) = T(F(\cdot \mid x))$, with T evaluated at the empirical df pertaining to $V_1, \ldots, V_k$.

The smoothness condition (3.6) from Section 3.1 on the joint density of (X, Y) turns out to be a handy condition also for the derivation of the bound established in the following result.

Theorem 3.5.2. *Under condition* (3.6) *and the assumption that the marginal density of X is positive at x, we have, uniformly for $n \in \mathbb{N}$ and $k \in \{1, \ldots, n\}$,*

$$H((V_1, \ldots, V_k), (W_1, \ldots, W_k)) = O(k^{1/2}(k/n)^{2/d}).$$

This result entails that any statistical procedure based on $V_1, \ldots, V_k$ approximately behaves like the corresponding one based on the iid vectors $W_1, \ldots, W_k$ with common df $F(\cdot \mid x)$. Within the error bound $O(k^{1/2}(k/n)^{2/d})$ (which does not depend on the dimension m of the covariate Y but on the dimension d of X), computations in regression analysis at one point may therefore be carried out within a classical statistical framework. For an example we refer to the next section.

The preceding result parallels the bound for the Hellinger distance between conditional empirical point processes and Poisson point processes established in Theorem 3.1.3.

Proof. Fix $r_0 > 0$ small enough. Since the Hellinger distance is bounded by $\sqrt{2}$, we obtain from the convexity theorem for the Hellinger distance (cf. Lemma 1.2.3)

$$H((V_1, \ldots, V_k), (W_1, \ldots, W_k))^2$$
$$\leq \int_0^{r_0} H(P_r^k, P_0^k)^2 \, \mathcal{L}(R_{k+1:n})(dr) + 2P(R_{k+1:n} \geq r_0)$$
$$\leq k \int_0^{r_0} H(P_r, P_0)^2 \, \mathcal{L}(R_{k+1:n})(dr) + 2P(R_{k+1:n} \geq r_0).$$

Hence, we only have to investigate the Hellinger distance between P_r and P_0 for $r \in (0, r_0)$.

The Lebesgue densities of P_0 and P_r are given by

$$h_0(y) = \frac{f(x, y)}{g(x)}, \qquad y \in \mathbb{R}^m,$$

and

$$h_r(y) = \frac{\int_{B(x,r)} f(z, y) \, dz}{\int_{B(x,r)} g(z) \, dz}, \qquad y \in \mathbb{R}^m,$$

respectively, where $g(z) := \int f(z, w) \, dw$ denotes the marginal density of X for z near x. By elementary computations we deduce from condition (3.6) the expansion

$$h_r^{1/2}(y) = h_0^{1/2}(y) \left(1 + O\left(r^2 R(y) \right) \right) \tag{3.13}$$

uniformly for $y \in \mathbb{R}^m$ and $0 < r \leq r_0$, where

$$\int R^2(y) f(x, y) \, dy < \infty.$$

We, consequently, obtain

$$H(P_r, P_0) = O(r^2),$$

uniformly for $0 < r < r_0$, and thus,

$$H((V_1, \ldots, V_k), (W_1, \ldots, W_k))^2$$
$$= O\left(k \int_0^{r_0} r^4 \mathcal{L}(R_{k+1:n})(dr) + P(R_{k+1:n} \geq r_0) \right).$$

By repeating those arguments which lead to expansion (3.13), it is easy to see that

$$F(r) := P(R \leq r) = P(X \in B(x, r)) = \text{volume } (B(x, r))(1 + O(r^2))$$
$$= c(d) r^d (1 + O(r^2)), \qquad 0 < r < r_0,$$

where $c(d)$ denotes the volume of the unit ball in $\mathbb{R}^d$. Thus, by Fubinis's theorem and the *quantile transformation* technique (see the proof of Theorem 2.2.2)

$$\int_0^{r_0} r^4 \, \mathcal{L}(R_{k+1:n})(dr) = E(R_{k+1:n}^4 \cdot 1_{[0,r_0]}(R_{k+1:n}))$$

$$= \int_0^{r_0} P(R_{k+1:n}^4 > r) \, dr$$

$$= \int_0^{r_0} P(R_{k+1:n} > r^{1/4}) \, dr$$

$$= \int_0^{r_0} P(F^{-1}(U_{k+1:n}) > r^{1/4}) \, dr$$

$$= \int_0^{r_0} P(U_{k+1:n} > F(r^{1/4})) \, dr$$

$$= \int_0^{r_0} P(U_{k+1:n} > c(d) r^{d/4} (1 + O(r^{1/2}))) \, dr$$

$$= \int_0^{r_0} P(U_{k+1:n}^{4/d} > c(d)^{4/d} r (1 + O(r^{1/2}))^{4/d}) \, dr$$

$$= O(E(U_{k+1:n}^{4/d})) = O(E(U_{k+1:n}^4)^{1/d}) = O((k/n)^{4/d})$$

uniformly for $k \in \{1, \ldots, n\}$, $n \in \mathbb{N}$, where $U_{1:n} \leq \ldots \leq U_{n:n}$ denote the order statistics pertaining to a sample of n independent and uniformly on (0,1) distributed rv. The inequality $E(U_{k+1:n}^{4/d}) \leq E(U_{k+1:n}^4)^{1/d}$ follows from Jensen's inequality, and the bound $E(U_{k+1:n}^4) = O((k/n)^4)$ is immediate from formula (1.7.4) in Reiss [385]. From the exponential bound for $P(U_{k+1:n} \geq \varepsilon)$ given in Lemma 3.1.1 in Reiss [385], we conclude that $P(R_{k+1:n} \geq r_0) = P(F^{-1}(U_{k+1:n}) \geq r_0) = P(U_{k+1:n} \geq F(r_0)) = O((k/n)^4)$ uniformly for $k \in \{1, \ldots, n\}$, $n \in \mathbb{N}$. This concludes the proof of Theorem 3.5.2. $\square$

3.6 Application: Optimal Accuracy of Estimators

In this section we will apply Theorem 3.5.2 to establish asymptotic optimal accuracy of the estimator sequence $\hat{\vartheta}_n(x)$ of a general regression functional $\vartheta(x) = T(F(\cdot \mid x))$, where $\hat{\vartheta}_n(x)$ is again simply the empirical counterpart of $\vartheta(x)$ with $F(\cdot \mid x)$ replaced by a sample df.

Such optimal rates of convergence have been established for the mean value functional $T_1(F) = \int t \, F(dt)$ by Stone ([428], [429]), for smooth functionals of the regression mean by Goldstein and Messer [177], and for the median functional $T_2(F) = F^{-1}(1/2)$ and, more generally, quantile functional by Truong [447] and Chaudhuri [63]. These results were established for the kernel estimator (with uniform kernel) and non-random bin width, leading to a random number of observations $V_1, V_2, \ldots$ Nearest neighbor competitors, based on a non-random number

$V_1, \ldots, V_k$, also achieve these optimal rates as shown by Bhattacharya and Mack [42] for T_1 and Bhattacharya and Gangopadhyay [41] for T_2, among others.

THE MODEL BIAS

Notice that the bound $k^{1/2}(k/n)^{2/d}$ in Theorem 3.5.2 does not depend on the dimension m of Y and that it converges to zero iff $k = k(n)$ satisfies $k/n^{4/(d+4)}$ $\to_{n\to\infty} 0$. If we choose $k(n)$ therefore of order $cn^{4/(d+4)}$ (independent of dimension m), the model error $O(k^{1/2}(k/n)^{2/d})$ becomes $O(c^{(d+4)/(2d)})$, uniformly for $c > 0$. This term $O(c^{(d+4)/(2d)})$ represents the non-vanishing bias for our model approximation $\mathcal{L}(W_1, \ldots, W_k)$ of $\mathcal{L}(V_1, \ldots, V_k)$. It can be regarded as an upper bound for the usually non-vanishing bias of any optimally tuned estimator of $T(F(\cdot \mid x))$, based on $V_1, \ldots, V_k$, for an arbitrary functional T.

LOCAL EMPIRICAL DISTRIBUTION FUNCTION

Denote by

$$F_{nk}(t|x) := k^{-1} \sum_{i \le k} 1_{(-\infty, t]}(V_i), \qquad t \in \mathbb{R}^m,$$

the (local) empirical df pertaining to the data $V_1, \ldots, V_k$ from $Y_1, \ldots, Y_n$, which are induced by the k nearest neighbors of x among $X_1, \ldots, X_n$. The natural nonparametric estimate of $\vartheta(x) = T(F(\cdot \mid x))$ is then the nearest neighbor (NN) estimate

$$\hat\vartheta_{nk}(x) := T(F_{nk}(\cdot \mid x)),$$

which is completely analogous to the kernel estimate defined in Section 3.2 but with a random sample size. Again we assume implicitly that $T : \mathcal{F} \to \mathbb{R}^l$ is a functional on a subspace $\mathcal{F}$ of the class of all df on $\mathbb{R}^m$ containing $F(\cdot \mid x)$, $F_{nk}(\cdot \mid x)$, $k \in \mathbb{N}$. In case of $T_1(F) = \int t\, F(dt)$ with $l = 1$, the estimator $\hat\vartheta_{nk}(x) = T_1(F_{nk}(\cdot \mid x))$ $= k^{-1} \sum_{i \le k} V_i$ is the local sample average; in case $T_2(F) = F^{-1}(q)$, the estimator $\hat\vartheta_{nk}(x)$ is the local sample quantile $F_{nk}(\cdot \mid x)^{-1}(q)$, $q \in (0, 1)$.

ASYMPTOTIC NORMALITY OF NN-ESTIMATES

Theorem 3.5.2 entails that we can approximate the distribution of $\hat\vartheta_{nk}(x)$ by that of

$$\hat\vartheta_k(x) := T(F_k(\cdot \mid x)),$$

where

$$F_k(t \mid x) := k^{-1} \sum_{i \le k} 1_{(-\infty, t]}(W_i), \qquad t \in \mathbb{R}^m,$$

is the empirical df pertaining to $W_1, \ldots, W_k$, being independent rv with common df $F(\cdot \mid x)$. We suppose implicitly that $F_k(\cdot \mid x)$, $k \in \mathbb{N}$, is also in the domain of the functional T. The following result, which is the nearest neighbor version of

Theorem 3.2.1, is immediate from Theorem 3.5.2. Note that this result is true for arbitrary dimension m of Y.

Proposition 3.6.1. *Suppose that* $\hat{\vartheta}_k(x) - \vartheta(x) = T(F_k(\cdot \mid x)) - T(F(\cdot \mid x))$ *is asymptotically normal that is,*

$$\sup_{t \in \mathbb{R}^l} \left| P\left(a_k \left(T(F_k(\cdot \mid x)) - T(F(\cdot \mid x)) \right) \le t \right) - N(\mu, \Sigma)\left((-\infty, t] \right) \right|$$

$$=: \ R(k) \longrightarrow_{k \to \infty} 0 \qquad (3.14)$$

for some norming sequence $0 < a_k \to \infty$ *as* $k \to \infty$, *where* $N(\mu, \Sigma)$ *denotes the normal distribution on* $\mathbb{B}^l$ *with mean vector* μ *and covariance matrix* Σ.

If condition (3.6) *is satisfied and the marginal density of* X *is positive at* x, *then asymptotic normality of* $\hat{\vartheta}_k(x) - \vartheta(x)$ *carries over to* $\hat{\vartheta}_{nk}(x) - \vartheta(x) = T(F_{nk}(\cdot \mid x)) - T(F(\cdot \mid x))$ *that is,*

$$\sup_{t \in \mathbb{R}^l} \left| P\left(a_k(\hat{\vartheta}_{nk}(x) - \vartheta(x)) \le t \right) - N(\mu, \Sigma)\left((-\infty, t] \right) \right|$$

$$= \ O\left(k^{1/2}(k/n)^{2/d} + R(k) \right).$$

Under suitable regularity conditions on the conditional df $F(\cdot \mid x)$, condition (3.14) is satisfied for a large class of statistical functionals including M, L and R estimators, with $a_k = k^{1/2}$ by the corresponding multivariate central limit theorems. In these cases we have typically the bound $R(k) = O(k^{-1/2})$ (cf. the monograph by Serfling [408]).

OPTIMAL ACCURACY

If a_k can be chosen as $k^{1/2}$, the choice for $k = k(n)$ of order $n^{4/(d+4)}$ roughly entails that the NN-estimator $\hat{\vartheta}_{nk}(x) = T(F_{nk}(\cdot \mid x))$ of $\vartheta(x) = T(F(\cdot \mid x))$ has accuracy of order $n^{-2/(d+4)}$ for a general functional T, independent of the dimension m of Y. As mentioned above, this is the well-known optimal (local) rate of convergence for the conditional mean as well as for the conditional median, both in dimensions $m = 1$ (and $l = 1$) for a twice continuously differentiable target function $\vartheta(x)$, roughly. These considerations indicate in particular that the bound $O(k^{1/2}(k/n)^{2/d})$ in Theorem 3.5.2 is sharp. The following result is immediate from Proposition 3.6.1.

Proposition 3.6.2. *Suppose that conditions* (3.6) *and* (3.14) *are satisfied with* $a_k = k^{1/2}$, *and that the marginal density of* X *is positive at* x. *Choose* $k = k(n) \sim c_1 n^{4/(d+4)}$. *Then we have, uniformly for* $c_1, c_2 > 0$,

$$\limsup_{n \to \infty} P\left(c_1^{1/2} n^{2/(d+4)} |\hat{\vartheta}_{nk}(x) - \vartheta(x)| > c_2 \right) = O(c_1^{(d+4)/(2d)} + c_2^r)$$

for any $r > 0$.

Note that if a_k has to be chosen of order smaller than $k^{1/2}$ in condition (3.14), then Proposition 3.6.1 entails that $\hat{\vartheta}_{nk}(x) = T(F_{nk}(\cdot \mid x))$ with k of order $n^{4/(d+4)}$ has asymptotic accuracy of order greater than $n^{-2/(d+4)}$; if a_k has to be chosen of order greater than $k^{1/2}$ it is vice versa. This indicates that the rate of order $k^{1/2}$ for a_k in condition (3.14) is necessary and sufficient for the estimator $\hat{\vartheta}_{nk}(x)$ to achieve the (optimal) local accuracy of order $n^{-2/(d+4)}$.

Part II

The IID Case:
Multivariate Extremes

Chapter 4

Basic Theory of Multivariate Maxima

In this chapter, we study the limiting distributions of componentwise defined maxima of iid d-variate rv. Such distributions are again max-stable as in the univariate case. Some technical results and first examples of max-stable df are collected in Section 4.1. In Section 4.2 and 4.3, we describe representations of max-stable df such as the de Haan-Resnick and the Pickands representation. Of special interest for the subsequent chapters will be the Pickands dependence function in Section 4.3 and the D-norm, which will be introduced in Section 4.4.

4.1 Limiting Distributions of Multivariate Maxima

Subsequently, arithmetic operations and order relations are meant componentwise; that is, e.g., $\boldsymbol{a} + \boldsymbol{b} = (a_1 + b_1, \ldots, a_d + b_d)$ for vectors $\boldsymbol{a} = (a_1, \ldots, a_d)$ and $\boldsymbol{b} = (b_1, \ldots, b_d)$. An interval $(\boldsymbol{a}, \boldsymbol{b}]$ is defined by $\bigtimes_{j \leq d}(a_j, b_j]$.

Recall that the df $F(\boldsymbol{x}) = Q(-\infty, \boldsymbol{x}]$ of a probability measure Q has the following properties:

(a) F is right-continuous: $F(\boldsymbol{x}_n) \downarrow F(\boldsymbol{x}_0)$ if $\boldsymbol{x}_n \downarrow \boldsymbol{x}_0$;

(b) F is normed: $F(\boldsymbol{x}_n) \uparrow 1$ if $x_{nj} \uparrow \infty$, $j = 1, \ldots, d$, and $F(\boldsymbol{x}_n) \downarrow 0$ if $\boldsymbol{x}_n \geq \boldsymbol{x}_{n+1}$ and $x_{nj} \downarrow -\infty$ for some $j \in \{1, \ldots, d\}$;

(c) F is Δ-monotone: For $\boldsymbol{a} \leq \boldsymbol{b}$,

$$
\Delta_{\mathbf{a}}^{\mathbf{b}} F = Q(\boldsymbol{a}, \boldsymbol{b}]
$$
$$
= \sum_{\mathbf{m} \in \{0,1\}^d} (-1)^{\left(d - \sum_{j \leq d} m_j\right)} F\left(b_1^{m_1} a_1^{1-m_1}, \ldots, b_d^{m_d} a_d^{1-m_d}\right) \geq 0.
$$

M. Falk et al., *Laws of Small Numbers: Extremes and Rare Events*, 3rd ed.,
DOI 10.1007/978-3-0348-0009-9_4, © Springer Basel AG 2011

Conversely, every function F, satisfying conditions (a)-(c), is the df of a probability measure Q. Usually, conditions (a) and (b) can be verified in a straightforward way. The Δ-monotonicity holds if, e.g., F is the pointwise limit of a sequence of df.

Let $\boldsymbol{X}_i = (X_{i1}, \ldots, X_{id})$, $i \leq n$, be iid d-variate rv with common df F. The d-variate maximum is defined by

$$\max_{i \leq n} \boldsymbol{X}_i := \left(\max_{i \leq n} X_{i1}, \ldots, \max_{i \leq n} X_{id} \right).$$

LIMITING DISTRIBUTIONS, MAX-STABILITY

It is an easy exercise to prove that

$$P\left(\max_{i \leq n} \boldsymbol{X}_i \leq \boldsymbol{x} \right) = F^n(\boldsymbol{x}).$$

Hence, the well-known formula for the df of a univariate maximum still holds in the multivariate framework. Limiting df are again called extreme value df (EVD). Such df can be characterized by the max-stability.

Again, a df G is max-stable if for each $n \in \mathbb{N}$,

$$G^n(\boldsymbol{d}_n + \boldsymbol{c}_n \boldsymbol{x}) = G(\boldsymbol{x})$$

for certain vectors $\boldsymbol{c}_n > 0$ and $\boldsymbol{d}_n$. If G is max-stable, then the marginal df also possess this property. This yields that a multivariate max-stable df is continuous (cf. Reiss [385], Lemma 2.2.6). Moreover, the components of the normalizing vectors are the normalizing constants in the univariate case.

If the G_j are univariate max-stable df, then

$$\min(G_1(x_1), \ldots, G_d(x_d))$$

is a max-stable df (representing the case of totally dependent rv). Moreover, for independent rv one obtains the max-stable df

$$\prod_{j \leq d} G_j(x_j).$$

One can prove that

$$\prod_{j \leq d} G_j(x_j) \ \leq \ G(\boldsymbol{x}) \ \leq \ \min(G_1(x_1), \ldots, G_d(x_d)) \qquad (4.1)$$

for every max-stable df G with margins G_j, see (4.37). Note that the right-hand side is the upper Fréchet bound which holds for every df; the proof is obvious. It follows that $G(\boldsymbol{x}) > 0$ if, and only if, $\boldsymbol{x} > \boldsymbol{\alpha}(G) := (\alpha(G_1), \ldots, \alpha(G_d))$.

In the following we primarily deal with max-stable df having reverse exponential margins $G_{2,1}(x) = e^x$, $x < 0$, which is the standard Weibull df with shape parameter $\alpha = -1$. This standardization in the univariate margins can always be achieved by means of a simple transformation: If G is max-stable with margins G_j, $j \leq d$, then

$$G\left(G_1^{-1}(G_{2,1}(x_1)), \ldots, G_d^{-1}(G_{2,1}(x_d))\right), \qquad \boldsymbol{x} < \boldsymbol{0}, \qquad (4.2)$$

defines a max-stable df with margins $G_{2,1}$.

Example 4.1.1 (Marshall-Olkin). Let Z_0, Z_1, Z_2 be independent standard reverse exponential rv; thus, $P(Z_i \leq x) = e^x = G_{2,1}(x)$ for $x < 0$. For each $\lambda \in (0,1)$ one obtains a bivariate max-stable df with margins $G_{2,1}$ by

$$P\left(\max\left(\frac{Z_j}{1-\lambda}, \frac{Z_0}{\lambda}\right) \leq x_j, \; j=1,2\right) = \exp\left((1-\lambda)(x_1 + x_2) + \lambda \min(x_1, x_2)\right),$$

where $x_j \leq 0$, $j = 1, 2$. If $\lambda = 0$ and $\lambda = 1$, then the bivariate df represent the cases of independent and totally dependent rv, respectively.

Because of its importance we are going to prove an extension of the foregoing example, see also Example 4.3.4.

Lemma 4.1.2. *For every $m \in \mathbb{N}$, let $Z_1, Z_2, \ldots, Z_m$ be iid rv with common standard reverse exponential df $G_{2,1}$. Let $a_{ij} > 0$ for $i \leq m$ and $j \leq d$. Then*

$$P\left(\max_{i \leq m} \frac{Z_i}{a_{ij}} \leq x_j, \; j = 1, \ldots, d\right) = \exp\left(\sum_{i \leq m} \min_{j \leq d} a_{ij} x_j\right), \qquad \boldsymbol{x} < \boldsymbol{0}, \qquad (4.3)$$

thus obtaining a d-variate max-stable df with exponential margins. If, in addition,

$$\sum_{i \leq m} a_{ij} = 1, \qquad j \leq d,$$

then the univariate margins are equal to $G_{2,1}$.

Proof. Note that $\max_{i \leq m} Z_i/a_{ij} \leq x_j$, $j \leq d$, if, and only if, $Z_i \leq \min_{j \leq d} a_{ij} x_j$, $i \leq m$; thus, (4.3) follows from the independence of $Z_1, \ldots, Z_m$. The max-stability is obvious. We see that the j-th marginal G_j is given by

$$G_j(x) = \exp\left(\left(\sum_{i \leq m} a_{ij}\right) x\right), \qquad x < 0,$$

and, hence, the assertion concerning the univariate margins holds. $\qquad \square$

In order to obtain a max-stable df on the right-hand side of (4.3) it suffices to assume that $\sum_{i \leq m} a_{ij} > 0$ for $j \leq d$. A continuation of this topic may be found in Section 4.3.

Because

$$\min_{i \leq n} \boldsymbol{X}_i = \max_{i \leq n}(-\boldsymbol{X}_i),$$

results for minima can be easily deduced from those for maxima.

Weak Convergence: The IID Case

Max-stable df are the limiting df of linearly normalized maxima. Recall that for a univariate df F the convergence

$$n(1 - F(b_n + a_n x)) =: n S_n(x) \to L(x) := -\log(G(x)), \qquad x > \alpha(G), \qquad (4.4)$$

as $n \to \infty$ implies that $F^n(b_n + a_n x) \to G(x)$ as $n \to \infty$. An extension to the d-variate case will be formulated as an inequality.

Let $\boldsymbol{X} = (X_1, \ldots, X_d)$ have the df F. For each non-void $K \subset \{1, \ldots, d\}$ define the marginal survivor function

$$S_K(\boldsymbol{x}) = P\{X_k > x_k, \, k \in K\}. \qquad (4.5)$$

Applying the well-known inclusion-exclusion formula

$$P\left(\bigcup_{j \le d} A_j\right) = \sum_{j \le d} (-1)^{j+1} \sum_{|K|=j} P\left(\bigcap_{k \in K} A_k\right) \qquad (4.6)$$

to $A_j = \{X_j > x_j\}$, one obtains the decomposition

$$1 - F = \sum_{j \le d} (-1)^{j+1} \sum_{|K|=j} S_K \qquad (4.7)$$

which will be crucial for the subsequent considerations. Thus, to establish the limiting distribution of maxima we may deal with survivor functions.

Lemma 4.1.3. *Let F_n be a d-variate df with univariate margins F_{nj}.*

(a) *Assume that F_{nj}^n converges weakly to a max-stable df G_{0j} for $j \le d$.*

(b) *Let $S_{n,K}$ be the survivor function of F_n corresponding to (4.5). Assume that, for each non-void $K \subset \{1, \ldots, d\}$,*

$$n\,S_{n,K}(\boldsymbol{x}) \to L_K(\boldsymbol{x}), \qquad \boldsymbol{x} > \boldsymbol{\alpha}, \qquad n \to \infty,$$

where the L_K are right-continuous functions and $\boldsymbol{\alpha} = (\alpha(F_{0j}))_{j \le d}$.

Let

$$G_0(\boldsymbol{x}) = \exp\left(\sum_{j \le d}(-1)^j \sum_{|K|=j} L_K(\boldsymbol{x})\right), \qquad \boldsymbol{x} > \boldsymbol{\alpha},$$

and $G_0(\boldsymbol{x}) = 0$ otherwise. Then

(i) *G_0 is a df with univariate margins G_{0j};*

(ii) *for some universal constant $C > 0$,*

$$|F_n^n(\boldsymbol{x}) - G_0(\boldsymbol{x})| \le \sum_K |nS_{n,K}(\boldsymbol{x}) - L_K(\boldsymbol{x})| + C/n, \qquad \boldsymbol{x} > \boldsymbol{\alpha},$$

where the summation runs over all non-void $K \subset \{1,\ldots,d\}$.

Proof. To prove (i) apply a slight extension of (4.1). Moreover,

$$\sup_{\mathbf{x}} |F_n^n(\boldsymbol{x}) - \exp(-n(1 - F_n(\boldsymbol{x})))| \le C/n$$

for some universal constant $C > 0$, and $-n(1 - F_n(\boldsymbol{x}))$ can be replaced by

$$\sum_{j \le d} (-1)^j \sum_{|K| = j} nS_{n,K}$$

according to (4.7). Substituting $nS_{n,K}$ by L_K, the proof of (ii) can easily be completed. For a detailed proof we refer to Section 7.2 in [385].   □

From Lemma 4.1.3 we know that the convergence of the functions $nS_{n,K}$ implies the convergence of F_n^n. For a converse conclusion in the multivariate framework see Galambos [167], Theorem 5.3.1.

Lemma 4.1.3 was formulated in such a manner that a triangular scheme of rv can also be dealt with. In the following example, the initial normal df depends on n via a correlation matrix $\Sigma(n)$; the limiting df H_Λ of the sample maxima of a triangular array of normal rv is max-stable, a property which does not necessarily hold under the conditions of Lemma 4.1.3 (we refer to the discussion about max-infinitely divisible df in the subsequent section).

Example 4.1.4 (Hüsler-Reiss). The rich structure of the family of multivariate normal distributions can be carried over to max-stable distributions. Let $(X_{1n},\ldots,X_{dn})$ be a vector of standard normal rv with df $F_{\Sigma(n)}$, where $\Sigma(n) = (\rho_{ij}(n))_{i,j \le d}$ is a non-singular correlation matrix. Let b_n be the unique positive solution of the equation $x = n\varphi(x)$, $x \ge 0$, where φ denotes the standard normal density. Note that $b_n^2 \sim 2\log(n)$.

Assume that

$$\left(\left((1 - \rho_{ij}(n))\log(n)\right)^{1/2} \right)_{i,j \le d} \to \Lambda := (\lambda_{ij})_{i,j \le d}, \qquad n \to \infty,$$

where $\lambda_{ij} \in (0,\infty)$ for $1 \le i, j \le d$ with $i \ne j$. Then

$$F_{\Sigma(n)}^n \left((b_n + x_j/b_n)_{j \le d}\right) \to H_\Lambda(\boldsymbol{x}), \qquad \boldsymbol{x} \in \mathbb{R}^d, \quad n \to \infty,$$

where

$$H_\Lambda(\boldsymbol{x}) = \exp\left(\sum_{k \le d} (-1)^k \sum_{1 \le j_1 < \cdots < j_k \le d} L_{j_1,\ldots,j_k}(\boldsymbol{x}) \right)$$

with

$$L_{j_1,\ldots,j_k}(\boldsymbol{x}) = \int_{x_{j_k}}^{\infty} S\left(\left(x_{j_i} - z + 2\lambda_{j_i j_k}^2\right)_{i \leq k-1} \mid \Gamma_{j_1,\ldots,j_k}\right) e^{-z}\, dz$$

and $S\left(\cdot \mid \Gamma_{j_1,\ldots,j_k}\right)$ is the survivor function of a $(k-1)$-variate normal vector with mean vector zero and covariance matrix

$$\Gamma_{j_1,\ldots,j_k} = 2\left(\lambda_{j_l j_k}^2 + \lambda_{j_m j_k}^2 - \lambda_{j_l j_m}^2\right)_{l,m \leq k-1}.$$

As the univariate margins of $F_{\Sigma(n)}^n\left((b_n + x_j/b_n)_{j \leq d}\right)$ are appropriately standardized standard normal df, it is clear that the univariate margins of their limit H_λ are Gumbel df G_3. We give some details in the bivariate case and include the cases of total dependence and independence. If

$$\left((1 - \rho(n))\log(n)\right) \to \lambda^2, \qquad n \to \infty,$$

for some $\lambda \in [0, \infty]$, then

$$F_{\rho(n)}^n(b_n + x/b_n, b_n + y/b_n) \to H_\lambda(x, y), \qquad n \to \infty,$$

where

$$H_\lambda(x, y) = \exp\left(-\Phi\left(\lambda + \frac{x-y}{2\lambda}\right)e^{-y} - \Phi\left(\lambda + \frac{y-x}{2\lambda}\right)e^{-x}\right). \qquad (4.8)$$

For $\lambda = \infty$ and $\lambda = 0$ the asymptotic independence and total dependence holds in the limit. If $\rho \in (-1, 1)$ is fixed, then the asymptotic independence holds.

An alternative representation of the foregoing max-stable df may be found in Joe's [274] paper which provides a broad discussion of parametric families of extreme value df (EVD) and their statistical inference. Another notable article is Tiago de Oliveira [445].

4.2 Representations and Dependence Functions

In contrast to the univariate case, multivariate max-stable df form a non-parametric family. In this section, we introduce different representations of max-stable df such as the representation by means of the exponent measure, the de Haan-Resnick representation and a certain spectral representation. The Pickands representation will be studied separately in Section 4.3.

THE MAX-INFINITE DIVISIBILITY

Contemporary multivariate EVT is mainly based on a characterization of max-infinitely divisible (max-id) df, established by Balkema and Resnick [24] for bivariate df.

A bivariate F is said to be *max-id* iff for any $n \in \mathbb{N}$ there is a bivariate df F_n such that $F = F_n^n$.

The following characterization is quite convenient. It follows from the fact that if the df F is the weak limit of F_n^n, $n \in \mathbb{N}$, where F_n, $n \in \mathbb{N}$, is a sequence of df on $\mathbb{R}^2$, then F is max-id, see Theorem 1 in Balkema and Resnick [24] or Proposition 5.1 in Resnick [393].

Lemma 4.2.1. *F is max-id iff F^t is a df for all $t > 0$.*

The following example provides a construction of max-id df. Actually, it will turn out that it provides a characterization of max-id df, see Theorem 4.2.3 below.

Example 4.2.2. Let μ be a σ-finite measure on $[-\infty, \infty)^2$ and define H on $\mathbb{R}^2$ by

$$
\begin{aligned}
H(x, y) &:= \mu \left(([-\infty, x] \times [-\infty, y])^{\complement} \right) \\
&= \mu \left([-\infty, \infty)^2 \backslash ([-\infty, x] \times [-\infty, y]) \right).
\end{aligned}
\tag{4.9}
$$

Consider now a Poisson process N on $(0, \infty) \times [-\infty, \infty)^2$ with intensity measure $\lambda \times \mu$, where λ denotes Lebesgue measure on $(0, \infty)$ and $\times$ the product measure. Denote by $(T_k, (X_k, Y_k))$, $k \in \mathbb{N}$, the points of the Poisson process N, i.e., $N(\cdot) = \sum_{k \in \mathbb{N}} \varepsilon_{(T_k, (X_k, Y_k))}(\cdot)$, where $\varepsilon_{\mathbf{u}}$ denotes the point measure with mass 1 at $\mathbf{u}$.

Put, for $t > 0$,

$$
\mathbf{Z}(t) := \left(\sup_{k \in \mathbb{N}} \{X_k : T_k \le t\}, \sup_{k \in \mathbb{N}} \{Y_k : T_k \le t\} \right) \in [-\infty, \infty)^2,
$$

with the convention $\sup \emptyset := -\infty$. Then we have, for $(x, y) \in \mathbb{R}^2$,

$$
\begin{aligned}
P\left(\mathbf{Z}(t) \le (x, y) \right) &= P \left(N \left((0, t] \times ([-\infty, x] \times [-\infty, y])^{\complement} \right) = 0 \right) \\
&= \exp \left(-(\lambda \times \mu) \left((0, t] \times ([-\infty, x] \times [-\infty, y])^{\complement} \right) \right) \\
&= \exp(-tH(x, y)).
\end{aligned}
$$

Provided that $\mathbf{Z}(t)$ is $\mathbb{R}^2$-valued a.s., we obtain from Lemma 4.2.1 that $F(x, y) := \exp(-H(x, y))$ is a max-id df on $\mathbb{R}^2$. To ensure that $\mathbf{Z}(t)$ is in $\mathbb{R}^2$ a.s., it is necessary that

$$
H(x_0, y_0) < \infty \quad \text{for some } (x_0, y_0) \in \mathbb{R}^2
\tag{4.10}
$$

and

$$
\mu(\mathbb{R} \times [-\infty, \infty)) = \mu([-\infty, \infty) \times \mathbb{R}) = \infty.
\tag{4.11}
$$

A σ-finite measure μ on $[-\infty, \infty)^2$ is called an *exponent measure* of the df $F = \exp(-H)$, if the conditions (4.9), (4.10) and (4.11) hold.

Example 4.2.2 shows that a df F is max-id if it has an exponent measure. The following characterization, which is due to Balkema and Resnick [24], shows that the converse implication is also true. For a proof we refer to Balkema and Resnick [24], Theorem 3, or to Resnick [393], Proposition 5.8.

Theorem 4.2.3 (Balkema and Resnick). *A df F on $\mathbb{R}^2$ is max-id iff it has an exponent measure.*

An exponent measure is not necessarily unique; just set $\mu(\{-\infty\} \times \{-\infty\}) > 0$. One may wonder why an exponent measure is defined on $[-\infty, \infty)^2$, whereas $(-\infty, \infty)^2$ would seemingly be a more natural choice. Take, for instance,

$$F(x, y) = \exp(x + y), \qquad x, y \le 0, \tag{4.12}$$

i.e., F is the df of the iid rv X, Y with $P(X \le x) = P(Y \le x) = \exp(x)$, $x \le 0$. Then F is max-id since F^t is the df of $(X/t, Y/t)$ for any $t > 0$. But there is no measure μ, defined on $\mathbb{R}^2$, such that

$$F(x, y) = \exp\left(-\mu\left(((-\infty, x] \times (-\infty, y])^{\complement}\right)\right), \qquad x, y \in \mathbb{R}. \tag{4.13}$$

Such a measure μ would have to satisfy

$$\mu\left(((-\infty, x] \times (-\infty, y])^{\complement}\right) = -x - y, \qquad x, y \le 0,$$

and, thus,

$$\begin{aligned}
&\mu((x_1, y_1] \times (x_2, y_2]) \\
&= \mu\left(((-\infty, x_1] \times (-\infty, y_2])^{\complement}\right) + \mu\left(((-\infty, y_1] \times (-\infty, x_2])^{\complement}\right) \\
&\quad - \mu\left(((-\infty, x_1] \times (-\infty, x_2])^{\complement}\right) - \mu\left(((-\infty, y_1] \times (-\infty, y_2])^{\complement}\right) \\
&= 0
\end{aligned}$$

for $x_1 < y_1 \le 0$, $x_2 < y_2 \le 0$, i.e., μ is the null-measure on $(-\infty, 0]^2$. Since $1 = F(0, 0) = \exp\left(-\mu\left(((-\infty, 0]^2)^{\complement}\right)\right)$, we have $\mu\left(((-\infty, 0]^2)^{\complement}\right) = 0$ as well and, thus, μ is the null-measure on $\mathbb{R}^2$. But this contradicts equation (4.13).

Consider, on the other hand, the measure μ on $[-\infty, 0]^2$, defined by

$$\mu(\{-\infty\} \times (x, 0]) = \mu((x, 0] \times \{-\infty\}) = -x, \qquad x \le 0,$$

and $\mu\left((-\infty, 0]^2\right) = 0 = \mu(\{(-\infty, \infty)\})$. Then μ has its complete mass on the set $(\{-\infty\} \times (-\infty, 0]) \cup ((-\infty, 0] \times \{-\infty\})$, μ is σ-finite and satisfies

$$\mu\left(([-\infty, x] \times [-\infty, y])^{\complement}\right) = \mu(\{-\infty\} \times (y, 0]) + \mu((x, 0] \times \{-\infty\}) = -x - y.$$

The investigations in this section can be extended from the bivariate case to an arbitrary dimension, where a d-variate exponent measure is σ-finite on $[-\infty,\infty)^d$ and satisfies d-variate versions of conditions (4.9), (4.10) and (4.11), see Vatan [452].

The following result is, thus, an immediate consequence of Theorem 4.2.3 by repeating the arguments in Example 4.2.2.

Corollary 4.2.4. *Each d-variate max-id df F can be represented as the component-wise supremum of the points of a Poisson process N with intensity measure $\lambda \times \mu$, where μ is equal to the exponent measure.*

The de Haan-Resnick Representation

In this section we describe the particular approach to multivariate extreme value theory as developed by de Haan and Resnick [193].

Recall that a df G on $\mathbb{R}^d$ is called max-stable iff for every $n \in \mathbb{N}$ there exist constants $a_{nj} > 0$, $b_{nj} \in \mathbb{R}$, $j \le d$, such that

$$G^n(a_{nj}x_j + b_{nj}, \, j \le d) = G(\mathbf{x}), \qquad \mathbf{x} = (x_1, \ldots, x_d) \in \mathbb{R}^d.$$

The df G is max-id and, thus, by Theorem 4.2.3 (d-variate version) it has an exponent measure μ on $[-\infty,\infty)^d = [-\infty,\infty)$, i.e., μ is σ-finite and satisfies

$$\mu\left(\mathbb{R}^{i-1} \times [-\infty,\infty) \times \mathbb{R}^{d-i}\right) = \infty, \qquad i \le d,$$

$$\mu\left([-\infty,\mathbf{x}_0]^{\complement}\right) < \infty \text{ for some } \mathbf{x}_0 \in \mathbb{R}^d,$$

$$G(\mathbf{x}) = \exp\left(-\mu\left([-\infty,\mathbf{x}]^{\complement}\right)\right), \qquad \mathbf{x} \in \mathbb{R}^d. \tag{4.14}$$

A max-stable df G on $\mathbb{R}^d$ is called *simple*, if each marginal df is $\exp\left(-x^{-1}\right)$, $x > 0$. Its exponent measure ν can be chosen such that $\nu\left([\mathbf{0},\infty)^{\complement}\right) = 0$. This can be seen as follows. Let μ be an exponent measure of G, define $M : [-\infty,\infty) \to [\mathbf{0},\infty)$ by $M(x_1, \ldots, x_d) := (\max(x_i, 0), \, i \le d)$, and let $\nu := \mu * M$ be the measure induced by ν and M, i.e., $\nu(B) = \mu(M^{-1}(B))$ for any Borel subset of $[\mathbf{0},\infty)$. Then we have in particular

$$G(\mathbf{x}) = \exp\left(-\nu\left([\mathbf{0},\mathbf{x}]^{\complement}\right)\right), \qquad \mathbf{x} \in \mathbb{R}^d. \tag{4.15}$$

This equation is not affected if we remove the point $\mathbf{0}$ from $[\mathbf{0},\infty)$, and restrict the measure ν to the punctuated set $E := [\mathbf{0},\infty)\backslash\{\mathbf{0}\}$. Then ν is uniquely determined.

Consider a d-variate simple max-stable df G. We, thus, have, for any $n \in \mathbb{N}$ and $\mathbf{x} \in \mathbb{R}^d$,

$$G^n(n\mathbf{x}) = G(\mathbf{x}).$$

This yields

$$G^{n/m}((n/m)\mathbf{x}) = G(\mathbf{x}), \qquad n, m \in \mathbb{N}, \, \mathbf{x} \in \mathbb{R}^d.$$

Choose now n, m such that $n/m \to t > 0$. Then the continuity of G implies

$$G^t(t\mathbf{x}) = G(\mathbf{x}), \qquad t > 0, \ \mathbf{x} \in \mathbb{R}^d. \tag{4.16}$$

From (4.15) and (4.16) we obtain that the exponent measure ν pertaining to G satisfies, for any $\mathbf{x} \in E$ and any $t > 0$,

$$\nu\left([\mathbf{0}, \mathbf{x}]^{\complement}\right) = t\nu\left([\mathbf{0}, t\mathbf{x}]^{\complement}\right) = t\nu\left(t[\mathbf{0}, \mathbf{x}]^{\complement}\right).$$

This equation can readily be extended to hold for all rectangles contained in E. For a set $B \subset E$ we write $tB := \{t\mathbf{x} : \mathbf{x} \in B\}$. The equality

$$\nu(tB) = \frac{1}{t}\nu(B), \tag{4.17}$$

thus, holds on a generating class closed under intersections and is, therefore, true for any Borel subset B of E (c.f. Exercise 3.1.3 in Resnick [393]).

Denote by $\|\mathbf{x}\|$ an arbitrary norm of $\mathbf{x} \in \mathbb{R}^d$. From (4.17) we obtain, for any $t > 0$ and any Borel subset A of the unit sphere $S_E := \{\mathbf{z} \in E : \|\mathbf{z}\| = 1\}$ in E,

$$\nu\left(\left\{\mathbf{x} \in E : \|\mathbf{x}\| \geq t, \frac{\mathbf{x}}{\|\mathbf{x}\|} \in A\right\}\right)$$
$$= \nu\left(\left\{t\mathbf{y} \in E : \|\mathbf{y}\| \geq 1, \frac{\mathbf{y}}{\|\mathbf{y}\|} \in A\right\}\right)$$
$$= \frac{1}{t}\nu\left(\left\{\mathbf{y} \in E : \|\mathbf{y}\| \geq 1, \frac{\mathbf{y}}{\|\mathbf{y}\|} \in A\right\}\right)$$
$$=: \frac{1}{t}\phi(A) \tag{4.18}$$

where ϕ is an *angular measure*.

Define the one-to-one function $T : E \to (0, \infty) \times S_E$ by $T(\mathbf{x}) := (\|\mathbf{x}\|, \mathbf{x}/\|\mathbf{x}\|)$, which is the transformation of a vector onto its *polar coordinates* with respect to the norm $\|\cdot\|$. From (4.18) we obtain that the measure $(\nu * T)(B) := \nu(T^{-1}(B))$, induced by ν and T, satisfies

$$(\nu * T)([t, \infty) \times A) = \nu\left(\left\{\mathbf{x} \in E : \|\mathbf{x}\| \geq t, \frac{\mathbf{x}}{\|\mathbf{x}\|} \in A\right\}\right)$$
$$= \frac{1}{t}\phi(A)$$
$$= \int_A \int_{[t,\infty)} r^{-2}\, dr\, d\phi(\mathbf{a})$$
$$= \int_{[t,\infty) \times A} r^{-2}\, dr\, d\phi(\mathbf{a})$$

and, hence,

$$(\nu * T)(dr, d\mathbf{a}) = r^{-2}dr\, d\phi(\mathbf{a}), \qquad r > 0, \mathbf{a} \in S_E, \tag{4.19}$$

in short notation. The exponent measure factorizes, therefore, across radial and angular components.

We have $\nu([\mathbf{0}, \mathbf{x}]^c) = \nu([\mathbf{0}, \mathbf{x}]^c \cap E) = (\nu * T)(T([\mathbf{0}, \mathbf{x}]^c \cap E))$ and, with the notation $\mathbf{z} = (z_1, \ldots, z_d)$ for an arbitrary vector $\mathbf{z} \in \mathbb{R}^d$,

$$
\begin{aligned}
T([\mathbf{0}, \mathbf{x}]^c \cap E) &= T(\{\mathbf{y} \in E : y_i > x_i \text{ for some } i \le d\}) \\
&= \{(r, \mathbf{a}) \in (0, \infty) \times S_E : ra_i > x_i \text{ for some } i \le d\} \\
&= \left\{(r, \mathbf{a}) \in (0, \infty) \times S_E : r > \min_{i \le d} x_i/a_i\right\}
\end{aligned}
$$

with the temporary convention $0/0 = \infty$.

Hence, we obtain from equation (4.19)

$$
\begin{aligned}
\nu([\mathbf{0}, \mathbf{x}]^c) &= (\nu * T)(T([\mathbf{0}, \mathbf{x}]^c \cap E)) \\
&= (\nu * T)\left(\left\{(r, \mathbf{a}) \in (0, \infty) \times S_E : r > \min_{i \le d} x_i/a_i\right\}\right) \\
&= \int_{S_E} \int_{(\min_{i \le d} x_i/a_i, \infty)} r^{-2} \, dr \, d\phi(\mathbf{a}) \\
&= \int_{S_E} \frac{1}{\min_{i \le d}(x_i/a_i)} \, d\phi(\mathbf{a}) \\
&= \int_{S_E} \max_{i \le d} \left(\frac{a_i}{x_i}\right) d\phi(\mathbf{a}),
\end{aligned}
$$

now with the convention $0/0 = 0$ in the last line.

We have, thus, established the following crucial result, due to de Haan and Resnick [193].

Theorem 4.2.5 (De Haan-Resnick Representation). *Any simple max-stable df G can be represented as*

$$
G(\mathbf{x}) = \exp\left(-\int_{S_E} \max_{i \le d} \left(\frac{a_i}{x_i}\right) d\phi(\mathbf{a})\right), \qquad \mathbf{x} \in [\mathbf{0}, \infty), \qquad (4.20)
$$

where the angular measure ϕ on S_E is finite and satisfies

$$
\int_{S_E} a_i \, d\phi(\mathbf{a}) = 1, \qquad i \le d. \qquad (4.21)
$$

Note that (4.21) is an immediate consequence of the fact that the marginals of G are $\exp(x^{-1})$, $x > 0$. The finiteness of ϕ follows from (4.21) and the fact that all norms on $\mathbb{R}^d$ are equivalent:

$$
\begin{aligned}
d &= \int_{S_E} \sum_{i \le d} a_i \, d\phi(\mathbf{a}) \\
&\ge \int_{S_E} \left(\sum_{i \le d} a_i^2\right)^{1/2} d\phi(\mathbf{a})
\end{aligned}
$$

$$\geq \text{const} \int_{S_E} \|\mathbf{a}\| \, d\phi(\mathbf{a})$$

$$= \text{const} \, \phi(S_E).$$

The reverse implication of Theorem 4.2.5 is also true, starting with a finite measure ϕ which satisfies (4.21). This can be deduced by following the preceding arguments in reverse order. The above derivation of Theorem 4.2.3 is taken from Section 5.4.1 of Resnick [393].

A SPECTRAL REPRESENTATION

Recall the well-known fact that a univariate rv X with arbitrary distribution Q can be obtained by putting

$$X := F^{-1}(U),$$

where U is uniformly on $[0, 1]$ distributed and F is the df of Q. This *probability integral transform* can be extended to any probability measure Q on an arbitrary complete and separable metric space S, equipped with the Borel σ-field: There exists a random element $\mathbf{f}$ from the interval $[0, 1]$, equipped with the Lebesgue-measure λ, into S, such that

$$Q = \lambda * \mathbf{f}; \tag{4.22}$$

see Theorem 3.2 in Billingsley [45].

This extension can readily be applied to the de Haan-Resnick representation (4.20) of a simple max-stable df G as follows. Put

$$Q := \frac{\phi}{\phi(S_E)}.$$

Then we obtain from (4.22) that there exists a rv $\mathbf{f} = (f_1, \ldots, f_d)$ on $[0, 1]$ such that $Q = \lambda * (f_1, \ldots, f_d)$. This implies the *spectral representation*

$$
\begin{aligned}
G(\mathbf{x}) &= \exp\left(-\int_{S_E} \max_{i \leq d}\left(\frac{a_i}{x_i}\right) d\phi(\mathbf{a})\right) \\
&= \exp\left(-\phi(S_E) \int_{S_E} \max_{i \leq d}\left(\frac{u_i}{x_i}\right) dQ(\mathbf{u})\right) \\
&= \exp\left(-\phi(S_E) \int_{[0,1]} \max_{i \leq d}\left(\frac{f_i(u)}{x_i}\right) du\right) \\
&= \exp\left(-\int_{[0,1]} \max_{i \leq d}\left(\frac{\widetilde{f}_i(u)}{x_i}\right) du\right),
\end{aligned} \tag{4.23}
$$

where the non-negative functions $\widetilde{f}_i(u) := \phi(S_E) f_i(u)$ satisfy $\int_{[0,1]} \widetilde{f}_i(u) \, du = 1$, $i \leq d$; see Corollary 5.2.9 for an extension. By transforming the margins of G, one immediately obtains corresponding representations of EVD with negative exponential or Gumbel margins; just use the transformations $0 > y \mapsto x = -1/y$ or $\mathbb{R} \ni z \mapsto x = \exp(z)$. The spectral representation was extended to max-stable *stochastic processes* by de Haan [187] and de Haan and Pickands [192].

THE BIVARIATE CASE

Next we consider the bivariate case $d = 2$. It is shown that the exponent measure ν of a bivariate simple max-stable df G can be represented by a univariate measure generating function on $[0, \pi/2]$. Choose a norm $\|\cdot\|$ on $\mathbb{R}^2$ and denote by

$$A(\vartheta) := \{(u, v) \in S_E : 0 \leq \arctan(v/u) \leq \vartheta\}$$

the set of those vectors (u, v) in S_E whose dihedral angle is less than ϑ, i.e., $v/u \leq \tan(\vartheta)$, $\vartheta \in [0, \pi/2]$. Then we have for the corresponding exponent measure ν by equation (4.18)

$$\begin{aligned}
\nu\Big(\Big\{(x, y) \in [0, \infty)^2 : \|(x, y)\| \geq t, \arctan\Big(\frac{y}{x}\Big) \in [0, \vartheta]\Big\}\Big) \\
= \nu\Big(\Big\{(x, y) \in [0, \infty)^2 : \|(x, y)\| \geq t, \frac{(x, y)}{\|(x, y)\|} \in A(\vartheta)\Big\}\Big) \\
= t^{-1}\phi(A(\vartheta)) \\
=: t^{-1}\Phi(\vartheta), \qquad 0 \leq \vartheta \leq \pi/2.
\end{aligned} \qquad (4.24)$$

Φ is the measure generating function of a finite measure on $[0, \pi/2]$, which is called *angular measure* as well. ν is obviously determined by the univariate function Φ, which may again be regarded as a dependence function. The estimation of Φ was dealt with by Einmahl et al. [120] and Drees and Huang [115]. Let $R_i = \left(X_i^2 + Y_i^2\right)^{1/2}$ and $\Theta_i = \mathrm{arctg}(X_i/Y_i)$ be the polar coordinates of (X_i, Y_i). Denote by $R_{1:n} \leq \cdots \leq R_{n:n}$ the order statistics of the rv R_i. Estimation of Φ can be based on those (X_i, Y_i) in a sample such that $\Theta_i \geq r$ with r being sufficiently large. With $k(n) \to \infty$ and $k(n)/n \to 0$ as $n \to \infty$, the estimator is of the form

$$\Phi_n(\vartheta) = \frac{1}{k(n)} \sum_{i \leq n} 1_{\{R_{n-k(n)+1:n} \leq R_i\}} 1_{[0, \vartheta]}(\Theta_i).$$

4.3 Pickands Representation and Dependence Function

In what follows we will mainly work with the Pickands representation of max-stable df and the pertaining Pickands dependence function D, introduced now.

THE PICKANDS REPRESENTATION

Extending Lemma 4.1.2 one may characterize the family of max-stable df with univariate margins $G_{2,1}$. Consider next a d-variate max-stable df G^* with reverse standard exponential marginals $\exp(x)$, $x \le 0$. Then,

$$G(\mathbf{x}) := G^*\left(-\frac{1}{x_1}, \ldots, -\frac{1}{x_d}\right), \tag{4.25}$$

is a simple max-stable df and, consequently, the de Haan-Resnick representation (4.20) implies the representation

$$
\begin{aligned}
G^*(\mathbf{x}) &= G\left(-\frac{1}{x_1}, \ldots, -\frac{1}{x_d}\right) \\
&= \exp\left(-\int_{S_E} \max_{i \le d}(-a_i x_i)\, d\phi(\mathbf{a})\right) \\
&= \exp\left(\int_{S_E} \min_{i \le d}(a_i x_i)\, d\phi(\mathbf{a})\right), \qquad \mathbf{x} \le \mathbf{0}.
\end{aligned} \tag{4.26}
$$

By choosing the L_1-norm $\|\mathbf{a}\|_1 = \sum_{i \le d} |a_i|$ for $\|\mathbf{a}\|$, S_E is the unit simplex S and, by equation (4.21),

$$\int_S a_j\, d\phi(\mathbf{a}) = 1, \qquad j \le d.$$

Hence (4.26) is the *Pickands representation* of G^* as established in Pickands [372].

Theorem 4.3.1 (Pickands). *A function G is a max-stable, d-variate df and has univariate margins $G_{2,1}$ if, and only if,*

$$G(\boldsymbol{x}) = \exp\left(\int_S \min_{j \le d}(u_j x_j)\, d\mu(\boldsymbol{u})\right), \qquad \boldsymbol{x} < 0, \tag{4.27}$$

where μ is a finite measure on the d-variate unit simplex

$$S = \left\{\boldsymbol{u} : \sum_{j \le d} u_j = 1,\ u_j \ge 0\right\}$$

having the property

$$\int_S u_j\, d\mu(\boldsymbol{u}) = 1, \qquad j \le d. \tag{4.28}$$

Proof. We only have to prove the if-part. We prove that G is Δ-monotone by showing that G is the pointwise limit of df. According to Lemma 4.1.2, it suffices to find a_{ijn} such that

$$\sum_{i \le m(n)} \min_{j \le d}(a_{ijn} x_j) \to \int_S \min_{j \le d}(u_j x_j)\, d\mu(\boldsymbol{u}), \qquad \boldsymbol{x} < 0, \quad n \to \infty. \tag{4.29}$$

Let

$$\mu_n = \sum_{i \le m(n)} \mu_n \{\boldsymbol{u}_{in}\} \, \varepsilon_{\boldsymbol{u}_{in}}$$

with $\boldsymbol{u}_{in} = (u_{ijn})_{j \le d}$, $n \in \mathbb{N}$, be a sequence of discrete measures on S, vaguely converging to μ. Then we have

$$\int_S \min_{j \le d} (u_j x_j) \, d\mu_n(\boldsymbol{u}) = \sum_{i \le m(n)} \min_{j \le d} (u_{ijn} x_j) \, \mu_n \{\boldsymbol{u}_{in}\}$$

and the desired relation (4.29) holds with $a_{ijn} = u_{ijn} \mu_n \{\boldsymbol{u}_{in}\}$. $\qquad\qquad\square$

Note that (4.28) implies that $\mu(S) = d$:

$$1 = \int_S u_d \, \mu(d\mathbf{u}) = \int_S 1 - \sum_{i \le d-1} u_i \, \mu(d\mathbf{u}) = \mu(S) - (d-1).$$

Finally, it can easily be seen that G is normed and has univariate margins $G_{2,1}$. Moreover, verify $G\left((x_i/n)_{i \le d}\right)^n = G(\boldsymbol{x})$ to show the max-stability.

THE PICKANDS DEPENDENCE FUNCTION

From Theorem 4.3.1 we deduce that a d-variate max-stable df G with reverse exponential margins $G_{2,1}$ can be rewritten in terms of the Pickands dependence function $D : R \to [0, \infty)$ where the domain of D is given by

$$R := \left\{ (t_1, \ldots, t_{d-1}) \in [0, 1]^{d-1} : \sum_{i \le d-1} t_i \le 1 \right\}. \qquad (4.30)$$

For $\mathbf{x} = (x_1, \ldots, x_d) \in (-\infty, 0\,]^d$, $\mathbf{x} \ne \mathbf{0}$, we have

$$G(\mathbf{x}) = \exp\left(\int_S \min(u_1 x_1, \ldots, u_d x_d) \, d\mu(\mathbf{u}) \right)$$

$$= \exp\left((x_1 + \cdots + x_d) \int_S \max\left(u_1 \frac{x_1}{x_1 + \cdots + x_d}, \ldots, u_d \frac{x_d}{x_1 + \cdots + x_d} \right) d\mu(\mathbf{u}) \right)$$

$$= \exp\left((x_1 + \cdots + x_d) D\left(\frac{x_1}{x_1 + \cdots + x_d}, \ldots, \frac{x_{d-1}}{x_1 + \cdots + x_d} \right) \right), \qquad (4.31)$$

where μ is the measure in Theorem 4.3.1 and

$$D(t_1, \ldots, t_{d-1})$$

$$:= \int_S \max\left(u_1 t_1, \ldots, u_{d-1} t_{d-1}, u_d \left(1 - \sum_{i \le d-1} t_i \right) \right) d\mu(\mathbf{u}) \qquad (4.32)$$

is the *Pickands dependence function*.

If the rv $(X_1, \ldots, X_d)$ follows the max-stable df G, then the cases $D(\mathbf{t}) = 1$ and $D(\mathbf{t}) = \max(t_1, \ldots, t_{d-1}, 1 - \sum_{i \le d-1} t_i)$, $\mathbf{t} \in R$, characterize the cases of independence and complete dependence of the rv $X_1, \ldots, X_d$.

Important Properties of Pickands Dependence Functions

(i) The dependence function D is obviously continuous with

$$D(\mathbf{e}_i) = 1, \quad 1 \le i \le d-1,$$

where $\mathbf{e}_i = (0, \ldots, 0, 1, 0, \ldots, 0) \in \mathbb{R}^{d-1}$ is the i-th unit vector in $\mathbb{R}^{d-1}$. The latter property is immediate from (4.28). Moreover, $D(\mathbf{0}) = 1$ as well. The vectors $\mathbf{e}_i$, $1 \le i \le d-1$, and $\mathbf{0} \in \mathbb{R}^{d-1}$ are the extreme points of the convex set R in (4.30).

(ii) In addition, $D(\mathbf{t}) \le 1$ for any $\mathbf{t} = (t_1, \ldots, t_{d-1}) \in R$ because, according to (4.28),

$$D(\mathbf{t}) = \int_S \max\left(u_1 t_1, \ldots, u_{d-1} t_{d-1}, u_d \left(1 - \sum_{i \le d-1} t_i \right) \right) d\mu(\mathbf{u})$$

$$\le \int_S u_1 t_1 + \cdots + u_{d-1} t_{d-1} + u_d \left(1 - \sum_{i \le d-1} t_i \right) d\mu(\mathbf{u}) = 1.$$

(iii) The function D is convex, that is, for $\mathbf{v}_1, \mathbf{v}_2 \in R$ and $\lambda \in [0,1]$:

$$D(\lambda \mathbf{v}_1 + (1-\lambda)\mathbf{v}_2) \le \lambda D(\mathbf{v}_1) + (1-\lambda) D(\mathbf{v}_2).$$

Writing $\mathbf{v}_i = (v_{i,1}, \ldots, v_{i,d-1})$, $i = 1, 2$, the convexity of D is immediate from the inequality

$$\max\left(u_1(\lambda v_{1,1} + (1-\lambda)v_{2,1}), \ldots, u_{d-1}(\lambda v_{1,d-1} + (1-\lambda)v_{2,d-1}), \right.$$

$$\left. u_d \left(1 - \sum_{i \le d-1} (\lambda v_{1,i} + (1-\lambda)v_{2,i}) \right) \right)$$

$$\le \lambda \max\left(u_1 v_{1,1}, \ldots, u_{d-1} v_{1,d-1}, u_d \left(1 - \sum_{i \le d-1} v_{1,i} \right) \right)$$

$$+ (1-\lambda) \max\left(u_1 v_{2,1}, \ldots, u_{d-1} v_{2,d-1}, u_d \left(1 - \sum_{i \le d-1} v_{2,i} \right) \right)$$

for arbitrary $\mathbf{u} = (u_1, \ldots, u_d) \in S$.

(iv) If $D(\mathbf{t}) = 1$ for an inner point $\mathbf{t} \in R$, then D is the constant function 1. This is an immediate consequence of the fact that a convex function on a convex subset $\mathcal{U}$ of a normed linear space, which attains a global maximum at an inner point of $\mathcal{U}$, is a constant, see, e.g. Roberts and Varberg [396, Theorem C, Section 51].

(v) We have, for arbitrary $\mathbf{t} \in R$,

$$D(\mathbf{t}) \geq \max\left(t_1, \ldots, t_{d-1}, 1 - \sum_{i \leq d-1} t_i\right) \geq \frac{1}{d}. \tag{4.33}$$

The first inequality is immediate from (4.28), the second one follows by putting $t_1 = \cdots = t_{d-1} = 1/d$. The minimum of the function D is attained at $(1/d, \ldots, 1/d) \in \mathbb{R}^{d-1}$:

$$D(\mathbf{t}) \geq D\left(\frac{1}{d}, \ldots, \frac{1}{d}\right) = \frac{1}{d} \int_S \max(u_1, \ldots, u_d) \, d\mu(\mathbf{u}),$$

if D satisfies the symmetry condition

$$D(t_1, \ldots, t_{d-1}) = D(s_1, \ldots, s_{d-1}) \tag{4.34}$$

for any subset $\{s_1, \ldots, s_{d-1}\}$ of $\{t_1, \ldots, t_d\}$, where $t_d := 1 - \sum_{i \leq d-1} t_i$. This follows from the inequality

$$\max(u_1, \ldots, u_d) = \max\left(u_1 \sum_{i \leq d} t_i, \ldots, u_d \sum_{i \leq d} t_i\right)$$
$$\leq \max(u_1 t_1, \ldots, u_d t_d) + \max(u_1 t_d, u_2 t_1, u_3 t_2, \ldots, u_d t_{d-1})$$
$$+ \max(u_1 t_{d-1}, u_2 t_d, u_3 t_1, \ldots, u_d t_{d-2}) + \cdots$$
$$+ \max(u_1 t_2, u_2 t_3, \ldots, u_{d-1} t_d, u_d t_1)$$

and, hence,

$$D\left(\frac{1}{d}, \ldots, \frac{1}{d}\right) \leq \frac{1}{d} D(t_1, \ldots, t_{d-1}) + \frac{1}{d} D(t_d, t_1, \ldots, t_{d-2})$$
$$+ \frac{1}{d} D(t_{d-1}, t_d, t_1, \ldots, t_{d-3}) + \cdots + \frac{1}{d} D(t_2, t_3, \ldots, t_d)$$
$$= D(t_1, \ldots, t_{d-1}).$$

Without the symmetry condition (4.34) on D, its minimum is not necessarily attained at $(1/d, \ldots, 1/d) \in \mathbb{R}^{d-1}$; a counterexample is given in (6.16).

(vi) If $D(1/d, \ldots, 1/d) = 1/d$ for an arbitrary dependence function D, then $D(\mathbf{t}) = \max(t_1, \ldots, t_d)$, $\mathbf{t} \in R$, where $t_d = 1 - \sum_{i \leq d-1} t_i$. This can easily be seen as follows. The equation

$$D\left(\frac{1}{d}, \ldots, \frac{1}{d}\right) = \frac{1}{d} \int_S \max(u_1, \ldots, u_d) \, d\mu(\mathbf{u}) = \frac{1}{d}$$

implies $\int_S \max(u_1, \ldots, u_d) \, d\mu(\mathbf{u}) = 1$ and, thus, for $i, j \leq d$,

$$
\begin{aligned}
0 &= \int_S \max(u_1, \ldots, u_d) - u_i \, d\mu(\mathbf{u}) \\
&\geq \int_S (u_j - u_i) 1_{\{u_j > u_i\}} \, d\mu(\mathbf{u}) \\
&\geq 0.
\end{aligned}
$$

This yields

$$
\begin{aligned}
\mu(\{\mathbf{u} \in S : u_j > u_i\}) &= 0, \qquad i, j \leq d, \\
\Longrightarrow \ \mu(\{\mathbf{u} \in S : u_i \neq u_j \text{ for some } i, j \leq d\}) &= 0 \\
\Longrightarrow \ \mu(\{\mathbf{u} \in S : u_1 = \cdots = u_d\}) &= \mu(S) = d
\end{aligned}
$$

and, hence,

$$
\begin{aligned}
D(\mathbf{t}) &= \int_S \max(u_1 t_1, \ldots, u_d t_d) \, d\mu(\mathbf{u}) \\
&= \int_{\{\mathbf{u} \in S : u_1 = \cdots = u_d\}} \max(u_1 t_1, \ldots, u_d t_d) \, d\mu(\mathbf{u}) \\
&= \max(t_1, \ldots, t_d) \int_S u_1 \, d\mu(\mathbf{u}) \\
&= \max(t_1, \ldots, t_d).
\end{aligned}
$$

(vii) Note that the symmetry condition (4.34) on D is equivalent to the condition that $X_1, \ldots, X_d$ are exchangeable, i.e., the distribution of $(X_{i_1}, \ldots, X_{i_d})$ is again G for any permutation $(i_1, \ldots, i_d)$ of $(1, \ldots, d)$.

(viii) The convex combination $D(\mathbf{t}) = (1 - \lambda)D_1(\mathbf{t}) + \lambda D_2(\mathbf{t})$ of two dependence functions D_1, D_2 is again a dependence function, $\lambda \in [0, 1]$. This is immediate by putting $\mu := (1 - \lambda)\mu_1 + \lambda\mu_2$, where μ_1, μ_2 are the measures on the simplex S corresponding to D_1 and D_2 in the definition (4.32) of a dependence function. The dependence function D is, thus, generated by μ. The functions $D(\mathbf{t}) = 1$ and $D(\mathbf{t}) = \max(t_1, \ldots, t_{d-1}, 1 - \sum_{i \leq d-1} t_i)$ are now extreme points of the convex set of all dependence functions.

(ix) The copula C of the EVD G with dependence function D is

$$
\begin{aligned}
C(\mathbf{u}) &= G(\log(u_1), \ldots, \log(u_d)) \\
&= \exp\left(\left(\sum_{i \leq d} \log(u_i) \right) D\left(\frac{\log(u_1)}{\sum_{i \leq d} \log(u_i)}, \ldots, \frac{\log(u_{d-1})}{\sum_{i \leq d} \log(u_i)} \right) \right)
\end{aligned}
$$

$$= \left(\prod_{i \le d} u_i \right)^{D\left(\frac{\log(u_1)}{\log\left(\prod_{i \le d} u_i\right)}, \dots, \frac{\log(u_{d-1})}{\log\left(\prod_{i \le d} u_i\right)} \right)}, \qquad \mathbf{u} \in (0,1)^d.$$

This copula, obviously, satisfies, for any $\lambda > 0$,

$$C\left(\mathbf{u}^\lambda\right) = C(\mathbf{u})^\lambda.$$

The case $D = 1$ yields the independence copula

$$C(\mathbf{u}) = \prod_{i \le d} u_i,$$

whereas $D(\mathbf{t}) = \max\left(t_1, \dots, t_{d-1}, 1 - \sum_{i \le d} t_i\right)$ yields the total dependence copula

$$C(\mathbf{u}) = \left(\prod_{i \le d} u_i \right)^{\frac{\min(\log(u_1), \dots, \log(u_d))}{\log\left(\prod_{i \le d} u_i\right)}}.$$

Recall that C is for each dependence function D a df on $(0,1)^d$ with uniform margins.

Properties (iv) and (vi) in the preceding list immediately imply the following characterization of independence and total dependence of the univariate margins of a multivariate EVD, which is due to Takahashi [438]. Note that an arbitrary multivariate EVD can be transformed to an EVD with standard negative exponential margins by just transforming the margins, see equation (5.47).

Theorem 4.3.2 (Takahashi). *Let G be an arbitrary d-dimensional EVD with margins G_j, $j \le d$. We have*

(i) $G(\mathbf{x}) = \prod_{j \le d} G_j(x_j)$ *for each* $\mathbf{x} = (x_1, \dots, x_d) \in \mathbb{R}^d$ *iff there exists one* $\mathbf{y} = (y_1, \dots, y_d) \in \mathbb{R}^d$ *with* $0 < G_j(y_j) < 1$, $j \le d$, *such that* $G(\mathbf{y}) = \prod_{j \le d} G_j(y_j)$.

(ii) $G(\mathbf{x}) = \min_{j \le d} G_j(x_j)$ *for each* $\mathbf{x} = (x_1, \dots, x_d) \in \mathbb{R}^d$ *iff there exists one* $\mathbf{y} = (y_1, \dots, y_d) \in \mathbb{R}^d$ *with* $0 < G_j(y_j) < 1$, $j \le d$, *such that* $G(\mathbf{y}) = G_1(y_1) = \dots = G_d(y_d)$.

The following result supplements the preceding one. It entails in particular that bivariate independence of the margins of a multivariate EVD is equivalent to complete independence of the margins.

Theorem 4.3.3. *Let G be an arbitrary d-dimensional EVD with one-dimensional margins G_j, $j \le d$. Suppose that for each bivariate margin $G_{(i,j)}$ of G there exists $\mathbf{y}_{(i,j)} = (y_{(i,j),1}, y_{(i,j),2}) \in \mathbb{R}^2$ with $0 < G_i(y_{(i,j),1}), G_j(y_{(i,j),2}) < 1$ such that $G_{(i,j)}(\mathbf{y}_{(i,j)}) = G_i(y_{(i,j),1})G_j(y_{(i,j),2})$. Then the margins of G are independent, i.e., $G(\mathbf{y}) = \prod_{j \le d} G_j(y_j)$, $\mathbf{y} \in \mathbb{R}^d$.*

Proof. Without loss of generality we can assume that G has standard negative exponential margins $G_j(x) = \exp(x)$, $x \leq 0$, and, thus, there exists a measure μ on $S = \left\{ \mathbf{u} \in [0,1]^d : \sum_{i \leq d} u_i = 1 \right\}$ with $\mu(S) = d$, $\int_S u_i \, d\mu(\mathbf{u}) = 1$, $i \leq d$, such that

$$G(\mathbf{x}) = \exp \left(\left(\sum_{i \leq d} x_i \right) \int_S \max_{j \leq d} \left(u_j \frac{x_j}{\sum_{i \leq d} x_i} \right) d\mu(\mathbf{u}) \right), \qquad \mathbf{x} \leq \mathbf{0} \in \mathbb{R}^d.$$

Since $G_{(i,j)}$ is a bivariate EVD with margins G_i, G_j, Takahashi's Theorem 4.3.2 implies that G_i and G_j are independent for each $1 \leq i, j \leq d$, $i \neq j$, i.e., $G_{(i,j)}(\mathbf{x}) = \exp(x_1 + x_2)$, $\mathbf{x} = (x_1, x_2) \leq \mathbf{0} \in \mathbb{R}^2$, and, therefore,

$$\int_S \max(u_i t, u_j(1-t)) \, d\mu(\mathbf{u}) = 1, \qquad t \in [0,1],\ 1 \leq i, j \leq d,\ i \neq j. \qquad (4.35)$$

This implies

$$1 = \int_S \max(u_i t, u_j(1-t)) \, d\mu(\mathbf{u}) \leq \int_S u_i t + u_j(1-t) \, d\mu(\mathbf{u}) = 1.$$

Putting $t = 1/2$, we obtain, for $1 \leq i, j \leq d$, $i \neq j$,

$$\int_S u_i + u_j - \max(u_i, u_j) \, d\mu(\mathbf{u}) = 0,$$

where the integrand is non-negative. This implies

$$\mu\left(\{\mathbf{u} \in S : u_i + u_j > \max(u_i, u_j)\}\right) = 0, \qquad 1 \leq i, j \leq d,\ i \neq j,$$

and, thus,

$$0 = \mu\left(\{\mathbf{u} \in S : u_i + u_j > \max(u_i, u_j) \text{ for some } 1 \leq i, j \leq d\}\right)$$
$$= \mu\left(\left(\cup_{i \leq d} \{\mathbf{e}_i\} \right)^{\complement} \right),$$

where $\mathbf{e}_i$ denotes the i-th unit vector in $\mathbb{R}^d$, yielding $\mu\left(\cup_{i \leq d} \{\mathbf{e}_i\} \right) = d$. Putting $t = 1/2$ in equation (4.35), we obtain

$$2 = \int_S \max(u_i, u_j) \, d\mu(\mathbf{u})$$
$$= \int_{\{\mathbf{e}_i, \mathbf{e}_j\}} \max(u_i, u_j) \, d\mu(\mathbf{u})$$
$$= \mu(\{\mathbf{e}_i\}) + \mu(\{\mathbf{e}_j\}), \qquad 1 \leq i, j \leq d,\ i \neq j.$$

But this implies $\mu(\{\mathbf{e}_i\}) = 1$, $i \leq d$, and, thus,

$$G(\mathbf{x}) = \exp \left(\sum_{i \leq d} x_i \right), \qquad \mathbf{x} \leq \mathbf{0} \in \mathbb{R}^d. \qquad \square$$

EXAMPLES OF PICKANDS DEPENDENCE FUNCTIONS

We discuss in detail two max-stable df in $\mathbb{R}^d$, namely, the Marshall-Olkin and the logistic (also known as negative logistic or Gumbel) df.

Example 4.3.4 (Marshall-Olkin df in $\mathbb{R}^d$). This is an extension of Example 4.1.1 and a special case of Example 4.1.2.

Let $Z_0, \ldots, Z_d$ be iid rv with common standard reverse exponential df. Put, for $\lambda \in (0, 1)$,

$$X_j := \max\left(\frac{Z_j}{1-\lambda}, \frac{Z_0}{\lambda}\right), \quad 1 \leq j \leq d.$$

Then $(X_1, \ldots, X_d)$ has a max-stable df with reverse exponential margins. Precisely, we have for $x_j \leq 0$, $1 \leq j \leq d$,

$$\begin{aligned} &P(X_j \leq x_j, 1 \leq j \leq d) \\ &= \exp\left((1-\lambda)(x_1 + \cdots + x_d) + \lambda\min(x_1, \ldots, x_d)\right) \\ &=: G_\lambda(x_1, \ldots, x_d). \end{aligned}$$

The df G_λ is a Marshall-Olkin [323] df in $\mathbb{R}^d$ with parameter $\lambda \in [0, 1]$, where $\lambda = 0$ is the case of independence of the margins, and $\lambda = 1$ is the case of complete dependence. The corresponding dependence function is

$$D_\lambda(t_1, \ldots, t_{d-1})$$

$$= 1 - \lambda + \lambda\max\left(t_1, \ldots, t_{d-1}, 1 - \sum_{i \leq d-1} t_i\right)$$

$$= 1 - \lambda\min\left(1 - t_1, \ldots, 1 - t_{d-1}, \sum_{i \leq d-1} t_i\right), \quad \mathbf{t} = (t_1, \ldots, t_{d-1}) \in R.$$

If we take $D_1(\mathbf{t}) = 1$ and $D_2(\mathbf{t}) = \max(t_1, \ldots, t_{d-1}, 1 - \sum_{i \leq d-1} t_i)$, then the convex combination $(1 - \lambda)D_1(\mathbf{t}) + \lambda D_2(\mathbf{t})$ of the two extremal cases of independence and complete dependence in the EVD model is just a Marshall-Olkin dependence function with parameter $\lambda \in [0, 1]$:

$$1 - \lambda + \lambda\max\left(t_1, \ldots, t_{d-1}, 1 - \sum_{i \leq d-1} t_i\right) = D_\lambda(\mathbf{t}).$$

Example 4.3.5 (Logistic df in $\mathbb{R}^d$). The logistic df ([183]), alternatively called negative logistic df or Gumbel df of type B, is defined by

$$G_\lambda(\mathbf{x}) = \exp\left(-\left(\sum_{i \leq d}(-x_i)^\lambda\right)^{1/\lambda}\right), \quad \lambda \geq 1.$$

It is max-stable with reverse exponential margins and has the dependence function

$$D_\lambda(t_1,\ldots,t_{d-1}) = \left(t_1^\lambda + \cdots + t_{d-1}^\lambda + \left(1 - \sum_{i \leq d-1} t_i\right)^\lambda \right)^{1/\lambda}.$$

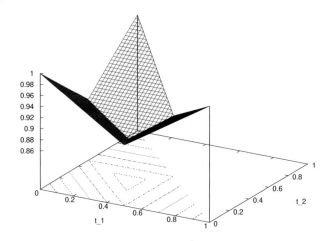

FIGURE 4.3.1. Dependence function $D(t_1,t_2) = 1 - \lambda\min(1 - t_1, 1 - t_2, t_1 + t_2)$ of the Marshall-Olkin df with $\lambda = .2$ in dimension $d = 3$, cf. Example 4.3.4.

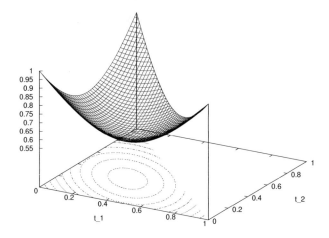

FIGURE 4.3.2. Dependence function $D(t_1,t_2) = (t_1^\lambda + t_2^\lambda + (1 - t_1 - t_2)^\lambda)^{1/\lambda}$ of the logistic df with $\lambda = 2$ in dimension $d = 3$, cf. Example 4.3.5.

We mention another df with independent margins in the limiting EVD.

Example 4.3.6 (Mardia's df in $\mathbb{R}^d$). Mardia's [318] df

$$H(\mathbf{x}) := \frac{1}{\sum_{i \leq d} \exp(-x_i) - (d - 1)}, \quad x_i \leq 0, \, 1 \leq i \leq d,$$

has reverse exponential margins, but it is not max-stable. Precisely, it satisfies

$$H^n\left(\frac{\mathbf{x}}{n}\right) = \frac{1}{\left(1 + \sum_{i \leq d}\left(\exp\left(-\frac{x_i}{n}\right) - 1\right)\right)^n} \longrightarrow_{n \to \infty} \exp\left(\sum_{i \leq d} x_i\right)$$

which is the EVD with independent reverse exponential margins.

4.4 The *D*-Norm

In this section we introduce quite a convenient representation of a d-dimensional EVD (and later a GPD) in terms of a norm on $\mathbb{R}^d$, called the D-norm.

The logistic distribution in Example 4.3.5 can obviously be written as

$$G_\lambda(\mathbf{x}) = \exp(-||\mathbf{x}||_\lambda),$$

where

$$||\mathbf{x}||_\lambda = \left(\sum_{i \leq d} |x_i|^\lambda\right)^{1/\lambda}$$

denotes the usual L_λ-norm on $\mathbb{R}^d$, $1 \leq \lambda \leq \infty$. Actually, it turns out that *any* EVD in (4.31) can be written as

$$G(\mathbf{x}) = \exp\left(\left(\sum_{i \leq d} x_i\right) D\left(\frac{x_1}{\sum_{i \leq d} x_i}, \ldots, \frac{x_{d-1}}{\sum_{i \leq d} x_i}\right)\right) = \exp\left(-||\mathbf{x}||_D\right),$$

where

$$||\mathbf{x}||_D := \left(\sum_{i \leq d} |x_i|\right) D\left(\frac{|x_1|}{\sum_{i \leq d} |x_i|}, \ldots, \frac{|x_{d-1}|}{\sum_{i \leq d} |x_i|}\right), \tag{4.36}$$

with the convention $||\mathbf{0}||_D = 0$. In fact, $|| \cdot ||_D$ defines a norm on $\mathbb{R}^d$, that is, for any $\mathbf{x}, \mathbf{y} \in \mathbb{R}^d$:

(i) $||\mathbf{x}||_D = 0 \Longleftrightarrow \mathbf{x} = \mathbf{0}$,

(ii) $||c\mathbf{x}||_D = |c| \, ||\mathbf{x}||_D \qquad$ for any $c \in \mathbb{R}$,

(iii) $||\mathbf{x} + \mathbf{y}||_D \leq ||\mathbf{x}||_D + ||\mathbf{y}||_D$.

The triangle inequality follows from the definition of D in (4.32):

$$\|\mathbf{x} + \mathbf{y}\|_D$$

$$= \left(\sum_{i \leq d} |x_i + y_i| \right) D \left(\frac{|x_1 + y_1|}{\sum_{i \leq d} |x_i + y_i|}, \ldots, \frac{|x_{d-1} + y_{d-1}|}{\sum_{i \leq d} |x_i + y_i|} \right)$$

$$= \left(\sum_{i \leq d} |x_i + y_i| \right) \int_S \max \left(u_1 \frac{|x_1 + y_1|}{\sum_{i \leq d} |x_i + y_i|}, \ldots, u_d \frac{|x_d + y_d|}{\sum_{i \leq d} |x_i + y_i|} \right) d\mu(\mathbf{u})$$

$$= \int_S \max \left(u_1 |x_1 + y_1|, \ldots, u_d |x_d + y_d| \right) d\mu(\mathbf{u})$$

$$\leq \int_S \max \left(u_1 |x_1|, \ldots, u_d |x_d| \right) d\mu(\mathbf{u})$$

$$\quad + \int_S \max \left(u_1 |y_1|, \ldots, u_d |y_d| \right) d\mu(\mathbf{u})$$

$$= \left(\sum_{i \leq d} |x_i| \right) \int_S \max \left(u_1 \frac{|x_1|}{\sum_{i \leq d} |x_i|}, \ldots, u_d \frac{|x_d|}{\sum_{i \leq d} |x_i|} \right) d\mu(\mathbf{u})$$

$$\quad + \left(\sum_{i \leq d} |y_i| \right) \int_S \max \left(u_1 \frac{|y_1|}{\sum_{i \leq d} |y_i|}, \ldots, u_d \frac{|y_d|}{\sum_{i \leq d} |y_i|} \right) d\mu(\mathbf{u})$$

$$= \|\mathbf{x}\|_D + \|\mathbf{y}\|_D.$$

We call $\|\mathbf{x}\|_D$ the *D-norm* on $\mathbb{R}^d$. From the inequalities

$$\max \left(t_1, \ldots, t_{d-1}, 1 - \sum_{i \leq d-1} t_i \right) \leq D(\mathbf{t}) \leq 1$$

for $\mathbf{t} = (t_1, \ldots, t_{d-1}) \in R$ we obtain for any $\mathbf{x} \in \mathbb{R}^d$ that its D-norm is always between the ∞-norm and the L_1-norm:

$$\|\mathbf{x}\|_\infty := \max(|x_1|, \ldots, |x_d|) \ \leq \ \|\mathbf{x}\|_D \ \leq \ \sum_{i \leq d} |x_i| =: \|\mathbf{x}\|_1. \qquad (4.37)$$

This, in turn, implies in particular that any EVD G with dependence function D satisfies the inequalities

$$\prod_{i \leq d} \exp(x_i) \leq G(\mathbf{x}) = \exp \left(-\|\mathbf{x}\|_D \right) \leq \exp \left(\min \left(x_1, \ldots, x_d \right) \right),$$

for $\mathbf{x} = (x_1, \ldots, x_d) \in (-\infty, 0]^d$, which is (4.1).

The inequalities (4.37) imply that a D-norm is *standardized*, i.e.,

$$\|\mathbf{e}_i\|_D = 1, \qquad i \leq d,$$

where $\mathbf{e}_i = (0, \ldots, 0, 1, 0, \ldots, 0)$ denotes the i-th unit vector in $\mathbb{R}^d$, $i \le d$.

The monotonicity of G implies further, for arbitrary $\mathbf{x} \le \mathbf{y} \le \mathbf{0}$,

$$\|\mathbf{x}\|_D \ge \|\mathbf{y}\|_D . \tag{4.38}$$

A norm $\|\cdot\|$ with this property will be called *monotone*. Note that this is equivalent with

$$\|\mathbf{x}\| \le \|\mathbf{y}\| , \qquad \mathbf{0} \le \mathbf{x} \le \mathbf{y}.$$

Takahashi's Theorem 4.3.2, for instance, can now be formulated as follows. Recall that the L_1-norm characterizes independence of the margins, whereas the ∞-norm characterizes complete dependence of the margins.

Theorem 4.4.1 (Takahashi). *We have the following equivalences:*

(i) $\|\cdot\|_D = \|\cdot\|_1 \iff \|\mathbf{y}\|_D = \|\mathbf{y}\|_1$ *for at least one* $\mathbf{y} \in \mathbb{R}^d$, *whose components are all different from 0;*

(ii) $\|\cdot\|_D = \|\cdot\|_\infty \iff \|(1, \ldots, 1)\|_D = \left\|\sum_{i \le d} \mathbf{e}_i\right\| = 1.$

In view of Takahashi's Theorem, the number

$$\varepsilon := \|(1, \ldots, 1)\|_D \in [1, d]$$

clearly measures the dependence structure of the margins of G, and we have in particular

$$\varepsilon = 1 \iff \|\cdot\|_D = \|\cdot\|_\infty \iff \text{complete dependence of the margins}$$

and

$$\varepsilon = d \iff \|\cdot\|_D = \|\cdot\|_1 \iff \text{independence of the margins}.$$

The number ε was introduced by Smith [421] as the *extremal coefficient*, defined as that constant which satisfies

$$G^*(x, \ldots, x) = F^\varepsilon(x), \qquad x \in \mathbb{R},$$

where G^* is an *arbitrary* d-dimensional EVD with identical margins $G_j^* = F$, $j \le d$. Without loss of generality we can transform the margins to the negative exponential distribution and obtain an EVD G with standard negative exponential margins and, thus,

$$G(x, \ldots, x) = \exp\left(x \left\|\sum_{i \le d} \mathbf{e}_i\right\|_D\right) = \exp(\varepsilon x), \qquad x \le 0,$$

yielding

$$\varepsilon = \|(1, \ldots, 1)\|_D = \left\|\sum_{i \le d} \mathbf{e}_i\right\|_D .$$

The obvious question *"When is an arbitrary norm $\|\cdot\|$ a D-norm?"* is answered by the following result due to Hofmann [222]. It is established by showing that

$$\nu([-\infty, \mathbf{x}]^{\complement}) := \|\mathbf{x}\|, \qquad \mathbf{x} \leq \mathbf{0},$$

defines an exponent measure on $[-\infty, \mathbf{0}] \setminus \{-\infty\}$ if and only if the norm $\|\cdot\|$ satisfies condition (4.39), which is essentially the Δ-monotonicity of a df.

Theorem 4.4.2 (Hofmann, 2009). *Let $\|\cdot\|$ be an arbitrary norm on $\mathbb{R}^d$. Then the function $G(\mathbf{x}) := \exp(-\|\mathbf{x}\|)$, $\mathbf{x} \leq \mathbf{0}$, defines a multivariate EVD if and only if the norm satisfies*

$$\sum_{\mathbf{m} \in \{0,1\}^d \,:\, m_i = 1,\, i \in K} (-1)^{d+1 - \sum_{j \leq d} m_j} \left\| \left(b_1^{m_1} a_1^{1-m_1}, \dots, b_d^{m_d} a_d^{1-m_d} \right) \right\| \geq 0 \quad (4.39)$$

for every non-empty $K \subset \{1, \dots, d\}$, $K \neq \{1, \dots, d\}$, and $-\infty < a_j \leq b_j \leq 0$, $1 \leq j \leq d$.

Note that the EVD $G(\mathbf{x}) = \exp(-\|\mathbf{x}\|)$, $\mathbf{x} \leq \mathbf{0}$, has *standard* reverse exponential margins if and only if $\|\mathbf{e}_i\| = 1$, $i \leq d$, i.e., if and only if the norm is standardized. In this case, the norm $\|\cdot\|$ is a D-norm.

The Bivariate Case

Putting $K = \{1\}$ and $K = \{2\}$, condition (4.39) reduces in the case $d = 2$ to

$$\|(b_1, b_2)\| \leq \min\left(\|(b_1, a_2)\|, \|(a_1, b_2)\|\right), \qquad \mathbf{a} \leq \mathbf{b} \leq \mathbf{0},$$

which, in turn, is obviously equivalent with

$$\|\mathbf{b}\| \leq \|\mathbf{a}\|, \qquad \mathbf{a} \leq \mathbf{b} \leq \mathbf{0},$$

i.e., the monotonicity of $\|\cdot\|$.

We, thus, obtain from Theorem 4.4.2 the following characterization in the bivariate case.

Lemma 4.4.3. *Take an arbitrary norm $\|\cdot\|$ on $\mathbb{R}^2$. Then*

$$G(\mathbf{x}) = \exp(-\|\mathbf{x}\|), \qquad \mathbf{x} \leq \mathbf{0},$$

defines an EVD in $\mathbb{R}^2$ if and only if the norm $\|\cdot\|$ is monotone.

The following lemma will be crucial for the characterization of a Pickands dependence function in the bivariate case. Together with the preceding lemma it entails, moreover, that in the bivariate case $G(\mathbf{x}) = \exp(-\|\mathbf{x}\|)$, $\mathbf{x} \leq \mathbf{0} \in \mathbb{R}^2$, defines an EVD with standard negative exponential margins, if and only if the norm $\|\cdot\|$ satisfies $\|\mathbf{x}\|_\infty \leq \|\mathbf{x}\| \leq \|\mathbf{x}\|_1$, $\mathbf{x} \geq \mathbf{0}$.

Lemma 4.4.4. *Let* $\|\cdot\|$ *be a norm on* $\mathbb{R}^d$. *If* $\|\cdot\|$ *is monotone and standardized, then we have, for* $\mathbf{0} \le \mathbf{x} \in \mathbb{R}^d$,

$$\|\mathbf{x}\|_\infty \le \|\mathbf{x}\| \le \|\mathbf{x}\|_1 .$$

For $d = 2$ *the converse statement is also true.*

Example 2.19 in Hofmann [221] shows that the preceding equivalence in $\mathbb{R}^2$ is not true for a general dimension d.

Proof. Let $\mathbf{0} \le \mathbf{x} = (x_1, \dots, x_d) \in \mathbb{R}^d$. Since the norm is standardized, we have by the triangle inequality

$$\|(x_1, \dots, x_d)\| \le \|(x_1, 0, \dots, 0)\| + \dots + \|(0, \dots, 0, x_d)\|$$
$$= x_1 + \dots + x_d$$
$$= \|(x_1, \dots, x_d)\|_1 .$$

Furthermore we obtain from the monotonicity of $\|\cdot\|$

$$\|(x_1, \dots, x_d)\| \ge \|(0, \dots, 0, x_i, 0 \dots, 0)\|$$
$$= x_i \|\mathbf{e}_i\|$$
$$= x_i, \qquad i \le d,$$

and, thus, $\|\mathbf{x}\| \ge \max(x_1, \dots, x_d) = \|\mathbf{x}\|_\infty$. Altogether we have $\|\mathbf{x}\|_\infty \le \|\mathbf{x}\| \le \|\mathbf{x}\|_1$.

Now let $d = 2$ and suppose that the norm satisfies $\|\mathbf{x}\|_\infty \le \|\mathbf{x}\| \le \|\mathbf{x}\|_1$ for $\mathbf{0} \le \mathbf{x}$. Then we have

$$1 = \|\mathbf{e}_i\|_\infty \le \|\mathbf{e}_i\| \le \|\mathbf{e}_i\|_1 = 1$$

and, thus, the norm is standardized.

Take $\mathbf{a} = (a_1, a_2)$, $\mathbf{b} = (b_1, b_2) \in \mathbb{R}^2$ with $\mathbf{0} \le \mathbf{a} \le \mathbf{b}$ and $\mathbf{0} < \mathbf{b}$. The condition $\|\mathbf{x}\|_\infty \le \|\mathbf{x}\|$ implies $b_i \le \max(b_1, b_2) = \|\mathbf{b}\|_\infty \le \|\mathbf{b}\|$ for $i = 1, 2$. From the triangle inequality we obtain

$$\|(a_1, b_2)\| = \left\| \frac{b_1 - a_1}{b_1} (0, b_2) + \frac{a_1}{b_1} (b_1, b_2) \right\|$$
$$\le \frac{b_1 - a_1}{b_1} \underbrace{\|(0, b_2)\|}_{=b_2 \le \|\mathbf{b}\|} + \frac{a_1}{b_1} \|(b_1, b_2)\|$$
$$\le \left(\frac{b_1 - a_1}{b_1} + \frac{a_1}{b_1} \right) \|(b_1, b_2)\|$$
$$= \|\mathbf{b}\|$$

and

$$
\begin{aligned}
\|\mathbf{a}\| &= \|(a_1, a_2)\| \\
&= \left\| \frac{b_2 - a_2}{b_2} (a_1, 0) + \frac{a_2}{b_2} (a_1, b_2) \right\| \\
&\leq \frac{b_2 - a_2}{b_2} \underbrace{\|(a_1, 0)\|}_{= a_1 \leq b_1 \leq \|\mathbf{b}\|} + \frac{a_2}{b_2} \underbrace{\|(a_1, b_2)\|}_{\leq \|\mathbf{b}\|, \text{ see above}} \\
&\leq \left(\frac{b_2 - a_2}{b_2} + \frac{a_2}{b_2} \right) \|\mathbf{b}\| \\
&= \|\mathbf{b}\|.
\end{aligned}
$$

Therefore the norm is monotone.   □

The preceding considerations can be utilized to characterize a Pickands dependence function in the bivariate case.

Proposition 4.4.5. *Consider an arbitrary function $D : [0, 1] \to [0, \infty)$ and put $\|(x, y)\|_D := (|x| + |y|) D (|x| / (|x| + |y|))$ for $x, y \in \mathbb{R}$ with the convention $\|\mathbf{0}\|_D = 0$. Then the following statements are equivalent.*

(i) $\|\cdot\|_D$ *is a monotone and standardized norm.*

(ii) $\|\cdot\|_D$ *is a norm that satisfies $\|\mathbf{x}\|_\infty \leq \|\mathbf{x}\|_D \leq \|\mathbf{x}\|_1$, $\mathbf{0} \leq \mathbf{x}$.*

(iii) $G(x, y) := \exp((x + y)D(x/(x + y)))$, $x, y \leq 0$, *defines a bivariate EVD with standard reverse exponential margins.*

(iv) *The function D is convex and satisfies $\max(t, 1 - t) \leq D(t) \leq 1$, $t \in [0, 1]$.*

(v) *The function D is convex and satisfies $\|\mathbf{x}\|_D \leq \|\mathbf{y}\|_D$ for $\mathbf{0} \leq \mathbf{x} \leq \mathbf{y}$ as well as $D(0) = D(1) = 1$.*

Proof. The equivalence of (i), (ii) and (iii) is a consequence of Lemmas 4.4.4 and 4.4.3. Next we show that (ii) and (iv) are equivalent. Suppose condition (ii) and choose $\lambda, t_1, t_2 \in [0, 1]$. The triangle inequality implies

$$
\begin{aligned}
D(\lambda t_1 + (1 - \lambda)t_2) &= \|\lambda(t_1, 1 - t_1) + (1 - \lambda)(t_2, 1 - t_2)\|_D \\
&\leq \lambda \|(t_1, 1 - t_1)\|_D + (1 - \lambda) \|(t_2, 1 - t_2)\|_D \\
&= \lambda D(t_1) + (1 - \lambda)D(t_2),
\end{aligned}
$$

i.e., D is a convex function. We have, moreover, for $t \in [0, 1]$,

$$
\max(t, 1 - t) = \|(t, 1 - t)\|_\infty \leq \|(t, 1 - t)\|_D = D(t) \leq \|(t, 1 - t)\|_1 = 1,
$$

which is (iv).

In what follows we show that (iv) implies (ii). The inequalities $\|\mathbf{x}\|_\infty \le \|\mathbf{x}\|_D \le \|\mathbf{x}\|_1$, $\mathbf{0} \le \mathbf{x} = (x, y)$, are obvious by putting $t = x/(x + y)$ in (iv). We also obtain $D(t) \ge 1/2$, $t \in [0, 1]$, and, thus, $\|\mathbf{x}\|_D = 0$ if and only if $\mathbf{x} = \mathbf{0}$ as well as $\|\lambda\mathbf{x}\|_D = |\lambda| \|\mathbf{x}\|_D$, $\lambda \in \mathbb{R}$, $\mathbf{x} \in \mathbb{R}^d$. The triangular inequality will follow from the subsequent considerations. The inequality $\max(t, 1 - t) \le D(t) \le 1$, $t \in [0, 1]$, implies, for $a, b \ge 0$, $a + b > 0$,

$$D\left(\frac{a}{a+b}\right) \ge \frac{b}{a+b} = \frac{b}{a+b}D(0) = \frac{b}{a+b}D(1)$$

as well as

$$D\left(\frac{a}{a+b}\right) \ge \frac{a}{a+b} = \frac{a}{a+b}D(0) = \frac{a}{a+b}D(1).$$

Hence we obtain for $0 \le (x_1, x_2) \le (y_1, y_2)$ with $x_1 + x_2 > 0$, $y_i > 0$, $i = 1, 2$,

$$
\begin{aligned}
D\left(\frac{x_1}{x_1+y_2}\right) &= D\left(\left(\frac{(y_1-x_1)y_2}{y_1(x_1+y_2)}\right) \cdot 0 + \left(\frac{(y_1+y_2)x_1}{y_1(x_1+y_2)}\right)\frac{y_1}{y_1+y_2}\right) \\
&\le \frac{(y_1-x_1)y_2}{y_1(x_1+y_2)}D(0) + \frac{(y_1+y_2)x_1}{y_1(x_1+y_2)}D\left(\frac{y_1}{y_1+y_2}\right) \\
&\le \frac{(y_1-x_1)(y_1+y_2)}{y_1(x_1+y_2)}D\left(\frac{y_1}{y_1+y_2}\right) + \frac{(y_1+y_2)x_1}{y_1(x_1+y_2)}D\left(\frac{y_1}{y_1+y_2}\right) \\
&= \frac{y_1+y_2}{x_1+y_2}D\left(\frac{y_1}{y_1+y_2}\right).
\end{aligned}
$$

Summarizing the preceding inequalities we obtain

$$
\begin{aligned}
D\left(\frac{x_1}{x_1+x_2}\right) &= D\left(\left(\frac{(y_2-x_2)x_1}{y_2(x_1+x_2)}\right) \cdot 1 + \left(\frac{(x_1+y_2)x_2}{y_2(x_1+x_2)}\right)\frac{x_1}{x_1+y_2}\right) \\
&\le \frac{y_2-x_2}{y_2(x_1+x_2)}x_1D(1) + \frac{x_2}{y_2}\frac{x_1+y_2}{x_1+x_2}D\left(\frac{x_1}{x_1+y_2}\right) \\
&\le \frac{y_2-x_2}{y_2(x_1+x_2)}y_1D(1) + \frac{x_2}{y_2}\frac{y_1+y_2}{x_1+x_2}D\left(\frac{y_1}{y_1+y_2}\right) \\
&\le \frac{y_2-x_2}{y_2}\frac{y_1+y_2}{x_1+x_2}D\left(\frac{y_1}{y_1+y_2}\right) + \frac{x_2}{y_2}\frac{y_1+y_2}{x_1+x_2}D\left(\frac{y_1}{y_1+y_2}\right) \\
&= \frac{y_1+y_2}{x_1+x_2}D\left(\frac{y_1}{y_1+y_2}\right).
\end{aligned}
$$

Multiplying by $(x_1 + x_2)$ shows the monotonicity $\|\mathbf{x}\|_D \le \|\mathbf{y}\|_D$, $\mathbf{0} \le \mathbf{x} \le \mathbf{y}$. Together with the convexity of D we will now see that $\|\cdot\|_D$ satisfies the triangular inequality for arbitrary $\mathbf{x}, \mathbf{y} \in \mathbb{R}^2$:

$$\|\mathbf{x} + \mathbf{y}\|_D$$

$$= (|x_1 + y_1| + |x_2 + y_2|) D\left(\frac{|x_1 + y_1|}{|x_1 + y_1| + |x_2 + y_2|}\right)$$

$$= \|(|x_1 + y_1|, |x_2 + y_2|)\|_D$$
$$\leq \|(|x_1| + |y_1|, |x_2| + |y_2|)\|_D$$
$$= (|x_1| + |x_2| + |y_1| + |y_2|)$$
$$\times D\left(\frac{|x_1| + |x_2|}{|x_1| + |x_2| + |y_1| + |y_2|}\frac{|x_1|}{|x_1| + |x_2|} + \frac{|y_1| + |y_2|}{|x_1| + |x_2| + |y_1| + |y_2|}\frac{|y_1|}{|y_1| + |y_2|}\right)$$
$$\leq (|x_1| + |x_2|) D\left(\frac{|x_1|}{|x_1| + |x_2|}\right) + (|y_1| + |y_2|) D\left(\frac{|y_1|}{|y_1| + |y_2|}\right)$$
$$= \|\mathbf{x}\|_D + \|\mathbf{y}\|_D.$$

Next we show that (iv) and (v) are equivalent. Suppose condition (iv). Then, obviously, $D(0) = D(1) = 1$. The monotonicity of $\|\cdot\|_D$ was established in the proof of the implication (iv) $\implies$ (ii). It, therefore, remains to show that (v) implies (iv). The convexity of D implies

$$D(t) = D((1-t) \cdot 0 + t \cdot 1) \leq (1-t)D(0) + tD(1) = t, \qquad t \in [0,1].$$

The monotonicity of $\|\cdot\|_D$ implies

$$(x_1 + x_2)D\left(\frac{x_1}{x_1 + x_2}\right) \leq (y_1 + y_2)D\left(\frac{y_1}{y_1 + y_2}\right), \qquad \mathbf{0} \leq \mathbf{x} \leq \mathbf{y}.$$

Choosing $x_1 \in [0,1]$ and putting $x_2 = 0$, $y_1 = x_1$, $y_2 = 1 - x_1$, we obtain from the above inequality

$$x_1 D(1) = x_1 \leq D(x_1).$$

Choosing $x_2 \in [0,1]$ and putting $x_1 = 0$, $y_1 = 1 - x_2$, $y_2 = x_2$, we obtain

$$x_2 D(0) = x_2 \leq D\left(1 - \frac{y_2}{y_1 + y_2}\right) = D(1 - x_2),$$

i.e., we have established (iv). $\square$

The equivalence of (iii) and (iv) in Proposition 4.4.5 is stated without proof in Deheuvels and Tiago de Oliveira [106]. An explicit proof is outlined in Beirlant et al. [32, Section 8.2.5]. The implication (iv) to (iii) under the additional condition that D is twice differentiable was established in Joe [275, Theorem 6.4].

Corollary 4.4.6. *Let $\|\cdot\|$ be an arbitrary norm on $\mathbb{R}^2$ and put $D(t) := \|(t, 1-t)\|$, $t \in [0,1]$. Then*

$$G(x,y) := \exp\left(-\|(x,y)\|\right)$$
$$= \exp\left((x+y)D\left(\frac{x}{x+y}\right)\right), \qquad x, y \leq 0,$$

defines an EVD with standard reverse exponential margins if and only if

$$\max(t, 1-t) \leq D(t) \leq 1, \qquad t \in [0,1]. \tag{4.40}$$

Proof. By Proposition 4.4.5 we only have to show the if-part. Precisely, we have to show that D is convex. But this is immediate from the triangular inequality:

$$
\begin{aligned}
D(&\lambda t_1 + (1-\lambda)t_2) \\
&= \|(\lambda t_1 + (1-\lambda)t_2, 1 - (\lambda t_1 + (1-\lambda)t_2))\| \\
&= \|\lambda(t_1, 1-t_1) + (1-\lambda)(t_2, 1-t_2)\| \\
&\leq \lambda \|(t_1, 1-t_1)\| + (1-\lambda)\|(t_2, 1-t_2)\| \\
&= \lambda D(t_1) + (1-\lambda)D(t_2), \qquad \lambda, t_1, t_2 \in [0,1]. \qquad \square
\end{aligned}
$$

Remark 4.4.7. Condition (4.40) is not necessarily satisfied by an arbitrary norm $\|\cdot\|$, even if it satisfies $\|(1,0)\| = \|(0,1)\| = 1$. Choose, for example, $\rho \in (-1,1)$ and put $\|(x,y)\|_\rho^2 = x^2 + 2\rho xy + y^2$, i.e., $\|\cdot\|_\rho$ is a generalized squared distance. Then we obtain for $\rho < -1/2$ that $\|(1/2,1/2)\|_\rho = (1/2 + \rho/2)^{1/2} < 1/2$ and, thus, $\|\cdot\|_\rho$ does not satisfy condition (4.40).

Proposition 4.4.5 can be utilized to characterize bivariate EVD in terms of compact and convex subsets of $\mathbb{R}^2$, thus, linking multivariate extreme value theory with convex geometry; we refer to Molchanov [341] for details. Take a compact and convex set $K \subset \mathbb{R}^2$, which is symmetric in the following sense: with $(x,y) \in K$ we require that $(-x,y) \in K$ and $(x,-y) \in K$ as well. We suppose in addition that $(1,0)$ and $(0,1)$ are boundary points of K.

Put $\|(0,0)\|_K := 0$ and, for $(x,y) \neq (0,0)$,

$$
\|(x,y)\|_K := \frac{1}{\sup\{t \in \mathbb{R} : t(x,y) \in K\}}.
$$

Then $\|\cdot\|_K$ defines a norm on $\mathbb{R}^2$, which satisfies condition (4.40) and, thus, $G_K(x,y) = \exp(-\|(x,y)\|_K)$, $x,y \leq 0$, is by Corollary 4.4.6 an EVD with standard reverse exponential margins. Note that K equals the unit ball with respect to this norm: $K = \{(x,y) \in \mathbb{R}^2 : \|(x,y)\|_K \leq 1\}$.

Take, on the other hand, $K = K_{\|\cdot\|_D} = \{(x,y) \in \mathbb{R}^2 : \|(x,y)\|_D \leq 1\}$ as the unit ball with respect to an arbitrary D-norm $\|\cdot\|_D$. Then K is a compact, convex and symmetric set with $\|(0,1)\|_D = \|(1,0)\|_D = 1$ and $\|(x,y)\|_K = \|(x,y)\|_D$. We have, thus, established the following characterization of bivariate EVD in terms of convex subsets of $\mathbb{R}^2$.

Corollary 4.4.8. *Let $G(x,y) = \exp(-\|(x,y)\|_D)$, $x,y \leq 0$, be any EVD with standard reverse exponential margins. Putting $K := \{(x,y) \in \mathbb{R}^2 : \|(x,y)\|_D \leq 1\}$ defines a one-to-one mapping from the set of bivariate EVD with standard reverse exponential margins into the set of compact, convex and symmetric subsets $K \subset \mathbb{R}^2$ such that $(-1,0)$ and $(0,1)$ are boundary points of K. We have, further, $\|(x,y)\|_D = \|(x,y)\|_K$, $(x,y) \in \mathbb{R}^2$.*

The preceding result can be utilized to generate bivariate EVD via compact subsets of $\mathbb{R}^2$. Take an arbitrary function $g : [0,1] \to [0,1]$, which is continuous and concave, and which satisfies $g(0) = 1$, $g(1) = 0$. The set $K :=$

$\{(x, y) \in \mathbb{R}^2 : |y| \leq g(|x|)\}$ then defines a convex, compact and symmetric sub-set of $\mathbb{R}^2$, and $(0, 1), (1, 0)$ are boundary points of K. Put for any $z \in [0, 1]$ $D(z) := \|(z, 1 - z)\|_K = 1/t_z$, where t_z is the unique solution of the equation $g(tz) = t(1 - z), t \geq 0$. Then $G(x, y) := \exp((x + y)D(x/(x + y)))$, $x, y \leq 0$, defines by Corollary 4.4.8 an EVD with reverse exponential margins.

With $g(z) := (1 - z^\lambda)^{1/\lambda}$, $\lambda \geq 1$, we obtain, for example, $t_z = (z^\lambda + (1 - z)^\lambda)^{-1/\lambda}$ and, thus, the set of logistic distributions

$$G(x, y) = \exp\left(-(|x|^\lambda + |y|^\lambda)^{1/\lambda}\right),$$

$x, y \leq 0$.

Molchanov [341] showed that max-stable rv in $[0, \infty)^d$ with unit Fréchet mar-gins are in one-to-one correspondence with convex sets K in $[0, \infty)^d$ called *max-zonoids*.

THE D-NORM IN ARBITRARY DIMENSION

In what follows we investigate the D-norm in $\mathbb{R}^d$. We start with a convex function and give a necessary and sufficient condition for the property that $\|\cdot\|_D$ actually defines a norm. Then we can use Theorem 4.4.2 to establish a necessary and suffi-cient condition for a convex function to be a Pickands dependence function. These considerations are taken from Hofmann [222]. Recall that a Pickands dependence function is defined on

$$R = \left\{(t_1, \ldots, t_{d-1}) \in [0, 1]^{d-1} : \sum_{j \leq d-1} t_j \leq 1\right\}.$$

Lemma 4.4.9. *Let $D : R \to (0, \infty)$ be a convex function. Then*

$$\|\mathbf{x}\|_D = \|\mathbf{x}\|_1 D\left(\frac{|x_1|}{\sum_{j \leq d} |x_j|}, \ldots, \frac{|x_{d-1}|}{\sum_{j \leq d} |x_j|}\right)$$

defines a norm on $\mathbb{R}^d$ iff, for $\mathbf{0} \leq \mathbf{x} \leq \mathbf{y}$, $\mathbf{x} \neq \mathbf{0}$,

$$D\left(\frac{x_1}{\sum_{j \leq d} x_j}, \ldots, \frac{x_{d-1}}{\sum_{j \leq d} x_j}\right) \leq \frac{\sum_{j \leq d} y_j}{\sum_{j \leq d} x_j} D\left(\frac{y_1}{\sum_{j \leq d} y_j}, \ldots, \frac{y_{d-1}}{\sum_{j \leq d} y_j}\right), \quad (4.41)$$

Proof. We, obviously, have $\|\lambda \mathbf{x}\|_D = |\lambda| \|\mathbf{x}\|_D$, $\lambda \in \mathbb{R}$, as well as $\|\mathbf{x}\|_D \geq 0$ and $\|\mathbf{x}\|_D = 0 \iff \mathbf{x} = \mathbf{0}$. The triangle inequality follows from the convexity of D, the triangle inequality of the absolute value and equation (4.41):

$$\|\mathbf{x} + \mathbf{y}\|_D$$

$$= \|\mathbf{x} + \mathbf{y}\|_1 D\left(\frac{|x_1 + y_1|}{\sum_{j \leq d} (|x_j + y_j|)}, \ldots, \frac{|x_{d-1} + y_{d-1}|}{\sum_{j \leq d} (|x_j + y_j|)}\right)$$

$$\leq \|\mathbf{x}+\mathbf{y}\|_1 \frac{\|\mathbf{x}\|_1 + \|\mathbf{y}\|_1}{\|\mathbf{x}+\mathbf{y}\|_1} D\left(\frac{|x_1|+|y_1|}{\sum_{j\leq d}(|x_j|+|y_j|)}, \ldots, \frac{|x_{d-1}|+|y_{d-1}|}{\sum_{j\leq d}(|x_j|+|y_j|)}\right)$$

$$= (\|\mathbf{x}\|_1 + \|\mathbf{y}\|_1) D\left(\frac{\sum_{j\leq d}|x_j|}{\sum_{j\leq d}(|x_j|+|y_j|)}\left(\frac{|x_1|}{\sum_{j\leq d}|x_j|}, \ldots, \frac{|x_{d-1}|}{\sum_{j\leq d}|x_j|}\right)\right.$$

$$\left. + \frac{\sum_{j\leq d}|y_j|}{\sum_{j\leq d}(|x_j|+|y_j|)}\left(\frac{|y_1|}{\sum_{j\leq d}|y_j|}, \ldots, \frac{|y_{d-1}|}{\sum_{j\leq d}|y_j|}\right)\right)$$

$$\leq (\|\mathbf{x}\|_1 + \|\mathbf{y}\|_1) \left(\frac{\sum_{j\leq d}|x_j|}{\sum_{j\leq d}(|x_j|+|y_j|)} D\left(\frac{|x_1|}{\sum_{j\leq d}|x_j|}, \ldots, \frac{|x_{d-1}|}{\sum_{j\leq d}|x_j|}\right)\right.$$

$$\left. + \frac{\sum_{j\leq d}|y_j|}{\sum_{j\leq d}(|x_j|+|y_j|)} D\left(\frac{|y_1|}{\sum_{j\leq d}|y_j|}, \ldots, \frac{|y_{d-1}|}{\sum_{j\leq d}|y_j|}\right)\right)$$

$$= \|\mathbf{x}\|_D + \|\mathbf{y}\|_D.$$

So we have established the if-part. Assume now that $\|\cdot\|_D$ defines a norm on $\mathbb{R}^d$. It is sufficient to prove equation (4.41) for $\mathbf{0} \leq \mathbf{x} \leq \mathbf{y}$, where $\mathbf{x}$ and $\mathbf{y}$ differ only in the k-th component. By iterating this step for every component one gets the general equation.

Set $\tilde{\mathbf{y}} := \mathbf{y} - 2y_k \mathbf{e}_k$, where $\mathbf{e}_k$ denotes the k-th unit vector in $\mathbb{R}^d$. The vector $\tilde{\mathbf{y}}$ differs from $\mathbf{y}$ only in the k-th component, and the absolute values of these components are identical. We, therefore, have $\|\mathbf{y}\|_D = \|\tilde{\mathbf{y}}\|_D$, and from the convexity of the D-norm we obtain

$$D\left(\frac{x_1}{\sum_{j\leq d} x_j}, \ldots, \frac{x_{d-1}}{\sum_{j\leq d} x_j}\right) = \frac{1}{\|\mathbf{x}\|_1} \|\mathbf{x}\|_D$$

$$= \frac{1}{\|\mathbf{x}\|_1} \left\|\frac{x_k + y_k}{2y_k}\mathbf{y} + \frac{y_k - x_k}{2y_k}\tilde{\mathbf{y}}\right\|_D$$

$$\leq \frac{1}{\|\mathbf{x}\|_1} \left(\frac{x_k + y_k}{2y_k}\|\mathbf{y}\|_D + \frac{y_k - x_k}{2y_k}\|\tilde{\mathbf{y}}\|_D\right)$$

$$= \frac{1}{\|\mathbf{x}\|_1} \|\mathbf{y}\|_D$$

$$= \frac{\sum_{j\leq d} y_j}{\sum_{j\leq d} x_j} D\left(\frac{y_1}{\sum_{j\leq d} y_j}, \ldots, \frac{y_{d-1}}{\sum_{j\leq d} y_j}\right). \qquad \square$$

Theorem 4.4.10. *Let D be a positive and convex function on R. Then D is a Pickands dependence function, i.e.,*

$$G(\mathbf{x}) := \exp\left(\left(\sum_{j\leq d} x_j\right) D\left(\frac{x_1}{\sum_{j\leq d} x_j}, \ldots, \frac{x_{d-1}}{\sum_{j\leq d} x_j}\right)\right), \qquad \mathbf{x} \leq \mathbf{0},$$

defines a d-dimensional EVD with standard converse exponential margins, if and only if

$$\sum_{\mathbf{m}\in\{0,1\}^d:\, m_j=1,j\in E} \left[(-1)^{d+1-\sum_{j\le d} m_j} \left(\sum_{j\le d}\left(-y_j^{m_j}x_j^{1-m_j}\right)\right) \right.$$
$$\left. \times D\left(\frac{y_1^{m_1}x_1^{1-m_1}}{\sum_{j\le d}\left(y_j^{m_j}x_j^{1-m_j}\right)}, \dots, \frac{y_{d-1}^{m_{d-1}}x_{d-1}^{1-m_{d-1}}}{\sum_{j\le d}\left(y_j^{m_j}x_j^{1-m_j}\right)}\right) \right] \ge 0 \quad (4.42)$$

for any $\mathbf{x}\le\mathbf{y}\le\mathbf{0}$ *and any subset* $E\subset\{1,\dots,d\}$, $E\ne\{1,\dots,d\}$, *and* D *satisfies* $D(\tilde{\mathbf{e}}_i)=D(\mathbf{0})=1$, $i\le d$, *where* $\tilde{\mathbf{e}}_i$ *denotes the i-th unit vector in* $\mathbb{R}^{d-1}$.

Proof. In what follows we show that condition (4.42) implies condition (4.41). For any $m\in\{1,\dots,d\}$ we apply condition (4.42) with $E=\{1,\dots,d\}\setminus\{m\}$ to the vectors $\sum_{i=1}^m x_i\mathbf{e}_i+\sum_{i=m+1}^d y_i\mathbf{e}_i$ and $\sum_{i=1}^{m-1}x_i\mathbf{e}_i+\sum_{i=m}^d y_i\mathbf{e}_i$ and obtain

$$\left\|\sum_{i=1}^m x_i\mathbf{e}_i+\sum_{i=m+1}^d y_i\mathbf{e}_i\right\|-\left\|\sum_{i=1}^{m-1}x_i\mathbf{e}_i+\sum_{i=m}^d y_i\mathbf{e}_i\right\|\ge 0.$$

Summation over m from 1 to d yields

$$0\le\sum_{m=1}^d\left(\left\|\sum_{i=1}^m x_i\mathbf{e}_i+\sum_{i=m+1}^d y_i\mathbf{e}_i\right\|-\left\|\sum_{i=1}^{m-1}x_i\mathbf{e}_i+\sum_{i=m}^d y_i\mathbf{e}_i\right\|\right)$$
$$=\sum_{m=1}^d\left(\left\|\sum_{i=1}^m x_i\mathbf{e}_i+\sum_{i=m+1}^d y_i\mathbf{e}_i\right\|\right)-\sum_{m=1}^d\left(\left\|\sum_{i=1}^{m-1}x_i\mathbf{e}_i+\sum_{i=m}^d y_i\mathbf{e}_i\right\|\right)$$
$$=\left\|\sum_{i=1}^d x_i\mathbf{e}_i\right\|+\sum_{m=1}^{d-1}\left(\left\|\sum_{i=1}^m x_i\mathbf{e}_i+\sum_{i=m+1}^d y_i\mathbf{e}_i\right\|\right)$$
$$-\sum_{m=2}^d\left(\left\|\sum_{i=1}^{m-1}x_i\mathbf{e}_i+\sum_{i=m}^d y_i\mathbf{e}_i\right\|\right)-\left\|\sum_{i=1}^d y_i\mathbf{e}_i\right\|$$
$$=\left\|\sum_{i=1}^d x_i\mathbf{e}_i\right\|+\sum_{m=1}^{d-1}\left(\left\|\sum_{i=1}^m x_i\mathbf{e}_i+\sum_{i=m+1}^d y_i\mathbf{e}_i\right\|\right)$$
$$-\sum_{m=1}^{d-1}\left(\left\|\sum_{i=1}^m x_i\mathbf{e}_i+\sum_{i=m+1}^d y_i\mathbf{e}_i\right\|\right)-\left\|\sum_{i=1}^d y_i\mathbf{e}_i\right\|$$
$$=\|\mathbf{x}\|-\|\mathbf{y}\|.$$

So $\|\cdot\|_D$ defines a norm with the properties from Lemma 4.4.9. Using the definition of the D-norm we see that

$$0 \leq \sum_{\mathbf{m} \in \{0,1\}^d : m_j = 1, j \in E} \left[(-1)^{d+1-\sum_{j \leq d} m_j} \left(\sum_{j \leq d} \left(-y_j^{m_j} x_j^{1-m_j} \right) \right) \right.$$

$$\left. D \left(\frac{y_1^{m_1} x_1^{1-m_1}}{\sum_{j \leq d} \left(y_j^{m_j} x_j^{1-m_j} \right)}, \dots, \frac{y_{d-1}^{m_{d-1}} x_{d-1}^{1-m_{d-1}}}{\sum_{j \leq d} \left(y_j^{m_j} x_j^{1-m_j} \right)} \right) \right]$$

$$= \sum_{\mathbf{m} \in \{0,1\}^d : m_j = 1, j \in E} \left[(-1)^{d+1-\sum_{j \leq d} m_j} \left\| \left(y_1^{m_1} x_1^{1-m_1}, \dots, y_{d-1}^{m_{d-1}} x_{d-1}^{1-m_{d-1}} \right) \right\|_D \right].$$

Application of Theorem 4.4.2 now completes the proof. $\qquad\qquad\qquad\qquad\square$

Chapter 5

Multivariate Generalized Pareto Distributions

In analogy to the univariate case, we introduce certain multivariate generalized Pareto df (GPD) of the form $W = 1 + \log(G)$ for the statistical modelling of multivariate exceedances, see Section 5.1. Various results around the multivariate peaks-over-threshold approach are compiled in Section 5.2. The peaks-over-threshold stability of a multivariate GPD is investigated in Section 5.3.

The special dependence of a multivariate EVD on its argument suggests the use of certain Pickands coordinates consisting of a *distance* and a *(pseudo)-angle*. Of decisive importance for our investigations will be a spectral decomposition of multivariate df into univariate ones based on the Pickands coordinates, see Sections 5.4. Using this approach, conditions and results can be carried over from the univariate setting to the multivariate one, see Section 5.5. Particularly, we study the domain of attraction of a multivariate EVD and compile results for df which belong to the δ-neighborhood of a multivariate GPD.

In addition, a given rv can be represented by the Pickands transform which consists of the random distance and the random angle. In Section 5.6 it will be shown that these rv are conditionally independent under a GPD. This result approximately holds under an EVD.

An important aspect for practical applications of multivariate GPD are simulations. In Section 5.7 we will present methods to simulate GPD. Sections 5.8 to 5.10 are dedicated to statistical inference in GP models, using the simulation as a first check of the practical applicability of the methods.

5.1 The Basics

Different from the univariate case, where $W(x) = 1 + \log(G(x))$, $\log(G(x)) \geq -1$ defines a df for any given max-stable df G and its multivariate version is not

M. Falk et al., *Laws of Small Numbers: Extremes and Rare Events*, 3rd ed.,
DOI 10.1007/978-3-0348-0009-9_5, © Springer Basel AG 2011

necessarily a df for dimensions $d > 2$, see Proposition 5.1.3. Yet, one finds df having such a form in the upper tail region, which is sufficient for our asymptotic considerations.

Therefore, we first introduce a *generalized Pareto (GP) function* pertaining to a max-stable G in $\mathbb{R}^d$ by

$$W(\mathbf{x}) := 1 + \log(G(\mathbf{x})) \tag{5.1}$$

$$= 1 + \left(\sum_{i \le d} x_i\right) D\left(\frac{x_1}{\sum_{i \le d} x_i}, \ldots, \frac{x_{d-1}}{\sum_{i \le d} x_i}\right), \qquad \log(G(\mathbf{x})) \ge -1.$$

Note that
$$W(\mathbf{x}) = 1 - ||\mathbf{x}||_D, \qquad \mathbf{x} \le \mathbf{0}, \, ||\mathbf{x}||_D \le 1,$$

where $||\mathbf{x}||_D$ is the D-norm of $\mathbf{x}$, see (4.36).

The GP function corresponding to the Marshall-Olkin EVD is for instance

$$W_\lambda(\mathbf{x}) = 1 + (1 - \lambda)(x_1 + \cdots + x_d) + \lambda \min(x_1, \ldots, x_d);$$

for the logistic df it is

$$W_\lambda(\mathbf{x}) = 1 - \left(\sum_{i \le d}(-x_i)^\lambda\right)^{1/\lambda}.$$

In what follows we introduce GPD which have the form $W = 1 + \log(G)$ in the upper tail region.

THE BIVARIATE CASE

For the dimension $d = 2$ the GP function itself is a GPD (mentioned by Kaufmann and Reiss [286]). Due to our special choice of EVD G with negative exponential margins, the margins of the bivariate GPD $W = 1 + \log(G)$ are equal to the uniform distribution on the interval $[-1, 0]$ which is a univariate GPD.

Lemma 5.1.1. *The bivariate GP function is a df.*

Proof. $W(\mathbf{x}) = \max(1 + \log(G(\mathbf{x})), 0)$, $\mathbf{x} \le \mathbf{0}$, is obviously continuous and normed. Its Δ-monotonicity follows from the Δ-monotonicity of G (see the beginning of Section 4.1): Let $r_n \downarrow 0$ be an arbitrary sequence of positive numbers converging to 0. Taylor expansion of exp at 0 implies, for $\mathbf{x} \le \mathbf{y} < \mathbf{0}$,

$$0 \le \Delta_{r_n\mathbf{x}}^{r_n\mathbf{y}} G$$
$$= G(r_n\mathbf{y}) - G(r_n(x_1, y_2)) - G(r_n(y_1, x_2)) + G(r_n\mathbf{x})$$
$$= (\exp(-r_n||\mathbf{y}||_D) - 1) + (1 - \exp(-r_n||(x_1, y_2)||_D))$$
$$\qquad + (1 - \exp(-r_n||(y_1, x_2)||_D)) + (\exp(-r_n||\mathbf{x}||_D) - 1)$$

$$= -r_n||\mathbf{y}||_D + r_n||(x_1, y_2)||_D + r_n||(y_1, x_2)||_D - r_n||\mathbf{x}||_D + O(r_n^2).$$

We, thus, obtain

$$0 \leq \lim_{n \to \infty} \frac{\Delta_{r_n \mathbf{x}}^{r_n \mathbf{y}} G}{r_n}$$
$$= 1 - ||\mathbf{y}||_D - (1 - ||(x_1, y_2)||_D) - (1 - ||(y_1, x_2)||_D) + (1 - ||\mathbf{x}||_D).$$

From the monotonicity of $|| \cdot ||_D$ (see (4.38)) one now concludes that

$$\Delta_{\mathbf{x}}^{\mathbf{y}} W = \max(1 - ||\mathbf{y}||_D, 0) - \max(1 - ||(x_1, y_2)||_D, 0)$$
$$- \max(1 - ||(y_1, x_2)||_D, 0) + \max(1 - ||\mathbf{x}||_D, 0) \geq 0.$$

$\square$

The support of such a bivariate GPD $W(x, y) = \max(1 - ||(x, y)||_D, 0) = \max(1 + (x + y)D(x/(x + y)), 0)$ is the set of those points $(x, y) \in (-\infty, 0]^2$, such that $(x + y)D(x/(x + y)) \geq -1$. Using the spectral decomposition introduced in Section 5.4, the support of W can conveniently be identified as the set of those points $(x, y) = c(z, 1 - z)$ with $z \in [0, 1]$ and $-1/D(z) \leq c \leq 0$. That is, we have $W(c(z, 1 - z)) = 1 + cD(z)$, $z \in [0, 1]$, $-1/D(z) \leq c \leq 1$, see Lemma 5.4.3.

Example 5.1.2 (Independent EVD rv). Let

$$G(x, y) = \exp(x + y), \qquad x, y \leq 0.$$

Then, $W = 1 + x + y = 1 - ||(x, y)||_1$ is the distribution of the rv $(U, -(1 + U))$, where U is uniformly distributed on $[-1, 0]$. Note that $(U, -(1 + U))$ is uniformly distributed on the line $\{(x, y) : x, y \leq 0, \ x + y = -1\}$.

This example indicates that the case of independent EVD rv has to be considered with particular caution. We will return to this case in Section 5.2 and give an interpretation.

GPD IN HIGHER DIMENSIONS

In dimension $d \geq 3$, the GP function is not necessarily a df. Take, for instance, $d = 3$ and $D = 1$. The GP function $W(\mathbf{x}) = \max(1 + x_1 + x_2 + x_3, 0)$, $\mathbf{x} \leq \mathbf{0}$, is not Δ-monotone, since the cube $[-1/2, 0]^3$ would get a negative probability:

$$\Delta_{(-\frac{1}{2}, -\frac{1}{2}, -\frac{1}{2})}^{\mathbf{0}} W = -\frac{1}{2}.$$

This example can be extended to an arbitrary dimension d. The following result is established by Hofmann [222], where $||\mathbf{x}||_\lambda = \left(\sum_{i \leq d} |x_i|^\lambda\right)^{1/\lambda}$ denotes the usual λ-norm of $\mathbf{x} \in \mathbb{R}^d$, $\lambda \geq 1$. This GP function coincides, however, in arbitrary dimension with the upper tail of a df in a neighborhood of $\mathbf{0}$, see below.

Proposition 5.1.3. *For any $\lambda \geq 1$, the GP function*

$$W_\lambda(\mathbf{x}) = \max\left(1 - \|\mathbf{x}\|_\lambda, 0\right), \qquad \mathbf{x} \leq \mathbf{0} \in \mathbb{R}^d,$$

does not define a df for $d \geq 3$.

We call a d-dimensional df W a (multivariate) GPD, if there is some EVD G with reverse exponential margins and dependence function D such that

$$W(\mathbf{x}) = 1 + \log(G(\mathbf{x})) = 1 - \|\mathbf{x}\|_D$$

for $\mathbf{x}$ in a neighborhood of $\mathbf{0}$. To shorten the notation, we will often write $W = 1 + \log(G)$ in this case.

To specify the notion "neighborhood of $\mathbf{0}$" we will establish that the cube $[-1/d, 0]^d$ is a suitable area for this neighborhood: We have for $\mathbf{x} \in [-1/d, 0]^d$ by the monotonicity and the standardization of the norm (see Lemma 4.4.9 and Theorem 4.4.10) with $\mathbf{e}_i$, $i \leq d$, denoting the standard unit vectors in $\mathbb{R}^d$:

$$\|\mathbf{x}\|_D \leq \left\|\left(\frac{1}{d}, \ldots, \frac{1}{d}\right)\right\|_D = \frac{1}{d}\|(1, \ldots, 1)\|_D \leq \frac{1}{d}\sum_{i=1}^d \|\mathbf{e}_i\|_D = \frac{1}{d} \cdot d = 1, \quad (5.2)$$

and, thus, $1 - \|\mathbf{x}\|_D \geq 0$ on $[-1/d, 0]^d$.

Furthermore, the Δ-monotonicity is satisfied by $1 - \|\cdot\|_D$ on $[-1/d, 0]^d$, since for $-1/d \leq a_i \leq b_i \leq 0$, $i \leq d$, we have

$$\Delta_{\mathbf{a}}^{\mathbf{b}}(1 - \|\cdot\|_D)$$

$$= \sum_{\mathbf{m} \in \{0,1\}^d} (-1)^{\left(d - \sum_{j \leq d} m_j\right)} \left(1 - \left\|\left(b_1^{m_1} a_1^{1-m_1}, \ldots, b_d^{m_d} a_d^{1-m_d}\right)\right\|_D\right)$$

$$= \sum_{\mathbf{m} \in \{0,1\}^d} (-1)^{\left(d+1 - \sum_{j \leq d} m_j\right)} \left\|\left(b_1^{m_1} a_1^{1-m_1}, \ldots, b_d^{m_d} a_d^{1-m_d}\right)\right\|_D$$

$$\geq 0$$

by Theorem 4.4.2. The characterization of a multivariate df at the beginning of Section 4.1, together with the preceding considerations, indicate that for every D-norm $\|\cdot\|_D$ there exists a df W on $(-\infty, 0]^d$, such that $W(\mathbf{x}) = 1 - \|\mathbf{x}\|_D$ on $[-1/d, 0]^d$. This is the content of the next theorem, which is established in Theorem 6.2.1 in Hofmann [222].

Theorem 5.1.4. *Let $\|\cdot\|_D$ be a d-dimensional D-Norm. Then there exists a d-variate df W with*

$$W(\mathbf{x}) = 1 - \|\mathbf{x}\|_D \ \text{for} \ \mathbf{x} \in [-1/d, 0]^d.$$

The following construction of a GPD is motivated by the representation $G(\mathbf{x}) = \exp\left(-\nu[-\infty, \mathbf{x}]^{\mathsf{C}}\right)$ of an EVD G by means of its exponent measure ν, see (4.14), and by a point process approximation w.r.t. the variational distance, see Section 7.1. Because $\nu\left([-\infty, \mathbf{t}]^{\mathsf{C}}\right) = -\log(G(\mathbf{t}))$, $\mathbf{t} < \mathbf{0}$, one gets by

$$Q_{\mathbf{t}} := \nu\left(\cdot \cap [-\infty, \mathbf{t}]^{\mathsf{C}}\right) / \nu\left([-\infty, \mathbf{t}]^{\mathsf{C}}\right) \tag{5.3}$$

a distribution which turns out to be a GPD. Another version of a GPD will be deduced from $Q_{\mathbf{t}}$ in Section 7.1. In the subsequent lemma, $Q_{\mathbf{t}}$ will also be characterized as the limit of certain conditional distributions of an EVD. A characterization of a GPD with underlying df being in the domain of attraction of an EVD is given in Theorem 5.2.4.

Lemma 5.1.5. *Let G be a d-dimensional EVD with reverse exponential margins and exponent measure ν.*

(i) *The df of $Q_{\mathbf{t}}$ is given by*

$$F_{\mathbf{t}}(\mathbf{x}) = Q_{\mathbf{t}}([-\infty, \mathbf{x}])$$
$$= \left(\log(G(\mathbf{x})) - \log(G(\min(\mathbf{t}, \mathbf{x})))\right)/a(\mathbf{t}), \quad \mathbf{x} \leq \mathbf{0},$$

where $a(\mathbf{t}) = -\log(G(\mathbf{t}))$.

(ii) *For $\mathbf{t}/a(\mathbf{t}) \leq \mathbf{x} \leq \mathbf{0}$ we have*

$$W_{\mathbf{t}}(\mathbf{x}) = F_{\mathbf{t}}(a(\mathbf{t})\mathbf{x}) = 1 + \log(G(\mathbf{x})).$$

(iii) *Let $\mathbf{X}$ be distributed according to G. Then, for $\mathbf{x} \leq \mathbf{0}$,*

$$\lim_{r \downarrow 0} P\left(\mathbf{X} \leq r\mathbf{x} \mid \mathbf{X} \in (-\infty, r\mathbf{t}]^{\mathsf{C}}\right) = F_{\mathbf{t}}(\mathbf{x}).$$

Proof. Assertion (i) follows from the equation

$$[-\infty, \mathbf{t}]^{\mathsf{C}} \cap [-\infty, \mathbf{x}] = [-\infty, \min(\mathbf{t}, \mathbf{x})]^{\mathsf{C}} \setminus [-\infty, \mathbf{x}]^{\mathsf{C}}$$

and the above representation of $-\log(G)$ by means of ν.
    To prove (ii) check that

$$F_{\mathbf{t}}(\mathbf{x}) = 1 + (\log(G(\mathbf{x}))/a(\mathbf{t}))$$

for every $\mathbf{x} \geq \mathbf{t}$ and utilize the equation $G^{s}(\mathbf{x}/s) = G(\mathbf{x})$ for $s > 0$.
    (iii) Because $\nu\left([-\infty, r\mathbf{x}]^{\mathsf{C}}\right) = r\nu\left([-\infty, \mathbf{x}]^{\mathsf{C}}\right)$ for every $r > 0$, and

$$[-\infty, r\mathbf{x}] \cap [-\infty, r\mathbf{t}]^{\mathsf{C}} = [-\infty, r\mathbf{x}] \setminus [-\infty, r\min(\mathbf{t}, \mathbf{x})],$$

one obtains from (4.14) and a Taylor expansion of exp that

$$P\left(\mathbf{X} \leq r\mathbf{x} \mid \mathbf{X} \in (-\infty, r\mathbf{t}]^{\complement}\right)$$

$$= \frac{\exp\left(-\nu\left([-\infty, r\mathbf{x}]^{\complement}\right)\right) - \exp\left(-\nu\left([-\infty, r\min(\mathbf{t}, \mathbf{x})]^{\complement}\right)\right)}{1 - \exp\left(-\nu\left([-\infty, r\mathbf{t}]^{\complement}\right)\right)}$$

$$\rightarrow_{r\downarrow 0} \frac{\nu\left([-\infty, \min(\mathbf{t}, \mathbf{x})]^{\complement}\right) - \nu\left([-\infty, \mathbf{x}]^{\complement}\right)}{\nu\left([-\infty, \mathbf{t}]^{\complement}\right)}$$

$$= F_{\mathbf{t}}(\mathbf{x}). \qquad \qquad \square$$

Characteristic properties of a multivariate GPD $W = 1 + \log(G)$ are, for example, that it is POT-stable, see (5.23) as well as Section 5.3, or that it yields the best attainable rate of convergence of extremes, equally standardized, see Theorem 5.4.7. Hence, these properties, which are well known for a univariate GPD (see Theorem 1.3.5 and Theorem 2.1.12), carry over to the multivariate case.

This discussion will be continued in Section 7.1 in conjunction with the point process approach to exceedances.

GP Functions and Quasi-Copulas

Shifting a GP function $W(\mathbf{x}) = 1 + \log(G(\mathbf{x}))$, $\mathbf{x} \in (-\infty, 0]^d$ with $\log(G(\mathbf{x})) \geq -1$, to the unit cube $[0, 1]^d$ by considering $\widetilde{W}(\mathbf{y}) := W(\mathbf{y} - \mathbf{1})$, $\mathbf{y} \in [0, 1]^d$, yields a copula in the case $d = 2$ by Lemma 5.1.1, but not necessarily in the case $d \geq 3$, see the discussion leading to Lemma 5.1.5.

It turns out, however, that $\widetilde{W}$ is in general a *quasi-copula* (Alsina et al. [11]). The notion of quasi-copulas was introduced to characterize operations on functions that cannot be derived from operations on rv. Due to a characterization of quasi-copulas by Genest et al. [173] in the bivariate case and Cuculescu and Theodorescu [86] in arbitrary dimensions, $\widetilde{W}$ is a quasi-copula since it satisfies the following three conditions for arbitrary $\mathbf{x} = (x_1, \ldots, x_d)$, $\mathbf{y} = (y_1, \ldots, y_d) \in [0, 1]^d$:

(i) $\widetilde{W}(\mathbf{x}) = 0$ if $x_i = 0$ for some i, and $\widetilde{W}(\mathbf{x}) = x_j$ if all $x_k = 1$ except x_j,

(ii) $\widetilde{W}(x_1, \ldots, x_d)$ is non-decreasing in each of its arguments,

(iii) $\widetilde{W}$ satisfies Lipschitz's condition, i.e.,

$$|\widetilde{W}(\mathbf{x}) - \widetilde{W}(\mathbf{y})| \leq ||\mathbf{x} - \mathbf{y}||_1.$$

The latter inequality follows from the representation (5.1) of W in the D-norm and (4.37). We have

$$\left| \widetilde{W}(\mathbf{x}) - \widetilde{W}(\mathbf{y}) \right| \le \left| \|\mathbf{x} - \mathbf{1}\|_D - \|\mathbf{y} - \mathbf{1}\|_D \right|$$
$$\le \|\mathbf{x} - \mathbf{y}\|_D$$
$$\le \|\mathbf{x} - \mathbf{y}\|_1 \, .$$

As a consequence we obtain, for example, that for any *track*

$$B = \{(F_i(t))_{i \le d} : 0 \le t \le 1\}$$

in the d-dimensional unit cube there exists a copula C_B, which coincides with $\widetilde{W}$ on B:

$$C_B(\mathbf{x}) = \widetilde{W}(\mathbf{x}), \qquad \mathbf{x} \in B.$$

By $F_1, \ldots, F_d$ we denote arbitrary univariate continuous df such that $F_i(0) = 0$, $F_i(1) = 1$, $i \le d$, see Cuculescu and Theodorescu [86].

We obtain, therefore, that for any track $B = \{(F_i(t))_{i \le d} : 0 \le t \le 1\}$ the function $F(t) := \widetilde{W}((F_i(t))_{i \le d})$ defines a univariate df on $[0, 1]$. The spectral decomposition of a GPD in the subsequent Lemma 5.4.3 is a specific example. In this case the tracks are lines and $F(t) = \widetilde{W}((F_i(t))_{i \le d}) = W((F_i(t) - 1)_{i \le d})$ is a uniform distribution for t large enough.

5.2 Multivariate Peaks-Over-Threshold Approach

Suppose that the d-dimensional df F is in the domain of attraction of an EVD G, i.e., there exist constants $\mathbf{a}_n > \mathbf{0}$, $\mathbf{b}_n \in \mathbb{R}^d$, $n \in \mathbb{N}$, such that

$$F^n(\mathbf{a}_n \mathbf{x} + \mathbf{b}_n) \to_{n \to \infty} G(\mathbf{x}), \qquad \mathbf{x} \in \mathbb{R}^d.$$

This is equivalent with convergence of the marginals *together* with convergence of the copulas:

$$C_F^n\left(\mathbf{u}^{1/n}\right) \to_{n \to \infty} C_G(\mathbf{u}) = G\left(\left(G_i^{-1}(u_i)\right)_{i \le d}\right), \qquad \mathbf{u} \in (0, 1)^d,$$

(Deheuvels [101, 102], Galambos [167], see Theorem 5.5.2) or, taking logarithms,

$$n\left(1 - C_F\left(\mathbf{u}^{1/n}\right)\right) \to_{n \to \infty} -\log(C_G(\mathbf{u})), \qquad \mathbf{u} \in (0, 1]^d.$$

Choosing $u_i := \exp(x_i)$, $x_i \le 0$, $i \le d$, we obtain

$$n(1 - C_F(\exp(\mathbf{x}/n))) \to_{n \to \infty} -\log(C_G(\exp(\mathbf{x}))) = l_G(\mathbf{x}), \qquad \mathbf{x} \le \mathbf{0},$$

where

$$l_G(\mathbf{x}) = -\log(C_G(\exp(\mathbf{x})))$$
$$= \left(-\sum_{i \le d} x_i\right) D\left(\frac{x_1}{\sum_{i \le d} x_i}, \ldots, \frac{x_{d-1}}{\sum_{i \le d} x_i}\right), \qquad \mathbf{x} \le \mathbf{0},$$

is the *stable tail dependence function* (Huang [234]) of G with corresponding Pickands dependence function D. Note that we have defined the function $l_G(\cdot)$ on $(-\infty, 0]^d$ instead of $[0, \infty)^d$, which is more common. The reason is equation (5.5).

Taylor expansion $\exp(\varepsilon) = 1 + \varepsilon + O(\varepsilon^2)$, $\varepsilon \to 0$, together with the fact that C_F has uniform margins implies

$$n(1 - C_F(\mathbf{1} + \mathbf{x}/n))) \to_{n \to \infty} l_G(\mathbf{x}), \qquad \mathbf{x} \le \mathbf{0},$$

or, in a continuous version,

$$t^{-1}(1 - C_F(\mathbf{1} + t\mathbf{x}))) \to_{t \downarrow 0} l_G(\mathbf{x}), \qquad \mathbf{x} \le \mathbf{0}, \tag{5.4}$$

see Section 4.2 in de Haan and de Ronde [194]. The stable tail dependence function is obviously homogeneous $t l_G(\mathbf{x}) = l_G(t\mathbf{x})$, $t \ge 0$, and, thus, (5.4) becomes

$$\frac{1 - C_F(\mathbf{1} + t\mathbf{x}) - l_G(t\mathbf{x})}{t} \to_{t \downarrow 0} 0.$$

Observe that

$$l_G(\mathbf{x}) = 1 - W(\mathbf{x}), \qquad \mathbf{x} \le \mathbf{0}, \tag{5.5}$$

where W is a multivariate GP function with uniform margins $W_i(x) = 1 + x$, $x \le 0$, $i \le d$, i.e.,

$$W(\mathbf{x}) = 1 + \log\left(\widetilde{G}(\mathbf{x})\right), \qquad \mathbf{x} \le \mathbf{0},$$

and $\widetilde{G}$ is a multivariate EVD with negative exponential margins $\widetilde{G}_i(x) = \exp(x)$, $x \le 0$, $i \le d$.

The preceding considerations together with elementary computations entail the following result, which is true for an EVD G with *arbitrary* margins, not necessarily negative exponential ones. By $\|\cdot\|$ we denote an arbitrary norm on $\mathbb{R}^d$.

Theorem 5.2.1. *An arbitrary df F is in the domain of attraction of a multivariate EVD G iff this is true for the univariate margins and if there exists a GPD W with uniform margins such that*

$$C_F(\mathbf{y}) = W(\mathbf{y} - \mathbf{1}) + o\left(\|\mathbf{y} - \mathbf{1}\|\right)$$

uniformly for $\mathbf{y} \in [0, 1]^d$.

The preceding result shows that the upper tail of the copula C_F of a df F can reasonably be approximated only by that of a GPD W with uniform margins.

Recall that an arbitrary copula C is itself a multivariate df, and each margin is the uniform distribution on $[0, 1]$. Putting in the preceding Theorem $F = C$, we obtain the following equivalences for an arbitrary copula.

Corollary 5.2.2. *A copula C is in the domain of attraction of an EVD G with standard negative exponential margins*

$\Longleftrightarrow$ *there exists a GP function W with standard uniform margins such that*

$$C(\mathbf{y}) = W(\mathbf{y} - \mathbf{1}) + o\left(\|\mathbf{y} - \mathbf{1}\|\right),$$

uniformly for $\mathbf{y} \in [0,1]^d$. In this case $W = 1 + \log(G)$.

$\Longleftrightarrow$ *there exists a norm $\|\cdot\|$ on $\mathbb{R}^d$ such that*

$$C(\mathbf{y}) = 1 - \|\mathbf{y} - \mathbf{1}\| + o\left(\|\mathbf{y} - \mathbf{1}\|\right),$$

uniformly for $\mathbf{y} \in [0,1]^d$. In this case $G(\mathbf{x}) = \exp\left(-\|\mathbf{x}\|\right)$, $\mathbf{x} \le \mathbf{0}$.

Proof. Recall from Section 4.4 that a GPD W with uniform margins can be written as

$$W(\mathbf{x}) = 1 - \|\mathbf{x}\|_D, \qquad \mathbf{x}_0 \le \mathbf{x} \le \mathbf{0},$$

where $\|\cdot\|_D$ is a D-norm on $\mathbb{R}^d$, i.e., $G(\mathbf{x}) = \exp\left(-\|\mathbf{x}\|_D\right)$, $\mathbf{x} \le \mathbf{0}$, defines an EVD on $\mathbb{R}^d$. If $C(\mathbf{y}) = W(\mathbf{y} - \mathbf{1}) + o\left(\|\mathbf{y} - \mathbf{1}\|\right)$, $\mathbf{y} \in [0,1]^d$, for some norm $\|\cdot\|$ on $\mathbb{R}^d$, then

$$C^n\left(\mathbf{1} + \frac{y}{n}\right) = \left(1 - \frac{1}{n}\|\mathbf{y}\|_D + o\left(\frac{1}{n}\|\mathbf{y}\|\right)\right)^n$$
$$\to_{n\to\infty} \exp\left(-\|\mathbf{y}\|_D\right) = G(\mathbf{y}), \qquad \mathbf{y} \le \mathbf{0}.$$

Together with Theorem 5.2.1 this implies Corollary 5.2.2. $\qquad\square$

In the final equivalence of Corollary 5.2.2, the norm can obviously be computed as

$$\|\mathbf{x}\| = \lim_{t\downarrow 0} \frac{1 - C(\mathbf{1} + t\mathbf{x})}{t} = l(\mathbf{x}), \qquad \mathbf{x} \le \mathbf{0},$$

i.e., *any* stable tail dependence function $l(\cdot)$ is, actually, a norm. The triangle inequality, satisfied by any norm, and the homogeneity of an arbitrary norm explain why $l(\cdot)$ is in general a convex and homogeneous function of order 1.

Take, for example, an arbitrary *Archimedean* copula

$$C_\varphi(\mathbf{u}) = \varphi^{-1}(\varphi(u_1) + \cdots + \varphi(u_d)), \qquad (5.6)$$

where the generator $\varphi : (0,\infty) \to [0,\infty)$ is a continuous function that is strictly decreasing on $(0,1]$, $\varphi(1) = 0$, $\lim_{x\downarrow 0} \varphi(x) = \infty$ and $\varphi^{-1}(t) = \inf\{x > 0 : \varphi(x) \le t\}$, $t \ge 0$.

If φ is differentiable from the left in $x = 1$ with left derivative $\varphi'(1-) \ne 0$, then

$$\frac{1 - C_\varphi(\mathbf{1} + t\mathbf{x})}{t} \to_{t\downarrow 0} \sum_{i \le d} |x_i| = \|\mathbf{x}\|_1, \qquad \mathbf{x} \le \mathbf{0}, \qquad (5.7)$$

i.e., each Archimedean copula with a generator φ as above is in the domain of attraction of the EVD $G(\mathbf{x}) = \exp\left(-\|\mathbf{x}\|_1\right)$, $\mathbf{x} \le \mathbf{0}$, with independent margins.

This concerns, for example, the Clayton and the Frank copula, which have generators $\varphi_C(t) = \vartheta^{-1}(t^{-\vartheta} - 1)$, $\varphi_F(t) = -\log\left((\exp(-\vartheta t) - 1)/(\exp(-\vartheta) - 1)\right)$, $\vartheta > 0$, but not the Gumbel copula with parameter $\lambda > 1$, which has generator $\varphi_G(t) = -(\log(t))^\lambda$, $\lambda \geq 1$, $0 < t \leq 1$.

In terms of rv, Theorem 5.2.1 becomes

Theorem 5.2.3. *Suppose that the d-dimensional rv* $\mathbf{X}$ *with df* F *is in the domain of attraction of an arbitrary EVD* G *with corresponding dependence function* D. *If* F *is continuous in its upper tail, then we have uniformly, for* $-\mathbf{1} < \mathbf{y} < \mathbf{0}$,

$$P\left(X_i \leq F_i^{-1}(1 + y_i), \, i \leq d\right) = W_G(\mathbf{y}) + o(\|\mathbf{y}\|),$$

where W_G *is a GPD whose upper tail is* $W_G(\mathbf{y}) = 1 + \left(\sum_{i \leq d} y_i\right) D\left(y_1 / \sum_{i \leq d} y_i, \ldots, y_{d-1} / \sum_{i \leq d} y_i\right)$.

We obtain, as a consequence, for the exceedance probabilities the approximation

$$P\left(\mathbf{X} \leq F^{-1}(\mathbf{1} + \mathbf{y}) \mid \mathbf{X} \not\leq F^{-1}(\mathbf{1} + \mathbf{y}_0)\right)$$

$$= \frac{P\left(\mathbf{X} \leq F^{-1}(\mathbf{1} + \mathbf{y}), \, \mathbf{X} \not\leq F^{-1}(\mathbf{1} + \mathbf{y}_0)\right)}{1 - P\left(\mathbf{X} \leq F^{-1}(\mathbf{1} + \mathbf{y}_0)\right)}$$

$$= \frac{P(\mathbf{X} \leq F^{-1}(\mathbf{1} + \mathbf{y})) - P(\mathbf{X} \leq F^{-1}(\mathbf{1} + \mathbf{y}_0)}{1 - P\left(\mathbf{X} \leq F^{-1}(\mathbf{1} + \mathbf{y}_0)\right)}$$

$$= \frac{W_G(\mathbf{y}) - W_G(\mathbf{y}_0) + o(\|\mathbf{y}_0\|)}{1 - W_G(\mathbf{y}_0) + o(\|\mathbf{y}_0\|)}$$

$$= \frac{W_G(\mathbf{y}) - W_G(\mathbf{y}_0)}{1 - W(\mathbf{y}_0)} + o(\|\mathbf{y}_0\|)$$

$$= P(\mathbf{Z} \leq \mathbf{y} \mid \mathbf{Z} \not\leq \mathbf{y}_0) + o(\|\mathbf{y}_0\|)$$

uniformly in $\mathbf{y}_0 \leq \mathbf{y} \leq \mathbf{0}$ as $\mathbf{y}_0 \to \mathbf{0}$, where the rv $\mathbf{Z}$ follows a GPD with upper tail $W_G(\mathbf{y})$, $\mathbf{y}_0 \leq \mathbf{y} \leq \mathbf{0}$.

Note that we have equality

$$P\left(\mathbf{X} \leq F^{-1}(\mathbf{1} + \mathbf{y}) \mid \mathbf{X} \not\leq F^{-1}(\mathbf{1} + \mathbf{y}_0)\right)$$
$$= P(\mathbf{Z} \leq \mathbf{y} \mid \mathbf{Z} \not\leq \mathbf{y}_0), \qquad \mathbf{y}_0 \leq \mathbf{y} \leq \mathbf{0},$$

if $\mathbf{X}$ follows a GPD and $\mathbf{y}_0$ is close to $\mathbf{0}$. This shows a first Peaks-over-Threshold stability of the class of multivariate GPD, which will be investigated in more detail in the separate Section 5.3.

The preceding result indicates that just as in the univariate case, also in the multivariate case exceedances above a high threshold can reasonably be approximated only by a GPD. This is made precise by the following result due to Rootzén and Tajvidi [401]. To state the theorem, let $\mathbf{X}$ be a d-dimensional rv with df F.

Further, let $\{\mathbf{u}(t) : t \in [1, \infty)\}$ be a d-dimensional curve starting at $\mathbf{u}(1) = \mathbf{0}$ and let $\boldsymbol{\sigma}(\mathbf{u}) = \boldsymbol{\sigma}(\mathbf{u}(t)) > \mathbf{0}$ be a function with values in $\mathbb{R}^d$.

While Lemma 5.1.5 shows that a multivariate GPD is the limit of certain conditional df of F being an EVD, Theorem 5.2.4 only requires F to be in the domain of attraction of an EVD.

Theorem 5.2.4 (Rootzén and Tajvidi, 2006).

(i) *Suppose that G is a d-dimensional EVD with $0 < G(\mathbf{0}) < 1$. If F is in the domain of attraction of G, then there exists an increasing continuous curve $\mathbf{u}$ with $F(\mathbf{u}(t)) \to 1$ as $t \to \infty$, and a function $\boldsymbol{\sigma}(\mathbf{u}) > 0$ such that*

$$P\left(\mathbf{X} \le \boldsymbol{\sigma}(\mathbf{u})\mathbf{x} + \mathbf{u} \mid \mathbf{X} \not\le \mathbf{u}\right) \to \frac{1}{-\log(G(\mathbf{0}))} \log\left(\frac{G(\mathbf{x})}{G(\min(\mathbf{x}, \mathbf{0}))}\right)$$

as $t \to \infty$ for all $\mathbf{x}$.

(ii) *Suppose that there exists an increasing continuous curve $\mathbf{u}$ with $F(\mathbf{u}(t)) \to 1$ as $t \to \infty$, and a function $\boldsymbol{\sigma}(\mathbf{u}) > 0$ such that, for $\mathbf{x} > \mathbf{0}$,*

$$P\left(\mathbf{X} \le \boldsymbol{\sigma}(\mathbf{u})\mathbf{x} + \mathbf{u} \mid \mathbf{X} \not\le \mathbf{u}\right) \to H(\mathbf{x})$$

for some df H as $t \to \infty$, where the marginals of H on $\mathbb{R}$ are non-degenerate. Then the above convergence holds for all $\mathbf{x}$ and there is a uniquely determined EVD G with $G(\mathbf{0}) = e^{-1}$ such that

$$H(\mathbf{x}) = \log\left(\frac{G(\mathbf{x})}{G(\min(\mathbf{x}, \mathbf{0}))}\right).$$

This G satisfies $G(\mathbf{x}) = \exp(H(\mathbf{x}) - 1)$ for $\mathbf{x} > \mathbf{0}$ and F is in the domain of attraction of G.

Note that

$$H(\mathbf{x}) := \log\left(\frac{G(\mathbf{x})}{G(\min(\mathbf{x}, \mathbf{0})}\right), \qquad \mathbf{x} \in \mathbb{R}^d, \tag{5.8}$$

defines a df, if G is an EVD with $G(\mathbf{0}) = e^{-1}$. This follows from the arguments in Lemma 5.1.5. In particular we obtain in this case

$$H(\mathbf{x}) = 1 + \log(G(\mathbf{x})), \qquad \mathbf{x} \ge \mathbf{0},$$

i.e., H is a GPD. The definition (5.8) of a GPD is due to Rootzén and Tajvidi [401]. Different from our definition of a GPD, it prescribes its values everywhere in $\mathbb{R}^d$. As a consequence, lower dimensional marginals of H as in (5.8) are not necessarily GPD again, see Rootzén and Tajvidi [401, Section 4].

THE CASES OF INDEPENDENCE AND COMPLETE DEPENDENCE

We will take a quick look at two special cases. The first can be interpreted with the help of Theorem 5.2.4. Take a GPD W with upper tail $W(\mathbf{x}) = 1 + \sum_{i \leq d} x_i = 1 - \|\mathbf{x}\|_1$, $\mathbf{x}_0 \leq \mathbf{x} \leq \mathbf{0}$. The corresponding EVD $G(\mathbf{x}) = \exp(-\|\mathbf{x}\|_1)$, $\mathbf{x} \leq \mathbf{0}$, has independent margins $G_i(x) = \exp(x)$, $x \leq 0$, $i \leq d$. The case $\|\cdot\|_D = \|\cdot\|_1$ is, therefore, referred to as the *independence case*. In the GPD setup, however, the behavior is different. Actually in this case no observations fall into an area close to $\mathbf{0}$, which can easily be seen by differentiating the df, resulting in the density 0 close to the origin (see Michel [332, Theorem 2] for details).

However, it is still justified to speak of this case as a case of independence with the following rational. Let $\mathbf{Y} = (Y_1, \ldots, Y_d)$ be a rv with df F and tail independent components Y_i. Suppose that the df of each Y_i is in the univariate domain of attraction of $\exp(x)$, $x \leq 0$. Then F is in the domain of attraction of $\exp(-\|\mathbf{x}\|_1)$, $\mathbf{x} \leq \mathbf{0}$, see Proposition 5.27 in Resnick [393].

Thus, by Theorem 5.2.4, we know that observations exceeding a high threshold have, after a suitable transformation of the margins, asymptotically the distribution $1 + \log(\exp(-\|\mathbf{x}\|_1)) = 1 - \|\mathbf{x}\|_1$ in the extreme area. So $W(\mathbf{x}) = 1 - \|\mathbf{x}\|_1$ is the asymptotic exceedance distribution of rv with tail independent components. This implies that rv with tail independent components have in the limit no observations close to the origin. Because of this we can, still, speak of $W(\mathbf{x}) = 1 - \|\mathbf{x}\|_1$ as the independence case.

In applications, one should check observations for tail independence before applying a GPD model to exceedances, to make sure that one is not in the case of independence.

We will give another interpretation of this GPD for the bivariate case in Section 6.1.

Next we will look at the other extreme case, the case of complete dependence $W(\mathbf{x}) = 1 - \|\mathbf{x}\|_\infty$. Here the margins are completely dependent. This can be seen as follows.

Lemma 5.2.5. *Let $X_1 < 0$ be uniformly distributed on $(-1, 0)$, and put $X_d := X_{d-1} := \cdots := X_1$. Then the joint df of $(X_1, \ldots, X_d)$ on the negative quadrant is $W(\mathbf{x}) = 1 - \|\mathbf{x}\|_\infty$ for $\mathbf{x} \leq \mathbf{0}$ with $\|\mathbf{x}\|_\infty \leq 1$ and equal to 0 elsewhere.*

Proof. Choose $\mathbf{x} = (x_1, \ldots, x_d) \in (-1, 0)^d$. Then

$$\begin{aligned}
W(\mathbf{x}) &= P(X_1 \leq x_1, \ldots, X_d \leq x_d) \\
&= P(X_1 \leq \min(x_1, \ldots, x_d)) \\
&= 1 + \min(x_1, \ldots, x_d) \\
&= 1 - \max(|x_1|, \ldots, |x_d|) \\
&= 1 - \|\mathbf{x}\|_\infty.
\end{aligned}$$

If one component of $\mathbf{x}$ is smaller than -1, we, obviously, have $W(\mathbf{x}) = 0$. $\square$

In case the rv $(X_1, \ldots, X_d)$ follows the EVD $\exp(-\|\mathbf{x}\|_\infty)$, we also have $X_1 = \cdots = X_d$ with probability 1. Therefore, this case is referred to as the case of *complete dependence*.

We have shown in Lemma 5.2.5 that in the case of complete dependence the GP function is a df on its entire support on which it is non-negative. This shows that Proposition 5.1.3 would not be valid for $\lambda = \infty$ and that there exists a multivariate case where the GP function is a df on its entire support.

THE GPD OF ASYMMETRIC LOGISTIC TYPE

In Section 5.1 we have shown the logistic GPD, which contains the above mentioned cases of independence and complete dependence. We will now present an extension of this model and use it to show that GPD are, in a certain sense, not uniquely determined when modelling threshold exceedances. The family of asymmetric logistic distributions was introduced in Tawn [440] for the extreme value case. It is derived there as a limiting distribution of componentwise maxima of storms recorded at different locations along a coastline.

Let $B := \mathcal{P}(\{1, \ldots, d\}) \backslash \{\emptyset\}$ be the power set of $\{1, \ldots, d\}$ containing all non-empty subsets, and let $\lambda_\Gamma \geq 1$ be arbitrary numbers for every $\Gamma \in B$ with $|\Gamma| > 1$ and $\lambda_\Gamma = 1$ for $|\Gamma| = 1$. Furthermore, let $0 \leq \psi_{j,\Gamma} \leq 1$, where $\psi_{j,\Gamma} = 0$ if $j \notin \Gamma$ and with the side condition $\sum_{\Gamma \in B} \psi_{j,\Gamma} = 1$ for $j = 1, \ldots, d$. Then a df with upper tail

$$W_{as}(x_1, \ldots, x_d) := 1 - \sum_{\Gamma \in B} \left\{ \sum_{j \in \Gamma} (-\psi_{j,\Gamma} x_j)^{\lambda_\Gamma} \right\}^{1/\lambda_\Gamma}, \qquad (5.9)$$

$x_i < 0$, $i = 1, \ldots, d$, close to 0, is called a GPD *of asymmetric logistic type*.

Due to the side conditions for the $\psi_{j,\Gamma}$ we have in this model $2^{d-1}(d+2) - (2d+1)$ free parameters, $2^d - d - 1$ for the various λ_Γ and the rest for the $\psi_{j,\Gamma}$, see Section 2 in Stephenson [425]. In the case $\psi_{j,\{1,\ldots,d\}} = 1$ for $j = 1, \ldots, d$ and $\lambda = \lambda_\Gamma \geq 1$, we have again the (symmetric) logistic distribution.

With $d = 2$ and the short notations $\psi_1 := \psi_{1,\{1,2\}}$, $\psi_2 := \psi_{2,\{1,2\}}$, $\lambda := \lambda_{\{1,2\}}$, formula (5.9) reduces to

$$W_{as}(x_1, x_2) = 1 + (1 - \psi_1)x_1 + (1 - \psi_2)x_2 - \left((-\psi_1 x_1)^\lambda + (-\psi_2 x_2)^\lambda\right)^{1/\lambda}.$$

In the case $d = 3$ we have

$$\begin{aligned}
W_{as}(x_1, x_2, x_3) &= 1 + (1 - \psi_1 - \psi_3 - \psi_7)x_1 + (1 - \psi_2 - \psi_5 - \psi_8)x_2 \\
&\quad + (1 - \psi_4 - \psi_6 - \psi_9)x_3 \\
&\quad - \left((-\psi_1 x_1)^{\lambda_1} + (-\psi_2 x_2)^{\lambda_1}\right)^{1/\lambda_1} \\
&\quad - \left((-\psi_3 x_1)^{\lambda_2} + (-\psi_4 x_3)^{\lambda_2}\right)^{1/\lambda_2}
\end{aligned}$$

$$- \left((-\psi_5 x_2)^{\lambda_3} + (-\psi_6 x_3)^{\lambda_3} \right)^{1/\lambda_3}$$
$$- \left((-\psi_7 x_1)^{\lambda_4} + (-\psi_8 x_2)^{\lambda_4} + (-\psi_9 x_3)^{\lambda_4} \right)^{1/\lambda_4} \qquad (5.10)$$

with the corresponding short notations for the $\psi_{j,\Gamma}$ and λ_Γ.

The following result can be shown by tedious but elementary calculations, see the proof of Lemma 2.3.18 in Michel [330].

Lemma 5.2.6. *The function*

$$w_{as}(\mathbf{x})$$

$$= \left(\prod_{i=1}^{d-1} (i\lambda_\Delta - 1) \right) \left(\prod_{i=1}^{d} \psi_{i,\Delta} \right)^{\lambda_\Delta} \left(\prod_{i=1}^{d} (-x_i) \right)^{\lambda_\Delta - 1} \left(\sum_{j=1}^{d} (-\psi_{j,\Delta} x_j)^{\lambda_\Delta} \right)^{\frac{1}{\lambda_\Delta} - d}$$

is the density of W_{as} for $\mathbf{x}_0 \leq \mathbf{x} < \mathbf{0}$.

Note that in the density in Lemma 5.2.6 only those parameters with index set $\Delta = \{1, \ldots, d\}$. In contrast to the extreme value case, the lower hierarchical parameters do not play a role close to the origin.

The next corollary follows from Lemma 5.2.6. It shows that different GPD can lead to the same conditional probability measure in the area of interest close to $\mathbf{0}$.

Corollary 5.2.7. *Let W_1 and W_2 be GPD such that there exists a neighborhood U of $\mathbf{0}$ (in the relative topology of the negative quadrant), such that*

$$P_{W_1}(B) = P_{W_2}(B)$$

for all Borel sets $B \subset U$. Then W_1 and W_2 and correspondingly their angular measures ν_1 and ν_2 may be different.

Proof. We will establish a counterexample in dimension $d = 3$. Let W_1 and W_2 be two trivariate GPD of asymmetric logistic type with identical parameters ψ_7, ψ_8, ψ_9 and λ_4 in the notation of (5.10) but with different parameter λ_1. Then we know from Lemma 5.2.6 that W_1 and W_2 have the same density close to the origin, i.e., $P_{W_1}(B) = P_{W_2}(B)$ for all Borel sets B close to the origin. Let $G_i = \exp(W_i - 1)$, $i = 1, 2$, be the corresponding extreme value distributions. The angular measures ν_1 and ν_2 belonging to G_1 and G_2 and, thus, W_1 and W_2 are given in Section 3.5.1 of Kotz and Nadarajah [293] in terms of their densities. These densities depend (on the lower boundaries of R_2) on the parameter λ_1 and are, thus, different for different λ_1, leading to $\nu_1 \neq \nu_2$. $\square$

The difference between the two angular measures in the proof of Corollary 5.2.7 is in the lower dimensional boundaries of the unit simplex. Both measures agree in the interior.

The proof of Corollary 5.2.7 can be carried over to arbitrary dimension $d > 3$, since its conclusion only uses free lower hierarchical parameters. As these exist in the asymmetric logistic case only for $d > 2$, we, thus, encounter here again an example of the fact that multivariate extreme value theory actually starts in dimension $d \geq 3$.

As a consequence of Corollary 5.2.7, it is possible to model exceedances by different GPD, which, however, lead to identical conditional probability measures close to the origin. For distributions $F_i \in \mathcal{D}(G_i)$, $i = 1, 2$ from the domains of attraction of different EVD $G_1 \neq G_2$ it may be possible to model exceedances over high thresholds of F_1 not only by W_1 but also by W_2 and vice versa.

A generalization of the asymmetric model to the generalized asymmetric model is given with the help of suitable norms in Hofmann [222], Section 5.5.

ANOTHER REPRESENTATION OF GPD AND EVD

The following result characterizes a GPD with standard uniform margins in terms of rv. For a proof we refer to Aulbach et al. [18]. It provides in particular an easy way to generate a multivariate GPD, thus extending the bivariate approach proposed by Buishand et al. [58] to an arbitrary dimension. Recall that an arbitrary multivariate GPD can be obtained from a GPD with ultimately uniform margins by just transforming the margins.

Proposition 5.2.8.

(i) *Let W be a multivariate GPD with standard uniform margins in a left neighborhood of $\mathbf{0} \in \mathbb{R}^d$. Then there exists a rv $\mathbf{Z} = (Z_1, \ldots, Z_d)$ with $Z_i \in [0, d]$ a.s., $E(Z_i) = 1$, $i \leq d$, and $\sum_{i \leq d} Z_i = d$ a.s. as well as a vector $(-1/d, \ldots, -1/d) \leq \mathbf{x}_0 < \mathbf{0}$ such that*

$$W(\mathbf{x}) = P\left(-U\left(\frac{1}{Z_1}, \ldots, \frac{1}{Z_d}\right) \leq \mathbf{x}\right), \qquad \mathbf{x}_0 \leq \mathbf{x} \leq \mathbf{0},$$

where the rv U is uniformly on $(0, 1)$ distributed and independent of $\mathbf{Z}$.

(ii) *The rv $-U(1/Z_1, \ldots, 1/Z_d)$ follows a GPD with standard uniform margins in a left neighborhood of $\mathbf{0} \in \mathbb{R}^d$ if U is independent of $\mathbf{Z} = (Z_1, \ldots, Z_d)$ and $0 \leq Z_i \leq c_i$ a.s. with $E(Z_i) = 1$, $i \leq d$, for some $c_1, \ldots, c_d \geq 1$.*

Note that the case of a GPD W with *arbitrary* uniform margins $W_i(x) = 1 - a_i x$ in a left neighborhood of $\mathbf{0}$ with scaling factors $a_i > 0$, $i \leq d$, immediately follows from the preceding result by substituting Z_i by $a_i Z_i$.

We remark that the distribution of the rv $\mathbf{Z}$ in part (i) of Proposition 5.2.8 is uniquely determined in the following sense. Let $\mathbf{T} = (T_1, \ldots, T_d)$ be another rv with values in $[0, d]^d$, $E(T_i) = 1$, $i \leq d$, $\sum_{i \leq d} T_i = d$ a.s., being independent of U and satisfying

$$-U\left(\frac{1}{Z_1}, \ldots, \frac{1}{Z_d}\right) =_D -U\left(\frac{1}{T_1}, \ldots, \frac{1}{T_d}\right). \tag{5.11}$$

Then we have

$$\mathbf{Z} =_D \mathbf{T}.$$

This can easily be seen as follows. Equation (5.11) implies

$$\frac{1}{U}\mathbf{Z} =_D \frac{1}{U}\mathbf{T} \implies \frac{1}{U}\frac{\mathbf{Z}}{\frac{\sum_{i\leq d} Z_i}{U}} =_D \frac{1}{U}\frac{\mathbf{T}}{\frac{\sum_{i\leq d} T_i}{U}} \implies \mathbf{Z} =_D \mathbf{T}.$$

If we drop the condition $\sum_{i\leq d} T_i = d$ a.s. and substitute for it the assumption $\sum_{i\leq d} T_i > 0$ a.s., then the above considerations entail

$$\mathbf{Z} =_D \frac{\mathbf{T}}{\frac{1}{d}\sum_{i\leq d} T_i}.$$

We, thus, obtain that $(T_1, \ldots, T_d)$ in (5.11) can be substituted for by

$$\widetilde{T}_i := \frac{T_i}{\frac{1}{d}\sum_{i\leq d} T_i}, \qquad i \leq d,$$

satisfying $\widetilde{T}_i \in [0, d]$, $E(\widetilde{T}_i) = 1$, $i \leq d$, $\sum_{i\leq d} \widetilde{T}_i = d$ as well as (5.11).

Proposition 5.2.8 entails the following representation of an arbitrary EVD with standard negative exponential margins. This result extends the spectral representation of an EVD in (4.23). It links, in particular, the set of copulas with the set of EVD.

Corollary 5.2.9.

(i) *Let G be an arbitrary EVD in $\mathbb{R}^d$ with standard negative exponential margins. Then there exists a rv $\mathbf{Z} = (Z_1, \ldots, Z_d)$ with $Z_i \in [0, d]$ a.s., $E(Z_i) = 1$, $i \leq d$, and $\sum_{i\leq d} Z_i = d$, such that*

$$G(\mathbf{x}) = \exp\left(\int \min_{i\leq d}(x_i Z_i)\, dP\right), \qquad \mathbf{x} \leq \mathbf{0}.$$

(ii) *Let, on the other hand, the rv $\mathbf{Z} = (Z_1, \ldots, Z_d)$ satisfy $Z_i \in [0, c_i]$ a.s. with $E(Z_i) = 1$, $i \leq d$, for some $c_1, \ldots, c_d \geq 1$. Then*

$$G(\mathbf{x}) := \exp\left(\int \min_{i\leq d}(x_i Z_i)\, dP\right), \qquad \mathbf{x} \leq \mathbf{0},$$

defines an EVD on $\mathbb{R}^d$ with standard negative exponential margins.

Proof. Let U be a uniformly on (0,1) distributed rv, which is independent of $\mathbf{Z}$. Both parts of the assertion follow from the fact that

$$W(\mathbf{x}) = P\left(-U\left(\frac{1}{Z_1}, \ldots, \frac{1}{Z_d}\right) \leq \mathbf{x}\right), \qquad \mathbf{x} \leq \mathbf{0},$$

defines a GPD on $\mathbb{R}^d$, i.e., $W = 1 + \log(G)$, which is in the domain of attraction of the EVD G with standard negative exponential margins: We have

$$W\left(\frac{\mathbf{x}}{n}\right)^n = (1 + \log\left(G\left(\frac{\mathbf{x}}{n}\right)\right)^n \to_{n\to\infty} G(\mathbf{x})$$

as well as, for n large,

$$
\begin{aligned}
W\left(\frac{\mathbf{x}}{n}\right)^n &= P\left(-U\left(\frac{1}{Z_1}, \ldots, \frac{1}{Z_d}\right) \le \frac{\mathbf{x}}{n}\right)^n \\
&= P\left(U \ge \frac{1}{n}\max_{i\le d}(-x_i Z_i)\right)^n \\
&= \left(\int P\left(U \ge \frac{1}{n}\max_{i\le d}(-x_i z_i) \mid Z_i = z_i, i \le d\right)(P * \mathbf{Z})(d\mathbf{z})\right)^n \\
&= \left(\int P\left(U \ge \frac{1}{n}\max_{i\le d}(-x_i z_i)\right)(P * \mathbf{Z})(d\mathbf{z})\right)^n \\
&= \left(1 - \int P\left(U \le \frac{1}{n}\max_{i\le d}(-x_i z_i)\right)(P * \mathbf{Z})(d\mathbf{z})\right)^n \\
&= \left(1 - \frac{1}{n}\int \max_{i\le d}(-x_i z_i)(P * \mathbf{Z})(d\mathbf{z})\right)^n \\
&\to_{n\to\infty} \exp\left(\int \min_{i\le d}(x_i Z_i)\, dP\right), \qquad \mathbf{x} \le \mathbf{0}. \qquad \square
\end{aligned}
$$

Let, for instance, C be an arbitrary d-dimensional copula, i.e., C is the df of a rv $\mathbf{S}$ with uniform margins $P(S_i \le s) = s$, $s \in (0,1)$, $i \le d$, (Nelsen [350]). Then $\mathbf{Z} := 2\mathbf{S}$ is a proper choice in part (ii) of Proposition 5.2.8 as well of Corollary 5.2.9. Precisely,

$$G(\mathbf{x}) := \exp\left(2\int_{[0,1]^d} \min(x_i u_i)\, C(d\mathbf{u})\right), \qquad \mathbf{x} \le \mathbf{0},$$

defines for an arbitrary copula C an EVD with standard negative exponential margins. This result maps the set of copulas in a natural way to the set of multivariate GPD and EVD, thus opening a wide range of possible scenarios.

MULTIVARIATE PIECING-TOGETHER

If X is a univariate rv with df F, then the df $F^{[x_0]}$ of X, conditional on the event $X > x_0$, is given by

$$F^{[x_0]}(x) = P(X \le x \mid X > x_0) = \frac{F(x) - F(x_0)}{1 - F(x_0)}, \qquad x \ge x_0,$$

where we require $F(x_0) < 1$. The POT approach shows that $F^{[x_0]}$ can reasonably be approximated only by a GPD with appropriate shape, location and scale parameter $W_{\gamma,\mu,\sigma}$, say,

$$
\begin{aligned}
F(x) &= (1 - F(x_0))F^{[x_0]}(x) + F(x_0) \\
&\approx (1 - F(x_0))W_{\gamma,\mu,\sigma}(x) + F(x_0), \qquad x \geq x_0.
\end{aligned}
$$

The piecing-together approach (PT) now consists in replacing the df F by

$$
F^*_{x_0}(x) = \begin{cases} F(x), & x \leq x_0, \\ (1 - F(x_0))W_{\gamma,\mu,\sigma}(x) + F(x_0), & x > x_0, \end{cases} \tag{5.12}
$$

where the shape, location and scale parameters γ, μ, σ of the GPD are typically estimated from given data. This modification aims at a more precise investigation of the upper end of the data.

Replacing F in (5.12) by the empirical df of n independent copies of X offers in particular a semiparametric approach to the estimation of high quantiles $F^{-1}(q) = \inf\{t \in \mathbb{R} : F(t) \geq q\}$ outside the range of given data, see, e.g., Section 2.3 of Reiss and Thomas [390].

For mathematical convenience we temporarily shift a copula to the interval $[-1,0]^d$ by shifting each univariate margin by -1. We call a df C_W on $[-1,0]^d$ a *GPD-copula* if each marginal df is the uniform distribution on $[-1,0]$ and C_W coincides close to zero with a GP function W as in equation (5.1), i.e., there exists $\mathbf{x}_0 < \mathbf{0}$ such that

$$
C_W(\mathbf{x}) = W(\mathbf{x}) = 1 - \|\mathbf{x}\|_D, \qquad \mathbf{x}_0 \leq \mathbf{x} \leq \mathbf{0},
$$

where the D-norm is standardized, i.e., $\|\mathbf{e}_i\|_D = 1$ for each unit vector in $\mathbb{R}^d$.

For later purposes we remark that a rv $\mathbf{V} \in [-1,0]^d$ following a GPD-copula can easily be generated using Proposition 5.2.8 as follows. Let U be uniformly on $(0,1)$ distributed and independent of the vector $\mathbf{S} = (S_1, \ldots, S_d)$, which follows an arbitrary copula on $[0,1]^d$. Then we have for $i \leq d$,

$$
P\left(-U\frac{1}{2S_i} \leq x\right) = \begin{cases} 1 + x, & \text{if } -\frac{1}{2} \leq x \leq 0, \\ \frac{1}{4|x|}, & \text{if } x < -\frac{1}{2}, \end{cases} =: H(x), \qquad x \leq 0,
$$

and, consequently,

$$
\mathbf{V} := \left(H\left(-\frac{U}{2S_1}\right) - 1, \ldots, H\left(-\frac{U}{2S_d}\right) - 1\right) = (V_1, \ldots, V_d)
$$

with

$$
V_i = \begin{cases} -\frac{U}{2S_i}, & \text{if } U \leq S_i, \\ \frac{S_i}{2U} - 1, & \text{if } U > S_i, \end{cases} \tag{5.13}
$$

follows by Proposition 5.2.8 a GPD-copula on $[-1,0]^d$.

The multivariate PT approach consists of two steps. In a first step, the upper tail of a given d-dimensional copula C, say, is cut off and substituted for by the upper tail of multivariate GPD-copula in a continuous manner. The result is again a copula, i.e., a d-dimensional distribution with uniform margins. The other step consists of the transformation of each margin of this copula by a given univariate df F_i^*, $1 \leq i \leq d$. This provides, altogether, a multivariate df with prescribed margins F_i^*, whose copula coincides in its central part with C and in its upper tail with a GPD-copula.

FITTING A GPD-COPULA TO A GIVEN COPULA

Let the rv $\mathbf{V} = (V_1, \ldots, V_d)$ follow a GPD-copula on $[-1,0]^d$. That is, $P(V_i \leq x) = 1 + x$, $-1 \leq x \leq 0$, is for each $i \leq d$ the uniform distribution on $[-1,0]$, and there exists $\mathbf{w} = (w_1, \ldots, w_d) < \mathbf{0}$ such that, for each $\mathbf{x} \in [\mathbf{w}, \mathbf{0}]$,

$$P(\mathbf{V} \leq \mathbf{x}) = 1 - \|\mathbf{x}\|_D,$$

where $\|\cdot\|_D$ is a standardized D-norm, i.e., it is a D-norm with the property $\|\mathbf{e}_i\|_D = 1$ for each unit vector $\mathbf{e}_i$ in $\mathbb{R}^d$, $1 \leq i \leq d$.

Let $\mathbf{Y} = (Y_1, \ldots, Y_d)$ follow an arbitrary copula C on $[-1,0]^d$ and suppose that $\mathbf{Y}$ is independent of $\mathbf{V}$. Choose a threshold $\mathbf{y} = (y_1, \ldots, y_d) \in [-1,0]^d$ and put

$$Q_i := Y_i 1_{(Y_i \leq y_i)} - y_i V_i 1_{(Y_i > y_i)}, \qquad i \leq d. \tag{5.14}$$

The rv $\mathbf{Q}$ then follows a GPD-copula on $[-1,0]^d$, which coincides with C on $\times_{i \leq d}[-1, y_i]$. This is the content of the following result. For a proof we refer to Aulbach et al. [18].

Proposition 5.2.10. *Suppose that $P(\mathbf{Y} > \mathbf{y}) > 0$. Each Q_i defined in (5.14) follows the uniform df on $[-1,0]$. The rv $\mathbf{Q} = (Q_1, \ldots, Q_d)$ follows a GPD-copula on $[-1,0]^d$, which coincides on $\times_{i \leq d}[-1, y_i]$ with C, i.e.,*

$$P(\mathbf{Q} \leq \mathbf{x}) = C(\mathbf{x}), \qquad \mathbf{x} \leq \mathbf{y}.$$

We have, moreover, with $x_i \in [\max(y_i, w_i), 0]$, $i \leq d$, for any non-empty subset K of $\{1, \ldots, d\}$,

$$P(Q_i \geq x_i, i \in K) = P(V_i \geq b_{i,K} x_i, i \in K),$$

where

$$b_{i,K} := \frac{P(Y_j > y_j, j \in K)}{-y_i} = \frac{P(Y_j > y_j, j \in K)}{P(Y_i > y_i)} \in (0, 1], \qquad i \in K.$$

The above approach provides an easy way to generate a rv $\mathbf{X} \in \mathbb{R}^d$ with prescribed margins F_i^*, $i \leq d$, such that $\mathbf{X}$ has a given copula in the central part of the data, whereas in the upper tail it has a GPD-copula.

Take $\mathbf{Q} = (Q_1, \ldots, Q_d)$ as in (5.14) and put $\widetilde{\mathbf{Q}} := (Q_1 + 1, \ldots, Q_d + 1)$. Then each component $\widetilde{Q}_i$ of $\widetilde{\mathbf{Q}}$ is uniformly distributed on $(0, 1)$ and, thus,

$$\mathbf{X} := (X_1, \ldots, X_d) := \left(F_1^{*-1}(\widetilde{Q}_1), \ldots, F_d^{*-1}(\widetilde{Q}_d) \right)$$

has the desired properties.

Combining the univariate *and* the multivariate PT approach now consists in defining F_i^* by choosing a threshold $u_i \in \mathbb{R}$ for each dimension $i \leq m$ as well as a univariate df F_i together with an arbitrary univariate GPD $W_{\gamma_i, \mu_i \sigma_i}$, and putting, for $i \leq d$,

$$F_i^*(x) := \begin{cases} F_i(x), & \text{if } x \leq u_i \\ (1 - F_i(u_i))W_{\gamma_i, \mu_i \sigma_i}(x) + F_i(u_i), & \text{if } x > u_i \end{cases} . \qquad (5.15)$$

This is typically done in a way such that F_i^* is a continuous function. This multivariate PT approach is utilized in [18] to operational loss data to evaluate the range of operational risk.

5.3   Peaks-Over-Threshold Stability of a GPD

Recall that a univariate GPD is in its standard form any member of

$$W(x) = 1 + \log(G(x)) = \begin{cases} 1 - (-x)^\alpha, & -1 \leq x \leq 0, & \text{(polynomial GPD)}, \\ 1 - x^{-\alpha}, & x \geq 1, & \text{(Pareto GPD)}, \\ 1 - \exp(-x), & x \geq 0, & \text{(exponential GPD)}, \end{cases}$$

where G is a univariate EVD.

According to Theorem 2.7.1, cf. also (2.15), a univariate GPD is characterized by its peaks-over-threshold (POT) stability: Let V be a rv which follows a univariate GPD W as in (5.3). Then we obtain for any x_0 with $W(x_0) \in (0, 1)$

$$P\left(V > tx_0 \mid V > x_0\right) = t^\alpha, \qquad t \in [0, 1], \qquad (5.16)$$

for a polynomial GPD,

$$P\left(V > tx_0 \mid V > x_0\right) = t^{-\alpha}, \qquad t \geq 1, \qquad (5.17)$$

for a Pareto GPD and

$$P\left(V > x_0 + t \mid V > x_0\right) = \exp(-t), \qquad t \geq 0, \qquad (5.18)$$

for the exponential GPD. By POT stability we mean that the above excess distributions are invariant to the choice of x_0.

The main contribution of this section is the multivariate extension of this result. It is, however, not obvious how to define a *multivariate* exceedance. Put, therefore, for any of the above three univariate cases, $A := (x_0, \infty)$. Then A satisfies the condition

$$x \in A \implies \begin{cases} t^{1/\alpha} x \in A, & t \in (0,1], \quad \text{in the polynomial case,} \\ t^{1/\alpha} x \in A, & t \geq 1, \quad \text{in the Pareto case,} \\ x + t \in A, & t \geq 0, \quad \text{in the exponential case.} \end{cases}$$

The preceding equations (5.16)-(5.18) can now be written as

$$P\left(t^{-1/\alpha} V \in A \mid V \in A\right) = t, \quad t \in (0,1], \text{ in the polynomial case,}$$

$$P\left(t^{-1/\alpha} V \in A \mid V \in A\right) = t^{-1}, \quad t \geq 1, \text{ in the Pareto case,}$$

$$P\left(V - t \in A \mid V \in A\right) = \exp(-t), \quad t \geq 0, \text{ in the exponential case.}$$

These different equations can be summarized as follows. Put, for an arbitrary univariate EVD G,

$$\psi(x) := \log(G(x)) = W(x) - 1, \quad 0 < G(x) < 1,$$

which defines a continuous and strictly monotone function with range $(-1, 0)$. Then we have, for $A = (x_0, 0)$ with $-1 \leq x_0 < 0$ and $P(V \in \psi^{-1}(A)) > 0$,

$$P(V \in \psi^{-1}(tA) \mid V \in \psi^{-1}(A)) = t, \quad t \in (0,1], \tag{5.19}$$

where the rv V follows the GPD W. This result is immediate from the fact that $\psi(V)$ follows the GPD $1 + x$, $-1 \leq x \leq 0$.

This POT stability of a univariate GPD will be extended to an arbitrary dimension. The following result is the multivariate analogue of equation (5.19).

Theorem 5.3.1. *Let $A \subset (\mathbf{x}_0, \mathbf{0}]$, $\mathbf{x}_0 \geq (-1/d, \ldots, -1/d)$, be a Borel set with the cone type property*

$$\mathbf{x} \in A \implies t\mathbf{x} \in A, \quad t \in (0,1]. \tag{5.20}$$

Suppose that the rv $\mathbf{V}$ follows an arbitrary GPD W with margins W_i, $i \leq d$, and put $\psi_i(x) = W_i(x) - 1$, $i \leq d$, $\Psi(\mathbf{x}) := (\psi_1(x_1), \ldots, \psi_d(x_d))$. If $P\left(\mathbf{V} \in \Psi^{-1}(A)\right) > 0$, then we have

$$P\left(\mathbf{V} \in \Psi^{-1}(tA) \mid \mathbf{V} \in \Psi^{-1}(A)\right) = t, \quad t \in [0,1].$$

Proof. We have

$$P\left(\mathbf{V} \in \Psi^{-1}(tA) \mid \mathbf{V} \in \Psi^{-1}(A)\right) = P\left(\Psi(\mathbf{V}) \in tA \mid \Psi(\mathbf{V}) \in A\right),$$

where $\widetilde{\mathbf{V}} := \Psi(\mathbf{V})$ follows a GPD with ultimate standard uniform margins $W_i(x) = 1+x$, $x_0 \leq x \leq 0$. From Proposition 5.2.8 we know that there exists a uniformly on $(0,1)$ distributed rv U and an independent rv $\mathbf{Z} = (Z_1, \ldots, Z_d)$ with $0 \leq Z_i \leq d$, $E(Z_i) = 1$, $i \leq d$, such that

$$P(\widetilde{\mathbf{V}} \leq \mathbf{x}) = P\left(-U\frac{1}{\mathbf{Z}} \leq \mathbf{x}\right), \qquad \mathbf{x}_0 \leq \mathbf{x} \leq \mathbf{0},$$

where $1/\mathbf{Z}$ is meant componentwise.

We obtain, consequently, for $t \in (0,1]$,

$$P\left(\frac{1}{t}\widetilde{\mathbf{V}} \in A\right) = P\left(-\frac{U}{t}\frac{1}{\mathbf{Z}} \in A\right)$$

$$= P\left(-\frac{U}{t}\frac{1}{\mathbf{Z}} \in A, U \leq t\right) + P\left(-\frac{U}{t}\frac{1}{\mathbf{Z}} \in A, U > t\right).$$

Note that the second probability vanishes as $-U/(t\mathbf{Z}) \notin A$ if $U > t$; recall that $1/\mathbf{Z} \geq (1/d, \ldots, 1/d)$, $\mathbf{x}_0 \geq (1/d, \ldots, 1/d)$ and that $A \subset (\mathbf{x}_0, \mathbf{0}]$. Conditioning on $U = u$ and substituting u by tu entails

$$P\left(-\frac{U}{t}\frac{1}{\mathbf{Z}} \in A, U \leq t\right) = \int_0^t P\left(\frac{u}{t\mathbf{Z}} \in A\right) du$$

$$= t\int_0^1 P\left(\frac{u}{\mathbf{Z}} \in A\right) du$$

$$= tP\left(\frac{U}{\mathbf{Z}} \in A\right)$$

$$= tP(\widetilde{\mathbf{V}} \in A).$$

We, thus, have established the equality

$$P\left(\frac{1}{t}\widetilde{\mathbf{V}} \in A\right) = tP(\widetilde{\mathbf{V}} \in A), \qquad 0 < t \leq 1.$$

As the set A has the property $\mathbf{x} \in A \implies t\mathbf{x} \in A$ for $t \in (0,1]$, we obtain the assertion:

$$P\left(\frac{1}{t}\widetilde{\mathbf{V}} \in A \,\Big|\, \widetilde{\mathbf{V}} \in A\right) = \frac{P\left(\frac{1}{t}\widetilde{\mathbf{V}} \in A, \widetilde{\mathbf{V}} \in A\right)}{P(\mathbf{V} \in A)}$$

$$= \frac{P\left(\frac{1}{t}\mathbf{V} \in A\right)}{P(\mathbf{V} \in A)}$$

$$= t, \qquad t \in (0,1]. \qquad \square$$

The following consequence of Theorem 5.3.1 generalizes results by Falk and Guillou [139] on the POT-stability of multivariate GPD by dropping the differentiability condition on the Pickands dependence function corresponding to the GPD W.

Corollary 5.3.2. *Let the rv* $\mathbf{V}$ *follow an arbitrary GPD* W *on* $\mathbb{R}^d$ *with ultimate univariate GPD margins* $W_1, \ldots, W_d$ *and let the the set* $A \subset \{\mathbf{x} \in \mathbb{R}^d : -1/d < W_i(x_i) - 1 \leq 0, i \leq d\}$ *satisfy*

$$\mathbf{x} \in A$$

$$\implies \begin{cases} \left(t^{1/\alpha_1}x_1, \ldots, t^{1/\alpha_d}x_d\right) \in A, & t \in (0,1], \\ & \quad \text{if } W_i(x) = 1 - |x|^{\alpha_i}, -1 \leq x \leq 0, i \leq d, \\ \left(t^{1/\alpha_1}x_1, \ldots, t^{1/\alpha_d}x_d\right) \in A, & t \geq 1, \text{if } W_i(x) = 1 - x^{-\alpha_i}, x \geq 1, i \leq d, \\ (x_1 + t, \ldots, x_d + t) \in A, & t \geq 0, \text{if } W_i(x) = 1 - \exp(-x), x \geq 0, i \leq d, \end{cases}$$

where $\alpha_i > 0$. *Then, if* $P(\mathbf{V} \in A) > 0$,

$$P\left(\left(t^{-1/\alpha_1}V_1, \ldots, t^{-1/\alpha_d}V_d\right) \in A \mid \mathbf{V} \in A\right)$$

$$= \begin{cases} t, & t \in (0,1], & \text{if } W_i(x) = 1 - |x|^{\alpha_i}, -1 \leq x \leq 0, i \leq d, \\ t^{-1}, & t \geq 1, & \text{if } W_i(x) = 1 - x^{-\alpha_i}, x \geq 1, i \leq d, \end{cases}$$

and

$$P((V_1 - t, \ldots, V_d - t) \in A \mid \mathbf{V} \in A) = \exp(-t), \qquad t \geq 0,$$

if $W_i(x) = 1 - \exp(-x)$, $x \geq 0$, $i \leq d$.

Proof. Put $\widetilde{A} := \{(\psi_1(a_1), \ldots, \psi_d(a_d)) : \mathbf{a} \in A\}$, where $\psi_i = W_i - 1$, $i \leq d$. Then $\widetilde{A} \subset ((-1/d, \ldots, -1/d), \mathbf{0}]$ satisfies

$$\mathbf{x} \in \widetilde{A} \implies t\mathbf{x} \in \widetilde{A}, \qquad 0 < t \leq 1,$$

and, thus, with $\Psi(\mathbf{x}) = (\psi_1(x_1), \ldots, \psi_d(x_d))$ we obtain from Theorem 5.3.1

$$P\left(\mathbf{V} \in \Psi^{-1}\left(t\widetilde{A}\right) \mid \mathbf{V} \in \Psi^{-1}\left(\widetilde{A}\right)\right) = t, \qquad t \in [0,1],$$

which implies the assertion for any of the three different cases. Note that $\{\mathbf{V} \in \Psi^{-1}\left(\widetilde{A}\right)\} = \{\mathbf{V} \in A\}$. $\qquad\square$

Example 5.3.3. Let $\mathbf{V}$ follow a GPD with ultimately standard uniform margins $W_i(x) = 1 + x$, $-1 \leq x \leq 0$, $i \leq d$. Then we obtain for any $c < 0$ close to 0 and arbitrary weights $\lambda_i > 0$, $i \leq d$,

$$P\left(\sum_{i \leq d} \lambda_i V_i > tc \mid \sum_{i \leq d} \lambda_i V_i > c\right) = t, \qquad t \in [0,1],$$

provided $P\left(\sum_{i \leq d} \lambda_i V_i > t\right) > 0$.

Proof. Set $A_c := \left\{ (-1/d, \ldots, -1/d) < \mathbf{a} \leq \mathbf{0} : \sum_{i \leq d} \lambda_i a_i > c \right\}$. Then the set A_c satisfies

$$\mathbf{x} \in A_c \implies t\mathbf{x} \in A_c, \qquad 0 < t \leq 1,$$

and, thus, the assertion is immediate from Corollary 5.3.2, applying the first case with $\alpha_1 = \cdots = \alpha_d = 1$.   □

We can interpret $\sum_{i \leq d} \lambda_i V_i$ as a linear portfolio with weights λ_i and risks V_i. A risk measure such as the expected shortfall $E\left(\sum_{i \leq d} \lambda_i V_i \mid \sum_{i \leq d} \lambda_i V_i > c \right)$, thus, fails in case of a multivariate GPD with ultimately uniform margins, as it is by the preceding example independent of the weights λ_i. For a further discussion we refer to [139].

The following characterization of a multivariate GPD with uniform margins, which requires no additional smoothness conditions on the underlying dependence function D, extends characterizations of a GPD as established in Theorem 2 and Proposition 6 in Falk and Guillou [139]. This result will suggest the definition of a statistic in Section 5.8, which tests for an underlying multivariate GPD. The conclusion " $\implies$ " is immediate from Corollary 5.3.2, first case. The reverse implication of Proposition 5.3.4 is established in Falk and Guillou [139], Proposition 6.

Proposition 5.3.4. *An arbitrary rv* $\mathbf{V} = (V_1, \ldots, V_d)$ *follows a GPD with uniform margins* $W_i(x) = 1 + a_i x$, $x_0 \leq x \leq 0$, *with some scaling factors* $a_i > 0$, $i \leq d$, *if, and only if, there exists* $\mathbf{x}_0 = (x_{0,1}, \ldots, x_{0,d}) < \mathbf{0}$ *with* $P(U_i > x_{0,i}) > 0$, $i \leq d$, *such that for any non-empty subset* $K \subset \{1, \ldots, d\}$ *of indices*

$$P(V_k > t x_k, k \in K) = t P(V_k > x_k, k \in K), \qquad t \in [0,1], \tag{5.21}$$

for any $\mathbf{x} = (x_1, \ldots, x_d) \in [\mathbf{x}_0, \mathbf{0}]$.

5.4   A Spectral Decomposition Based on Pickands Coordinates

Motivated by the special dependence of G and W on $\mathbf{x}$ and D we introduce the Pickands coordinates pertaining to $\mathbf{x}$. A consequence of this definition will be a decomposition of a multivariate df H into a family of univariate df. This decomposition also suggests estimators of the Pickands dependence function D under EVD and GPD.

Recall that the Pickands dependence function D is defined on the set

$$R = \left\{ (t_1, \ldots, t_{d-1}) \in [0,1]^{d-1} : \sum_{i \leq d-1} t_i \leq 1 \right\},$$

see (4.30).

PICKANDS COORDINATES IN $\mathbb{R}^d$

Any vector $\mathbf{x} = (x_1, \ldots, x_d) \in (-\infty, 0]^d$ with $\mathbf{x} \neq 0$ can be uniquely represented as

$$\mathbf{x} = (x_1 + \cdots + x_d) \left(\frac{x_1}{x_1 + \cdots + x_d}, \ldots, \frac{x_{d-1}}{x_1 + \cdots + x_d}, 1 - \frac{x_1 + \cdots + x_{d-1}}{x_1 + \cdots + x_d} \right)$$

$$=: c \left(z_1, \ldots, z_{d-1}, 1 - \sum_{i \leq d-1} z_i \right),$$

where $c < 0$ and $\mathbf{z} = (z_1, \ldots, z_{d-1}) \in R$ are the *Pickands coordinates* of $\mathbf{x}$. They are similar to polar coordinates in $\mathbb{R}^d$, but using the $\|\cdot\|_1$-norm in place of the usual $\|\cdot\|_2$-norm. The vector $\mathbf{z}$ represents the angle and the number c the distance of $\mathbf{x}$ from the origin. Therefore, $\mathbf{z}$ and c are termed *angular* and *(pseudo-)radial* component (Nadarajah [343]).

SPECTRAL DECOMPOSITIONS OF DISTRIBUTION FUNCTIONS

Let H be an arbitrary df on $(-\infty, 0]^d$ and put, for $\mathbf{z} \in R$ and $c \leq 0$,

$$H_{\mathbf{z}}(c) := H\left(c\left(z_1, \ldots, z_{d-1}, 1 - \sum_{i \leq d-1} z_i \right) \right).$$

With $\mathbf{z}$ being fixed, $H_{\mathbf{z}}$ is a univariate df on $(-\infty, 0]$. This can easily be seen as follows. Let $\mathbf{U} = (U_1, \ldots, U_d)$ be a rv with df H. Then we have

$$H_{\mathbf{z}}(c) = \begin{cases} P\left(\max\left(\max_{i:z_i>0} \frac{U_i}{z_i}, \frac{U_d}{1-\sum_{i \leq d-1} z_i} \right) \leq c \right), & 0 < \sum_{i \leq d-1} z_i < 1, \\ P\left(\max_{i:z_i>0} \frac{U_i}{z_i} \leq c \right), & \text{if} \quad \sum_{i \leq d-1} z_i = 1, \\ P(U_d \leq c), & \sum_{i \leq d-1} z_i = 0. \end{cases}$$

Note that

$$H_{\mathbf{z}}(c) = P\left(\max_{i \leq d} \frac{U_i}{z_i} \leq c \right),$$

where $z_d := 1 - \sum_{i \leq d-1} z_i$, if H is continuous at $\mathbf{0}$.

The df H is obviously uniquely determined by the family

$$\mathcal{P}(H) := \{H_{\mathbf{z}} : \mathbf{z} \in R\}$$

of the univariate *spectral df* $H_{\mathbf{z}}$. This family $\mathcal{P}(H)$ of df is the *spectral decomposition* of H.

First we study two examples of spectral representations, namely those for EVD and GPD. In these cases the spectral df are univariate EVD and, respectively, GPD (merely in the upper tail if $d \geq 3$).

Lemma 5.4.1. *For a max-stable df G with reverse exponential margins we have*

$$G_{\mathbf{z}}(c) = \exp\big(c\,D(\mathbf{z})\big), \quad c \leq 0,\, \mathbf{z} \in R,$$

and, thus, $\mathcal{P}(G)$ is the family of reverse exponential distributions with parameter $D(\mathbf{z})$, $\mathbf{z} \in R$.

Proof. The assertion is obvious from the equation

$$G_{\mathbf{z}}(c) = G\bigg(c\Big(z_1, \ldots, z_{d-1},\, 1 - \sum_{i \leq d-1} z_i\Big)\bigg) = \exp\big(c\,D(z_1, \ldots, z_{d-1})\big). \qquad \square$$

In the subsequent lemma it is pointed out that the converse implication also holds true.

Lemma 5.4.2. *Let H be an arbitrary df on $(-\infty, 0]^d$ with spectral decomposition*

$$H_{\mathbf{z}}(c) = \exp\big(cg(\mathbf{z})\big), \qquad c \leq 0,\, \mathbf{z} \in R,$$

where $g : R \to (0, \infty)$ is an arbitrary function with $g(\mathbf{0}) = g(\mathbf{e}_i) = 1$ and $\mathbf{e}_i$ is the i-th unit vector in $\mathbb{R}^{d-1}$. Then H is an EVD with reverse exponential margins and dependence function g.

Proof. H is max-stable because

$$H^n\bigg(\frac{c}{n}\Big(z_1, \ldots, z_{d-1},\, 1 - \sum_{i \leq d-1} z_i\Big)\bigg) = H_{\mathbf{z}}^n\Big(\frac{c}{n}\Big) = \exp\big(cg(z)\big)$$

$$= H\bigg(c\Big(z_1, \ldots, z_{d-1},\, 1 - \sum_{i \leq d-1} z_i\Big)\bigg)$$

and it has reverse exponential margins. The assertion is now a consequence of the Pickands representation of H and the preceding lemma. $\qquad \square$

Corresponding to Lemma 5.4.1 we deduce that the spectral df of a GPD is equal to a uniform df (in a neighborhood of 0).

Lemma 5.4.3. *Let $W = 1 + \log(G)$ be a GPD. For c in a neighborhood of 0*

$$W_{\mathbf{z}}(c) = 1 + c\,D(\mathbf{z}),$$

and, thus, each member $W_{\mathbf{z}}$ of $\mathcal{P}(W)$ coincides in its upper tail with the uniform df on the interval $(-1/D(\mathbf{z}), 0)$, $\mathbf{z} \in R$.

The following result is just a reformulation of Lemma 5.4.1 and Lemma 5.4.3. Put for $\mathbf{z} = (z_1, \ldots, z_{d-1}) \in R$, $c \leq 0$ and $\mathbf{U} = (U_1, \ldots, U_d)$ with df H,

$$
M_{\mathbf{z},\mathbf{U}} := \begin{cases} \max\left(\max_{i:z_i>0}\dfrac{U_i}{z_i}, \dfrac{U_d}{1-\sum_{i\leq d-1}z_i}\right), & \text{if } 0 < \sum_{i\leq d-1}z_i < 1, \\[2ex] \max_{i:z_i>0}\dfrac{U_i}{z_i}, & \text{if } \sum_{i\leq d-1}z_i = 1, \\[2ex] U_d, & \text{if } \sum_{i\leq d-1}z_i = 0, \end{cases}
$$

$$
= \max_{i\leq d}\frac{U_i}{z_d} \tag{5.22}
$$

almost surely, if H is continuous at $\mathbf{0}$, where $z_d = 1 - \sum_{i\leq d-1}z_i$. Recall that this condition is satisfied for EVD and GPD.

Corollary 5.4.4. *We have for* $\mathbf{z} \in R$ *and* $c < 0$ *that*

$$
P(M_{\mathbf{z},\mathbf{U}} \leq c) = \begin{cases} \exp(cD(\mathbf{z})), & \text{if } H = G, \\ 1 + cD(\mathbf{z}), & \text{if } H = W \text{ and } c \text{ close to } 0. \end{cases}
$$

ESTIMATION OF THE PICKANDS DEPENDENCE FUNCTION

The rv $|M_{\mathbf{z},\mathbf{U}}|$ is, thus, exponential distributed with parameter $D(\mathbf{z})$ if $H = G$, and its df coincides near 0 with the uniform df on $(0, 1/D(\mathbf{z}))$ if $H = W$. This suggests in the case $H = G$ the following estimator of $D(\mathbf{z})$, based on n independent copies $\mathbf{U}_1, \ldots, \mathbf{U}_n$ of $\mathbf{U}$.

Put, for $\mathbf{z} \in R$,

$$
\hat{D}_{n,EV}(\mathbf{z}) := \frac{1}{\frac{1}{n}\sum_{i\leq n}|M_{\mathbf{z},\mathbf{U}_i}|}
$$

The estimator is motivated by the usual estimation of the parameter of an exponential distribution by the reciprocal of the mean of the observations.

The following result is now immediate from the central limit theorem applied to $\hat{D}_{n,EV}(\mathbf{z})$. In the case $d = 2$, the estimator $\hat{D}_{n,EV}(\mathbf{z})$ is known as Pickands ([372]) estimator. A functional central limit theorem and a law of the iterated logarithm for $\hat{D}_{n,EV}(\mathbf{z})$ was established in the bivariate case by Deheuvels [103]. For further literature on non-parametric estimation of the dependence function for a multivariate extreme value distribution, we refer to Zhang et al. [471].

Lemma 5.4.5. *We have, for* $\mathbf{z} \in R$ *as* $n \to \infty$,

$$
n^{1/2}(\hat{D}_{n,EV}(\mathbf{z}) - D(\mathbf{z})) \longrightarrow_D N(0, D^2(\mathbf{z})).
$$

Suppose that $\mathbf{U}$ has a GPD. Conditional on the assumption that $M_{\mathbf{z},\mathbf{U}} \geq c_0 > -1$, $M_{\mathbf{z},\mathbf{U}}/c_0$ is uniformly distributed on $(0, 1)$,

$$
P(M_{\mathbf{z},\mathbf{U}} \geq uc_0 \mid M_{\mathbf{z},\mathbf{U}} \geq c_0) = u, \qquad u \in (0, 1),
$$

if c_0 is close to 0. Consider only those observations among $M_{\mathbf{z},\mathbf{U}_1}/c_0, \ldots, M_{\mathbf{z},\mathbf{U}_n}/c_0$

with $M_{\mathbf{z},\mathbf{U}_i}/c_0 \leq 1$. Denote these exceedances by $M_1, \ldots, M_{K(n)}$, where $K(n) = \sum_{i \leq n} 1(M_{\mathbf{z},\mathbf{U}_i}/c_0 \leq 1)$. By Theorem 1.3.1, $M_1, M_2, \ldots$ are independent and uniformly on $(0,1)$ distributed rv. The points $(i/(K(n) + 1), M_{i:K(n)})$, $1 \leq i \leq K(n)$, then should be close to the line $\{(u, u) : u \in [0, 1]\}$. This plot offers a way to check graphically, whether $\mathbf{U}$ has actually a GPD. We will exploit this relationship in Section 5.8 to derive tests for checking the GPD assumption of real data and thereby finding appropriate thresholds for the POT approach.

If $\mathbf{U}$ has an EVD, then we obtain from a Taylor expansion of exp at 0,

$$P(M_{\mathbf{z},\mathbf{U}} \geq uc_0 \mid M_{\mathbf{z},\mathbf{U}} \geq c_0) = \frac{1 - \exp(uc_0 D(\mathbf{z}))}{1 - \exp(c_0 D(\mathbf{z}))} = u(1 + O(c_0))$$

uniformly for $\mathbf{z} \in R$ and $u \in (0, 1)$ as $c_0 \uparrow 0$. A uniform-uniform plot of the exceedances $M_1, \ldots, M_{K(n)}$ would then be close to the identity only for c_0 converging to 0.

FURTHER POT-STABILITY OF W

Pickands coordinates offer an easy way to show further POT-stability of a GPD $W = 1 + \log(G)$; see also Section 5.3. Choose $\mathbf{z} = (z_1, \ldots, z_{d-1}) \in R$, put $z_d := 1 - \sum_{i \leq d-1} z_i \in (0, 1)$, and let $0 > r_n \to 0$ be a sequence of arbitrary negative numbers converging to 0 as $n \to \infty$. Let $\mathbf{X} = (X_1, \ldots, X_d)$ have GPD W. Then we obtain with $0 > c > -1/D(\mathbf{z})$ for the conditional distribution if n is large,

$$P\left(X_i \geq r_n z_i |c| D(\mathbf{z}) \text{ for some } 1 \leq i \leq d \mid X_i \geq r_n z_i \text{ for some } 1 \leq i \leq d\right)$$
$$= \frac{1 - W(r_n |c| D(\mathbf{z})(z_1, \ldots, z_d))}{1 - W(r_n(z_1, \ldots, z_d))}$$
$$= \frac{r_n |c| (z_1 + \cdots + z_d) D^2(\mathbf{z})}{r_n(z_1 + \cdots + z_d) D(\mathbf{z})}$$
$$= -cD(\mathbf{z})$$
$$= 1 - W(c\mathbf{z})$$
$$= P(X_i \geq cz_i \text{ for some } 1 \leq i \leq d). \tag{5.23}$$

The following variant of POT-stability is also satisfied by a GPD $W = 1 + \log(G)$ with dependence function D.

Lemma 5.4.6. *Put $k = dD(1/d, \ldots, 1/d)$ and choose $t_i \in [-1/k, 0)^d$, $i \leq d$, such that $\kappa := P(X_i \geq t_i, i \leq d) > 0$, where $(X_1, \ldots, X_d)$ has df W, and $W(\mathbf{x}) = 1 + \log(G(\mathbf{x}))$ for $\mathbf{x} \in \times_{i \leq d}[t_i, 0]$.*
Then we have for $0 \geq \kappa s_i \geq t_i$, $i \leq d$,

$$P(X_i \geq \kappa s_i, i \leq d \mid X_i \geq t_i, i \leq d) = P(X_i \geq s_i, i \leq d).$$

Proof. The assumption is easy to check:

$$P(X_i \geq \kappa s_i,\, i \leq d \mid X_i \geq t_i,\, i \leq d)$$

$$= \frac{P(X_i \geq \kappa s_i,\, i \leq d)}{P(X_i \geq t_i,\, i \leq d)}$$

$$= \frac{1}{\kappa} \Delta^{\mathbf{0}}_{(\kappa s_1,\ldots,\kappa s_d)} W$$

$$= \frac{1}{\kappa} \sum_{\mathbf{m} \in \{0,1\}^d} (-1)^{d - \sum_{j \leq d} m_j} W\left(0^{m_1}(\kappa s_1)^{1-m_1}, \ldots, 0^{m_d}(\kappa s_d)^{1-m_d}\right)$$

$$= \sum_{\mathbf{m} \in \{0,1\}^d} (-1)^{d - \sum_{j \leq d} m_j} W\left(0^{m_1} s_1^{1-m_1}, \ldots, 0^{m_d} s_d^{1-m_d}\right)$$

$$= P(X_i \geq s_i,\, i \leq d).$$

Note that $\sum_{\mathbf{m} \in \{0,1\}^d} (-1)^{d - \sum_{j \leq d} m_j} = 0$ and that the constant function $D = 1$ is, therefore, excluded from the above considerations, since we have in this case $\Delta^{\mathbf{0}}_{(t_1,\ldots,t_d)} W = 0$. $\qquad\square$

For further results on the asymptotic distribution of bivariate excesses using copulas we refer to Wüthrich [467] and Juri and Wüthrich [279].

The Best Attainable Rate of Convergence

In Theorem 2.1.11 and 2.1.12 we showed that the univariate GPD are characterized by the fact that they yield the best rate of convergence of the upper extremes in a sample, equally standardized. In the sequel we extend this result to arbitrary dimensions.

Let $\mathbf{U}^{(1)}, \mathbf{U}^{(2)}, \ldots$ be a sequence of independent copies of a rv $\mathbf{U}$ with df H, which realizes in $(-\infty, 0]^d$. Then, $M_{\mathbf{z},\mathbf{U}^{(1)}}, M_{\mathbf{z},\mathbf{U}^{(2)}}, \ldots$ defines for any $\mathbf{z} \in R$ a sequence of univariate rv with common df $H_{\mathbf{z}}(c) = H(c(z_1, \ldots, z_{d-1}, 1 - \sum_{i \leq d-1} z_i))$, $c \leq 0$.

Denote for any $n \in \mathbb{N}$ by $M_{\mathbf{z},1:n} \leq M_{\mathbf{z},2:n} \leq M_{\mathbf{z},n:n}$ the ordered values of $M_{\mathbf{z},\mathbf{U}^{(1)}}, \ldots, M_{\mathbf{z},\mathbf{U}^{(n)}}$, so that $M_{\mathbf{z},n-k+1:n}$ is the k-th largest order statistic, $k = 1, \ldots, n$.

Suppose that the df H is a GPD $H = 1 + \log(G)$. Then we obtain by elementary arguments

$$\sup_{\mathbf{z} \in R} \sup_{c \leq 0} \left| P\left(n M_{\mathbf{z},n-k+1:n} \leq \frac{c}{D(\mathbf{z})} \right) - P\left(\sum_{j \leq k} \xi_j \leq c \right) \right| = O\left(\frac{k}{n}\right), \qquad (5.24)$$

where $\xi_1, \xi_2, \ldots$ are iid rv on $(-\infty, 0)$ with common standard reverse exponential distribution, cf Theorem 2.1.11.

But also the reverse conclusion holds, as the subsequent Corollary 5.4.8 shows by putting $g(x) = Cx$. The following results extend Theorem 2.1.12 and Corollary 2.1.13 to arbitrary dimensions. Suppose that the df H is continuous near $\mathbf{0}$ and that $H_{\mathbf{z}}(c)$ is a strictly increasing function in $c \in [c_0, 0]$ for all $\mathbf{z} \in R$ and some $c_0 < 0$.

Theorem 5.4.7. *Suppose that there exist norming constants $a_{\mathbf{z},n} > 0$, $b_{\mathbf{z},n} \in \mathbb{R}$ such that*

$$\Delta(n,k) := \sup_{\mathbf{z} \in R} \sup_{c \in \mathbb{R}} \left| P((M_{\mathbf{z},n-k+1:n} - b_{\mathbf{z},n})/a_{\mathbf{z},n} \leq c) - P\left(\sum_{j \leq k} \xi_j \leq c \right) \right| \rightarrow_{n \to \infty} 0$$

for any sequence $k = k(n) \in \{1, \ldots, n\}$, $n \in \mathbb{N}$, with $k/n \rightarrow_{n \to \infty} 0$. Then there exist positive integers $a_1, a_2, \ldots, a_d$ such that

$$H(\mathbf{x}) = W\left(\frac{x_1}{a_1}, \ldots, \frac{x_d}{a_d} \right)$$

for all $\mathbf{x} = (x_1, \ldots, x_d)$ in a neighborhood of $\mathbf{0}$, where $W = 1 + \log(G)$ is a GPD.

The following consequence is obvious.

Corollary 5.4.8. *If there exist norming constants $a_{\mathbf{z},n} > 0$, $b_{\mathbf{z},n} \in \mathbb{R}$ such that*

$$\sup_{\mathbf{z} \in R} \sup_{c \in \mathbb{R}} \left| P((M_{\mathbf{z},n-k+1:n} - b_{\mathbf{z},n})/a_{\mathbf{z},n} \leq c) - P\left(\sum_{j \leq k} \xi_j \leq c \right) \right| \leq g(k/n),$$

where $g : [0,1] \to \mathbb{R}$ satisfies $\lim_{x \to 0} g(x) = 0$, then the conclusion of Theorem 5.4.7 holds.

Proof. By repeating the arguments in the proof of the main result in Falk [129], one shows that there exists $c_0 < 0$ such that, for any $\mathbf{z} \in R$ and any $0 \geq c, c' \geq c_0$,

$$\frac{1 - H_{\mathbf{z}}(c)}{c} = \frac{1 - H_{\mathbf{z}}(c')}{c'}.$$

This implies the representation

$$H_{\mathbf{z}}(c) = 1 + cA(\mathbf{z}), \qquad \mathbf{z} \in R, \ 0 \geq c \geq c_0,$$

where

$$A(\mathbf{z}) := \frac{1 - H_{\mathbf{z}}(c')}{|c'|}$$

is independent of c'. The function $A(\mathbf{z})$ will not be a Pickands dependence function as it does, for example, not necessarily satisfy $A(\mathbf{e}_i) = 1 = A(\mathbf{0})$ where $\mathbf{e}_i$ is the i-th unit vector in $\mathbb{R}^{d-1}$ and $\mathbf{0} \in \mathbb{R}^{d-1}$ as well.

Put, for $1 \leq i \leq d - 1$,

$$a_i := \frac{1}{A(\mathbf{e}_i)} \quad \text{and} \quad a_d := \frac{1}{A(\mathbf{0})}.$$

Then we obtain, for $\mathbf{x} = (x_1, \ldots, x_d) \neq \mathbf{0}$ in a neighborhood of $\mathbf{0}$,

$$
\begin{aligned}
&H(a_1 x_1, \ldots, a_d x_d) \\
&= 1 + (a_1 x_1 + \cdots + a_d x_d) A\left(\frac{a_1 x_1}{\sum_{i \leq d} a_i x_i}, \ldots, \frac{a_{d-1} x_{d-1}}{\sum_{i \leq d} a_i x_i}\right) \\
&= 1 + (x_1 + \cdots + x_d) D\left(\frac{x_1}{\sum_{i \leq d} x_i}, \ldots, \frac{x_{d-1}}{\sum_{i \leq d} x_i}\right) \\
&=: H^*(x_1, \ldots, x_d),
\end{aligned}
$$

where, with $z_d := 1 - \sum_{i \leq d-1} z_i$,

$$D(\mathbf{z}) := (a_1 z_1 + \cdots + a_{d-1} z_{d-1} + a_d z_d) A\left(\frac{a_1 z_1}{\sum_{i \leq d} a_i z_i}, \ldots, \frac{a_{d-1} z_{d-1}}{\sum_{i \leq d} a_i z_i}\right)$$

for any $\mathbf{z} \in R$.

Note that we have $D(\mathbf{e}_i) = 1 = D(\mathbf{0})$ and that for $\mathbf{x} = (x_1, \ldots, x_d) \in (-\infty, 0]^d$, $\mathbf{x} \neq \mathbf{0}$,

$$H^*\left(\frac{\mathbf{x}}{n}\right)^n \to_{n \to \infty} \exp\left((x_1 + \cdots + x_d) D\left(\frac{x_1}{\sum_{i \leq d} x_i}, \ldots, \frac{x_{d-1}}{\sum_{i \leq d} x_i}\right)\right) =: G(\mathbf{x}).$$

G is, therefore, an EVD with standard reverse exponential margins and, thus, D is by the Pickands representation of an EVD necessarily a dependence function. This completes the proof of Theorem 5.4.7. $\qquad\qquad \square$

5.5   Multivariate Domains of Attraction, Spectral Neighborhoods

The question now suggests itself, whether one can establish a *necessary and sufficient* condition for $H \in \mathcal{D}(G)$ in terms of the spectral decomposition of H. This question leads to the subsequent Theorem 5.5.3. Likewise, we extend the concept of a δ-neighborhood of a GPD from the univariate case to higher dimensions by using the spectral decomposition.

THE DOMAIN OF ATTRACTION

The Gnedenko-de Haan Theorem 2.1.1 provides necessary and sufficient conditions for a univariate df to belong to the domain of attraction of a univariate EVD. The concept is less straightforward for the multivariate case; for various characterizations of the domain of attraction of a multivariate EVD we refer to Section 3.2 of

Kotz and Nadarajah [293] and to Section 5.4.2 of Resnick [393]. At first we want to provide an extension of the univariate domain of attraction definition (see (1.20) in Section 1.3) to the multivariate case.

Definition 5.5.1. *Let F be a d-dimensional df. Then we say F belongs to the domain of attraction of an EVD G, abbr. $F \in \mathcal{D}(G)$, if there exists constants $\boldsymbol{a}_n > 0$, $\boldsymbol{b}_n \in \mathbb{R}^d$, $n \in \mathbb{N}$, such that*

$$F^n(\boldsymbol{a}_n \boldsymbol{x} + \boldsymbol{b}_n) \to_{n \to \infty} G(\boldsymbol{x}), \qquad \boldsymbol{x} \in \mathbb{R}^d .$$

A useful theorem regarding the characterization of a multivariate domain of attraction goes back to Deheuvels [102] and reads as follows.

Theorem 5.5.2 (Deheuvels). *Let $\boldsymbol{X}$, $\boldsymbol{X}_1, \dots, \boldsymbol{X}_n$ be i.i.d d-dimensional rv with common df F. Then*

$$F^n(\boldsymbol{a}_n \boldsymbol{x} + \boldsymbol{b}_n) \to_{n \to \infty} G(\boldsymbol{x}) , \qquad \boldsymbol{x} \in \mathbb{R}^,$$

for some constants $\boldsymbol{a}_n > 0$, $\boldsymbol{b}_n \in \mathbb{R}^d$, $n \in \mathbb{N}$, if, and only if, each margin F_i of F converges to the univariate margin $G_i(x)$ of G, and the convergence of the copulas holds

$$C_F^n(\boldsymbol{u}^{1/n}) \to_{n \to \infty} C_G(\boldsymbol{u}), \qquad \boldsymbol{u} \in (0,1)^d. \tag{5.25}$$

Proof. See Deheuvels [102] or Galambos [167].   □

For the following approach regarding the spectral decomposition, it is useful to recall the Gnedenko-de Haan Theorem 2.1.1 in the case $d = 1$ and $G(x) = \exp(x)$, $x \le 0$: We have $H \in \mathcal{D}(G)$ iff $\omega(H) = \sup\{x \in \mathbb{R} : H(x) < 1\} < \infty$ and

$$\lim_{c \uparrow 0} \frac{1 - H(\omega(H) + ct)}{1 - H(\omega(H) + c)} = t, \qquad t > 0.$$

Now we are ready to state our result for a general dimension d. By $H_1, \dots, H_d$ we denote the marginal df of H and by $H^{\boldsymbol{\omega}}(\mathbf{x}) := H(\boldsymbol{\omega}(H) + \mathbf{x})$, $\mathbf{x} \in (-\infty, 0]^d$, the shifted df if $\boldsymbol{\omega}(H) = (\omega(H_1), \dots, \omega(H_d))$ are finite numbers. A proof is given in Falk [137].

Theorem 5.5.3.

(i) *Suppose that $H \in \mathcal{D}(G)$ and that $H_1 = \dots = H_d$. Then we have $\omega(H_1) < \infty$ and*

$$\forall \mathbf{z} \in R : \quad \lim_{c \uparrow 0} \frac{1 - H_{\mathbf{z}}^{\omega}(ct)}{1 - H_{\mathbf{z}}^{\omega}(c)} = t, \qquad t > 0, \tag{5.26}$$

and

$$\forall \mathbf{z}_1, \mathbf{z}_2 \in R : \quad \lim_{c \uparrow 0} \frac{1 - H_{\mathbf{z}_1}^{\omega}(c)}{1 - H_{\mathbf{z}_2}^{\omega}(c)} = \frac{A(\mathbf{z}_1)}{A(\mathbf{z}_2)} \tag{5.27}$$

for some positive function $A : R \to (0, \infty)$.

(ii) *Suppose that $\omega(H_i) < \infty$, $1 \le i \le d$, and that (5.26) and (5.27) hold. Then we have $H \in \mathcal{D}(G)$.*

If we assume identical margins $H_1 = \cdots = H_d$ of H, then we obtain from the preceding result the characterization of the domain of attraction $H \in \mathcal{D}(G)$ in terms of the spectral decomposition of H:

$$H \in \mathcal{D}(G) \iff \omega(H_1) < \infty \text{ and (5.26) and (5.27) hold.}$$

Theorem 5.5.3 can easily be extended to an EVD with (reverse) Weibull or Fréchet margins as follows. Suppose that $G_{\alpha_1,\dots,\alpha_d}$ is an EVD with i-th marginal

$$G_i(x) = \exp(\psi_{\alpha_i}(x)), \qquad 1 \le i \le d,$$

where

$$\psi_\alpha(x) := \begin{cases} -(-x)^\alpha, & x < 0, \text{ if } \alpha > 0, \\ -x^\alpha, & x > 0, \text{ if } \alpha < 0, \end{cases}$$

defining, thus, the family of (reverse) Weibull and Fréchet df $\exp(\psi_\alpha(x))$. Note that

$$G_{\alpha_1,\dots,\alpha_d}\left(\psi_{\alpha_1}^{-1}(x_1),\dots,\psi_{\alpha_d}^{-1}(x_d)\right) = G_{1,\dots,1}(x_1,\dots,x_d), \quad x_i < 0, \ 1 \le i \le d,$$

where $G_{1,\dots,1} = G$ has reverse exponential margins. Let H be an arbitrary d-dimensional df and put with $\boldsymbol{\psi} = (\psi_{\alpha_1},\dots,\psi_{\alpha_d})$ for $x_i < 0, 1 \le i \le d$,

$$H^{\boldsymbol{\psi}}(x_1,\dots,x_d) := H\left(\omega_1 + \psi_{\alpha_1}^{-1}(x_1),\dots,\omega_d + \psi_{\alpha_d}^{-1}(x_d)\right)$$

where $\omega_i = 0$ if $\alpha_i < 0$ and $\omega_i = \omega(H_i)$ if $\alpha_i > 0$. Then we have

$$H^{\boldsymbol{\psi}} \in \mathcal{D}\left(G_{1,\dots,1}\right) \iff H \in \mathcal{D}\left(G_{\alpha_1,\dots,\alpha_d}\right)$$

and Theorem 5.5.3 can be applied.

THE SPECTRAL NEIGHBORHOOD OF A GPD

The spectral decomposition provides a comparatively simple sufficient condition for an arbitrary multivariate df H to belong to the domain of attraction of an EVD G with reverse exponential margins.

Theorem 5.5.4. *Suppose that for any $\mathbf{z} \in R$,*

$$1 - H_{\mathbf{z}}(c) = |c|g(\mathbf{z})(1 + o(1)), \qquad c \uparrow 0, \tag{5.28}$$

for some function g with $g(\mathbf{e}_i) = 1 = g(\mathbf{0})$, $1 \le i \le d - 1$. Then, $g(\mathbf{z}) =: D(\mathbf{z})$ is a Pickands dependence function and the df H is in the domain of attraction of the EVD G with standard reverse exponential margins and dependence function D. Precisely, we have

$$H_{\mathbf{z}}^n\left(\frac{c}{n}\right) = H^n\left(\frac{c}{n}\left(z_1,\dots,z_{d-1}, 1 - \sum_{i \le d-1} z_i\right)\right) \longrightarrow_{n\to\infty} \exp(cD(\mathbf{z})), \qquad c \le 0.$$

Before we prove Theorem 5.5.4 we have to add some remarks and definitions. Note that condition (5.28) is by Hôpital's rule satisfied if $H_{\mathbf{z}}$ has a positive derivative $h_{\mathbf{z}}(c) = (\partial/\partial c)H_{\mathbf{z}}(c)$ for $c < 0$ close to 0 and any $\mathbf{z} = (z_1, \ldots, z_{d-1}) \in R$ such that for some function g with $g(\mathbf{e}_i) = 1 = g(\mathbf{0})$, $1 \le i \le d-1$,

$$h_{\mathbf{z}}(c) = g(\mathbf{z})(1 + o(1)). \tag{5.29}$$

Recall that the spectral decomposition of a GPD W with Pickands dependence function D can be written as $1 - W_z(c) = |c|D(z)$ for $c < 0$ close to 0. Therefore, condition (5.28) is equivalent to the condition that H *belongs to the spectral neighborhood of the GPD W*, that is,

$$1 - H_{\mathbf{z}}(c) = (1 - W_z(c))(1 + o(1)), \qquad c \uparrow 0, \ \mathbf{z} \in R. \tag{5.30}$$

This condition is related to condition (4.4) in the univariate case.

Likewise, condition (5.29) is equivalent to

$$h_{\mathbf{z}}(c) = D(\mathbf{z})(1 + o(1)) \qquad c \uparrow 0, \ \mathbf{z} \in R. \tag{5.31}$$

In this case we say that H *belongs to the differentiable spectral neighborhood of the GPD W*.

If we weaken the condition $g(\mathbf{e}_i) = 1 = g(\mathbf{0})$ in Theorem 5.5.4 to $g(\mathbf{e}_i) > 0$, $1 \le i \le d-1$, $g(\mathbf{0}) > 0$, then we obtain

$$H^n\left(\frac{\mathbf{x}}{n}\right) \longrightarrow_{n \to \infty} \exp\left(\left(\sum_{i \le d} x_i\right) g\left(\frac{x_1}{\sum_{i \le d} x_i}, \ldots, \frac{x_{d-1}}{\sum_{i \le d} x_i}\right)\right) =: G(\mathbf{x}),$$

where G is a max-stable df; i.e., $G^n(\mathbf{x}/n) = G(\mathbf{x})$. But it does not necessarily have standard reverse exponential margins, since

$$G(0, \ldots, 0, x_i, 0, \ldots, 0) = \begin{cases} \exp(x_i g(\mathbf{e}_i)), & \text{if } i \le d-1, \\ \exp(x_d g(\mathbf{0})), & \text{if } i = d. \end{cases}$$

In this case we obtain, however, for $\mathbf{x} \in (-\infty, 0]^d$ with $a_i := 1/g(\mathbf{e}_i)$, $1 \le i \le d-1$, $a_d := 1/g(\mathbf{0})$,

$$H^n\left(\frac{1}{n}(a_1 x_1, \ldots, a_d x_d)\right)$$

$$\to_{n \to \infty} \exp\left(\left(\sum_{i \le d} a_i x_i\right) g\left(\frac{a_1 x_1}{\sum_{i \le d} a_i x_i}, \ldots, \frac{a_{d-1} x_{d-1}}{\sum_{i \le d} a_i x_i}\right)\right)$$

$$= \exp\left(\left(\sum_{j \le d} x_j\right)\left(\sum_{i \le d} a_i \frac{x_i}{\sum_{j \le d} x_j}\right)\right.$$

$$\left. \times g\left(\frac{a_1 x_1/\sum_{j \le d} x_j}{\sum_{i \le d} a_i x_i/\sum_{j \le d} x_j}, \ldots, \frac{a_{d-1} x_{d-1}/\sum_{j \le d} x_j}{\sum_{i \le d} a_i x_i/\sum_{j \le d} x_j}\right)\right)$$

$$= \exp\left(\left(\sum_{i\le d} x_i\right) D\left(x_1 / \sum_{i\le d} x_i, \ldots, x_{d-1} / \sum_{i\le d} x_i\right)\right)$$

$$=: G^*(x_1, \ldots, x_d),$$

where with $z_d := 1 - \sum_{i\le d-1} z_i$,

$$D(z_1, \ldots, z_{d-1}) := \left(\sum_{i\le d} a_i z_i\right) g\left(\frac{a_1 z_1}{\sum_{i\le d} a_i z_i}, \ldots, \frac{a_{d-1} z_{d-1}}{\sum_{i\le d} a_i z_i}\right), \qquad \mathbf{z} \in R,$$

is a Pickands dependence function and G^* is max stable with standard reverse exponential margins.

Proof. We have by condition (5.28) for any $c < 0$

$$\lim_{n\to\infty} \frac{H_{\mathbf{z}}(c/n) - 1}{c/n} = g(\mathbf{z}),$$

which yields

$$\lim_{n\to\infty} n(H_{\mathbf{z}}(c/n) - 1) = cg(\mathbf{z}).$$

From the expansion $\log(1 + \varepsilon) = \varepsilon + O(\varepsilon^2)$, as $\varepsilon \to 0$, we obtain

$$H_{\mathbf{z}}^n\left(\frac{c}{n}\right) = \exp\left(n \log\left(1 + \left(H_{\mathbf{z}}\left(\frac{c}{n}\right) - 1\right)\right)\right)$$

$$= \exp\left(n\left(H_{\mathbf{z}}\left(\frac{c}{n}\right) - 1\right) + O\left(\frac{1}{n}\right)\right) \xrightarrow{n\to\infty} \exp(cg(\mathbf{z})).$$

We, therefore, have, for any $\mathbf{x} = (x_1, \ldots, x_d) \in (-\infty, 0]^d$,

$$H^n\left(\frac{\mathbf{x}}{n}\right) \xrightarrow{n\to\infty} \exp\left(\left(\sum_{i\le d} x_i\right) g\left(\frac{x_1}{\sum_{i\le d} x_i}, \ldots, \frac{x_{d-1}}{\sum_{i\le d} x_i}\right)\right) =: G(\mathbf{x}) \qquad (5.32)$$

with the convention that the right-hand side equals 1 if $\mathbf{x} = \mathbf{0}$.

From the fact that $H^n(0, \ldots, 0, x_i/n, 0, \ldots, 0)$ converges to $\exp(x_i)$, $x_i \le 0$, $1 \le i \le d$, one concludes that g is continuous: We have for arbitrary $\mathbf{x} = (x_1, \ldots, x_d)$, $\mathbf{y} = (y_1, \ldots, y_d) \in (-\infty, 0]^d$,

$$\left| H^n\left(\frac{\mathbf{x}}{n}\right) - H^n\left(\frac{\mathbf{y}}{n}\right) \right|$$

$$\le \sum_{i\le d} \left| H^n\left(0, \ldots, 0, \frac{x_i}{n}, 0, \ldots, 0\right) - H^n\left(0, \ldots, 0, \frac{y_i}{n}, 0, \ldots, 0\right) \right|,$$

see Lemma 2.2.6 in Reiss [385]. By putting $y_i = x_i + \varepsilon_i$, $1 \le i \le d$, where $\varepsilon_1, \ldots, \varepsilon_d$ are small and $\varepsilon_i \le 0$ if $x_i = 0$, we obtain, thus, by (5.32)

$$\left| \exp\left(\left(\sum_{i \leq d} (x_i + \varepsilon_i) \right) g\left(\frac{x_1 + \varepsilon_1}{\sum_{i \leq d} (x_i + \varepsilon_i)}, \dots, \frac{x_{d-1} + \varepsilon_{d-1}}{\sum_{i \leq d} (x_i + \varepsilon_i)} \right) \right) \right.$$

$$\left. - \exp\left(\left(\sum_{i \leq d} x_i \right) g\left(\frac{x_1}{\sum_{i \leq d} x_i}, \dots, \frac{x_{d-1}}{\sum_{i \leq d} x_i} \right) \right) \right|$$

$$\leq \sum_{i \leq d} |\exp(x_i) - \exp(x_i + \varepsilon_i)|.$$

Letting $\varepsilon_i \to 0$, $1 \leq i \leq d$, this inequality implies that $g(\mathbf{z})$, $\mathbf{z} \in R$, is a continuous function. Hence, $G(\mathbf{x})$ is by Lemma 7.2.1 in Reiss [385] a df on $(-\infty, 0]^d$. It is obviously max-stable with standard reverse exponential margins and, hence, it coincides with its Pickands representation in (5.32), which completes the proof. $\square$

In the subsequent lines we modify the concept of a spectral neighborhood of a GPD.

A Spectral δ-Neighborhood of a GPD

Using the spectral decomposition, we can easily extend the definition of δ-neighborhoods of a univariate GPD to arbitrary dimensions. We say that the df H *belongs to the spectral δ-neighborhood* of the GPD W if it is continuous in a neighborhood of $\mathbf{0} \in \mathbb{R}^d$ and satisfies uniformly for $\mathbf{z} \in R$ the expansion

$$1 - H_{\mathbf{z}}(c) = \big(1 - W_{\mathbf{z}}(c) \big) \big(1 + O(|c|^{\delta}) \big) \tag{5.33}$$

for some $\delta > 0$ as $c \uparrow 0$. The EVD G with reverse exponential margins is, for example, in the spectral δ-neighborhood of the corresponding GPD W with $\delta = 1$. Because $D(\mathbf{z}) \geq 1/d$ for any $\mathbf{z} \in R$ we have

$$1 - G_{\mathbf{z}}(c) = 1 - \exp\big(c\, D(\mathbf{z}) \big)$$

$$= c\, D(\mathbf{z}) + O(c^2)$$

$$= c\, D(\mathbf{z}) \big(1 + O(c) \big)$$

$$= \big(1 - W_{\mathbf{z}}(c) \big) \big(1 + O(c) \big).$$

Mardia's df is, for example, in the δ-neighborhood with $\delta = 1$ of the GPD W with dependence function $D(\mathbf{z}) = 1$, $\mathbf{z} \in R$.

Note that by putting $\mathbf{z} = \mathbf{e}_i$, $1 \leq i \leq d-1$, and $\mathbf{z} = \mathbf{0}$, equation (5.33) implies that the univariate margins of the df H are in the spectral δ-neighborhood of the uniform distribution on $(-1, 0)$:

$$P(U_i > c) = |c| \big(1 + O(|c|^{\delta}) \big), \quad 1 \leq i \leq d.$$

The following result extends the characterization of δ-neighborhoods of a univariate GPD in Theorem 2.2.5 in terms of the rate of convergence of extremes to arbitrary dimensions. A proof for the case $d = 2$ is given in Falk and Reiss [152]; this proof can easily be extended to general $d \geq 2$. For related results on the rate of convergence of multivariate extremes in terms of probability metrics we refer to Omey and Rachev [359], de Haan and Peng [191] and the literature cited therein.

Theorem 5.5.5. *Let H be a d-dimensional df.*

(i) *If H is for some $\delta \in (0, 1]$ in the spectral δ-neighborhood of the GPD $W = 1 + \log(G)$, then we have*

$$\sup_{\mathbf{x} \in (-\infty, 0]^d} \left| H^n\left(\frac{\mathbf{x}}{n}\right) - G(\mathbf{x}) \right| = O(n^{-\delta}).$$

(ii) *Suppose that $H_{\mathbf{z}}(c)$ as defined in (5.33) is differentiable with respect to c in a neighborhood of 0 for any $\mathbf{z} \in R$, i.e., $h_{\mathbf{z}}(c) := (\partial/\partial c)H_{\mathbf{z}}(c)$ exists for $c \in (-\varepsilon, 0)$ and any $\mathbf{z} \in R$. Suppose, moreover, that $H_{\mathbf{z}}$ satisfies the von Mises condition*

$$\frac{-c\,h_{\mathbf{z}}(c)}{1 - H_{\mathbf{z}}(c)} =: 1 + \eta_{\mathbf{z}}(c) \to_{c\uparrow 0} 1, \quad \mathbf{z} \in R,$$

with remainder term $\eta_{\mathbf{z}}$ satisfying

$$\sup_{\mathbf{z} \in R} \left| \int_c^0 \frac{\eta_{\mathbf{z}}(t)}{t}\, dt \right| \to_{c\uparrow 0} 0.$$

If

$$\sup_{\mathbf{x} \in (-\infty, 0]^d} \left| H^n\left(\frac{\mathbf{x}}{n}\right) - G(\mathbf{x}) \right| = O(n^{-\delta})$$

for some $\delta \in (0, 1]$, then H is in the spectral δ-neighborhood of the GPD $W = 1 + \log(G)$.

For Mardia's distribution

$$H_{\mathbf{z}}(c) = \frac{1}{\sum_{i \leq d-1} \exp(-c\mathbf{z}_i) + \exp\left(c\left(\sum_{i \leq d-1} z_i - 1\right)\right) - (d-1)}$$

we have, for example,

$$1 - H_{\mathbf{z}}(c) = -c\bigl(1 + O(c)\bigr)$$

and

$$h_{\mathbf{z}}(c) = 1 + O(c)$$

uniformly for $\mathbf{z} \in R$ as $c \uparrow 0$ and, thus,

$$\eta_{\mathbf{z}}(c) = \frac{-c h_{\mathbf{z}}(c)}{1 - H_{\mathbf{z}}(c)} - 1 = O(c),$$

uniformly for $\mathbf{z} \in R$. The conditions of Theorem 5.5.5 are, therefore, satisfied by Mardia's distribution with $G(\mathbf{x}) = \exp(\sum_{i \leq d} x_i)$, $\mathbf{x} = (x_1, \ldots, x_d) \in (-\infty, 0]^d$, and $\delta = 1$.

An Estimator of D

An equivalent formulation of condition (5.33) in terms of $M_{\mathbf{z},\mathbf{U}}$ is

$$P(M_{\mathbf{z},\mathbf{U}} > c) = cD(\mathbf{z})(1 + O(|c|^{\delta}))$$

uniformly for $\mathbf{z} \in R$ as $c \uparrow 0$. This suggests as an estimator of $D(\mathbf{z})$, based on n independent copies $\mathbf{U}_1, \ldots, \mathbf{U}_n$ of $\mathbf{U}$, the relative frequency

$$\hat{D}_{n,c}(\mathbf{z}) := \frac{1}{cn} \sum_{i \le n} 1(M_{\mathbf{z},\mathbf{U}_i} > c).$$

We have

$$E(\hat{D}_{n,c}(\mathbf{z})) = D(\mathbf{z})(1 + O(|c|^{\delta}))$$

and

$$\mathrm{Var}(\hat{D}_{n,c}(\mathbf{z})) = \frac{D(\mathbf{z})}{nc}(1 + O(|c|^{\delta})).$$

The asymptotic normality of $\hat{D}_{n,c}(\mathbf{z})$ is now a consequence of the Moivre-Laplace theorem.

Lemma 5.5.6. *If $c = c_n < 0$ satisfies $n|c| \to \infty$, $n|c|^{1+2\delta} \to 0$ as $n \to \infty$, then we have*

$$(n|c|)^{1/2}(\hat{D}_{n,c}(\mathbf{z}) - D(\mathbf{z})) \longrightarrow_D N(0, D(\mathbf{z})),$$

provided the df of $\mathbf{U}$ is in the spectral δ-neighborhood of the GPD with dependence function D.

The function $\hat{D}_{n,c}(\mathbf{z})$ is neither continuous nor convex in $\mathbf{z}$. Convex estimators of the dependence function for bivariate EVD are studied, for example, by Tiago de Oliveira [444], Deheuvels and Tiago de Oliveira [106], Hall and Tajvidi [203] and Jiménez et al. [272]. Kernel and parametric estimators of the dependence function were studied by Smith et al. [423] and Abdous et al. [1]. We refer to Section 3.6 of Kotz and Nadarajah [293] for a thorough review of statistical estimation in multivariate EVD models. For GPD models we present some statistical estimation procedures in Sections 5.8 to 5.10 and 6.6.

The following corollary provides a sufficient condition for H to belong to the spectral δ-neighborhood of a GPD W.

Corollary 5.5.7. *Suppose in addition to the assumptions of the preceding result that, for some $\delta \in (0, 1]$,*

$$h_{\mathbf{z}}(c) = g(\mathbf{z})(1 + O(|c|^{\delta})) \tag{5.34}$$

as $c \uparrow 0$ uniformly for $\mathbf{z} \in R$. Then H is in the spectral δ-neighborhood of the GPD W with Pickands dependence function g and the conclusion of Theorem 5.5.4 applies.

Proof. The assertion is immediate from the expansion

$$1 - H_{\mathbf{z}}(c) = \int_{(c,0)} h_{\mathbf{z}}(u) \, du = |c| g(\mathbf{z}) \big(1 + O(|c|^{\delta}) \big).$$

$\square$

If condition (5.34) holds then H *belongs to the differentiable spectral δ-neighborhood of the GPD W.* Notice that, necessarily, $g(\mathbf{z}) = D(\mathbf{z})$.

For Mardia's distribution we have, for example, $h_{\mathbf{z}}(c) = 1 + O(c)$ uniformly for $\mathbf{z} \in R$ as $c \uparrow 0$. Consequently, by Corollary 5.5.7, Mardia's distribution is in the spectral δ-neighborhood with $\delta = 1$ of the GPD W with dependence function $D(\mathbf{z}) = 1$, $\mathbf{z} \in R$, and Theorem 5.5.5 on the speed of convergence applies.

5.6   The Pickands Transform

Let $\mathbf{U} := (U_1, \ldots, U_d)$ be an arbitrary rv, which takes values in $(-\infty, 0]^d$, and denote its df by H. Suppose that H has continuous partial derivatives of order d near $\mathbf{0} \in \mathbb{R}^d$. Then

$$h(x_1, \ldots, x_d) := \frac{\partial^d}{\partial x_1 \cdots \partial x_d} H(x_1, \ldots, x_d)$$

is a density of H in a neighborhood of $\mathbf{0}$ (see e.g. Bhattacharya and Rao [43], Theorem A.2.2). Define the transformation $T : (-\infty, 0]^d \setminus \{\mathbf{0}\} \to R \times (-\infty, 0)$ by

$$T(\mathbf{x}) := \left(\frac{x_1}{x_1 + \cdots + x_d}, \ldots, \frac{x_{d-1}}{x_1 + \cdots + x_d}, x_1 + \cdots + x_d \right) =: (\mathbf{z}, c), \qquad (5.35)$$

which is the transformation of $\mathbf{x} = (x_1, \ldots, x_d)$ onto its Pickands coordinates $\mathbf{z} = (z_1, \ldots, z_{d-1})$ and c, see equation (4.31) for the definition of the unit simplex R. This mapping is one-to-one with the inverse function

$$T^{-1}(\mathbf{z}, c) = c \left(z_1, \ldots, z_{d-1}, 1 - \sum_{i \le d-1} z_i \right). \qquad (5.36)$$

It turns out that the *Pickands transform* of $\mathbf{U}$ onto its Pickands coordinates

$$(\mathbf{Z}, C) := T(\mathbf{U})$$

has some characteristic features which make it a very useful tool for the investigation of d-variate POT models.

Nadarajah [343] uses this representation of $\mathbf{U}$ to provide analytical results on the tail behavior of a bivariate df in the domain of attraction of a bivariate max-stable distribution G. Moreover, Coles [71], Section 8.3.2., applies Pickands coordinates to show that the intensity of the limiting Poisson point process of the sequence of point processes $N_n = \sum_{i \leq n} \varepsilon_{n\mathbf{U}^{(i)}}$, $n \in \mathbb{N}$, factorizes across radial and angular components. By $\mathbf{U}^{(1)}, \mathbf{U}^{(2)}, \ldots$ we denote independent copies of $\mathbf{U}$, which has df G with reverse exponential margins. This result goes back to de Haan [188]; see Corollary 5.6.7 below for an extension. Capéraà et al. [59] investigate a non-parametric estimation procedure for G, which is based on the angular components $\mathbf{Z}^{(i)}$ of $\mathbf{U}^{(i)}$, $1 \leq i \leq n$. For further applications such as the generation of pseudo rv with distribution G we refer to Section 3 of Kotz and Nadarajah [293].

The Pickands Transform of a GPD Random Vector

The subsequent lemma provides the density of the Pickands transform in case of a GPD with a smooth dependence function D. For a proof of the following result see Falk and Reiss [155].

Lemma 5.6.1. *Consider a GPD $W = 1 + \log(G)$ with a Pickands dependence function D having continuous partial derivatives of order d. Let $\mathbf{U} = (U_1, \ldots, U_d)$ be a rv with df W. The Pickands transform $T(\mathbf{U}) = (Z, C)$ then has on $R \times (c_0, 0)$, with $c_0 < 0$ close to 0, a density $f(\mathbf{z}, c)$, which is independent of c. We have*

$$f(\mathbf{z}, c) = |c|^{d-1} \left(\frac{\partial^d}{\partial x_1 \cdots \partial x_d} W \right) \left(T^{-1}(\mathbf{z}, c) \right)$$

$$= \varphi(\mathbf{z}) \quad \text{for } \mathbf{z} \in R, \ c \in (c_0, 0).$$

As with many assertions before, the restriction that c_0 has to be close enough to 0 is due to the fact that we have

$$W(\mathbf{x}) = 1 + \log(G(\mathbf{x})) = 1 + \left(\sum_{i \leq d} x_i \right) D \left(\frac{x_1}{\sum_{i \leq d} x_i}, \ldots, \frac{x_{d-1}}{\sum_{i \leq d} x_i} \right)$$

only for $\mathbf{x}$ close to 0 if $d \geq 3$, see Lemma 5.1.5.

It turns out that the angular component $\mathbf{Z}$ and the radial component C are independent, conditional on $C > c_0$. Moreover, C is on $(-1, 0)$ uniformly distributed and, conditional on $C > c_0$, $\mathbf{Z}$ has the density $f(\mathbf{z}) := \varphi(\mathbf{z}) / \int_R \varphi(\mathbf{y}) \, d\mathbf{y}$ on R.

Theorem 5.6.2. *Suppose that $\mathbf{U} = (U_1, \ldots, U_d)$ follows a GPD W, where the dependence function D has continuous partial derivatives of order d and the density*

$$\varphi(\mathbf{z}) = |c|^{d-1}(\partial^d/(\partial x_1 \cdots \partial x_d)W)(T^{-1}(\mathbf{z}, c)) \tag{5.37}$$

of the Pickands transform gives positive mass on R:

$$\zeta := \int_R \varphi(\mathbf{z}) \, d\mathbf{z} > 0. \tag{5.38}$$

Then for $c_0 < 0$ close to 0 we have

(i) *Conditional on $C = U_1 + \cdots + U_d > c_0$, the angular component $\mathbf{Z} = (U_1/C, \ldots, U_{d-1}/C)$ and the radial component C of the Pickands transform are independent.*

(ii) *C follows on $(c_0, 0)$ a uniform distribution, precisely*

$$P(C > c) = \zeta|c|, \qquad c_0 \leq c \leq 0.$$

and, thus,

$$P(C \geq c \mid C > c_0) = \frac{|c|}{|c_0|}, \qquad c_0 \leq c \leq 0.$$

(iii) *Conditional on $C > c_0$, $\mathbf{Z}$ has the density*

$$f(\mathbf{z}) = \frac{\varphi(\mathbf{z})}{\zeta}, \quad \mathbf{z} \in R.$$

Proof. We have for $c \in (c_0, 0)$,

$$P(C > c) = P(C > c, \mathbf{Z} \in R)$$

$$= \int_{(c,0)} \int_R f(\mathbf{z}, u) \, d\mathbf{z} \, du$$

$$= \int_{(c,0)} \int_R \varphi(\mathbf{z}) \, d\mathbf{z} \, du = |c|\zeta,$$

which proves (ii). Further we have for any Borel measurable set $B \subset R$ by (i)

$$P(\mathbf{Z} \in B | C > c_0) = \frac{P(\mathbf{Z} \in B, C > c_0)}{P(C > c_0)}$$

$$= \frac{1}{|c_0|\zeta} \int_B \int_{(c_0,0)} \varphi(\mathbf{z}) \, dc \, d\mathbf{z} = \frac{1}{\zeta} \int_B \varphi(\mathbf{z}) \, d\mathbf{z},$$

which proves (iii). Finally, we have for $c_0 < c < 0$,

$$P(\mathbf{Z} \in B, C \geq c | C > c_0) = \frac{P(\mathbf{Z} \in B, C \geq c)}{P(C > c_0)}$$

$$= \frac{1}{\zeta|c_0|} \int_B \int_{(c,0)} \varphi(\mathbf{z}) \, dc \, d\mathbf{z}$$

$$= \frac{1}{\zeta} \int_B \varphi(\mathbf{z}) \, d\mathbf{z} \quad \frac{|c|}{|c_0|}$$

$$= P(\mathbf{Z} \in B | C \geq c_0) P(C \geq c | C \geq c_0)$$

by (ii) and (iii). This completes the proof of the theorem. $\qquad\square$

The function φ defined in (5.37) is called *Pickands density*. Remark that the Pickands density is the density of a probability measure only after the division by ζ (in case $\zeta > 0$), else it is the density of a measure, which assigns the mass ζ to the simplex R.

Note that the constant dependence function $D = 1$ is not included in the preceding Theorem 5.6.2 since in this case $\partial^d/(\partial x_1 \cdots \partial x_d)W = \partial^d/(\partial x_1 \cdots \partial x_d)(1 + \sum_{i \le d} x_i) = 0$ and, thus, $\int_R \varphi(\mathbf{z})\, d\mathbf{z} = 0$. The converse implication

$$\int_R \varphi(\mathbf{z})\, d\mathbf{z} = 0 \Rightarrow D = 1$$

does not hold in general. Consider, e.g., the Pickands dependence function

$$D(z_1, z_2) = \left(z_1^\lambda + z_2^\lambda\right)^{1/\lambda} + (1 - z_1 - z_2).$$

It pertains to the df $H_\lambda(x_1, x_2, x_3) = G_\lambda(x_1, x_2)G(x_3)$ where G_λ is the logistic df with $\lambda > 1$, cf. Example 4.3.5, and $G(x) = \exp(x)$, $x < 0$, i.e., we have

$$H_\lambda(x_1, x_2, x_3) = \exp\left(-\left((-x_1)^\lambda + (-x_2)^\lambda\right)^{1/\lambda} + x_3\right).$$

Obviously, $D \ne 1$, but for the GPD W_λ belonging to H_λ, i.e.,

$$W_\lambda(x_1, x_2, x_3) = 1 - \left((-x_1)^\lambda + (-x_2)^\lambda\right)^{1/\lambda} + x_3,$$

we have $\partial^3/(\partial x_1 \partial x_2 \partial x_3)W_\lambda = 0$ and, thus, $\int_R \varphi(\mathbf{z})\, d\mathbf{z} = 0$. However, we can establish relationships to the pairwise bivariate Pickands dependence functions

$$D_{rs}(z) := D(z\mathbf{e}_r + (1 - z)\mathbf{e}_s), \quad z \in [0, 1], \tag{5.39}$$

where $\mathbf{e}_r$ and $\mathbf{e}_s$ are the r-th and s-th unit vectors in $\mathbb{R}^{d-1}$ and $\mathbf{e_d} := \mathbf{0} \in \mathbb{R}^{d-1}$, $r, s \in \{1, \ldots, d\}$. To justify the definition in (5.39) let $\mathbf{X} = (X_1, \ldots, X_d)$ be a d-variate random vector whose df is a d-variate EVD G with Pickands dependence function D. Then the bivariate marginal df of the random vector (X_r, X_s), $r, s \in \{1, \ldots, d\}$, $r \ne s$, is a bivariate EVD with the above Pickands dependence function D_{rs}.

Lemma 5.6.3. *Let φ be the Pickands density of a d-variate GPD W with Pickands dependence function D as given in Lemma 5.4.1. Then we have*

(i) $D = 1 \iff D_{rs} = 1$ *for every pair $r, s \in \{1, \ldots, d\}$,*

(ii) $\int_R \varphi(\mathbf{z})\, d\mathbf{z} = 0 \iff D_{rs} = 1$ *for at least one pair $r, s \in \{1, \ldots, d\}$.*

Proof. The necessity of $D = 1$ in part (i) follows directly from the definition (5.39) of the pairwise bivariate Pickands dependence functions. The sufficiency can be deduced from Theorem 4.3.3, which entails the equivalence of complete and pairwise bivariate independence of the margins of a multivariate EVD, by using the representation (4.31) of an EVD in terms of the Pickands dependence function. For a proof of part (ii) we refer to Lemma 1.2 in Frick and Reiss [162]. □

From Lemma 5.6.3 it follows again that $D = 1$ implies $\int_R \varphi(\mathbf{z})\, d\mathbf{z} = 0$. And obviously we have

$$\int_R \varphi(\mathbf{z})\, d\mathbf{z} = 0 \Longleftrightarrow D = 1 \tag{5.40}$$

in the bivariate case and if D satisfies the symmetry condition (4.34) since all the pairwise bivariate Pickands dependence functions are equal in this case.

Moreover, we deduce from Lemma 5.6.3 that $\int_R \varphi(\mathbf{z})\, d\mathbf{z} > 0$ stands for tail dependence since it implies $D \neq 1$. The case $\int_R \varphi(\mathbf{z})\, d\mathbf{z} = 0$ represents tail independence in at least one bivariate marginal distribution.

DIFFERENTIABLE δ-NEIGHBORHOODS OF PICKANDS TRANSFORMS

We can easily extend Theorem 5.6.2 to certain differentiable δ-neighborhoods of GPD. Let $\mathbf{U} = (U_1, \ldots, U_d)$ be a rv such that the corresponding Pickands transform $(\mathbf{Z}, C)$ has for some $c_0 < 0$ a density $f(\mathbf{z}, c)$ on $R \times (c_0, 0)$. Suppose that the density satisfies for some $\delta > 0$ the expansion

$$f(\mathbf{z}, c) = \varphi(\mathbf{z}) + O(|c|^\delta) \tag{5.41}$$

uniformly for $\mathbf{z} \in R$. Then we say that the df H of $\mathbf{U}$ is in the *differentiable δ-neighborhood of the GPD W with dependence function D*.

The max-stable df G is, for example, in the differentiable δ-neighborhood of $W = 1 + \log(G)$ with $\delta = 1$ (Falk and Reiss [155]). The df of $\mathbf{Z}$ in the bivariate case with underlying max-stable df G has been derived by Ghoudi et al. [179]; under the additional assumption that the second derivative of the dependence function D exists, the density of $(\mathbf{Z}, C)$ was computed by Deheuvels [103].

Theorem 5.6.4. *Suppose that the df H of $\mathbf{U}$ is for some $\delta > 0$ in the differentiable δ-neighborhood of the GPD W with dependence function D. Suppose that $\zeta = \int_R \varphi(\mathbf{z})\, d\mathbf{z} > 0$. Conditional on $C > c_0$, the corresponding transform $(\mathbf{Z}, C/c_0)$ then has a density $f_{c_0}(\mathbf{z}, c)$ which satisfies*

$$f_{c_0}(\mathbf{z}, c) = \frac{\varphi(\mathbf{z})}{\zeta} + O(|c_0|^\delta)$$

uniformly on $R \times (0, 1)$ as $c_0 \uparrow 0$.

Under the conditions of the previous theorem, $\mathbf{Z}$ and C/c_0 are asymptotically for $c_0 \uparrow 0$ independent, conditional on $C > c_0$, where $\mathbf{Z}$ has in the limit the density $f(\mathbf{z}) = \varphi(\mathbf{z})/\zeta$ and C/c_0 is uniformly on $(0, 1)$ distributed. This will be formulated in the subsequent corollary. Note that the variational distance between probability measures equals $1/2$ times the L_1-distance between their densities with respect to a dominating measure, see, for example, Lemma 3.3.1 in Reiss [385].

Corollary 5.6.5. *Let* $\mathbf{Z}^*, C^*$ *be independent rv,* $\mathbf{Z}^*$ *having density* f *on* R *and* C^* *being uniformly distributed on* $(0,1)$. *Denote by* $\mathcal{B}$ *the* σ-*field of Borel sets in* $R \times [0,1]$. *Then we obtain from Theorem 5.6.4 that*

$$\sup_{B \in \mathcal{B}} \left| P\big((\mathbf{Z}, C/c_0) \in B \mid C > c_0\big) - P\big((\mathbf{Z}^*, C^*) \in B\big) \right|$$

$$= \frac{1}{2} \int_R \int_{[0,1]} |f_{c_0}(\mathbf{z}, c) - f(\mathbf{z})| \, dc \, d\mathbf{z} = O(|c_0|^\delta).$$

The asymptotic independence of $\mathbf{Z}$ and C, conditional on $C > c_0$ explains, why the intensity measure of the limiting Poisson process of the sequence of point processes of exceedances in Corollary 5.6.7 below factorizes across radial and angular components, which was first observed by de Haan [188], see also Section 8.3.2 of Coles [71].

Expansions of Pickands densities of Finite Length with Regularly Varying Functions

The first-order condition (5.41) characterizing the differentiable δ-neighborhood of a GPD was refined to a higher-order condition by Frick and Reiss [162] who use an expansion of $f(\mathbf{z}, c)$ again with $\varphi(\mathbf{z})$ as a leading term.

Let $\mathbf{U} = (U_1, \ldots, U_d)$ be an arbitrary rv on $(-\infty, 0]^d$, whose Pickands transform has a density $f(\mathbf{z}, c)$ on $R \times (c_0, 0)$ for $c_0 < 0$ close to 0. Assume that

$$f(\mathbf{z}, c) = \varphi(\mathbf{z}) + \sum_{j=1}^{k} B_j(c)\tilde{A}_j(\mathbf{z}) + o(B_k(c)), \quad c \uparrow 0, \qquad (5.42)$$

uniformly for $\mathbf{z} \in R$ for some $k \in \mathbb{N}$, where the $\tilde{A}_j : R \to \mathbb{R}$, $j = 1, \ldots, k$, are integrable functions. In addition, we require that the functions $B_j : (-\infty, 0) \to (0, \infty)$, $j = 1, \ldots, k$, satisfy

$$\lim_{c \uparrow 0} B_j(c) = 0 \qquad (5.43)$$

and

$$\lim_{c \uparrow 0} \frac{B_j(ct)}{B_j(c)} = t^{\beta_j}, \quad t > 0, \beta_j > 0. \qquad (5.44)$$

Without loss of generality, let $\beta_1 < \beta_2 < \cdots < \beta_k$. We say that the density $f(\mathbf{z}, c)$ satisfies an *expansion of length* $k + 1$ if the conditions (5.42)-(5.44) hold. Recall that in analogy to (2.17) a function fulfilling condition (5.44) is regularly varying in 0 with β_j being the exponent of variation. According to Resnick [393] one can always represent a β-varying function as $|c|^\beta L(c)$, where L is slowly varying in 0 meaning that the exponent of variation is zero. The functions $B_j(c) = |c|^{\beta_j}$, $j = 1, \ldots, k$, e.g., satisfy the preceding conditions.

Due to the properties of slowly varying functions the density in (5.42) also satisfies

$$f(\mathbf{z}, c) = \varphi(\mathbf{z}) + \sum_{j=1}^{\kappa} B_j(c)\tilde{A}_j(\mathbf{z}) + o(B_\kappa(c)), \quad c \uparrow 0, \tag{5.45}$$

for any $1 \le \kappa \le k$. With regard to the testing problem in Section 6.5 the existence of an index j such that $\int_R \tilde{A}_j(\mathbf{z})\, d\mathbf{z} \ne 0$ is essential. Then it is appropriate to choose κ as

$$\kappa = \min\left\{ j \in \{1, \dots, k\} : \int_R \tilde{A}_j(\mathbf{z})\, d\mathbf{z} \ne 0 \right\}. \tag{5.46}$$

If $k = 1$, we write B and $\tilde{A}$ instead of B_1 and $\tilde{A}_1$, respectively, and denote the exponent of variation of B by β.

The Pickands density of an EVD G satisfies the expansion

$$f(\mathbf{z}, c) = \varphi(\mathbf{z}) + \sum_{j=1}^{\infty} |c|^j \tilde{A}_j(\mathbf{z})$$

uniformly for $\mathbf{z} \in R$, where $c \in (c_0, 0)$ with $c_0 < 0$ close to 0. Because the $\tilde{A}_j$ are uniformly bounded on the simplex R the expansions can be reduced to an expansion of arbitrary finite length.

The d-variate standard normal distribution $\mathcal{N}(\mathbf{0}, \Sigma)$ with positive definite correlation matrix Σ, transformed to reverse exponential margins possesses a Pickands density that satisfies the expansion

$$f_\Sigma(\mathbf{z}, c) = B(c)\tilde{A}(\mathbf{z}) + o(B(c)), \quad c \uparrow 0,$$

with

$$B(c) = |c|^{\sum_{i,j=1}^{d} \sigma_{ij} - 1} L(c),$$

$$L(c) = (-\log|c|)^{\sum_{i,j=1}^{d} \sigma_{ij}/2 - d/2},$$

and

$$\tilde{A}(\mathbf{z}) = (\det \Sigma)^{-1/2} (4\pi)^{\sum_{i,j=1}^{d} \sigma_{ij}/2 - d/2} \prod_{i,j=1}^{d} (z_i z_j)^{(\sigma_{ij} - \delta_{ij})/2},$$

where $I_d = (\delta_{ij})_{i,j=1,\dots,d}$ and $\Sigma^{-1} = (\sigma_{ij})_{i,j=1,\dots,d}$. The function B is regularly varying in 0 with the exponent of variation $\beta = \sum_{i,j=1}^{d} \sigma_{ij} - 1 > 0$, cf. Example 3 in Frick and Reiss [162].

Expansions of Pickands densities can be used to characterize the dependence structure of the underlying rv. Particularly, the first term $\varphi(\mathbf{z})$ distinguishes between tail dependence and (marginal) tail independence according to Lemma 5.6.3. In Section 6.1 we establish a relationship to spectral expansions, and in Section 6.5 expansions of Pickands densities are used to test the tail dependence in arbitrary dimensions.

The POT Approach Based on the Pickands Transform

Let now $\mathbf{U}^{(i)} = (U_1^{(i)}, \ldots, U_d^{(i)})$, $1 \leq i \leq n$, be iid rv with common df H, which is in a differentiable δ-neighborhood of a GPD with $\zeta = \int_R \varphi(\mathbf{z}) \, d\mathbf{z} > 0$. Denote by $(\mathbf{Z}^{(i)}, C^{(i)})$, $1 \leq i \leq n$, the corresponding Pickands transforms.

Fix a threshold $c_0 < 0$ and consider only those observations among the sample $(\mathbf{Z}^{(i)}, C^{(i)}/c_0)$, $1 \leq i \leq n$, with $C^{(i)} > c_0$. We denote these by $(\widetilde{\mathbf{Z}}^{(j)}, \widetilde{C}^{(j)}/c_0)$, $1 \leq j \leq K(n)$, where $K(n) = \sum_{j \leq n} 1(C^{(j)} > c_0)$ is binomial $B(n, p(c_0))$ distributed with parameters n and $p(c_0) = P(C^{(j)} > c_0) = |c_0|(\zeta + O(|c_0|^\delta))$.

We obtain from Theorem 1.3.1 that the exceedances $(\widetilde{\mathbf{Z}}^{(j)}, \widetilde{C}^{(j)}/c_0)$, $j = 1, 2, \ldots$, are independent copies of $(\widetilde{\mathbf{Z}}, \widetilde{C}/c_0)$, which realizes in $R \times [0, 1]$ and has the distribution $P((\mathbf{Z}, C/c_0) \in \cdot | C > c_0)$. From Corollary 5.6.5 we deduce in the sequel that the empirical point process of the exceedances can be approximated in variational distance within the bound $O(n|c_0|^{1+\delta})$ by the empirical point process of $(\mathbf{Z}_j^*, C_j^*)$, $1 \leq j \leq K(n)$, which are independent copies of $(\mathbf{Z}^*, C^*)$ defined in Corollary 5.6.5.

We represent the sample $(\widetilde{\mathbf{Z}}^{(j)}, \widetilde{C}^{(j)}/c_0)$, $1 \leq j \leq K(n)$, by means of the empirical point process

$$N_{n,c_0} := \sum_{j \leq K(n)} \varepsilon_{(\widetilde{\mathbf{Z}}^{(j)}, \widetilde{C}^{(j)}/c_0)}.$$

The empirical process N_{n,c_0} is a random element in the set $\mathbb{M}(R \times [0, 1])$ of all finite point measures on $(R \times [0, 1], \mathcal{B})$, equipped with the smallest σ-field $\mathcal{M}(R \times [0, 1])$ such that for any $B \in \mathcal{B}$ the projection $\mathbb{M}(R \times [0, 1]) \ni \mu \mapsto \mu(B)$ is measurable; see the discussion around (1.3) in Section 1.1.

Let $(\mathbf{Z}_1^*, C_1^*), (\mathbf{Z}_2^*, C_2^*), \ldots$ be independent copies of $(\mathbf{Z}^*, C^*)$, which are also independent of $K(n)$. Denote by

$$N_{n,c_0}^* := \sum_{j \leq K(n)} \varepsilon_{(\mathbf{Z}_j^*, C_j^*)}$$

the point process pertaining to $(\mathbf{Z}_1^*, C_1^*), \ldots, (\mathbf{Z}_{K(n)}^*, C_{K(n)}^*)$. By $d(\xi, \eta)$ we denote the variational distance between two random elements ξ, η in an arbitrary measurable space $(\mathbb{M}, \mathcal{M})$:

$$d(\xi, \eta) = \sup_{M \in \mathcal{M}} |P(\xi \in M) - P(\eta \in M)|.$$

Theorem 5.6.6. *Suppose that the df of* $\mathbf{U}$ *is for some* $\delta > 0$ *in the differentiable* δ-*neighborhood of a GPD with dependence function* D. *We assume that* $\zeta = \int_R \varphi(\mathbf{z}) \, d\mathbf{z} > 0$. *Then we have*

$$d(N_{n,c_0}, N_{n,c_0}^*) = O(n|c_0|^{1+\delta}).$$

Proof. From Corollary 1.2.4 together with Corollary 5.6.5 we obtain

$$d(N_{n,c_0}, N_{n,c_0}^*) \le E(K(n))d((\widetilde{\mathbf{Z}}, \widetilde{C}/c_0), (\mathbf{Z}^*, C^*)) = O(n|c_0|^{1+\delta}). \qquad \square$$

We replace now the binomial distributed rv $K(n)$ by a Poisson distributed one $\tau(n)$ with parameter $E(\tau(n)) = n|c_0|\zeta$, which is stochastically independent of $(\mathbf{Z}_1^*, C_1^*), (\mathbf{Z}_2^*, C_2^*), \dots$ From the triangular inequality, Lemmata 1.2.1, 1.2.2, 3.1.4 and Theorem 5.6.6 we derive the bound

$$d\Big(N_{n,c_0}, \sum_{j \le \tau(n)} \varepsilon_{(\mathbf{Z}_j^*, C_j^*)}\Big) = O\Big(n|c_0|^{1+\delta} + |c_0| + (n|c_0|)^{1/2}|c_0|^\delta\Big).$$

Choose $L < 0$ and put $c_0 = L/n$. Then we obtain

$$d(N_{n,L/n}, N_L^*) = O(n^{-\delta}),$$

where

$$N_L^* := \sum_{j \le \tau(n)} \varepsilon_{(\mathbf{Z}_j^*, C_j^*)}$$

is a Poisson process with intensity measure

$$\nu(B) = E(N_L^*(B)) = E(\tau(n))P((\mathbf{Z}^*, C^*) \in B)$$
$$= |L|\zeta \int_B f(\mathbf{z})\, dc\, d\mathbf{z} = |L| \int_B \varphi(\mathbf{z})\, dc\, d\mathbf{z}, \qquad B \in \mathcal{B},$$

which is independent of n.

Since the function $T(\mathbf{x}) = (\mathbf{z}, c)$ is one-to-one, we obtain now the following result, which goes back to de Haan [188]; see also Section 8.3.2 of Coles [71]. It provides in addition a bound for the rate of convergence of

$$N_n^L := \sum_{j \le K(n)} \varepsilon_{(n/L)\widetilde{\mathbf{U}}^{(j)}} = \sum_{j \le K(n)} \varepsilon_{T^{-1}(\widetilde{\mathbf{Z}}^{(j)}, \widetilde{C}^{(j)}/(L/n))}$$

to the Poisson process

$$N_T^{*L} := \sum_{j \le \tau(n)} \varepsilon_{T^{-1}(\mathbf{Z}_j^*, C_j^*)},$$

whose intensity measure factorizes across radial and angular components. By $\widetilde{\mathbf{U}}^{(1)}, \widetilde{\mathbf{U}}^{(2)}, \dots$ we denote those observations in the sample $\mathbf{U}^{(1)}, \mathbf{U}^{(2)}, \dots, \mathbf{U}^{(n)}$ whose radial Pickands coordinates satisfy $C^{(i)} \ge L/n$.

Corollary 5.6.7. *With the preceding notation we have under the conditions of Theorem 5.6.6, for any $L < 0$,*

$$d(N_n^L, N_T^{*L}) = d(N_{n,L/n}, N_L^*) = O(n^{-\delta}).$$

THE PICKANDS TRANSFORM FOR A GENERAL EVD

In the sequel we will introduce the Pickands transform for a d-dimensional EVD G with arbitrary univariate EVD margins. The family of non-degenerate *univariate* EVD can be parametrized by $\alpha \in \mathbb{R}$ with

$$G_\alpha(x) = \begin{cases} \exp\left(-(-x)^\alpha\right), & x \le 0 \\ 1, & x > 0 \end{cases} \qquad \text{for } \alpha > 0,$$

$$G_\alpha(x) = \begin{cases} 0, & x \le 0 \\ \exp(-x^\alpha), & x > 0 \end{cases} \qquad \text{for } \alpha < 0$$

and

$$G_0(x) := \exp(-e^{-x}), \qquad x \in \mathbb{R},$$

being the family of (reverse) Weibull, Fréchet and the Gumbel distribution; see Section 2.2. Note that G_1 is the standard reverse exponential df.

We denote in what follows by $G_{\boldsymbol{\alpha}}$ with $\boldsymbol{\alpha} = (\alpha_1, \dots, \alpha_d) \in \mathbb{R}^d$ a d-dimensional max-stable df, whose i-th univariate margin is equal to G_{α_i}, $i \le d$. The corresponding GPD is any df $W_{\boldsymbol{\alpha}}$ such that for $\mathbf{x}$ with $G_{\boldsymbol{\alpha}}(\mathbf{x})$ in a neighborhood of $\mathbf{1}$

$$W_{\boldsymbol{\alpha}}(\mathbf{x}) = 1 + \log\left(G_{\boldsymbol{\alpha}}(\mathbf{x})\right).$$

The univariate margins of $W_{\boldsymbol{\alpha}}$ coincide in their upper tails with those of the usual one-dimensional GPD, see Section 1.3. The df G with reverse exponential margins and the corresponding GPD, which we considered above, would now be denoted by $G_{(1,\dots,1)}$ and $W_{(1,\dots,1)}$.

The following auxiliary function will be crucial for our further investigation. Put, for $x \in \mathbb{R}$ with $0 < G_{\alpha_i}(x) < 1$,

$$\psi_{\alpha_i}(x) := \log\left(G_{\alpha_i}(x)\right)$$

$$= \begin{cases} -(-x)^{\alpha_i}, & x < 0, & \text{if } \alpha_i > 0 \\ -x^{\alpha_i}, & x > 0, & \text{if } \alpha_i < 0 \\ -e^{-x}, & x \in \mathbb{R}, & \text{if } \alpha_i = 0. \end{cases}$$

Note that ψ_{α_i} is a strictly monotone and continuous function, whose range is $(-\infty, 0)$.

The next result will be crucial for the definition of the Pickands transform for arbitrary $G_{\boldsymbol{\alpha}}$ and $W_{\boldsymbol{\alpha}}$.

Lemma 5.6.8. *Suppose that the rv $\mathbf{X} = (X_1, \dots, X_d)$ has df $G_{\boldsymbol{\alpha}}$. Put*

$$U_i := \psi_{\alpha_i}(X_i), \qquad 1 \le i \le d.$$

Then $\mathbf{U} = (U_1, \dots, U_d)$ has df $G_{(1,\dots,1)}$.

Proof. Since $G_{\alpha_i}(X_i)$ is uniformly distributed on $(0,1)$, it is obvious that $U_i = \log(G_{\alpha_i}(X_i))$ has df $\exp(x) = G_1(x)$, $x < 0$. It remains to show that the df of $(U_1, \ldots, U_d)$, H say, is max-stable. But this follows from the fact that G_α is max-stable with $G_{\alpha_i}^n(\psi_{\alpha_i}^{-1}(x/n)) = G_{\alpha_i}(\psi_{\alpha_i}^{-1}(x))$: We have for $x_i < 0$, $1 \le i \le d$,

$$
\begin{aligned}
H(x_1/n, \ldots, x_d/n)^n &= P\big(U_i \le x_i/n, 1 \le i \le d\big)^n \\
&= P\big(X_i \le \psi_{\alpha_i}^{-1}(x_i/n), 1 \le i \le d\big)^n \\
&= P\big(X_i \le \psi_{\alpha_i}^{-1}(x_i), 1 \le i \le d\big) = H(x_1, \ldots, x_d). \qquad \square
\end{aligned}
$$

Lemma 5.6.8 can also be formulated as

$$
G_\alpha\big(\psi_{\alpha_1}^{-1}(x_1), \ldots, \psi_{\alpha_d}^{-1}(x_d)\big) = G_{(1,\ldots,1)}(x_1, \ldots, x_d), \qquad x_i < 0, \, i \le d. \tag{5.47}
$$

The max-stability of G_α is, thus, preserved by the transformation of each univariate marginal onto the reverse exponential distribution.

Corollary 5.6.9. *Suppose that* $\mathbf{X} = (X_1, \ldots, X_d)$ *has common df* W_α. *Put*

$$
U_i := \psi_{\alpha_i}(X_i), \qquad 1 \le i \le d.
$$

Then $\mathbf{U} = (U_1, \ldots, U_d)$ *has common df* $W_{(1,\ldots,1)}$.

Proof. We have for $c_0 < x_i < 0$, $i \le d$, c_0 close to 0:

$$
\begin{aligned}
P(U_i \le x_i, i \le d) &= P\big(X_i \le \psi_{\alpha_i}^{-1}(x_i), 1 \le i \le d\big) \\
&= 1 + \log\Big(G_\alpha\big(\psi_{\alpha_1}^{-1}(x_1), \ldots, \psi_{\alpha_d}^{-1}(x_d)\big)\Big) \\
&= 1 + \log\big(G_{(1,\ldots,1)}(x_1, \ldots, x_d)\big) = W_{(1,\ldots,1)}(x_1, \ldots, x_d). \quad \square
\end{aligned}
$$

The preceding result can also be formulated as

$$
W_\alpha\big(\psi_{\alpha_1}^{-1}(x_1), \ldots, \psi_{\alpha_d}^{-1}(x_d)\big) = W_{(1,\ldots,1)}(x_1, \ldots, x_d), \qquad c_0 < x_i < 0, \, i \le d,
$$

c_0 close to 0.

As a consequence of the above results we can represent an arbitrary G_α as

$$
\begin{aligned}
G_\alpha(x_1, \ldots, x_d) &\\
&= G_{(1,\ldots,1)}\big(\psi_{\alpha_1}(x_1), \ldots, \psi_{\alpha_d}(x_d)\big) \\
&= \exp\Big(\Big(\sum_{i \le d} \psi_{\alpha_i}(x_i)\Big) D\Big(\frac{\psi_{\alpha_1}(x_1)}{\sum_{i \le d} \psi_{\alpha_i}(x_i)}, \ldots, \frac{\psi_{\alpha_{d-1}}(x_{d-1})}{\sum_{i \le d} \psi_{\alpha_i}(x_i)}\Big)\Big) \\
&=: \exp\big(c_\alpha D(\mathbf{z}_\alpha)\big),
\end{aligned}
$$

where D is a dependence function as defined in (4.32). Equally, we have

$$
W_\alpha(x_1, \ldots, x_d) = W_{(1,\ldots,1)}\big(\psi_{\alpha_1}(x_1), \ldots, \psi_{\alpha_d}(x_d)\big) = 1 + c_\alpha D(\mathbf{z}_\alpha),
$$

whenever $c_{\boldsymbol{\alpha}} = \sum_{i \leq d} \psi_{\alpha_i}(x_i)$ is close enough to 0.

The functions $c_{\boldsymbol{\alpha}}$ and $\mathbf{z}_{\boldsymbol{\alpha}}$ now provide the Pickands transform of a rv $\mathbf{X} = (X_1, \ldots, X_d)$ with df $G_{\boldsymbol{\alpha}}$ or $W_{\boldsymbol{\alpha}}$: Put

$$
C_{\boldsymbol{\alpha}} := \sum_{i \leq d} \psi_{\alpha_i}(X_i),
$$

$$
\mathbf{Z}_{\boldsymbol{\alpha}} := \Big(\frac{\psi_{\alpha_1}(X_1)}{C_{\boldsymbol{\alpha}}}, \ldots, \frac{\psi_{\alpha_{d-1}}(X_{d-1})}{C_{\boldsymbol{\alpha}}} \Big).
$$

By the fact that $U_i = \psi_{\alpha_i}(X_i)$, $1 \leq i \leq d$, have joint df $G_{(1,\ldots,1)}$ or $W_{(1,\ldots,1)}$, it is obvious that Theorems 5.6.2, 5.6.4 and Corollary 5.6.5 on the (asymptotic) distribution and independence of the angular and radial component of the Pickands transform in the case $\boldsymbol{\alpha} = (1, \ldots, 1)$ immediately apply to $C_{\boldsymbol{\alpha}}$ and $\mathbf{Z}_{\boldsymbol{\alpha}}$ with arbitrary $\boldsymbol{\alpha}$. Put for $G_{\boldsymbol{\alpha}}$,

$$
G_{\boldsymbol{\alpha}}^{\psi}(\mathbf{x}) := G_{\boldsymbol{\alpha}}\big(\psi_{\alpha_1}^{-1}(x_1), \ldots, \psi_{\alpha_d}^{-1}(x_d)\big),
$$

where $x_i < 0$, $1 \leq i \leq d$. From Lemma 5.6.8 we know that $G_{\boldsymbol{\alpha}}^{\psi} = G_{(1,\ldots,1)}$ and, hence, we obtain from Lemma 5.4.1, for the spectral decomposition of $G_{\boldsymbol{\alpha}}^{\psi}$,

$$
\big(G_{\boldsymbol{\alpha}}^{\psi}\big)_{\mathbf{z}}(c) = G_{\boldsymbol{\alpha}}\Big(\psi_{\alpha_1}^{-1}(cz_1), \ldots, \psi_{\alpha_{d-1}}^{-1}(cz_{d-1}), \psi_{\alpha_d}^{-1}\Big(c\Big(1 - \sum_{i \leq d-1} z_i\Big)\Big)\Big)
$$

$$
= \exp\big(cD(\mathbf{z})\big), \qquad c < 0, \ \mathbf{z} \in R. \tag{5.48}
$$

$\mathcal{P}(G_{\boldsymbol{\alpha}}^{\psi})$ is, thus, the family of reverse exponential distributions with parameter $D(\mathbf{z})$, $\mathbf{z} \in R$. Equally, we obtain for

$$
W_{\boldsymbol{\alpha}}^{\psi}(x_1, \ldots, x_d) := W_{\boldsymbol{\alpha}}\big(\psi_{\alpha_1}^{-1}(x_1), \ldots, \psi_{\alpha_d}^{-1}(x_d)\big)
$$

from Corollary 5.6.9 and Lemma 5.4.3,

$$
\big(W_{\boldsymbol{\alpha}}^{\psi}\big)_{\mathbf{z}}(c) = W_{\boldsymbol{\alpha}}\Big(\psi_{\alpha_1}^{-1}(cz_1), \ldots, \psi_{\alpha_d}^{-1}\Big(c\Big(1 - \sum_{i \leq d-1} z_i\Big)\Big)\Big)
$$

$$
= 1 + cD(z_1, \ldots, z_{d-1}), \qquad c \geq c_0.
$$

The members of the family $\mathcal{P}(W_{\boldsymbol{\alpha}}^{\psi})$ coincide, thus, in their upper tails with the family of uniform distributions on the interval $(-1/D(\mathbf{z}), 0)$, $z \in R$. The preceding results in this section now carry over to $\mathcal{P}(G_{\boldsymbol{\alpha}}^{\psi})$ and $\mathcal{P}(W_{\boldsymbol{\alpha}}^{\psi})$.

Example 5.6.10 (Hüsler-Reiss). *Consider the bivariate Hüsler-Reiss EVD with parameter $\lambda \in [0, \infty)$,*

$$
H_{\lambda}(x_1, x_2) = \exp\Big(-\Phi\Big(\lambda + \frac{x_1 - x_2}{2\lambda}\Big) e^{-x_2} - \Phi\Big(\lambda + \frac{x_2 - x_1}{2\lambda}\Big) e^{-x_1}\Big),
$$

for x_1, $x_2 \in \mathbb{R}$, where Φ denotes the standard normal df. H_λ is max-stable with Gumbel margins, i.e., we have $H_\lambda = G_{(0,0)}$. From (5.48) we obtain, for $c < 0$ and $z \in [0, 1]$,

$$
\begin{aligned}
&H_\lambda\left(\psi_0^{-1}(cz), \psi_0^{-1}(c(1-z))\right) \\
&= H_\lambda\left(-\log(z) - \log(|c|), -\log(1-z) - \log(|c|)\right) \\
&= \exp\left(c\left(\Phi\left(\lambda + \frac{\log((1-z)/z)}{2\lambda}\right)(1-z) + \Phi\left(\lambda + \frac{\log(z/(1-z))}{2\lambda}\right)z\right)\right) \\
&=: \exp\left(c\, D_\lambda(z)\right),
\end{aligned}
$$

where $D_\lambda(z)$, $z \in [0, 1]$, is a dependence function as given in (4.32) with $d = 2$.

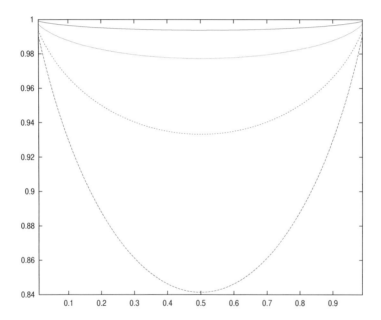

FIGURE 5.6.1. Dependence functions of the bivariate Hüsler-Reiss df with parameters $\lambda = 1, 1.5, 2, 2.5$; from bottom to top.

5.7   Simulation Techniques

A crucial point for further investigation of multivariate GPD and their practical application to real data sets is the need for simulations. Possible applications of the simulation methods, which will be presented in this section, are of course Monte-Carlo and bootstrapping methods. Another usage is a first check of new statistical testing or estimation procedures in multivariate GPD models, as it will be done in Sections 5.8 to 5.10 and 6.6.

Not much attention has been paid to the simulation of multivariate GPD so far. Michel [330] and [331] are among the very few contributions to this topic and are the basis of the simulation techniques which we are going to present. Hofmann [222] adds an algorithm for the generation of a rv from a nested logistic GPD in dimension $d = 3$. Clearly, the representation $\mathbf{X} = U/\mathbf{Z}$ in Proposition 5.2.8 of a GPD rv $\mathbf{X}$ entails its simulation via the uniformly distributed on $(0, 1)$ rv U and the independent rv $\mathbf{Z}$. But it is by no means obvious, how to simulate this way a target GPD such as a logistic one.

A summary of the work done on the simulation of multivariate EVD can be found in Stephenson [425].

Simulation of GPD with Bounded Pickands Density

The algorithm introduced here is based on the so-called rejection method and the transformation to Pickands coordinates (5.35). The method presented here will be applicable for low dimensions only due to computational reasons, but it has the advantage of being able to simulate a broad variety of GPD, namely those with a bounded Pickands density.

We shortly describe the rejection method in general. Suppose we want to simulate a distribution F on a compact set $A \subset \mathbb{R}^d$. Let $g(\mathbf{x}) = mf(\mathbf{x})$ be a constant multiple of the density f of F. We require that g is bounded by some number M. Then Algorithm 5.7.1 describes the *rejection method* for the generation of one rv with df F.

Algorithm 5.7.1.

1. *Generate a rv $\mathbf{X} = (X_1, \ldots, X_d)$, uniformly distributed on A.*

2. *Generate a random number Y independent of $\mathbf{X}$, which is uniformly distributed on $[0, M]$.*

3. *Return $\mathbf{X}$ if $Y \leq g(\mathbf{X})$, else go to 1.*

It is obvious that Algorithm 5.7.1 can be very inefficient, since a lot of points might have to be generated to get a "useable" one.

A natural scheme for the simulation of a GPD W is the application of the rejection method to W. However, even in the most common cases like the logistic case, the density of a GPD is not bounded, see Section 4.2 of Michel [331]. Therefore, the rejection method is not directly applicable. In many cases, however, a detour via Pickands coordinates can be helpful.

The idea is to generate the Pickands coordinates $\mathbf{Z}$ and C of a rv, which follows W, separately. These are by Theorem 5.6.2 independent under $C > c_0$ and, in addition, C is uniformly distributed on $(c_0, 0)$ and, thus, easy to simulate.

For the generation of $\mathbf{Z}$ we can apply the rejection method if its density f and, thus, the Pickands density $\varphi = \zeta \cdot f$ are bounded. This is for example the case with the logistic GPD, see Theorem 2.4 in Michel [331], but also for many other cases like the asymmetric logistic model and large parts of the nested logistic model, see Section 2.3 of Michel [330]. Since $\mathbf{Z}$ lies in the $d-1$-dimensional unit simplex R, one needs to simulate the uniform distribution on R. This can be done by the following algorithm, which is investigated in Corollary 4.4 of Michel [331].

Algorithm 5.7.2. *Set $k:=0$ and, for $i=d-1,\dots,1$, do*

1. *Generate a uniformly distributed number x_i on $(0,1)$, independent of x_j, $i+1 \le j \le d-1$.*

2. *Compute $u_i := (1-k)\left(1-(1-x_i)^{1/i}\right)$.*

3. *Put $k := k + u_i$.*

Return the vector $(u_1,\dots,u_d-1)$.

To generate now the rv $\mathbf{Z}$, one uses the Pickands density φ for the application of the rejection method, since for the use of f the number ζ would have to be calculated, which is possible only approximately and only with great numerical effort. In the end one has to invert the Pickands transformation to get the desired rv. The algorithm below implements these considerations and is the desired simulation algorithm.

Algorithm 5.7.3.

1. *Generate a vector $(z_1,\dots,z_{d-1})$, which has density f, with Algorithm 5.7.1 applied to the Pickands density φ, where the uniform distribution on the unit simplex R is done by Algorithm 5.7.2.*

2. *Generate, independent of $(z_1,\dots,z_{d-1})$, a number c, uniformly distributed on $(c_0,0)$.*

3. *Return the vector $\left(cz_1,\dots,cz_{d-1}, c - c\sum_{i=1}^{d-1} z_i\right)$.*

Remark that Algorithm 5.7.3 only simulates a GPD under the condition $C > c_0$. We will show later, how this condition can dropped to simulate unconditional GPD.

Experimental and theoretical investigations of Algorithm 5.7.3 in Section 3.2 of Michel [330] show that runtimes explode for large d. Thus, the algorithm is in general only suited for low dimensions.

A Special Case: Simulation of GPD of Logistic Type

In the special case of the logistic GPD a simulation algorithm can be used which does not have the runtime disadvantages of Algorithm 5.7.3. It is based on the Shi transformation (Shi [413]), which is a variant of the Pickands transformation for the logistic case. It will be introduced next.

The mapping $P : (0, \infty) \times (0, \pi/2)^{d-1} \to (0, \infty)^d$ with

$$
P(r, \psi_1, \ldots, \psi_{d-1})
$$
$$
= r \left(\cos(\psi_1), \cos(\psi_2)\sin(\psi_1), \ldots, \cos(\psi_{d-1}) \prod_{j=1}^{d-2} \sin(\psi_j), \prod_{j=1}^{d-1} \sin(\psi_j) \right) \quad (5.49)
$$

is the *polar transformation*, and its inverse defines the *polar coordinates* r, $\boldsymbol{\psi} = (\psi_1, \ldots, \psi_{d-1})$ in $(0, \infty)^d$. The following facts are well known (see, for example, Mardia et al. [319], Section 2.4): the mapping P is one-to-one, infinitely often differentiable and satisfies the equation

$$
1 = ||P(1, \psi_1, \ldots, \psi_{d-1})||_2^2 = \sum_{i=1}^{d-1} \cos^2(\psi_i) \prod_{j=1}^{i-1} \sin^2(\psi_j) + \prod_{j=1}^{d-1} \sin^2(\psi_j), \quad (5.50)
$$

i.e., the function $P(1, \psi_1, \ldots, \psi_{d-1})$ is a one-to-one mapping from $(0, \pi/2)^{d-1} \subset \mathbb{R}^{d-1}$ onto the intersection of $(0, \infty)^d$ with the unit sphere of $\mathbb{R}^d$ with respect to the Euclidian $|| \cdot ||_2$-norm.

Lemma 5.7.4. *The mapping* $T : (0, \pi/2)^{d-1} \to (0, \infty)^d$, *defined by*

$$
T(\psi_1, \ldots, \psi_{d-1})
$$
$$
:= \left(\cos^2(\psi_1), \cos^2(\psi_2)\sin^2(\psi_1), \ldots, \cos^2(\psi_{d-1}) \prod_{j=1}^{d-2} \sin^2(\psi_j), \prod_{j=1}^{d-1} \sin^2(\psi_j) \right),
$$

maps the cube $(0, \pi/2)^{d-1}$ *one-to-one and infinitely often differentiable onto the simplex* $S = \left\{ \mathbf{x} \in (0,1)^d : \sum_{i \le d} x_i = 1 \right\}$, *i.e., to the unit circle in* $(0, \infty)^d$, *minus the unit vectors, with regard to the* $|| \cdot ||_1$-*norm.*

Proof. The function $x \mapsto x^2$ maps the interval $(0,1)$ one-to-one onto itself, thus the bijectivity and differentiability of T follow from the corresponding properties of the polar transformation. Let $(x_1, \ldots, x_d) = T(\psi_1, \ldots, \psi_{d-1})$. By (5.50) the relation $\sum_{i=1}^d x_i = 1$ directly follows. □

We have

$$
\left\| \left(\frac{(-x_1)^\lambda}{||\mathbf{x}||_\lambda^\lambda}, \ldots, \frac{(-x_d)^\lambda}{||\mathbf{x}||_\lambda^\lambda} \right) \right\|_1 = 1
$$

for $\mathbf{x} \in (-\infty, 0)^d$ and $\lambda \geq 1$ and, thus, $\left((-x_1)^\lambda / \|\mathbf{x}\|_\lambda^\lambda, \ldots, (-x_d)^\lambda / \|\mathbf{x}\|_\lambda^\lambda\right) \in$ S. This point has a representation with regard to the transformation T. More precisely, there exist a uniquely determined $(\psi_1, \ldots, \psi_{d-1}) \in (0, \pi/2)^{d-1}$ with

$$
\left(\frac{(-x_1)^\lambda}{\|\mathbf{x}\|_\lambda^\lambda}, \ldots, \frac{(-x_d)^\lambda}{\|\mathbf{x}\|_\lambda^\lambda}\right)
$$
$$
= \left(\cos^2(\psi_1), \cos^2(\psi_2)\sin^2(\psi_1), \ldots, \cos^2(\psi_{d-1}) \prod_{j=1}^{d-2} \sin^2(\psi_j), \prod_{j=1}^{d-1} \sin^2(\psi_j)\right).
$$

By taking the λ-th root, multiplying with $-\|\mathbf{x}\|_\lambda$ and putting $c := \|\mathbf{x}\|_\lambda$ one arrives at

$$
\mathbf{x} = (x_1, \ldots, x_d)
$$
$$
= ST_\lambda(c, \psi_1, \ldots, \psi_{d-1})
$$
$$
:= -c\Bigg(\cos^{2/\lambda}(\psi_1), \cos^{2/\lambda}(\psi_2)\sin^{2/\lambda}(\psi_1), \ldots, \cos^{2/\lambda}(\psi_{d-1}) \prod_{j=1}^{d-2} \sin^{2/\lambda}(\psi_j),
$$
$$
\prod_{j=1}^{d-1} \sin^{2/\lambda}(\psi_j)\Bigg).
$$

This transformation is called the *Shi transformation ST_λ*. The transformation $ST_\lambda : (0, \infty) \times (0, \pi/2)^{d-1} \to (-\infty, 0)^d$ is one-to-one and infinitely often differentiable. The components of the vector $(c, \psi_1, \ldots, \psi_{d-1}) := ST_\lambda^{-1}(x_1, \ldots, x_d)$ are the *Shi coordinates* of $(x_1, \ldots, x_d)$, where c is called the *radial component* and $\boldsymbol{\psi} := (\psi_1, \ldots, \psi_{d-1})$ is the *angular component*. Note that $c = \|\mathbf{x}\|_\lambda$. The corresponding random Shi coordinates of a rv will be denoted as usual with upper case letters.

Note that in the case $\lambda = 2$ the Shi transformation is up to sign the polar transformation from (5.49). For $\lambda = 1$ we have a variant of the inverse Pickands transformation, where the angular component has an additional parametrization with regard to the cube $(0, \pi/2)^{d-1}$.

The Shi transformation was originally introduced in Shi [413] and was used by Stephenson [425] to simulate the rv following an extreme value distribution of logistic type.

With the following theorem we establish a basis for a simulation algorithm for the logistic GPD W_λ. As a helpful notation let $B_r^\lambda := \{\mathbf{x} \in (-\infty, 0)^d : \|\mathbf{x}\|_\lambda < r\}$, $r > 0$, be the *ball* in $(-\infty, 0)^d$ of radius r with respect to the $\|\cdot\|_\lambda$-norm, centered at the origin.

Theorem 5.7.5. *Let $\mathbf{X} = (X_1, \ldots, X_d)$ follow a logistic GPD W_λ with parameter $\lambda > 1$. Choose a number $c_0 > 0$, such that $W_\lambda(\mathbf{x}) = 1 - \|\mathbf{x}\|_\lambda$ on $B_{c_0}^\lambda$. Then the* *rv* $(C, \Psi_1, \ldots, \Psi_{d-1}) = ST_\lambda^{-1}(\mathbf{X})$ *has, under the condition* $\mathbf{X} \in B_{c_0}^\lambda$, *on* $(0, c_0) \times (0, \pi/2)^{d-1}$ *the density*

$$f(c, \psi_1, \ldots, \psi_{d-1}) = f(\psi_1, \ldots, \psi_{d-1}) = \prod_{i=1}^{d-1} \left(2i - \frac{2}{\lambda}\right) \cos(\psi_i) \sin^{2(d-i)-1}(\psi_i).$$

Additionally, f has positive mass on $(0, \pi/2)^{d-1}$:

$$\eta := \int_{\left(0, \frac{\pi}{2}\right)^{d-1}} f(\psi_1, \ldots, \psi_{d-1}) \, d(\psi_1, \ldots, \psi_{d-1}) = \prod_{i=1}^{d-1} \left(\frac{i - \frac{1}{\lambda}}{d - i}\right) > 0.$$

Furthermore, we have, conditional on $C = \|\mathbf{X}\|_\lambda < c_0$:

(i) *The random Shi coordinates* $C, \Psi_1, \ldots, \Psi_{d-1}$ *are independent.*

(ii) *The rv C is on* $(0, c_0)$ *uniformly distributed.*

(iii) *The angular component* Ψ_i *has the df*

$$F_i(\psi) := \sin^{2(d-i)}(\psi), \qquad 0 \le \psi \le \pi/2,$$

with the corresponding quantile function $F_i^{-1}(u) = \arcsin\left(u^{1/(2(d-i))}\right)$, $0 < u < 1$, $i = 1, \ldots, d-1$.

Proof. The proof follows the lines of the proof of Theorem 5.6.2. The somewhat lengthy details are worked out in Section 3.1 of Michel [330]. □

Theorem 5.7.5 is analogous to Theorem 5.6.2, but with the restriction to the logistic case and using the Shi transformation. While in Theorem 5.6.2 the independence of the angular and the radial component of Pickands coordinates for general GPD could be shown, Theorem 5.7.5 states in addition the mutual independence of the angular components. We are also able to specify their distributions precisely. But recall that we are restricting ourselves to the logistic case.

In what follows, we will apply Theorem 5.7.5 to derive an algorithm for the simulation of GPD of logistic type in general dimension.

Algorithm 5.7.6.

1. *Generate U_1 uniformly on* $(0, c_0)$ *and* $U_2, \ldots, U_d$ *uniformly on* $(0, 1)$, *all mutually independent.*

2. *Compute* $\Psi_i := F_i^{-1}(U_{i+1})$ *for* $i = 1, \ldots, d-1$.

3. *Return the vector* $(X_1, \ldots, X_d) = ST_\lambda(U_1, \Psi_1, \ldots, \Psi_{d-1})$.

Example 5.7.7. *Figure* 5.7.1 *illustrates results of Algorithm* 5.7.6 *for* $d = 2$ *and* 3 *and miscellaneous* λ. *In each plot* 1000 *points were generated. The generated points arrange themselves in a sort of* d-*dimensional cone, whose peak lies in the origin and whose center is the line* $x_i = x_j$, $i, j = 1, \ldots, d$, *the bisector of the negative quadrant. The lower end is naturally bounded by* $\|\mathbf{x}\|_\lambda = c_0$. *The parameter* λ *describes the width of the cone. For* λ *close to* 1 *it is opened very wide, for larger* λ *it becomes more narrow.*

These plots will inspire us in Section 5.10 *to develop testing methods for logistic GP models.*

Chapter 7 of Hofmann [222] shows a generalization of the Shi transformation to the nested logistic model and uses it to develop an analogous simulation algorithm in dimension $d = 3$. Other generalizations of the Shi transformation are to date unknown to the authors.

SIMULATION OF UNCONDITIONAL GPD

Previously in this section we have introduced simulations of GPD, which are only able to simulate conditional GPD in a neighborhood of the origin. One encounters a problem, if unconditionally GP distributed rv are to be simulated. In this situation, the POT-stability of a GPD as formulated in Lemma 5.4.6, will enable us to overcome this problem.

Since a Pickands dependence function D satisfies $D(\mathbf{z}) \leq 1$ for all $\mathbf{z} \in R$, we conclude $k := dD(1/d, \ldots, 1/d) \leq d$ and, with the choice of $t_i \in [-1/d, 0)$, the corresponding assumptions in Lemma 5.4.6 are fulfilled by Theorem 5.1.4 for all GPD W.

Corollary 5.7.8. *Let the rv* $\mathbf{Y}$ *follow an arbitrary GPD* W. *Put* $\kappa := P(\mathbf{Y} \geq \mathbf{t})$, *where* $\mathbf{t} = (t, \ldots, t) \in (-1/d, 0)^d$ *and suppose that* $\kappa > 0$. *Furthermore, let the rv* $\mathbf{X}$ *be conditionally GP distributed, i.e.,* $P(\mathbf{X} \geq \mathbf{x}) = \bar{W}(\mathbf{x})/\kappa$ *for* $\mathbf{x} \geq \mathbf{t}$, *where we denote by* $\bar{W}$ *the survivor function of* W. *Then* $\mathbf{Y}$ *and* $\mathbf{X}/\kappa$ *are close to* $\mathbf{0}$ *identically distributed and, thus,* $\mathbf{X}/\kappa$ *is GP distributed.*

Proof. By Theorem 5.1.4 we can assume that $W(\mathbf{x}) = 1 + \log(G(\mathbf{x}))$ for $\mathbf{x} \geq \mathbf{t}$. Thus, we have by Lemma 5.4.6

$$P(\mathbf{X} \geq \mathbf{x}) = \frac{\bar{W}(\mathbf{x})}{\kappa} = \frac{P(\mathbf{Y} \geq \mathbf{x})}{P(\mathbf{Y} \geq \mathbf{t})} = P(\mathbf{Y} \geq \mathbf{x} \mid \mathbf{Y} \geq \mathbf{t})$$

$$= P\left(\mathbf{Y} \geq \frac{\mathbf{x}}{\kappa}\kappa \mid \mathbf{Y} \geq \mathbf{t}\right) = P\left(\mathbf{Y} \geq \frac{\mathbf{x}}{\kappa}\right) = P(\kappa\mathbf{Y} \geq \mathbf{x})$$

Thus $\mathbf{X}$ and $\kappa\mathbf{Y}$ are identically distributed close to 0 and, therefore, also $\mathbf{Y}$ and $\mathbf{X}/\kappa$. $\square$

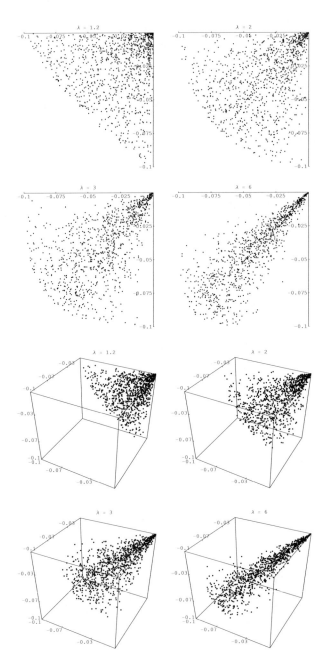

FIGURE 5.7.1. Simulated logistic GPD rv for $d = 2$ (top) and $d = 3$ (bottom), for $\lambda = 1.2$, 2, 3, 6 each time.

Corollary 5.7.8 provides a straightforward manner of getting GPD rv from conditionally GPD rv. It suffices to simulate conditional GPD, as done in Algorithms 5.7.6 or 5.7.3. A division by κ turns them into unconditionally GPD rv, since only the distribution close to the origin is crucial for the definition of a GPD.

Algorithm 5.7.9.

1. *Generate n rv $x_1, \ldots, x_n$ in the cube $K_t := (-t, 0)^d$, which are conditionally distributed by a GPD (for example with the help of Algorithm 5.7.6 for the logistic type or Algorithm 5.7.3 for other cases, rejecting those rv, which are outside K_t). Choose t, for example, as $-1/(2d)$.*

2. *Compute $\kappa = \bar{W}(t, \ldots, t)$ and $y_i := x_i/\kappa$, $i = 1, \ldots, n$.*

3. *Return $y_1, \ldots, y_n$.*

5.8 Testing the GPD Assumption, Threshold Selection

When applying GPD models to actual data sets, the first task is to check whether a GPD model really fits the data, at least in the upper tail. Hand in hand with this problem comes the question, well known from the univariate case: what is the appropriate choice of a threshold, over which one can model the observations as coming from a GPD?

By making suggestions how to handle these problems, the contribution of this section is twofold: First we develop a non-asymptotic and exact level-α test based on the single-sample t-test, which checks whether multivariate data are actually generated by a multivariate GPD. The idea for this test was already described in the remarks after Lemma 5.4.5. Its performance is evaluated by theoretical considerations and by simulations using the simulation algorithms of Section 5.7. This procedure is also utilized for deriving a Gauss-test based threshold selection in multivariate POT models.

TESTING FOR A MULTIVARIATE GPD

Let $\mathbf{U} = (U_1, \ldots, U_d)$ be an arbitrary rv in $(-\infty, 0)^d$ with $P(\mathbf{U} > \mathbf{x}) > 0$ for all $\mathbf{x} = (x_1, \ldots, x_d) < \mathbf{0}$ close to $\mathbf{0}$. Note that $\max(U_k/x_k, k \in K) < t \iff U_k/x_k < t, k \in K \iff U_k > tx_k, k \in K$, for any non-empty set $K \subset \{1, \ldots, d\}$ and $t \in [0, 1]$. The following characterization of a multivariate GPD with uniform margins is just a reformulation of Proposition 5.3.4,

Corollary 5.8.1. *The rv $\mathbf{U}$ follows a GPD with uniform margins iff for any $\mathbf{x} = (x_1, \ldots, x_d) < \mathbf{0}$ close to $\mathbf{0}$,*

$$P\left(\max\left(\frac{U_k}{x_k}, k \in K\right) < t \mid \max\left(\frac{U_k}{x_k}, k \in K\right) < 1\right) = t, \qquad t \in [0, 1],$$

for any non-empty subset $K \subset \{1, \ldots, d\}$.

By putting $x_k = x < 0$ for $k = 1, \ldots, d$, we obtain from Corollary 5.8.1 that for a GPD rv $\mathbf{U}$ and x close to 0 we have

$$P\left(\|\mathbf{U}\|_\infty < t\,|x| \mid \|\mathbf{U}\|_\infty < |x|\right) = t, \qquad t \in [0,1], \tag{5.51}$$

i.e., the $\|\cdot\|_\infty$-norm of $\mathbf{U}$ is uniformly distributed, conditional on exceeding some small threshold near zero. Similar results to (5.51) have been shown for the $\|\cdot\|_1$-norm in Theorem 5.6.2 and, in addition, for the usual λ-norm in case of a logistic dependence function in Theorem 5.7.5.

The theoretical results derived above can now be utilized to develop a testing procedure for multivariate GPD.

Suppose that we observe n independent copies $\mathbf{U}_1, \ldots, \mathbf{U}_n$ of a rv $\mathbf{U}$, not necessarily following a GPD. Corollary 5.8.1 suggests the following procedure to test for an underlying GPD with uniform margins: choose $\mathbf{x} < \mathbf{0}$ close to $\mathbf{0}$, determine the exceedance vectors $\mathbf{U}_j > \mathbf{x}$ and test the maxima of the vectors $\mathbf{U}_j/\mathbf{x}$ for the uniform distribution on $(0,1)$, where division is meant componentwise.

Denote the exceedances in the order of their outcome by $\mathbf{V}_1, \ldots, \mathbf{V}_{K(n)}$ and by

$$M_m := \max\left(\frac{V_{m,1}}{x_1}, \ldots, \frac{V_{m,d}}{x_d}\right), \qquad m \leq K(n),$$

the largest value of the scaled components of the m-th exceedance $\mathbf{V}_m$. Then, by Theorem 1.3.1, $M_1, M_2, \ldots$ are, for $\mathbf{x}$ close to $\mathbf{0}$, independent and uniformly on $(0,1)$ distributed rv if $\mathbf{U}$ follows a GPD with uniform margins and they are independent of their total number $K(n)$. To test for this uniform distribution we could use any goodness-of-fit test such as the single-sample Kolmogorov-Smirnov test statistic

$$
\begin{aligned}
\mathrm{KS} &:= K(n)^{1/2} \sup_{0 \leq t \leq 1} \left| \frac{1}{K(n)} \sum_{m \leq K(n)} 1_{[0,t]}(M_m) - t \right| \\
&= \max_{m \leq K(n)} \left(\max\left(M_{m:K(n)} - \frac{m-1}{K(n)}, \frac{m}{K(n)} - M_{m:K(n)} \right) \right),
\end{aligned} \tag{5.52}
$$

where $M_{1:K(n)} \leq M_{2:K(n)} \leq \cdots \leq M_{K(n):K(n)}$ denote the ordered values of $M_1, \ldots, M_{K(n)}$. The asymptotic distribution of KS is known, so that a p-value can also be computed, see e.g. Sheskin [412], pages 241 ff.

A goodness-of-fit test, however, commonly uses a limit distribution of the test statistic and, therefore, requires a sample size which is not too small. This condition contradicts the fact that we will typically observe only a few exceedances above a high threshold. A non-asymptotic test is, therefore, preferred such as the following suggestion.

We transform $M_1, \ldots, M_{K(n)}$ by the inverse Φ^{-1} of the standard normal df $\Phi(x) = (2\pi)^{-1/2} \int_{-\infty}^{x} \exp(-y^2/2)\, dy$ and, thus, obtain iid standard normal rv

$$Z_m := \Phi^{-1}(M_m), \qquad m \leq K(n),$$

under the null hypothesis of an underlying GPD with uniform margins. The next result is an immediate consequence.

Theorem 5.8.2. *Let* $\mathbf{U}_1, \ldots, \mathbf{U}_n$ *be independent copies of a rv* $\mathbf{U}$ *that follows a GPD with uniform margins. Suppose that* $P(\mathbf{U} > \mathbf{x}) > 0$ *for all* $\mathbf{x} < \mathbf{0}$ *close to* $\mathbf{0}$. *The test statistic of the single-sample t test*

$$t_{K(n)} = \frac{\frac{1}{K(n)^{1/2}} \sum_{m \leq K(n)} Z_m}{\left(\frac{1}{K(n)-1} \left(\sum_{m \leq K(n)} Z_m^2 - \frac{1}{K(n)} \left(\sum_{m \leq K(n)} Z_m \right)^2 \right) \right)^{1/2}} \tag{5.53}$$

then follows a t distribution with $k - 1$ *degrees of freedom, conditional on* $k = K(n) \geq 2$.

We reject the null hypothesis that the rv $\mathbf{U}$ follows a GPD with uniform margins if $|t_{K(n)}|$ gets too large or, equivalently, if the corresponding p-value

$$p = 2\left(1 - T_{K(n)-1}(|t_{K(n)}|)\right) \tag{5.54}$$

gets too small, typically if $p \leq 0.05$. By T_k we denote the df of the t distribution with k degrees of freedom. Note that the preceding test is a conditional one, given the number $K(n)$ of exceedances.

Note, on the other hand, that the above test for a multivariate GPD with uniform margins is not fail-safe. Consider $\mathbf{X} = (X_1, \ldots, X_d)$, where the X_i are iid rv with df $F(y) = 1 - (-y)^{1/d}$, $-1 \leq y \leq 0$. Then we have for any $\mathbf{x} \in [-1, 0)^d$

$$P(\mathbf{X} > t\mathbf{x} \mid \mathbf{X} > \mathbf{x}) = t, \qquad t \in [0, 1],$$

but the df of $\mathbf{X}$ is not a GPD with uniform margins. To safeguard oneself against such a counterexample one could test the univariate excesses for a one-dimensional uniform distribution on $(0, 1)$ as well.

Performance of the Test

To study the performance of the above test we consider in what follows n independent copies $\mathbf{X}_1, \ldots, \mathbf{X}_n$ of a rv $\mathbf{X} = (X_1, \ldots, X_d)$, which realizes in $(-\infty, 0)^d$ and where the df F satisfies for $t \in (0, 1)$ and $\mathbf{x} \leq \mathbf{0}$, $\mathbf{x} \neq \mathbf{0}$, in a left neighborhood $\mathcal{U}$ of $\mathbf{0}$ the expansion

$$\frac{\frac{d}{dt}(1 - F(t\mathbf{x}))}{\frac{d}{dt}(1 - W(t\mathbf{x}))} = 1 + J\left(\frac{\mathbf{x}}{\|\mathbf{x}\|_1}\right)(1 - W(t\mathbf{x}))^\delta + r(t, \mathbf{x}), \tag{5.55}$$

with some $\delta > 0$, where W is the df of a GPD with uniform margins, $J(\cdot)$ is
a function on the set $S := \{\mathbf{z} \leq \mathbf{0} : \|\mathbf{z}\|_1 = 1\}$ of directions in $(-\infty, 0)^d$ and the
function $r(\cdot, \cdot)$ satisfies uniformly for $t \in (0, 1)$ and $\mathbf{x} \in \mathcal{U}$ the expansion $r(t, \mathbf{x}) = o\left((1 - W(t\mathbf{x}))^\delta\right)$.

Notice that condition (5.55) implies by l'Hôpital's rule that F and W are tail
equivalent if the function K is bounded, i.e.,

$$\lim_{\mathbf{x} \uparrow 0} \frac{1 - F(\mathbf{x})}{1 - W(\mathbf{x})} = 1.$$

Condition (5.55) is a condition on the spectral decomposition of F, see Sections 5.4 and 5.5. It essentially requires that the df F is in the differentiable
δ-spectral neighborhood of W, i.e., the tracks $F(|t| \mathbf{x})$, $t \leq 0$, belong to the wide
class of Hall [202]. It is, for example, satisfied with $\delta = 1$ and $J = 1$ if F is an
EVD G with negative exponential margins and $W = 1 + \log(G)$.

Choose a non-empty subset $L \subset \{1, \ldots, d\}$, $\mathbf{x} \in \mathcal{U}$, $\mathbf{x} < \mathbf{0}$, and put $\mathbf{x}_L = (\widetilde{x}_{1,L}, \ldots, \widetilde{x}_{d,L})$ with $\widetilde{x}_{i,L} = x_i$ if $i \in L$ and $\widetilde{x}_{i,L} = 0$ elsewhere, and set $\widetilde{\mathbf{x}}_L := \left(\widetilde{x}_{1,L}/\sum_{i \in L} x_i, \ldots, \widetilde{x}_{d-1,L}/\sum_{i \in L} x_i\right)$. The following lemma is established in Falk
and Michel [147], Lemma 4.1.

Lemma 5.8.3. *Suppose that the rv* $\mathbf{X}$ *satisfies condition* (5.55). *Choose* $\mathbf{x} < \mathbf{0}$
in $\mathcal{U}$, *a sequence of numbers* $c_n \downarrow 0$ *as* $n \to \infty$, *and put* $\mathbf{x}_n := c_n \mathbf{x}$, $n \in \mathbb{N}$. *If*
$A(\mathbf{x}) := \sum_{j \leq d} (-1)^{j+1} \sum_{|L|=j} \|\mathbf{x}_L\|_1 D(\widetilde{x}L) > 0$, *then we obtain, uniformly for*
$t \in [0, 1]$,

$$P(\mathbf{X} > t\mathbf{x}_n \mid \mathbf{X} > \mathbf{x}_n)$$
$$= t + \frac{c_n^\delta}{\delta + 1} \left(t(t^\delta - 1) \frac{B(\mathbf{x})}{A(\mathbf{x})} + \left(t^{1+\delta}(1 - t) + t(1 - t^{1+\delta})\right) o(1) \right),$$

where $B(\mathbf{x}) := \sum_{j \leq d} (-1)^{j+1} \sum_{|L|=j} \|\mathbf{x}_L\|_1^{1+\delta} D^{1+\delta}(\widetilde{x}L) K(\mathbf{x}_L/\|\mathbf{x}_L\|_1)$.

Note that the condition $A(\mathbf{x}) > 0$ in Lemma 5.8.3 is a rather weak one, as
we have $A(\mathbf{x}) \geq 0$ anyway.

Consider now the exceedances $\mathbf{X}_i > \mathbf{x}_n = c_n \mathbf{x}$ among $\mathbf{X}_1, \ldots, \mathbf{X}_n$. Denote these by $\mathbf{V}_1^{(n)}, \ldots, \mathbf{V}_{K(n)}^{(n)}$, where the number $K(n)$ is binomial distributed
$B(n, p(n))$ with $p(n) = P(\mathbf{X} > \mathbf{x}_n) = c_n(A(\mathbf{x}) + O(c_n^\delta))$, and $K(n)$ is independent
of the exceedances $\mathbf{V}_1^{(n)}, \mathbf{V}_2^{(n)}, \ldots$ Then, by Lemma 5.8.3,

$$M_m^{(n)} = \max\left(\frac{V_{m,1}^{(n)}}{c_n x_1}, \ldots, \frac{V_{m,d}^{(n)}}{c_n x_d}\right), \qquad m \leq K(n),$$

are iid with df

$$F^{(n)}(t) := \frac{P(\mathbf{X} > t\mathbf{x}_n)}{P(\mathbf{X} > \mathbf{x}_n)}$$

$$= t\left(1 + \frac{c_n^\delta}{1+\delta}\left((t^\delta - 1)\frac{B(\mathbf{x})}{A(\mathbf{x})} + \left(t^\delta(1-t) + (1-t^{1+\delta})\right)o(1)\right)\right)$$

for $0 \le t \le 1$. Denote by $F(t) = t$, $0 \le t \le 1$, the df of the uniform distribution on $(0,1)$. Note that $E\left(\Phi^{-1}\left(M_1^{(n)}\right)\right) = \int_0^1 \Phi^{-1}(t)\,F^{(n)}(dt)$ and that $\int_0^1 \Phi^{-1}(t)\,F(dt) = \int_0^1 \Phi^{-1}(t)\,dt = 0$. Integration by parts together with the substitution $t \mapsto \Phi(y)$ implies

$$E\left(\Phi^{-1}\left(M_1^{(n)}\right)\right) = \int_0^1 \Phi^{-1}(t)\,(F^{(n)} - F)(dt)$$

$$= \int_0^1 \Phi^{-1}(t)\frac{d}{dt}(F^{(n)}(t) - t)\,dt$$

$$= -\int_0^1 \frac{d}{dt}(\Phi^{-1}(t))(F^{(n)}(t) - t)\,dt$$

$$= -\int_0^1 \frac{1}{\varphi(\Phi^{-1}(t))}(F^{(n)}(t) - t)\,dt$$

$$= -\int_{-\infty}^\infty \frac{1}{\varphi(y)}(F^{(n)}(\Phi(y)) - \Phi(y))\varphi(y)\,dy$$

$$= \int_{-\infty}^\infty \frac{c_n^\delta}{1+\delta}\Phi(y)(1-\Phi^\delta(y))\frac{B(\mathbf{x})}{A(\mathbf{x})}\,dy + o(c_n^\delta)$$

$$= \frac{c_n^\delta}{1+\delta}\frac{B(\mathbf{x})}{A(\mathbf{x})}\int_{-\infty}^\infty \Phi(y)(1-\Phi^\delta(y))\,dy + o(c_n^\delta)$$

$$= \text{const } c_n^\delta + o(c_n^\delta).$$

The single-sample t test computed from the transformed exceedances $Z_m^{(n)} = \Phi^{-1}(M_m^{(n)})$, $1 \le m \le K(n)$,

$$t_{K(n)} = \frac{\frac{1}{K(n)^{1/2}}\sum_{m \le K(n)} Z_m^{(n)}}{\left(\frac{1}{K(n)-1}\left(\sum_{m \le K(n)} Z_m^{(n)2} - \frac{1}{K(n)}\left(\sum_{m \le K(n)} Z_m^{(n)}\right)^2\right)\right)^{1/2}}$$

$$= \frac{\frac{1}{K(n)^{1/2}}\sum_{m \le K(n)}(Z_m^{(n)} - E(Z_1^{(n)}))}{\left(\frac{1}{K(n)-1}\left(\sum_{m \le K(n)} Z_m^{(n)2} - \frac{1}{K(n)}\left(\sum_{m \le K(n)} Z_m^{(n)}\right)^2\right)\right)^{1/2}}$$

$$+ \frac{K(n)^{1/2}E(Z_1^{(n)})}{\left(\frac{1}{K(n)-1}\left(\sum_{m \le K(n)} Z_m^{(n)2} - \frac{1}{K(n)}\left(\sum_{m \le K(n)} Z_m^{(n)}\right)^2\right)\right)^{1/2}}$$

will, thus, converge to infinity in probability and detect that the underlying df is not a GPD, if $K(n)^{1/2}E(Z_1^{(n)})$ tends to infinity. Since $K(n)/(np(n)) \to 1$ in probability if $np(n) \sim nc_n \to \infty$ as $n \to \infty$, the deviation from a GPD is detected if $(nc_n)^{1/2}c_n^\delta \to \infty$, i.e., if $nc_n^{1+2\delta} \to \infty$ as $n \to \infty$.

SIMULATION OF THE TEST

We want to illustrate the performance of the above test for an underlying GPD by showing some simulations. The test actually tests for a (shifted) GPD copula, and its performance seems to be reasonably good if a threshold $\mathbf{x} = (x, x, \ldots, x)$ with identical entries is used.

Each of the following plots displays the points $(j/100, p_{j/100})$, $-100 \le j \le -1$ with linear interpolation, where $p_{j/100}$ is the p-value defined in (5.54) based on the threshold $\mathbf{x}_T = (j/100, \ldots, j/100)$. Note that within a graphic the same sample is used for the 100 tests, only the threshold is chosen differently each time. Since only very few observations are dropped when raising the threshold from $j/100$ to $(j+1)/100$, these 100 interpolation points suffice to get stable results.

By the choice of the threshold with identical components, the results of various tests and different thresholds can elegantly be plotted into one graphic with the thresholds on the horizontal and the p-values on the vertical axis. The thick line in the graphics represents the p-values using the t-test (5.54), the thin line is the corresponding p-value using the KS-test statistic (5.52). For reference an additional horizontal line is drawn at the 5%-level. Recall that a p-value below 5% typically leads to a rejection of the null hypothesis.

To provide the number $K(n)$ of observations the test with threshold $j/100$ uses, a dashed line connects the points $(j/100, n^{-1}\sum_{i=1}^{n}\mathbf{1}_{(\mathbf{x}_T, \mathbf{0})}(\mathbf{x}_i))$ in each plot, i.e., the relative number of points exceeding the corresponding threshold. The number of observations exceeding the thresholds -0.8, -0.6, -0.4 and -0.2 is also given in the second line of the labelling of the horizontal axis.

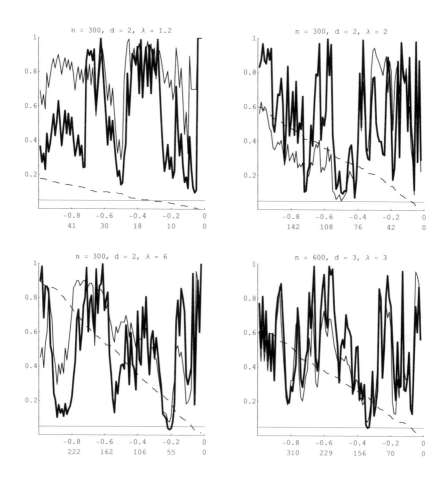

FIGURE 5.8.1. Plots for logistic GP distributed rv.

Figure 5.8.1 shows plots with simulated data from a logistic GPD $W(\mathbf{x}) = 1 - \|\mathbf{x}\|_\lambda$ with different values of λ, sample sizes n and dimensions d. The data were generated by Algorithms 5.7.6 and 5.7.9. We see that only in a very few cases the true hypothesis of GPD rv is rejected. The few rejections are not surprising, since each plot displays the outcome of 100 tests (on the same data set but with different thresholds). Both, the t-test and the KS-test, therefore, keep their prescribed levels.

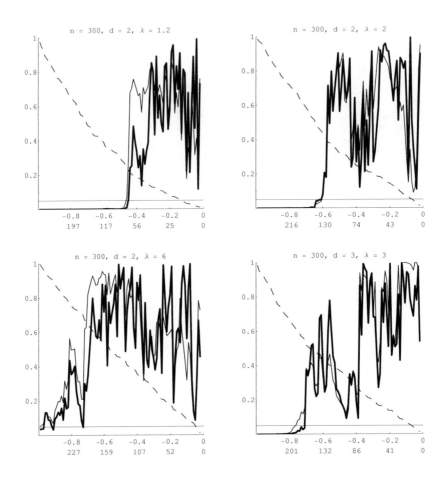

FIGURE 5.8.2. Plots for rv following a logistic extreme value copula.

Figure 5.8.2 shows the results of the tests with underlying EVD $G(\mathbf{x}) = \exp(-\|\mathbf{x}\|_\lambda)$, $\mathbf{x} \le \mathbf{0}$, transformed to uniform margins, thus simulating an extreme value copula. The data were generated with the help of Algorithm 1.1 from Stephenson [425]. Since GPD and EVD are tail equivalent, the tests do not detect a deviation from the null hypothesis close to 0, but only away from the origin. Note also the dependence of the power of the test on the parameter λ.

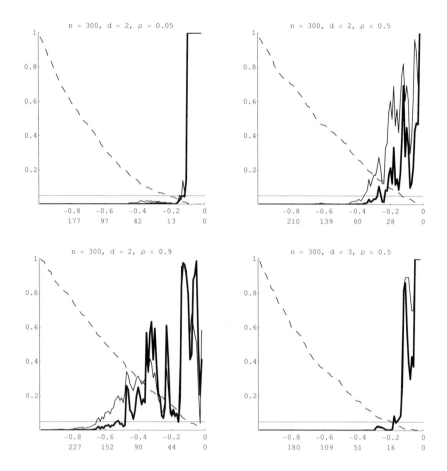

FIGURE 5.8.3. Plots for rv following a normal copula.

In Figure 5.8.3 normal vectors with constant correlation coefficient ρ for all bivariate components, which are transformed to uniform margins on $(-1, 0)$ are used, i.e., the data are generated by a shifted normal copula. Since a shifted multivariate normal copula is in the domain of attraction of $\exp(-||\mathbf{x}||_1)$, $\mathbf{x} \leq \mathbf{0}$, the appropriate GPD model fitting the exceedances would be $W(\mathbf{x}) = 1 - ||\mathbf{x}||_1$, which is, however, a peculiar case. With this GPD underlying the data no observation should be close to the origin, see Section 5.2, and thus a GPD should not be detected in our simulations. This can be observed in Figure 5.8.3 for low correlations, as the null hypothesis is not rejected only in those cases with a very small number of exceedances. A high coefficient of correlation, however, decreases the power of our test. Such a phenomenon is also discussed in Coles [71], Section 8.4. The KS test seems to have here less power than the t-test as well.

A *t*-Test Based Threshold Selection in Multivariate POT Models

The choice of an appropriate threshold is very much an open matter, even in the univariate case. The selection is typically supported visually by a diagram such as the mean excess plot, see for example Davison and Smith [94]. In what follows we will utilize the test for a multivariate GPD developed previously to derive a *t*-test based multivariate threshold selection rule.

We consider in this section a rv $\mathbf{X} = (X_1, \ldots, X_d)$ that is in the domain of attraction of an EVD G with negative exponential margins, i.e., there exist vectors $\mathbf{a}_n > \mathbf{0}$, $n \in \mathbb{N}$, with $\mathbf{a}_n \downarrow_{n \to \infty} \mathbf{0}$, and a vector $\mathbf{b} \in \mathbb{R}^d$ such that the df F of $\mathbf{X}$ satisfies

$$F^n(\mathbf{a}_n\mathbf{x} + \mathbf{b}) \to_{n \to \infty} G(\mathbf{x}), \qquad \mathbf{x} \in \mathbb{R}^d. \tag{5.56}$$

By $\mathbf{xy} = (x_i y_i)_{i \leq d}$ we denote as usual the componentwise product of two vectors $\mathbf{x}, \mathbf{y} \in \mathbb{R}^d$. The vector $\mathbf{b}$ is the upper and finite endpoint of the support of F. For convenience we assume that $\mathbf{b}$ is known and we set it to $\mathbf{0}$.

Note that a rv $\mathbf{X}$ with continuous marginal df F_i, $i \leq d$, is in the domain of attraction of an EVD with arbitrary margins implies that the transformed vector $(F_i(X_i) - 1)_{i \leq d}$ is in the domain of attraction of an EVD G with negative standard exponential margins, see Resnick [393], Proposition 5.10. The df of the transformed rv then satisfies (5.56) with $\mathbf{b} = \mathbf{0}$ and $\mathbf{a}_n = (1/n, \ldots, 1/n)$.

Taking logarithms we obtain from (5.56),

$$n(1 - F(\mathbf{a}_n\mathbf{x})) \to_{n \to \infty} -\log(G(\mathbf{x})), \qquad \mathbf{x} \leq \mathbf{0}. \tag{5.57}$$

Define the decreasing and continuous function $\mathbf{a} : [1, \infty) \to (0, \infty)^d$ by $\mathbf{a}(n) := \mathbf{a}_n$, $n \in \mathbb{N}$, and by linear interpolation for $n < s < n + 1$. It then follows from (5.57) that

$$s(1 - F(\mathbf{a}(s)\mathbf{x})) \to_{s \to \infty} -\log(G(\mathbf{x})), \qquad \mathbf{x} \leq \mathbf{0}.$$

Observe that $-\log(G(\mathbf{x})) = 1 - W(\mathbf{x})$ for $\mathbf{x}$ near $\mathbf{0}$, where W denotes a GPD corresponding to G. From equation (4.7) we obtain, for $t \in [0, 1]$,

$$P(\mathbf{X} > t\mathbf{a}(s)\mathbf{x}) = \sum_{j \leq d} (-1)^{j+1} \sum_{|L|=j} (1 - P(X_k \leq ta_k(s)x_k, \ k \in L))$$

and, thus, for $\mathbf{x} < \mathbf{0}$ near $\mathbf{0}$ with Proposition 5.3.4,

$$
\begin{aligned}
P\left(\mathbf{X} > t\mathbf{a}(s)\mathbf{x} \mid \mathbf{X} > \mathbf{a}(s)\mathbf{x}\right) \\
= \frac{\sum_{j \leq d}(-1)^{j+1}\sum_{|L|=j} s\left(1 - F(t\mathbf{a}_{n,L}\mathbf{x}_L)\right)}{\sum_{j \leq d}(-1)^{j+1}\sum_{|L|=j} s\left(1 - F(\mathbf{a}_{n,L}\mathbf{x}_L)\right)} \\
\to_{s \to \infty} \frac{\sum_{j \leq d}(-1)^{j+1}\sum_{|L|=j}(1 - W(t\mathbf{x}_L))}{\sum_{j \leq d}(-1)^{j+1}\sum_{|L|=j}(1 - W(\mathbf{x}_L))} \\
= t.
\end{aligned}
\tag{5.58}
$$

This means that the df of an exceedance $\mathbf{X} > \mathbf{x}_T$ above a threshold $\mathbf{x}_T$ close to $\mathbf{0}$ approaches the uniform distribution on $(0,1)$.

Suppose that we have n independent copies $\mathbf{X}_1, \ldots, \mathbf{X}_n$ of $\mathbf{X}$ and that we are interested in some upper tail analysis of $\mathbf{X}$. To this end we choose a threshold $\mathbf{x}_T < \mathbf{0}$ close to $\mathbf{0}$ and consider only those observations among the sample which exceed this threshold. We denote these exceedances by $\mathbf{Y}_1 \ldots, \mathbf{Y}_{K(n)}$, which can be handled as independent copies of a rv $\mathbf{Y}$. From (5.58) we obtain that the df

$$P(\mathbf{Y} > t\mathbf{x}_T) = P(\mathbf{X} > t\mathbf{x}_T \mid \mathbf{X} > \mathbf{x}_T) \approx t$$

is close to the uniform df on $(0,1)$. The selection of the threshold $\mathbf{x}_T$ such that the above approximation is close enough for practical purposes is an obvious problem. Choosing the threshold very close to $\mathbf{0}$ improves the above approximation but reduces the number $K(n)$ of exceedances, which is a trade-off situation. In the univariate case $d = 1$ the selection of a threshold x_T can, for example, be supported visually by a diagram such as the empirical mean excess function

$$\hat{e}(x_T) = \frac{1}{K(n)} \sum_{j \leq K(n)} \frac{Y_j}{x_T} \approx E\left(\frac{Y}{x_T}\right) \approx \frac{1}{2}$$

as proposed by Davison and Smith [94], which should be close to the constant $1/2$ if x_T is close enough to 0 and $K(n)$ is large enough, see, e.g. Section 2.2 of Reiss and Thomas [390]. The considerations in this section can also be utilized for a t-test based threshold selection $\mathbf{x}_T$ as described in the sequel.

Denote again by

$$M_m = \max\left(\frac{Y_{m,1}}{x_{T,1}}, \ldots, \frac{Y_{m,d}}{x_{T,d}}\right), \qquad m \leq K(n),$$

the largest component of the excess vector $(Y_{m,1}/x_{T,1}, \ldots, Y_{m,d}/x_{T,d})$. Then the rv $M_1, M_2, \ldots$ are independent copies of a univariate rv M with df

$$P(M \leq t) = P(\mathbf{X} > t\mathbf{x}_T \mid \mathbf{X} > \mathbf{x}_T) \approx t, \qquad t \in [0,1],$$

if $\mathbf{x}_T$ is close enough to $\mathbf{0}$.

The single-sample t-test statistic $t_{K(n)}$ as given in (5.53) of the transformed rv $Z_m = \Phi^{-1}(M_m)$, $m \leq K(n)$, can then be utilized for a threshold selection: Fix $\mathbf{x}_0 < \mathbf{0}$ and put $\mathbf{x}(c) := c\mathbf{x}_0$, $c > 0$. One could use the smallest $c =: c_T > 0$ such that the p-value of $\left|t_{K(n)}\right|$ for the corresponding threshold $\mathbf{x}_T = c_T\mathbf{x}_0$ is bigger than some prescribed value such as 0.05. This threshold selection can be used to decide above which threshold an approximation of the underlying df by a GPD is justified. We refer to Section 7 of Falk and Michel [147] for an application of this rule to a real data set.

5.9   Parametric Estimation Procedures

Since searching for appropriate models in a non-parametric family like the GPD family can be too ambitious, one way out is to look for models in certain parametric subfamilies. Once one has decided for one such subfamily, one has to identify the corresponding model parameters. This is usually done by estimation methods.

In this section we present a short overview of several methods for parametric estimation in GPD models, which use decompositions of the corresponding rv with the help of different versions of the Pickands coordinates. The estimators are compared to each other with simulated data sets.

We will present two ML methods based on the angular density and one which uses the Pickands the angular density. Relative frequencies will be needed for another estimation procedure. Since the overview will be a short summary, the interested reader is referred to Michel [330] and [333] for details.

The Pickands Transformation Reloaded

Recall that with $R = \left\{ \mathbf{x} \in (0,1)^{d-1} : \sum_{i \leq d-1} x_i < 1 \right\}$ being the open and $\bar{R}$ the closed unit simplex in $\mathbb{R}^{d-1}$, the Pickands dependence function $D : \bar{R} \to [0,1]$ of an arbitrary GPD W with standard uniform margins can be written as

$$D(t_1, \ldots, t_{d-1}) \tag{5.59}$$

$$= \int_{\bar{R}} \max \left(u_1 t_1, \ldots, u_{d-1} t_{d-1}, \left(1 - \sum_{i=1}^{d-1} u_i \right) \left(1 - \sum_{i=1}^{d-1} t_i \right) \right) \nu(d\mathbf{u}),$$

where ν is the angular measure on $\bar{R}$, with characteristic properties

$$\nu\left(\bar{R} \right) = d \quad \text{and} \quad \int_{\bar{R}} u_i \nu(d\mathbf{u}) = 1, \qquad 1 \leq i \leq d-1, \tag{5.60}$$

see Section 4.3.

By $d^* := \nu(R)$ we denote the mass of ν in the interior of $\bar{R}$; recall that by (5.60) we have $\nu\left(\bar{R} \right) = d$ and, thus, $0 \leq d^* \leq d$. If the measure ν, restricted to R_{d-1}, possesses a density, we denote it by l and call it the *angular density*. In the literature it is also common to call the angular measure/density the *spectral measure/density*, see for example Einmahl et al. [121].

We have already introduced the Pickands coordinates as an important tool for analyzing a GPD in Section 5.6, since they decompose a GPD rv into two conditionally independent components, under the condition that the radial component exceeds some high value. The conditional distribution of the radial component is a uniform distribution, the angular component has the density $f(\mathbf{z}) := \varphi(\mathbf{z})/\zeta$, where φ is the Pickands density.

We define now a variant of the Pickands coordinates, which will also prove very useful.

The transformation $T_F : (-\infty, 0)^d \to R \times (-\infty, 0)$, defined by

$$T_F(x) := \left(\frac{\frac{1}{x_1}}{\frac{1}{x_1} + \cdots + \frac{1}{x_d}}, \ldots, \frac{\frac{1}{x_{d-1}}}{\frac{1}{x_1} + \cdots + \frac{1}{x_d}}, \frac{1}{x_1} + \cdots + \frac{1}{x_d} \right) \qquad (5.61)$$

$$=: (z_1, \ldots, z_{d-1}, c),$$

is the transformation to Pickands coordinates $\mathbf{z} := (z_1, \ldots, z_{d-1})$ and c *with respect to Fréchet margins*, where $\mathbf{z}$ is again called the angular component and c the radial component.

The transformation (5.61) is closely related to the Pickands transformation (5.35). We have, in addition, applied the transformation $0 < y \mapsto 1/y$ which conveys exponentially distributed rv to Fréchet rv, which is a common marginal transformation in extreme value theory. We, therefore, use the symbol T_F.

The distributions of the radial and angular components of the Pickands coordinates with respect to Fréchet margins are asymptotically known. To explain these, we need some additional notation.

By

$$K_s := \left\{ \mathbf{x} \in (-\infty, 0)^d : \|\mathbf{x}\|_\infty < s \right\}, \qquad s > 0, \qquad (5.62)$$

we denote the (open) cube with edge length s in the negative quadrant. For $r, s > 0$, let

$$Q_{r,s} := \left\{ \mathbf{z} \in R : T_F^{-1}(\mathbf{z}, -r) \in K_s \right\} \qquad (5.63)$$

be the set of angular components of the Pickands transformation with respect to Fréchet margins of the points in the cube K_s, whose radial component has the value $-r$. This set can also be written as

$$Q_{r,s} = \left\{ \mathbf{z} \in R : z_i > \frac{1}{rs}, i \le d - 1, \sum_{i \le d-1} z_i < 1 - \frac{1}{rs} \right\}.$$

Put for $r, s > 0$,

$$\chi(r, s) := \int_{Q_{r,s}} l(\mathbf{z}) \, d\mathbf{z}. \qquad (5.64)$$

Then we have $\chi(r, s) \uparrow_{r \to \infty} d^*$, since $Q_{r,s} \uparrow_{r \to \infty} R$.

One can show that

$$P\left(\mathbf{Z} \in B \mid C = -r, \mathbf{Z} \in Q_{r,s}\right) = \frac{1}{\chi(r, s)} \int_{B \cap Q_{r,s}} l(\mathbf{z}) \, d\mathbf{z} \qquad (5.65)$$

with $\mathbf{Z}$ and C being the random Pickands coordinates with respect to Fréchet margins corresponding to the GPD W, and B is some Borel set in R. One can, further, show that

$$\sup_{B \in \mathbb{B}_{d-1} \cap R} \left| P\left(\mathbf{Z} \in B \mid \mathbf{X} \in A_{r,s}\right) - \int_B \frac{l(\mathbf{z})}{d^*} \, d\mathbf{z} \right| = O\left(d^* - \chi(r, s)\right), \qquad (5.66)$$

where $\mathbb{B}_{d-1}$ is the Borel-σ-field in $\mathbb{R}^{d-1}$ and

$$A_{r,s} := \left\{ \mathbf{x} \in K_s : c = \sum_{i \leq d} \frac{1}{x_i} < -r \right\}. \qquad (5.67)$$

Altogether, the angular Pickands coordinate $\mathbf{Z}$ with respect to Fréchet margins has asymptotically, for $r \to \infty$, the conditional density $l(\mathbf{z})/d^*$, which is a scaled version of the angular density. For the highly technical proofs of the preceding results we refer to Theorems 5.1.6 and 5.1.7 of Michel [330]. The rate of convergence of the approximation depends on the rate of convergence of χ to d^*. In cases close to independence of the margins, this convergence is typically very slow, whereas in cases close to complete dependence of the margins this convergence will be quite fast, see Example 5.1.5 of Michel [330].

ML Estimation with the Angular Density

We introduce two ML based estimation procedures for a parametric family of GPD. For general information on the ML method we refer to Section 2.6 of Coles [71] and Section 4.2 of Serfling [408].

In what follows, we assume that we have n independent copies $\widetilde{\mathbf{X}}^{(1)}, \ldots, \widetilde{\mathbf{X}}^{(n)}$ of a rv $\widetilde{\mathbf{X}}$, which follows a GPD $W_{\lambda_1,\ldots,\lambda_k}$ from a k-parametric family, which satisfies $W_{\lambda_1,\ldots,\lambda_k}(\mathbf{x}) = 1 + \log(G(\mathbf{x}))$ for $\mathbf{x} \in K_s$, where G is an EVD with standard negative exponential margins. Let $W_{\lambda_1,\ldots,\lambda_k}$ have the angular density $l_{\lambda_1,\ldots,\lambda_k}$, and suppose that $d^*_{\lambda_1,\ldots,\lambda_k} > 0$. To keep the notation as simple as possible we set $\boldsymbol{\lambda} := (\lambda_1, \ldots, \lambda_k)$, where $\boldsymbol{\lambda} \in \Lambda \subset \mathbb{R}^k$.

We further consider only the copies with $\left\|\widetilde{\mathbf{X}}^{(i)}\right\|_\infty < s$ and denote by $\widetilde{\mathbf{Z}}^{(i)}$ and $\widetilde{C}^{(i)}$ the corresponding random Pickands coordinates with respect to Fréchet margins, $i = 1, \ldots, n$. We choose a threshold $r > 0$ and consider only those observations $\widetilde{\mathbf{X}}^{(i)}$ with $\widetilde{C}^{(i)} < -r$, i.e., $\widetilde{\mathbf{X}}^{(i)} \in A_{r,s}$. We denote these by $\mathbf{X}^{(1)}, \ldots, \mathbf{X}^{(m)}$. They are independent from each other and from the random number $m = K(n)$, see Theorem 1.3.1.

The corresponding Pickands coordinates $\mathbf{Z}^{(i)}$ with respect to Fréchet margins have a density that is not exactly known, but which is close to $l_{\boldsymbol{\lambda}}(\mathbf{z})/d^*_{\boldsymbol{\lambda}}$, see (5.66). This is a suitable approach for an ML estimation of $\boldsymbol{\lambda}$: Determine $\hat{\boldsymbol{\lambda}}_{m,r}$ such that the expression

$$\Upsilon(\boldsymbol{\lambda}) := \log\left(\prod_{i=1}^m \frac{l_{\boldsymbol{\lambda}}\left(\mathbf{Z}^{(i)}\right)}{d^*_{\boldsymbol{\lambda}}} \right) = \sum_{i=1}^m \log\left(\frac{l_{\boldsymbol{\lambda}}\left(\mathbf{Z}^{(i)}\right)}{d^*_{\boldsymbol{\lambda}}} \right)$$

is maximized in $\boldsymbol{\lambda}$.

Asymptotic consistency, normality and efficiency of ML estimators have been extensively studied; see, for example Pfanzagl [367]. We, however, do not use the *exact* ML procedure, since we do not insert the observations into their density, but into a function, which is only *close* to it. We, therefore, refer to this approach as the *asymptotic* ML method. The asymptotic distribution of the above defined asymptotic MLE $\hat{\boldsymbol{\lambda}}_{m,r}$ is, under suitable regularity conditions for $m \to \infty$ and $r \to \infty$, again the normal distribution with mean $\boldsymbol{\lambda}$ and covariance matrix $\mathbf{V}_{\boldsymbol{\lambda}}^{-1} := (v_{\boldsymbol{\lambda},j_1,j_2})_{j_1,j_2=1,\ldots,k}^{-1}$, where

$$v_{\boldsymbol{\lambda},j_1,j_2} := \int_R \frac{d_{\boldsymbol{\lambda}}^*}{l_{\boldsymbol{\lambda}}(\mathbf{z})} \left(\frac{\partial}{\partial \lambda_{j_1}} \frac{l_{\boldsymbol{\lambda}}(\mathbf{z})}{d_{\boldsymbol{\lambda}}^*} \right) \left(\frac{\partial}{\partial \lambda_{j_2}} \frac{l_{\boldsymbol{\lambda}}(\mathbf{z})}{d_{\boldsymbol{\lambda}}^*} \right) d\mathbf{z}.$$

The matrix $\mathbf{V}_{\boldsymbol{\lambda}}$ is the Fisher information matrix. The estimator is, thus, asymptotically efficient in the sense that the covariance matrix of its limiting normal distribution for $m \to \infty$, $r \to \infty$ is the inverse of the Fisher information matrix, see Section 4.1.3 in Serfling [408]. The exact regularity conditions and the proof of the asymptotic normality are somewhat technical and lengthy. They follow the approach presented in Section 4.2 of Serfling [408]. For the details we refer to Theorem 6.1.2 of Michel [330].

As stated above, the approximation of the density of the $\mathbf{Z}^{(i)}$ by $l_{\boldsymbol{\lambda}}/d_{\boldsymbol{\lambda}}^*$ is quite crude for fixed r, if one is close to the independence case, due to the slow convergence of χ in this case. Under the additional condition $C^{(i)} = -r_i$ and $\mathbf{Z}^{(i)} \in Q_{r_i,s}$, we know from (5.65) that the $\mathbf{Z}^{(i)}$ have the density

$$\tilde{l}_{\boldsymbol{\lambda},r_i}(\mathbf{z}) = \begin{cases} \frac{l_{\boldsymbol{\lambda}}(\mathbf{z})}{\chi_{\boldsymbol{\lambda}}(r_i,s)} & \text{for } \mathbf{z} \in Q_{r_i,s}, \\ 0 & \text{else.} \end{cases}$$

This can be used for a conditional approach of a MLE of $\boldsymbol{\lambda} = (\lambda_1, \ldots, \lambda_k)$ by choosing $\hat{\boldsymbol{\lambda}}_{m,r}$ such that the expression

$$\tilde{\Upsilon}(\boldsymbol{\lambda}) := \log \left(\prod_{i=1}^m \tilde{l}_{\boldsymbol{\lambda},r_i}\left(\mathbf{Z}^{(i)}\right) \right) = \sum_{i=1}^m \log \left(\tilde{l}_{\boldsymbol{\lambda},r_i}\left(\mathbf{Z}^{(i)}\right) \right)$$

is maximized in $\boldsymbol{\lambda}$. Since we are using conditional densities, we refer to this method as the *conditional* ML method.

The conditional ML method with $\tilde{\Upsilon}(\boldsymbol{\lambda})$ also leads to an estimator, which is asymptotically normal under suitable regularity conditions for $m \to \infty$, $r \to \infty$, with the same covariance matrix $\mathbf{V}_{\boldsymbol{\lambda}}^{-1}$ as above; we refer to Theorem 6.1.4 of Michel [330] for the somewhat lengthy details.

MLE WITH THE PICKANDS DENSITY

We present a different ML approach. As before, we assume that we have n independent copies $\tilde{\mathbf{X}}^{(1)}, \ldots, \tilde{\mathbf{X}}^{(n)}$ of a rv $\tilde{\mathbf{X}}$, which follows a GPD $W_{\boldsymbol{\lambda}}$ from a k-parametric family and the usual representation $W_{\boldsymbol{\lambda}}(\mathbf{x}) = 1 + \log(G_{\boldsymbol{\lambda}}(\mathbf{x}))$ with some EVD $G_{\boldsymbol{\lambda}}$ on K_s. By $\varphi_{\boldsymbol{\lambda}}$ we denote the corresponding Pickands density in the parametric model and by $\zeta_{\boldsymbol{\lambda}}$ its integral as given in (5.38). This time we denote by $\widetilde{\mathbf{Z}}^{(i)}$ and $\widetilde{C}^{(i)}$ the corresponding *standard* Pickands coordinates, $i = 1, \ldots, n$. We choose a threshold $r < 0$ close enough to 0, and consider only those observations with $\widetilde{C}^{(i)} > r$, representing again the extreme observations. We denote these by $\mathbf{X}^{(1)}, \ldots, \mathbf{X}^{(m)}$. They are by Theorem 1.3.1 independent of the random number $m = K(n)$, and have the density $\varphi_{\boldsymbol{\lambda}}(\mathbf{z})/\zeta_{\boldsymbol{\lambda}}$, independent of r, see Theorem 5.6.2. Again, we can do a ML estimation of $\boldsymbol{\lambda}$ by choosing $\hat{\boldsymbol{\lambda}}_m$ such that the expression

$$\Omega(\boldsymbol{\lambda}) := \log\left(\prod_{i=1}^{m} \frac{\varphi_{\boldsymbol{\lambda}}\left(\mathbf{Z}^{(i)}\right)}{\zeta_{\boldsymbol{\lambda}}}\right) = \sum_{i=1}^{m} \log\left(\varphi_{\boldsymbol{\lambda}}\left(\mathbf{Z}^{(i)}\right)\right) - m\log(\zeta_{\boldsymbol{\lambda}})$$

is maximized in $\boldsymbol{\lambda}$.

Since here, in contrast to the previous section, we insert the observations into their exact densities, the proof of the asymptotic normality (this time for $m \to \infty$ only) of the MLE under suitable regularity conditions follows from the corresponding standard literature. The asymptotic normal distribution has mean $\boldsymbol{\lambda}$ and covariance matrix $\mathbf{U}_{\boldsymbol{\lambda}}^{-1} := (u_{\boldsymbol{\lambda}, j_1, j_2})^{-1}_{j_1, j_2 = 1, \ldots, k}$ with

$$u_{\boldsymbol{\lambda}, j_1, j_2} := E_{\boldsymbol{\lambda}}\left(\left(\frac{\partial \log\left(\frac{\varphi_{\boldsymbol{\lambda}}(\mathbf{z})}{\zeta_{\boldsymbol{\lambda}}}\right)}{\partial \lambda_{j_1}}\right)\left(\frac{\partial \log\left(\frac{\varphi_{\boldsymbol{\lambda}}(\mathbf{z})}{\zeta_{\boldsymbol{\lambda}}}\right)}{\partial \lambda_{j_2}}\right)\right).$$

The estimation is again asymptotically efficient. Since it is based on the Pickands density we refer to it as the *Pickands* ML method.

ESTIMATION VIA RELATIVE FREQUENCIES

In the following, we present another estimation approach in parametric multivariate GPD models. The idea for this method results from the fact that only the the *number of observations*, which fall into a certain area, can be asymptotically sufficient for the parameters of the model, see Falk [136].

We assume again that we have independent and identically distributed rv $\mathbf{X}_1, \ldots, \mathbf{X}_n$, which follow a GPD $W_{\lambda_1, \ldots, \lambda_k}$ whose angular density $l_{\lambda_1, \ldots, \lambda_k}(\mathbf{z})$ is continuously differentiable in $\lambda_1, \ldots, \lambda_k$ and $d^*_{\lambda_1, \ldots, \lambda_k} > 0$. For simplicity of notation, we put again $\boldsymbol{\lambda} := (\lambda_1, \ldots, \lambda_k)$, where $\boldsymbol{\lambda} \in \Lambda \subseteq \mathbb{R}^k$. We, furthermore, assume that the parameter space Λ is an open non-empty subset of $\mathbb{R}^k$.

For $v > d$ put

$$Q_v := \left\{\mathbf{z} \in R : z_i > \frac{1}{v}, \ i \leq d-1, \ \sum_{i \leq d-1} z_i < 1 - \frac{1}{v}\right\}.$$

The restriction $v > d$ ensures that the set Q_v is not empty. We have $Q_{r,s} = Q_{rs}$ for the set $Q_{r,s}$ introduced in (5.63). We put, furthermore,

$$B_{r,v} := \left\{\mathbf{x} \in (-\infty, 0)^d : c < -r, \ \mathbf{z} \in Q_v\right\},$$

where c and $\mathbf{z}$ are the Pickands coordinates of $\mathbf{x}$ with respect to Fréchet margins.

It is crucial in the following that $B_{r,v} \subset K_s$ holds for $0 < s < 1$. Therefore, s has to be chosen such that $W_{\boldsymbol{\lambda}}$ possesses the representation (5.1) on the set K_s. To ensure that $B_{r,v} \subset K_s$, the numbers v and r have to be chosen such that the inequality $v < sr$ holds. The set $B_{r,v}$ is illustrated in the bivariate case in Figure 5.9.1. The parameter v reflects the angle of $B_{r,v}$. It is small for v close to d, and it converges to a right angle for $v \to \infty$.

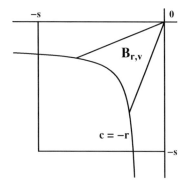

FIGURE 5.9.1. The set $B_{r,v}$ in the bivariate case.

By applying the transformation theorem to the Pickands coordinates with respect to Fréchet margins, we obtain

$$P(\mathbf{X}_1 \in B_{r,v}) = \int_{-\infty}^{-r} \int_{Q_v} c^{-2} l_{\boldsymbol{\lambda}}(\mathbf{z})\, d\mathbf{z}\, dc = \frac{\chi_{\boldsymbol{\lambda}}(v)}{r}$$

where $\chi_{\boldsymbol{\lambda}}(v) := \int_{Q_v} l_{\boldsymbol{\lambda}}(\mathbf{z})\, d\mathbf{z}$. The proof is given in Lemma 5.1.1 of Michel [330]. By

$$h(B_{r,v}) = \frac{1}{n} \sum_{i=1}^{n} 1_{B_{r,v}}(\mathbf{X}_i)$$

we denote the relative frequency of the occurrence of the event $B_{r,v}$ where 1_A is the usual indicator function of a set A. By the law of large numbers, this relative frequency converges for $n \to \infty$ to the probability of occurrence $\chi_{\boldsymbol{\lambda}}(v)/r$.

Choose now v_j, $j = 1, \ldots, k$, such that $d < v_1 < \ldots < v_k < sr$ and define the function $\boldsymbol{\psi} : \Lambda \to [0,1]^k$ by

$$\boldsymbol{\psi}(\boldsymbol{\lambda}) = \left(\frac{\chi_{\boldsymbol{\lambda}}(v_1)}{r}, \ldots, \frac{\chi_{\boldsymbol{\lambda}}(v_k)}{r} \right).$$

In what follows we assume that $\boldsymbol{\psi}$ is one-to-one. We estimate the parameter $\boldsymbol{\lambda}$ by a solution $\hat{\boldsymbol{\lambda}}$ of the equation

$$\boldsymbol{\psi}(\boldsymbol{\lambda}) = (h(B_{r,v_1}), \ldots, h(B_{r,v_k})). \tag{5.68}$$

For this we have to assume that

$$(h(B_{r,v_1}), \ldots, h(B_{r,v_k})) \in \mathrm{Im}(\boldsymbol{\psi}), \tag{5.69}$$

with $\mathrm{Im}(\boldsymbol{\psi})$ being the range of the function $\boldsymbol{\psi}$, else $\hat{\boldsymbol{\lambda}}$ is arbitrarily defined. Condition (5.69) holds by the law of large numbers with a probability converging to 1 as n increases, i.e.,

$$P\left((h(B_{r,v_1}), \ldots, h(B_{r,v_k})) \notin \mathrm{Im}(\boldsymbol{\psi})\right) \to_{n\to\infty} 0,$$

if $\psi(\boldsymbol{\lambda})$ is not an boundary element of the range, which we assume. For real data sets, it can happen that (5.69) is not satisfied. One could then modify the parameters v_i if possible, see Algorithm 5.9.1 at the end of this section, where we try to find appropriate v_i in this case.

By the law of large numbers we get

$$\boldsymbol{\psi}(\hat{\boldsymbol{\lambda}}) = (h(B_{r,v_1}), \ldots, h(B_{r,v_k})) \to_{n\to\infty} \left(\frac{\chi_{\boldsymbol{\lambda}}(v_1)}{r}, \ldots, \frac{\chi_{\boldsymbol{\lambda}}(v_k)}{r}\right) = \boldsymbol{\psi}(\boldsymbol{\lambda}).$$

The function $\boldsymbol{\psi}$ is, by the assumptions on $l_{\boldsymbol{\lambda}}$, continuously differentiable and, thus, we conclude from $\boldsymbol{\psi}(\hat{\boldsymbol{\lambda}}) \to_{n\to\infty} \boldsymbol{\psi}(\boldsymbol{\lambda})$ and that $\boldsymbol{\psi}$ is assumed one-to-one, the convergence $\hat{\boldsymbol{\lambda}} \to \boldsymbol{\lambda}$ for $n \to \infty$.

The asymptotic normality of the estimator can also be shown under suitable regularity conditions for $n \to \infty$. The asymptotic covariance matrix is

$$\left(\mathbf{J}_{\boldsymbol{\psi}(\boldsymbol{\lambda})}\right)^{-1} \boldsymbol{\Sigma} \left(\mathbf{J}^T_{\boldsymbol{\psi}(\boldsymbol{\lambda})}\right)^{-1}, \tag{5.70}$$

where $\mathbf{J}_{\boldsymbol{\psi}(\boldsymbol{\lambda})}$ is the Jacobian of ψ at $\boldsymbol{\lambda}$ and $\boldsymbol{\Sigma} = (\varsigma_{ij})_{1 \le i,j \le k}$ with

$$\varsigma_{ij} := \frac{\chi_{\boldsymbol{\lambda}}\left(v_{\min(i,j)}\right)}{r} - \frac{\chi_{\boldsymbol{\lambda}}(v_i)\chi_{\boldsymbol{\lambda}}(v_j)}{r^2}, \qquad 1 \le i,j \le k.$$

For the complete result with the exact conditions and the quite technical proof we refer to Section 6.3 of Michel [330].

It is often the case that the variance (for $k = 1$) or the determinant of the covariance matrix (for $k > 1$) can be minimized with respect to $\mathbf{v} := (v_1, \ldots, v_k)$ while $\boldsymbol{\lambda}$ is fixed. This would be the optimal $\mathbf{v}$, on which the estimation procedure should be based. Since this optimum, however, depends directly on the parameter, which is to be estimated, it cannot be computed in practice. To get approximations of the optimal $\mathbf{v}$ the following iterative algorithm comes naturally.

Algorithm 5.9.1. *Let $i = 1$, $\eta > 0$ small, and $I \in \mathbb{N}$.*

1. *Determine $\boldsymbol{\lambda}^{(0)}$ by the asymptotic ML method, maximizing Υ.*

2. *Determine $\mathbf{v}^{(i)} = \left(v_1^{(i)}, \ldots, v_k^{(i)}\right) \in (d, rs]^k$ such that the determinant of the asymptotic covariance matrix (5.70) with underlying parameter $\boldsymbol{\lambda}^{(i-1)}$ becomes minimal.*

3. *Determine the estimator* $\boldsymbol{\lambda}^{(i)}$ *through (5.68) with parameter* $\mathbf{v}^{(i)}$.

4. *If*

$$\left| \frac{\boldsymbol{\lambda}^{(i-1)} - \boldsymbol{\lambda}^{(i)}}{\boldsymbol{\lambda}^{(i-1)}} \right| \leq \eta$$

or $i = I$, *return* $\boldsymbol{\lambda}^{(i)}$, *else increase* i *by 1 and go to 2.*

The first estimation of $\boldsymbol{\lambda}^{(0)}$ can also be done by some other estimation procedure, but maximizing Υ by the asymptotic ML method is typically the computationally fastest way.

The choice of the break off parameters η and I depends on the underlying problem and cannot be specified in general.

COMPARISON OF THE ESTIMATION PROCEDURES

We compare the preceding procedures by applying them to identical simulated data sets. We examine the following procedures:

- Asymptotic ML: estimation of $\boldsymbol{\lambda}$ by maximizing Υ.

- Conditional ML: estimation of $\boldsymbol{\lambda}$ by maximizing $\widetilde{\Upsilon}$.

- Pickands ML: estimation of $\boldsymbol{\lambda}$ by maximizing Ω.

- Simple Iteration: estimation of $\boldsymbol{\lambda}$ by Algorithm 5.9.1, where $I = 1$.

- Multiple Iteration: estimation of $\boldsymbol{\lambda}$ by Algorithm 5.9.1, where $\eta = 0.01$ and $I = 10$.

Since asymptotic ML, conditional ML, simple and multiple iteration are based on Pickands coordinates with respect to Fréchet margins, whereas Pickands ML is based on standard Pickands coordinates, the estimation procedures use different (random) sample sizes. Thus a comparison with identical sample sizes becomes difficult. The corresponding thresholds have to be adjusted, so that approximately the same number of observations exceeds the corresponding threshold for each of the above estimators. We did this by setting

$$r_P = -1.5 \cdot \frac{d^2}{r_F}, \tag{5.71}$$

where r_F is the threshold for Pickands coordinates with respect to Fréchet margins, and r_P is the threshold for standard Pickands coordinates.

Before we present the simulation results, we want to add some short considerations on the computational efficiency of the methods in the logistic case, which will be the basis of our simulations.

For the asymptotic ML method we only have to maximize a quickly evaluable function, which is usually done with an iteration procedure and is very efficient with the corresponding software packages. From an efficiency point of view, this is the only method that is usable in dimensions ≥ 5, since no numerical evaluation of an integral is required, if an analytical expression of d_λ^* is known.

For the conditional ML method we also have to maximize a function, which includes, however, various integrals, which are typically accessible only by numerical methods. This makes the conditional ML method numerically quite inefficient.

For the Pickands ML method, we have to maximize a function with just one numerical integral (φ_λ) in general. In most practical cases, asymptotic ML is faster due to the known analytical representation of d_λ^*.

For the simple iteration we have to do an asymptotic ML estimation first, then, to determine $\mathbf{v}$, we have to maximize a function, which contains two numerical integrals and one numerical derivative. Subsequently, we have to solve an equation numerically which contains one integral. Thus, this method is more costly than the asymptotic ML and the Pickands ML method, but a lot less numerical integrals have to be determined than with the conditional ML method. Although, for the multiple iteration, the simple Iteration has to be executed repeatedly, it is less costly than conditional ML.

The five procedures introduced above are now compared by their results on identical simulated data sets. For fixed dimension d and parameter λ, a sample of size n of observations following a multivariate logistic GPD is simulated by Algorithm 5.7.9 and 5.7.6. This sample corresponds to the observations in a real data set after the transformation to uniform margins. The estimation $\hat{\lambda}$ for λ can now be done by the five methods. This procedure of creating *one* sample by simulation and estimating the parameter λ by the *five* methods introduced above is now repeated 100 times. This leaves us with 100 estimations of the same parameter λ for each method. To be able to present the results graphically, boxplots have been drawn to display the 100 estimations for each method. These boxplots can then be used to visually compare the estimations, for example with respect to biasedness and variability.

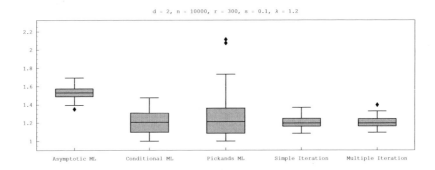

FIGURE 5.9.2. Comparison of estimators for logistic data with $\lambda = 1.2$.

The parameters are first set to $d = 2$, $n = 10000$, $r = 300$, $s = 0.1$ and $\lambda = 1.2$, the threshold for the Pickands ML method is computed as in (5.71). The resulting graphic is shown in Figure 5.9.2.

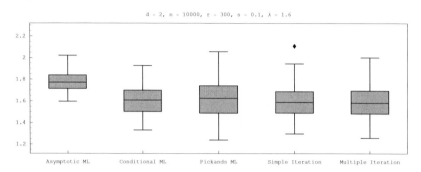

FIGURE 5.9.3. Comparison of estimators for logistic data with $\lambda = 1.6$.

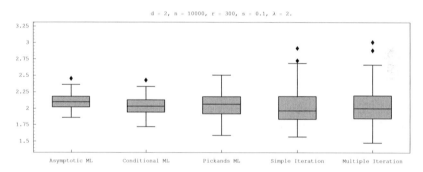

FIGURE 5.9.4. Comparison of estimators for logistic data with $\lambda = 2$.

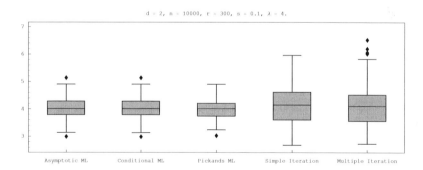

FIGURE 5.9.5. Comparison of estimators for logistic data with $\lambda = 4$.

To be able to compare the behavior of the estimators for different underlying logistic GPD, we have generated analogous graphics for $\lambda = 1.6$, 2 and 4. They are displayed in Figures 5.9.3 to 5.9.5.

The conditional ML and the Pickands ML do not show big differences, with conditional ML having slightly lower variation. They seem to work quite well in all cases. Very noticeable is the large bias of the asymptotic ML method for λ close to one. The asymptotic ML is, thus, an estimation procedure, which is asymptotically efficient and works fine for cases of high dependence, but it is biased for fixed sample sizes when approaching the independence case and, thus, it is not reliable there.

The two iteration methods hardly differ in their behavior. The iteration procedures work quite well close to the independence case $\lambda = 1$, where they have the smallest variabilities. Close to the dependence case the procedures however have large variabilities and the ML methods should be preferred. Further simulations in the trivariate logistic case show an identical behavior to the bivariate case.

To summarize: The conditional ML method seems to yield the best results. But since it requires quite a long time for computation, it might not be the best one suited for practical purposes, especially in dimensions larger than 2. In these cases one should use one of the other estimation procedures.

An application of these methods to a real data set is provided in Section 7 of Michel [333].

5.10   Testing in Logistic GPD Models

Together with the estimation of parameters in multivariate GPD models comes the question of testing for parameters in these models. This seems to be a relatively uncharted area in the investigations of the multivariate GPD. To start the investigations we restrict ourselves in this section to the logistic GPD models and introduce some basic testing procedures. We develop two testing scenarios which are closely related to the usual one sample Gauss-test and two sample t-test.

THE BASIC SETUP

The characteristic feature of logistic GPD rv can be best formulated in terms of the angular Pickands coordinate. Look for example at the simulation results in Figure 5.7.1. The generated points arrange themselves in a sort of d-dimensional cone, whose peak lies in the origin and whose center is the line $x_i = x_j, i, j = 1, \ldots, d$, the bisector of the negative quadrant. The parameter λ describes the width of the cone. For λ close to 1 it is opened very wide, for larger λ it becomes more narrow.

The width of the cone can also be described by the variance of the angular Pickands coordinates of the observations. This variance is the characteristic feature of the logistic GPD with parameter λ. The center of the cone corresponds to the maximum of the Pickands density (see Figure 6.6.2 for some graphs of bivariate logistic Pickands densities), which is also the expectation of the angular component, as we will see below.

The above facts will be used to test for special values of the underlying parameter λ. Before we will define the test statistics, we have to establish some auxiliary results.

Theorem 5.10.1. *Let* $\mathbf{X}$ *follow an exchangeable GPD model, i.e.,* $(X_1, \ldots, X_d)$ *and* $(X_{\sigma(1)}, \ldots, X_{\sigma(d)})$ *have for any permutation* σ *of* $(1, \ldots, d)$ *the same GPD. Consider the angular Pickands coordinates* $Z_i = X_i/(X_1 + \cdots + X_d)$ *under the condition that* $X_1 + \cdots + X_d > c_0$ *for some* $c_0 < 0$ *close enough to* 0. *Then we have* $E(Z_i) = 1/d$, $i \le d$.

Proof. Due to the exchangeability of the componentes of the rv $\mathbf{X}$ we have, close to 0,

$$1 = E\left(\frac{X_1 + \cdots + X_d}{X_1 + \cdots + X_d}\right) = E\left(\sum_{i=1}^{d} \frac{X_i}{\sum_{j=1}^{d} X_j}\right) = \sum_{i=1}^{d} E\left(\frac{X_i}{\sum_{j=1}^{d} X_j}\right)$$

$$= dE\left(\frac{X_i}{\sum_{j=1}^{d} X_j}\right) = dE(Z_i)$$

and, thus, the assertion. $\qquad\qquad\square$

Note that the logistic model is an exchangeable model and, thus, Theorem 5.10.1 is applicable in this case.

We will start constructing our tests for the bivariate case. We have $E(Z) = 1/2$ for the bivariate logistic GPD, with the Pickands density

$$\varphi_\lambda(z) = (\lambda - 1)z^{\lambda-1}(1-z)^{\lambda-1}\left(z^\lambda + (1-z)^\lambda\right)^{1/\lambda - 2}, \qquad z \in [0,1].$$

according to Theorem 2.4 of Michel [331]. Denoting its integral by ζ_λ, we put

$$\sigma_\lambda^2 := \operatorname{Var}(Z) = E((Z - 1/2)^2) = E\left(Z^2\right) - \frac{1}{4} = \int_0^1 z^2 \frac{\varphi_\lambda(z)}{\zeta_\lambda}\,dz - \frac{1}{4}$$

and

$$\theta_\lambda^2 := \operatorname{Var}\left(\left(Z - \frac{1}{2}\right)^2\right) = E\left(\left(Z - \frac{1}{2}\right)^4\right) - E^2\left(\left(Z - \frac{1}{2}\right)^2\right)$$

$$= \int_0^1 \left(z - \frac{1}{2}\right)^4 \frac{\varphi_\lambda(z)}{\zeta_\lambda}\,dz - \sigma_\lambda^4.$$

Both numbers are finite due to $z^2\varphi_\lambda(z) \le \varphi_\lambda(z)$ and $(z - 1/2)^4\varphi_\lambda(z) \le \varphi_\lambda(z)$ for $z \in [0,1]$, $\int_0^1 \varphi_\lambda(z)/\zeta_\lambda\,dz = 1$. The two numbers are shown in Figure 5.10.1, depending on λ.

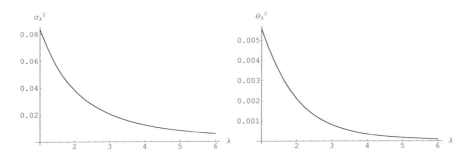

FIGURE 5.10.1. Graphs of σ_λ^2 and θ_λ^2.

Both of them are monotonically decreasing and converging to 0 for $\lambda \to \infty$. In addition, one can easily see by the dominated convergence theorem that

$$\lim_{\lambda \to 1} \sigma_\lambda^2 = \frac{1}{12} \quad \text{and} \quad \lim_{\lambda \to 1} \theta_\lambda^2 = \frac{1}{180}.$$

A ONE-SAMPLE TEST

We start with the case, where we have one bivariate sample following a logistic GPD and we have some null-hypothesis $\lambda = \lambda_0$ about the underlying parameter of the bivariate logistic GPD, which we want to test. The idea for the test presented in the following is to compare the empirical variance of the angular Pickands coordinate with its theoretical counterpart σ_λ^2, under the hypothesis $\lambda = \lambda_0$. This corresponds to the well-known situation of the one-sample Gauss-test, where we have one sample following a normal distribution with known variance and we have some null-hypothesis $\mu = \mu_0$ about the mean of the normal distribution in the sample.

Let $\left(X_1^{(i)}, X_2^{(i)}\right)$, $i \le n$, be n independent copies of the rv (X_1, X_2), which follows a logistic GPD with parameter $\lambda > 1$. Consider the angular Pickands components of those $m = K(n)$ rv with $X_1^{(i)} + X_2^{(i)} > c_0$. Denote these by Z_i, $i \le m$. The rv

$$\eta_{m,\lambda} := \frac{1}{m^{1/2}} \sum_{i=1}^{m} \frac{\left(Z_i - \frac{1}{2}\right)^2 - \sigma_\lambda^2}{\theta_\lambda}$$

is by the central limit theorem for $m \to \infty$ standard normal distributed.

To test the hypothesis that the parameter $\lambda = \lambda_0$ actually underlies our initial observations, we compute η_{m,λ_0} and reject the hypothesis, if $|\eta_{m,\lambda_0}|$ is too large, i.e., if the asymptotic p-value $1 - (2\Phi(|\eta_{m,\lambda_0}|) - 1)$ is too small, typically smaller than 0.05.

To investigate the behavior of the test and especially to examine the normal approximation of the distribution of η_{m,λ_0}, we did a lot of simulations. For one setting of the parameters m, λ and λ_0, the corresponding logistic GPD rv were simulated with Algorithm 5.7.6 and the corresponding p-values were computed. The null-hypothesis was rejected if the p-value was below 0.05. This was repeated 1000 times and the corresponding relative frequency of rejections computed. The procedure was repeated for different combinations of the parameters m, λ and λ_0. The results are displayed in Figure 5.10.2.

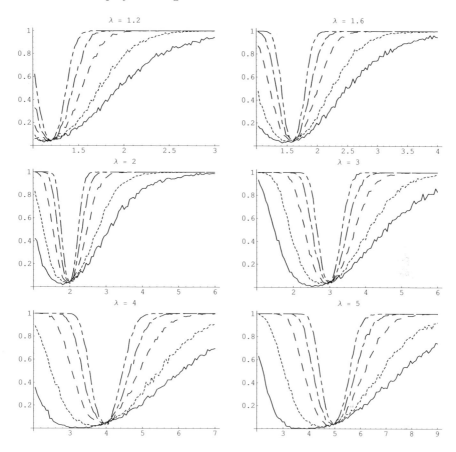

FIGURE 5.10.2. Simulations of the test for different parameters and sample sizes ($m = 10$ solid lines, $m = 20$ small dashes, $m = 50$ long dashes, $m = 100$ long and small dashes, $m = 200$ long and two small dashes). The vertical axis gives the relative frequency of rejection, the horizontal axis the tested parameter λ_0.

The test alway keeps the level, as the true null-hypothesis $\lambda = \lambda_0$ is never rejected in more than 5% of the cases. The test is, however, not quite powerful enough for a sample size of 10, as its p-values fall below the 5% rejection frequency for lots of λ_0 deviating from the true parameter. This can also be seen for $m = 20$ and large λ. Thus, in these cases, the normal approximation is not close enough to the actual distribution of the test statistic $\eta_{m,\lambda}$, so that a sample size of more than 20 should be used to make the test sufficiently powerful. A sample size of 50 already gives quite reasonable results.

The test statistic $\eta_{m,\lambda}$ is the usual Gauss-test statistic based on $(Z_i - 1/2)^2$ and, therefore, also one-sided tests can easily be derived testing the hypotheses $\lambda \leq \lambda_0$ or $\lambda \geq \lambda_0$.

A TWO-SAMPLE TEST

Next we investigate a two-sample problem by comparing two independent logistic GPD samples with the hypothesis that the same (but unknown) parameter λ_0 is underlying both samples. The idea here is to compare the empirical variance of the angular Pickands coordinates in the two samples. This corresponds to the well-known two-sample t-test comparing the means of two samples with unknown but identical variation.

Let $(X_1^{(i)}, X_2^{(i)})$, $i \leq n$, be independent and identically distributed logistic GPD rv with underlying parameter λ, and let $(\widetilde{X}_1^{(i)}, \widetilde{X}_2^{(i)})$, $i \leq \tilde{n}$, be independent and identically distributed logistic GPD rv with underlying parameter $\tilde{\lambda}$. We require, in addition, that the two sets of rv are independent.

Consider the angular Pickands components of those m rv with $X_1^{(i)} + X_2^{(i)} > c_0$ and, correspondingly, the angular Pickands components of those $\tilde{m}$ rv with $\widetilde{X}_1^{(i)} + \widetilde{X}_2^{(i)} > \tilde{c}_0$. Denote these by Z_i, $i \leq m$, and $\widetilde{Z}_i$, $i \leq \tilde{m}$, respectively. Put $Q_i := (Z_i - 1/2)^2$, $i \leq m$, and $\widetilde{Q}_i := (\widetilde{Z}_i - 1/2)^2$, $i \leq \tilde{m}$. We consider the usual two-sample t-test statistic

$$\eta_{m,\tilde{m}} := \frac{\frac{1}{m}\sum_{i=1}^{m} Q_i - \frac{1}{\tilde{m}}\sum_{i=1}^{\tilde{m}} \widetilde{Q}_i}{S\sqrt{\frac{1}{m} + \frac{1}{\tilde{m}}}},$$

where S is the usual pooled variance of the Q_i and $\widetilde{Q}_i$, see Section 2.3 of Falk et al. [145]. Assume that $m/(m + \tilde{m}) \to \gamma$ for $m, \tilde{m} \to \infty$. Then the statistic $\eta_{m,\tilde{m}}$ is asymptotically normally distributed under the hypothesis $\lambda = \tilde{\lambda}$. By elementary calculations and the central limit theorem it can be shown that $\theta_\lambda/S \to 1$ for $m, \tilde{m} \to \infty$ and if $\lambda = \tilde{\lambda}$. We can rewrite the test statistic as

$$\eta_{m,\tilde{m}} = \underbrace{\frac{\theta_\lambda}{S}}_{\to 1} \left(\underbrace{\underbrace{\sqrt{\frac{\tilde{m}}{m + \tilde{m}}}}_{\to \sqrt{1-\gamma}} \underbrace{\frac{1}{\sqrt{m}} \sum_{i=1}^{m} \frac{Q_i - \sigma_\lambda^2}{\theta_\lambda}}_{\to_D N(0,1)}}_{\to_D N(0,1-\gamma)} - \underbrace{\underbrace{\sqrt{\frac{m}{m + \tilde{m}}}}_{\to \sqrt{\gamma}} \underbrace{\frac{1}{\sqrt{\tilde{m}}} \sum_{i=1}^{\tilde{m}} \frac{\widetilde{Q}_i - \sigma_\lambda^2}{\theta_\lambda}}_{\to_D N(0,1)}}_{\to_D N(0,\gamma)} \right)$$

$$\to_D N(0,1).$$

The asymptotic p-value for testing the two-sided hypothesis $\lambda = \widetilde{\lambda}$ is, thus, $1 - \left(2\Phi \left(\left| \eta_{m,\widetilde{m}} \right| \right) - 1 \right)$. As stated before, testing the one-sided hypotheses $\lambda \leq \widetilde{\lambda}$ or $\lambda \geq \widetilde{\lambda}$ is also possible.

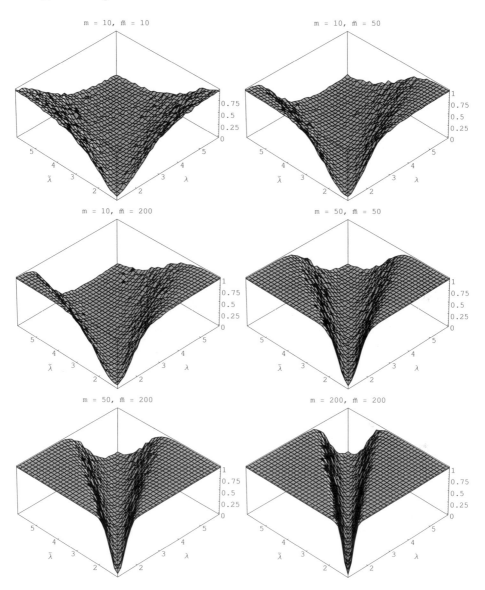

FIGURE 5.10.3. Simulations of the two-sample test with various parameters and various sample sizes. The vertical axes show the relative frequencies of rejections, the two planar axes show the underlying parameters λ and $\widetilde{\lambda}$.

Along the same lines as above, simulations were computed to investigate the behavior of the test and the normal approximation of the distribution of $\eta_{m,\tilde{m}}$. The results are displayed in Figure 5.10.3.

The test keeps its prescribed level as the true hypothesis is never rejected in more than 5% of the cases. The test performs, however, poorly if one of the two samples has a size of only 10 in each sample. A sample size of 50 yields good results.

Multivariate Generalizations

In the general d-dimensional case, the logistic Pickands density is given by

$$\varphi_\lambda(z_1,\ldots,z_{d-1}) = \left(\prod_{i=1}^{d-1}(i\lambda-1)\right)\frac{\left(\prod_{i=1}^{d-1}z_i\right)^{\lambda-1}\left(1-\sum_{i=1}^{d-1}z_i\right)^{\lambda-1}}{D_\lambda(z_1,\ldots,z_{d-1})^{d\lambda-1}} \qquad (5.72)$$

see Theorem 2.7 of Michel [331], with D_λ being the logistic dependence function, see Example 4.3.5.

We know by Theorem 5.10.1 that the expectation of the angular Pickands coordinate is $(1/d,\ldots,1/d)$. In analogy to the bivariate case, we define the deviation of the angular component from its expectation by

$$\sigma_\lambda^2 := E\left(\left\|\mathbf{Z}-\left(\frac{1}{d},\ldots,\frac{1}{d}\right)\right\|^2\right) = \int_R\left\|\mathbf{z}-\left(\frac{1}{d},\ldots,\frac{1}{d}\right)\right\|^2\frac{\varphi_\lambda(\mathbf{z})}{\zeta_\lambda}\,d\mathbf{z}$$

and we denote its variance by

$$\theta_\lambda^2 := \mathrm{Var}\left(\left\|\mathbf{Z}-\left(\frac{1}{d},\ldots,\frac{1}{d}\right)\right\|^2\right)$$

$$= E\left(\left\|\mathbf{Z}-\left(\frac{1}{d},\ldots,\frac{1}{d}\right)\right\|^4\right) - E^2\left(\left\|\mathbf{Z}-\left(\frac{1}{d},\ldots,\frac{1}{d}\right)\right\|^2\right)$$

$$= \int_R\left\|\mathbf{z}-\left(\frac{1}{d},\ldots,\frac{1}{d}\right)\right\|^4\frac{\varphi_\lambda(\mathbf{z})}{\zeta_\lambda}\,d\mathbf{z} - \sigma_\lambda^4,$$

where $R = \left\{\mathbf{x}\in(0,1)^{d-1}:\sum_{i\le d-1}x_i < 1\right\}$ is the unit simplex. Both numbers are finite, since the norm is always bounded by 1.

For the one-sample test, the scenario is now as follows: Let $(X_1^{(i)},\ldots,X_d^{(i)})$, $i \le n$, be independent copies of $(X_1,\ldots,X_d)$ following a logistic GPD. Consider the angular Pickands components of those $m = K(n)$ rv with $X_1^{(i)}+\cdots+X_d^{(i)} > c_0$. Denote these by $\mathbf{Z}_i = (Z_1^{(i)},\ldots,Z_d^{(i)})$, $i \le m$. The rv

$$\eta_{m,\lambda} := \frac{1}{m^{1/2}}\sum_{i\le m}\frac{\left\|\mathbf{Z}_i-\left(\frac{1}{d},\ldots,\frac{1}{d}\right)\right\|^2-\sigma_\lambda^2}{\theta_\lambda}$$

is for $m \to \infty$ asymptotically standard normal distributed. This entails the computation of an approximate p-value.

In the same way, the two-sample test statistic can be generalized to

$$\eta_{m,\widetilde{m}} := \frac{\frac{1}{m}\sum_{i=1}^{m}Q_i - \frac{1}{\widetilde{m}}\sum_{i=1}^{\widetilde{m}}\widetilde{Q}_i}{S\sqrt{\frac{1}{m}+\frac{1}{\widetilde{m}}}},$$

where S is the usual pooled variance of $Q_i = \|\mathbf{Z}_i - (1/d,\ldots,1/d)\|^2$ and $\widetilde{Q}_i = \left\|\widetilde{\mathbf{Z}}_i - (1/d,\ldots,1/d)\right\|^2$.

Chapter 6

The Pickands Approach in the Bivariate Case

The restriction to bivariate rv enables the study of their distributions in much greater detail. We introduce, for example, a certain measure generating function M, see Section 6.1, and prove that the pertaining Pickands dependence function D is absolutely continuous, see Lemma 6.2.1 and the subsequent discussion. This property is unknown in higher dimensions. We also introduce an expansion of order $k + 1$, $k \in \mathbb{N}$, of the spectral df in the bivariate case, see Section 6.1, which turns out to be quite useful in a testing problem.

The tail dependence parameter, which measures the tail dependence between two rv, will be introduced in Section 6.1 and computed for rv in a certain neighborhood of an EVD. The more general tail dependence function will be investigated in Section 6.4.

Results about the Pickands transform of bivariate GPD and EVD rv will be reformulated and strengthened in Section 6.3. This will include the case of Marshall-Olkin rv.

Tests for tail independence of two rv are investigated in Section 6.5. It turns out that the radial component of the Pickands transform is a powerful tool to discriminate between tail independence and tail dependence; an extension to higher dimensions is indicated.

Another speciality of the bivariate case is the ability to estimate the angular density in GP models via the angular component of the Pickands coordinates, shown in Section 6.6.

6.1 Preliminaries

Recall from Section 4.3 that a bivariate EVD G with reverse standard exponential margins has the representation

M. Falk et al., *Laws of Small Numbers: Extremes and Rare Events*, 3rd ed.,
DOI 10.1007/978-3-0348-0009-9_6, © Springer Basel AG 2011

$$G(x, y) = \exp\left((x + y)D\left(\frac{x}{x + y}\right)\right), \qquad x, y \le 0,$$

where $D : [0, 1] \to [0, 1]$ is the Pickands dependence function

$$D(z) = \int_S \max\left(uz, v(1 - z)\right) d\mu(u, v). \tag{6.1}$$

From equations (4.31) and (4.32) with $d = 2$ recall that

$$S = \{(u, v) : u + v = 1, \, u, v \ge 0\}$$

is the unit simplex in $\mathbb{R}^2$, and μ is an arbitrary measure on S with the properties

$$\mu(S) = 2, \quad \int_S u \, d\mu(u, v) = \int_S v \, d\mu(u, v) = 1.$$

The Dependence Function D in the Bivariate Case

We show that the measure μ on the simplex S in $\mathbb{R}^2$ can be replaced by a measure ν on the interval $[0, 1]$, which we call the angular measure. In Section 6.2 we will outline the relationship between the Pickands dependence function D and the measure generating function M of ν.

Denote by $\pi_2(x, y) := y$ the projection onto the second component of an arbitrary vector $(x, y) \in \mathbb{R}^2$ and put

$$\nu(B) := \pi_2\mu(B) = \mu(\pi_2^{-1}(B))$$

for any Borel set B in $[0, 1]$. Then ν is a measure on $[0, 1]$ with $\nu([0, 1]) = 2$ and

$$\int_{[0,1]} v \, d\nu(v) = \int_S v \, d\mu(u, v) = 1 = \int_{[0,1]} 1 - v \, d\nu(v)$$

and D becomes

$$D(z) = \int_S \max(uz, v(1 - z)) \, d\mu(u, v)$$

$$= \int_{[0,1]} \max((1 - v)z, v(1 - z)) \, d\nu(v).$$

Let ν be, on the other hand, an arbitrary measure on $[0, 1]$ with $\nu([0, 1]) = 2$ and $\int_{[0,1]} v \, d\nu(v) = 1$. Define $\pi : [0, 1] \to S$ by $\pi(v) := (1 - v, v)$ and put, for an arbitrary Borel set B in S,

$$\mu(B) := \nu(\pi^{-1}(B)).$$

Then μ is a measure on S with $\mu(S) = 2$ and $\int_S v \, d\mu(u, v) = \int_{[0,1]} v \, d\nu(v) = 1$. The Pickands dependence function corresponding to μ is

$$D(z) = \int_S \max(uz, v(1-z)) \, d\mu(u,v)$$

$$= \int_{[0,1]} \max((1-v)z, v(1-z)) \, d\nu(v).$$

We summarize the preceding considerations in the following result.

Lemma 6.1.1. *A bivariate function G is a max-stable df with univariate standard reverse exponential margins if, and only if, it has the representation*

$$G(x,y) = \exp\left((x+y)D\left(\frac{x}{x+y}\right)\right), \qquad x, y \le 0, \ (x,y) \ne (0,0),$$

where $D : [0,1] \to [0,1]$ is the Pickands dependence function

$$D(z) = \int_{[0,1]} \max((1-u)z, u(1-z)) \, d\nu(u)$$

and ν is an arbitrary measure on $[0,1]$ with $\nu([0,1]) = 2$ and $\int_{[0,1]} u \, d\nu(u) = 1$.

The measure generating function M of the preceding angular measure ν will be investigated in Section 6.2. The following properties of D were already established for a general dimension d in Section 4.3.

Lemma 6.1.2. *We have for an arbitrary dependence function $D : [0,1] \to \mathbb{R}$:*

(i) *D is continuous and convex.*

(ii) *$1/2 \le \max(z, 1-z) \le D(z) \le 1$.*

(iii) *$D(1/2) = \min_{z \in [0,1]} D(z)$, if $D(z) = D(1-z)$, $z \in [0,1]$.*

(iv) *$D(z) = 1$, $z \in [0,1]$, and $D(z) = \max(z, 1-z)$, $z \in [0,1]$, are the cases of independence and complete dependence of the margins of the EVD $G(x,y) = \exp((x+y)D(x/(x+y)))$.*

THE TAIL DEPENDENCE PARAMETER

Starting with the work by Geffroy [168], [169] and Sibuya [415], X and Y with joint EVD G are said to be *tail independent* or *asymptotically independent* if the *tail dependence parameter*

$$\chi := \lim_{c \uparrow 0} P(Y > c \mid X > c) \tag{6.2}$$

equals 0. Note that $\chi = 2(1 - D(1/2))$ and, thus, the convexity of $D(z)$ implies that $\chi = 0$ is equivalent to the condition $D(z) = 1$, $z \in [0,1]$. We refer to Section 6.4 for a discussion of a certain tail dependence function δ with $\delta(1/2) = \chi$.

The attention given to statistical properties of asymptotically independent distributions is largely a result of a series of articles by Ledford and Tawn [310], [311] and [312]. Coles et al. [74] give a synthesis of the theory. For a directory of coefficients of tail dependence such as χ we refer to Heffernan [215].

Let the rvs U, V have dfs F_U and F_V and let

$$\chi(q) := P\big(V > F_V^{-1}(q) \mid U > F_U^{-1}(q)\big) \tag{6.3}$$

be the tail dependence parameter at level $q \in (0,1)$; see Reiss and Thomas [389], (2.57). We have

$$\chi(q) = \chi + O(1-q)$$

if (U,V) follows a bivariate EVD. Also

$$\chi(q) = \chi = 2(1 - D(1/2)), \qquad q \geq 1/2,$$

for bivariate GPD, see (9.24) and (10.8) of Reiss and Thomas [389]. Recall that in the bivariate case, a GPD can be defined as $W(x,y) = 1 + \log(G(x,y))$ for $\log(G(x,y)) \geq -1$, see Lemma 5.1.1.

Tail independence $\chi = 0$ is, therefore, characterized for a GPD $W(x,y) = 1 + (x+y)D(x/(x+y))$ by $\chi(q) = 0$ for large values of q or, equivalently, by $D(z) = 1$, $z \in [0,1]$, and, hence, by $W(x,y) = 1 + x + y$. Note, however, that $W(x,y) = 1 + x + y$ is the df of $(U, -1-U)$, i.e., we have tail independence $\chi = 0$ for a GPD iff we have complete dependence $V = -1 - U$, which seems to be a bit weird.

This *exact tail independence* of $W(x,y) = 1 + x + y$ can, however, easily be explained as follows: Consider independent copies $(U_i, V_i) = (U_i, -1 - U_i)$, $i \leq n$, $n \geq 2$, of $(U,V) = (U, -1-U)$ with df W. Then we have for the vector of componentwise maxima

$$\big(\max_{i \leq n} U_i, \max_{i \leq n} V_i\big) = \big(\max_{i \leq n} U_i, \max_{i \leq n}(-1 - U_i)\big) = (\xi_1, \xi_2),$$

where ξ_1, ξ_2 are independent rv. This is due to the equivalence

$$U_i > U_j \iff -1 - U_i < -1 - U_j.$$

This is another interpretation of the case of independence for GPDs, additional to our remarks in Section 5.2.

By taking the limit of $\chi(q)$, as $q \uparrow 1$, in (6.3) one can easily extend the definition of the tail dependence parameter to random variables with arbitrary joint distribution functions.

Towards Residual Tail Dependence

Looking at the notion of tail independence and tail dependence, one may intuitively say that there is tail dependence in (x,y) if both components x and y are simultaneously large. Otherwise one may speak of tail independence.

Translating this idea to relative frequencies of data (x_i, y_i), $i \geq 1$, one may say that there is tail dependence if the frequency of both values x_i and y_i being large, relative to the frequency of x_i being large, is bounded away from zero. Within the stochastic model this leads again to the conditional probability $P(Y > c \mid X > c)$ in (6.2), which is bounded away from zero if X and Y are tail dependent. Let us discuss this idea by regarding bivariate data from standard normal distributions with various correlation coefficients ρ, cf. Figure 6.1.1. If $\rho = 0$, the data are obviously tail independent. For $\rho = 0.7$ and $\rho = 0.9$ there seems to be a stronger tail dependence. Finally, if $\rho = -0.7$, the tail independence is stronger than for $\rho = 0$.

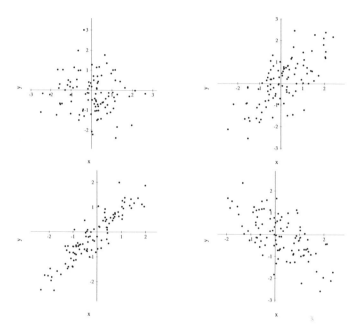

FIGURE 6.1.1. Plots of bivariate normal samples with $\rho = 0$ (top left), $\rho = 0.7$, (top right), $\rho = 0.9$ (bottom left) and $\rho = -0.7$ (bottom right).

If, however, one considers a standard normally distributed random vector (X, Y) with $\rho < 1$ and marginal df Φ, one can show that $P(Y > \Phi^{-1}(q) | X > \Phi^{-1}(q)) \to 0$, as $q \uparrow 1$. Thus, we have tail independence in this case in contrast to the above intuition. The reason is that there may be a residual tail dependence in the data even if they are tail independent. This type of tail dependence can be captured by using the dependence measure $\bar{\chi}$ introduced by Coles et al. [74], namely

$$\bar{\chi} = \lim_{q \uparrow 1} \left(\frac{2 \log P\{U > F_U^{-1}(q)\}}{\log P\{V > F_V^{-1}(q), U > F_U^{-1}(q)\}} - 1 \right)$$

where U and V are random variables with dfs F_U and F_V. It may be of interest in its own right that for rvs X and Y with reverse exponential dfs the preceding parameter $\bar{\chi}$ is given by

$$\bar{\chi} = \lim_{c\uparrow 0} \left(\frac{2\log P\{X > c\}}{\log P\{Y > c, X > c\}} - 1 \right),$$

cf. Frick et al. [161]. In the standard normal case we have $\bar{\chi} = \rho$.

DIFFERENTIABLE SPECTRAL EXPANSIONS OF FINITE LENGTH WITH REGULARLY VARYING FUNCTIONS

We strengthen the condition that a df $H(x,y)$, x, $y \leq 0$, belongs to the differentiable spectral neighborhood, respectively, to the differentiable spectral δ-neighborhood of a GPD W with Pickands dependence function D, cf. conditions (5.31) and (5.34).

Let again $H_z(c) = H(c(z, 1 - z))$, $z \in [0,1]$, $c \leq 0$, be the spectral decomposition of an arbitrary df $H(x,y)$, x, $y \leq 0$, see Section 5.4. We assume in the following that the partial derivatives

$$h_z(c) := \frac{\partial}{\partial c} H_z(c) \quad \text{and} \quad \tilde{h}_z(c) := \frac{\partial}{\partial z} H_z(c) \tag{6.4}$$

of $H_z(c)$ exist for c close to 0 and any $z \in [0,1]$, and that they are continuous. In addition, we require that $h_z(c)$ satisfies the expansion

$$h_z(c) = D(z) + \sum_{j=1}^{k} B_j(c) A_j(z) + o(B_k(c)), \quad c \uparrow 0, \tag{6.5}$$

uniformly for $z \in [0,1]$ for some $k \in \mathbb{N}$, where D is a Pickands dependence function and the $A_j : [0,1] \to \mathbb{R}$, $j = 1, \ldots, k$, are integrable functions. In addition, we require that the functions $B_j : (-\infty, 0) \to (0, \infty)$, $j = 1, \ldots, k$, satisfy

$$\lim_{c\uparrow 0} B_j(c) = 0 \tag{6.6}$$

and

$$\lim_{c\uparrow 0} \frac{B_j(ct)}{B_j(c)} = t^{\beta_j}, \quad t > 0, \beta_j > 0. \tag{6.7}$$

Without loss of generality, let $\beta_1 < \beta_2 < \cdots < \beta_k$. We say that the df H satisfies a *differentiable spectral expansion of length* $k+1$ if the conditions (6.5)-(6.7) hold. The functions B_j are again regularly varying in 0 with exponent of variation β_j, $j = 1, \ldots, k$, cf. condition (5.44). As in the case of Pickands densities, cf. (5.45), one can reduce the spectral expansion in (6.5) to an expansion of length $\kappa + 1$ for any $1 \leq \kappa \leq k$, i.e.,

$$h_z(c) = D(z) + \sum_{j=1}^{\kappa} B_j(c) A_j(z) + o(B_\kappa(c)), \quad c \uparrow 0.$$

With regard to the testing problem in Section 6.5, the existence of an index j such that

$$(2 + \beta_j) \int_0^1 A_j(z)\, dz - A_j(0) - A_j(1) \neq 0, \qquad (6.8)$$

is essential. Then it is appropriate to choose κ as

$$\kappa = \min \left\{ j \in \{1, \ldots, k\} : (2 + \beta_j) \int_0^1 A_j(z)\, dz - A_j(0) - A_j(1) \neq 0 \neq 0 \right\}. \quad (6.9)$$

We also refer to the discussion of the existence of κ in the lines following Corollary 6.5.1.

If $k = 1$, we write B and A instead of B_1 and A_1, respectively, and denote the exponent of variation of B by β. We remark that the special case of a differentiable spectral expansion of length 2 with $B(c) = c$ was investigated in the second edition of this book.

According to Theorem 5.5.4 one could equivalently replace the dependence function $D(z)$ in (6.5) by some function $a : [0, 1] \to [0, \infty)$ with $a(0) = a(1) = 1$. Then $a(z)$ is actually a Pickands dependence function $D(z)$. From Theorem 5.5.4 one obtains that $H(x, y)$ is in the bivariate domain of attraction of the EVD G with this dependence function D and

$$\lim_{c \uparrow 0} P(Y > c \mid X > c) = 2(1 - D(1/2)).$$

Thus, we have tail independence of X and Y iff $D(z) = 1$, $z \in [0, 1]$. In this case, the residual tail dependence can be captured by

$$\bar{\chi} = \frac{1 - \beta_1}{1 + \beta_1}. \qquad (6.10)$$

Therefore, we also call β_1 the *residual tail dependence parameter*.

We discuss some examples. Assume that $D'(z)$ is continuous. Because a GPD W has the spectral decomposition $W_z(c) = 1 + cD(z)$ for c close to 0 we know that W satisfies the conditions (6.4) and (6.5) with $w_z(c) = D(z)$, $\tilde{w}_z(c) = cD'(z)$ and $A_j(z) = 0$, $j \geq 1$.

In addition, an EVD G has the spectral decomposition $G_z(c) = \exp(cD(z))$ and, hence, it satisfies condition (6.4) and (6.5) with

$$g_z(c) = \exp(cD(z))D(z) = D(z) + cD(z)^2 + o(c), \qquad (6.11)$$

and

$$\tilde{g}_z(c) = \exp(cD(z))cD'(z).$$

Mardia's df $H(x, y) = (\exp(-x) + \exp(-y) - 1)^{-1}$, $x, y \leq 0$, satisfies, for instance, a differentiable spectral expansion of length 2 with $D(z) = 1$ and $A(z) = 2 - z^2 - (1 - z)^2$.

The standard normal distribution with correlation coefficient $\rho \in (0,1)$ transformed to $[-1,0]$-uniform margins satisfies the expansion

$$h_z(c) = 1 + B(c)A(z) + o(B(c)), \qquad c \uparrow 0,$$

with

$$B(c) = |c|^{\frac{2}{1+\rho}-1} L(c),$$

where

$$L(c) = (1+\rho)^{\frac{3}{2}}(1-\rho)^{-\frac{1}{2}}(4\pi)^{-\frac{\rho}{1+\rho}}(-\log|c|)^{-\frac{\rho}{1+\rho}},$$

and

$$A(z) = -\frac{2}{1+\rho}(z(1-z))^{\frac{1}{1+\rho}}.$$

The function B is regularly varying in 0 with the exponent of variation

$$\beta = \frac{2}{1+\rho} - 1 \in (0,1) \tag{6.12}$$

for $\rho \in (0,1)$. Plugging β into (6.10) one receives again $\bar{\chi} = \rho$. For further details we refer to Frick et al. [161].

The following lemma states that an expansion of finite length of a Pickands density can be deduced from a spectral expansion of finite length under certain conditions.

Lemma 6.1.3. *Let H be the distribution function of a bivariate random vector $\mathbf{X} = (X_1, X_2)$ with values in $(-\infty, 0]^2$ satisfying the spectral expansion (6.5) uniformly for $z \in [0,1]$, where the Pickands dependence function D and the A_j, $j = 1, \ldots, k$, are twice continuously differentiable.*

(i) *Putting*

$$\tilde{A}_j(z) = -\beta_j A_j(z) - \frac{\beta_j}{1+\beta_j} A_j'(z)(1-2z) + \frac{1}{1+\beta_j} A_j''(z)z(1-z), \tag{6.13}$$

where β_j is the exponent of variation of the function B_j, one gets

$$\int_0^1 \tilde{A}_j(z)\,dz = -(2+\beta_j)\int_0^1 A_j(z)\,dz + A_j(0) + A_j(1) \tag{6.14}$$

for $j = 1, \ldots, k$.

(ii) *If the remainder term*

$$R(z,c) := h_z(c) - D(z) - \sum_{j=1}^{k} B_j(c)A_j(z)$$

is positive and differentiable in c, then the density of the Pickands transform
$(Z, C) = T(\mathbf{X})$ satisfies the expansion

$$f(z, c) = \varphi(z) + \sum_{j=1}^{k} B_j(c)\tilde{A}_j(z) + o(B_k(c)), \quad c \uparrow 0, \qquad (6.15)$$

uniformly for $z \in [0, 1]$ with $\varphi(z) = D''(z)z(1 - z)$ and $\tilde{A}_j$ as in (6.13). The
regularly varying functions B_j are the same as in expansion (6.5).

(iii) *If the parameter κ of the spectral expansion in (6.9) exists, the parameter κ*
   of the Pickands density in (5.46) exists, too, and both are the same.

Proof. Part (i) can easily be deduced by partial integration. For the proof of part
(ii) we refer to Lemma 2.1 in Frick and Reiss [162]. The assertion of part (iii)
follows directly from part (i). $\square$

The concepts of differentiable spectral expansions and expansions of Pickands
densities are of interest in their own right, applications may be found in Section
6.5 in conjunction with a testing problem.

6.2   The Measure Generating Function M

The restriction to the bivariate case enables the representation of an arbitrary
dependence function D in terms of the measure generating function corresponding
to the measure ν, see Lemma 6.1.1. The derivation of several characteristics of D
such as its absolute continuity will be a consequence.

ANOTHER REPRESENTATION OF THE PICKANDS DEPENDENCE FUNCTION

The following representation of a dependence function in the bivariate case will be
crucial for our subsequent considerations.

Lemma 6.2.1. *Let ν be an arbitrary measure on $[0, 1]$ with $\nu([0, 1]) = 2$ and*
$\int_{[0,1]} u\,\nu(du) = 1$. Denote by $M(z) := \nu([0, z])$, $z \in [0, 1]$, the corresponding mea-
sure generating function. Then we have for the dependence function D correspond-
ing to ν (cf. Lemma 6.1.1) the representation

$$D(z) = 1 - z + \int_{[0,z]} M(x)\,dx, \qquad 0 \le z \le 1.$$

Proof. According to Lemma 6.1.1 and straightforward calculations we have

$$D(z) = \int_{[0,1]} \max((1 - u)z, u(1 - z))\,d\nu(u)$$

$$= z \int_{[0,z]} (1 - u)\,d\nu(u) + (1 - z) \int_{(z,1]} u\,d\nu(u)$$

$$= zM(z) - z + 1 - \int_{[0,z]} u\, \nu(du).$$

Now Fubini's theorem implies

$$\int_{[0,z]} u\, \nu(du) = \int_{[0,z]} \int_{[0,z]} 1_{[0,u)}(x)\, dx\, d\nu(u)$$

$$= \int_{[0,z]} \int_{[0,z]} 1_{(x,z]}(u)\, d\nu(u)\, dx$$

$$= \int_{[0,z]} M(z) - M(x)\, dx$$

$$= zM(z) - \int_{[0,z]} M(x)\, dx.$$

This representation yields the assertion

$$D(z) = 1 - z + \int_{[0,z]} M(x)\, dx. \qquad \qquad \square$$

Note that M coincides with the angular distribution in Section 5.9 in the bivariate case. The preceding result yields the following consequences.

(i) The function $F(z) := D(z) - 1 + z$ defines for an arbitrary dependence function D a probability df on $[0,1]$, whose Lebesgue density is M:

$$F(z) = \int_{[0,z]} M(x)\, dx, \qquad z \in [0,1].$$

(ii) Recall that a Lebesgue density is uniquely determined almost everywhere. Since M is continuous from the right, the representation of $D(z)$ in (i) implies that M and, thus, the measure ν is uniquely determined by D.

The Marshall-Olkin dependence function

$$D_\lambda(z) = 1 - \lambda \min(z, 1 - z)$$

is, for example, generated by the measure ν, which has mass 2λ at $z = 1/2$ and mass $1 - \lambda$ at 0 and 1.

(iii) The representation

$$D(z) = 1 - z + \int_{[0,z]} M(x)\, dx$$

implies, moreover, that D is absolutely continuous with derivative $D'(z) := M(z) - 1$:

$$D(z_2) - D(z_1) = \int_{[z_1,z_2]} M(x) - 1\, dx, \qquad 0 \le z_1 \le z_2 \le 1.$$

We obtain, therefore, that $D''(z)$ exists if, and only if $M'(z)$ exists, and in this case these derivatives coincide. This was observed by Smith [417].

The measure generating function corresponding to the logistic dependence function with parameter $\lambda \geq 1$,

$$D(z) = \left(z^\lambda + (1-z)^\lambda\right)^{1/\lambda}$$

is, for example,

$$
\begin{aligned}
M(z) &= 1 + D'(z) \\
&= 1 + \frac{z^{\lambda-1} - (1-z)^{\lambda-1}}{\left(z^\lambda + (1-z)^\lambda\right)^{1-1/\lambda}}, \qquad z \in [0,1].
\end{aligned}
$$

This function is continuous with $M(0) = 0$.

The measure generating function pertaining to the independence case $D(z) = 1$, $z \in [0,1]$, is $M(z) = 1$, $z \in [0,1)$, $M(1) = 2$. The corresponding measure ν has mass 1 at each of the points 0 and 1. The measure generating function pertaining to the complete dependence case $D(z) = \max(z, 1-z)$ is $M(z) = 0, z \in [0, 1/2)$, and $M(z) = 2$, $z \in [1/2, 1]$, i.e., the corresponding measure ν has mass 2 at $1/2$.

ESTIMATION OF THE MEASURE GENERATING FUNCTION M

The representation

$$M(z) = D'(z) + 1$$

offers an easy way to estimate the measure generating function M by means of an estimator of D, which we introduced in Section 5.4 in general dimension.

Let $(U_1, V_1), \ldots, (U_n, V_n)$ be independent copies of (U, V), whose df H is in the δ-neighborhood of the GPD W with dependence function D. Choose an auxiliary parameter $c < 0$ and consider, for $z \in [0, 1)$,

$$
\hat{D}_{n,c}(z) = \frac{1}{c}\left(1 - \frac{1}{n}\sum_{i=1}^{n} 1\left(U_i \leq cz, V_i \leq c(1-z)\right)\right)
$$

$$
= \frac{1}{cn}\sum_{i=1}^{n} 1\left(U_i > cz \text{ or } V_i > c(1-z)\right).
$$

If we let $c = c_n$ tend to 0 with n such that $n|c| \to \infty$, $n|c|^{1+2\delta} \to 0$ as n increases, then we obtain from Lemma 5.5.6

$$
(n|c|)^{1/2}\left(\hat{D}_{n,c}(z) - D(z)\right) \longrightarrow_D N\left(0, D(z)\right).
$$

The idea now suggests itself to use

$$T_{n,c,h}(z) := \frac{\hat{D}_{n,c}(z+h) - \hat{D}_{n,c}(z)}{h}$$

with $h > 0$ for the estimation of the derivative from the right $D'(z)$ of $D(z)$. We have

$$P\Big(U > cz \text{ or } V > c(1-z)\Big)$$
$$= 1 - H(c(z, 1-z)))$$
$$= \Big(1 - W(c(z, 1-z))\Big)\Big(1 + O(|c|^{\delta})\Big)$$
$$= |c|D(z)\Big(1 + O(|c|^{\delta})\Big)$$

and, consequently,

$$E\Big(T_{n,c,h}(z)\Big) = \frac{1}{hc}\Big(H\Big(c(z, 1-z)\Big) - H\Big(c(z+h, 1-z-h)\Big)\Big)$$
$$= \frac{1}{h}\Big(D(z+h) - D(z)\Big) + O\left(\frac{|c|^{1+\delta}}{h}\right),$$

provided that c is small enough.

The variance of $T_{n,c,h}(z)$ satisfies

$$nhc\mathrm{Var}\Big(T_{n,c,h}(z)\Big)$$
$$= \frac{1}{hc}\Bigg(E\Big(\Big(\mathbb{1}(U \le cz, V \le c(1-z))$$
$$- \mathbb{1}(U \le c(z+h), V \le c(1-z-h))\Big)^2\Big)$$
$$- \Big(H(c(z, 1-z)) - H(c(z+h, 1-z-h))\Big)^2\Bigg)$$
$$= \frac{1}{hc}\Bigg(H\Big(c(z, 1-z)\Big) + H\Big(c(z+h, 1-(z+h))\Big)$$
$$- 2H\Big(c(z, 1-(z+h))\Big)$$
$$- \Big(H\Big(c(z, 1-z)\Big) - H\Big(c(z+h, 1-(z+h))\Big)\Big)^2\Bigg)$$
$$= \frac{1}{hc}\Bigg(cD(z) + cD(z+h) - 2c(1-h)D\left(\frac{z}{1-h}\right) + O(|c|^{1+\delta})$$

$$- \left(- cD(z+h) + cD(z) + O(|c|^{1+\delta}) \right)^2 \Bigg)$$

$$= 2(1-h) \frac{D(z) - D\left(\frac{z}{1-h}\right)}{h} + 2D(z) + \frac{D(z+h) - D(z)}{h} + O\left(\frac{|c|^\delta}{h} + ch\right)$$

$$\longrightarrow_{h,c \to 0} D'(z)(1-2z) + 2D(z)$$

$$= 1 + (1-2z)M(z) + 2\int_{[0,z]} M(x)\,dx =: \sigma^2(z),$$

provided that $|c|^\delta/h \to 0$.

The following result is now an immediate consequence of the central limit theorem.

Lemma 6.2.2. *Suppose that the df H of (U,V) belongs to the spectral δ-neighborhood of the GPD W. If we choose $c = c_n \to 0$, $h = h_n \to 0$ with $n|c|h \to \infty$, $|c|^\delta/h \to 0$ as n increases, then we obtain, for $z \in [0,1)$,*

$$(n|c|h)^{1/2}\left(T_{n,c,h}(z) - E(T_{n,c,h}(z)) \right) \longrightarrow_D N(0, \sigma^2(z)).$$

Lemma 6.2.1 implies that for small c,

$$E\left(T_{n,c,h}(z) \right) - D'(z) = \frac{D(z+h) - D(z)}{h} - D'(z) + O\left(\frac{|c|^{1+\delta}}{h}\right)$$

$$= \frac{1}{h}\int_{(z,z+h]} M(x) - M(z)\,dx + O\left(\frac{|c|^{1+\delta}}{h}\right).$$

The following result is now immediate from Lemma 6.2.2.

Theorem 6.2.3. *Suppose that the df H of (U,V) belongs to the spectral δ-neighborhood of the GPD W with dependence function $D(z) = 1 - z + \int_{[0,z]} M(x)\,dx$, $z \in [0,1]$. Choose $z \in [0,1)$ and suppose that $M(z+\varepsilon) - M(z) = O(\varepsilon^\alpha)$, $\varepsilon > 0$, for some $\alpha > 1/2$. Then the estimator*

$$\hat{M}_n(z) := T_{n,c,h}(z) + 1$$

of $M(z)$ is asymptotically normal:

$$(n|c|h)^{1/2}\left(\hat{M}_n(z) - M(z) \right) \longrightarrow_D N\left(0, \sigma^2(z)\right),$$

provided that $c = c_n \to 0$, $h = h_n \to 0$ satisfy $|c|^\delta/h \to 0$, $n|c|h \to \infty$, $n|c|h^{2\alpha} \to 0$ as $n \to \infty$.

6.3 The Pickands Transform in the Bivariate Case

In this section we want to reformulate and strengthen results about the Pickands transform of bivariate GPD and EVD rv and include the case of Marshall-Olkin rv.

THE DISTRIBUTION OF THE DISTANCE $C = U + V$

We start with a result about the distance $C = U + V$ pertaining to an EVD rv.

Lemma 6.3.1. *Let (U, V) be an EVD rv with Pickands dependence function D. We have, for $c < 0$ close to 0,*

$$P(U + V \leq c) = \exp(c) - c \int_{[0,1]} \exp(cD(z)) \left(D(z) + D'(z)(1 - z) \right) dz.$$

Proof. The following arguments are taken from Ghoudi et al. [179]. The conditional df of $U + V$, given $U = u < 0$, is, for c close to 0 ,

$$
\begin{aligned}
&P(U + V \leq c \mid U = u) \\
&= P(V \leq c - u \mid U = u) \\
&= \lim_{\varepsilon \downarrow 0} \frac{P(V \leq c - u, U \in [u, u + \varepsilon])}{P(U \in [u, u + \varepsilon])} \\
&= \lim_{\varepsilon \downarrow 0} \frac{G(u + \varepsilon, c - u) - G(u, c - u)}{\varepsilon} \cdot \frac{\varepsilon}{\exp(u + \varepsilon) - \exp(u)} \\
&= \frac{G(u, c - u)}{\exp(u)} \left(D\left(\frac{u}{c}\right) + D'\left(\frac{u}{c}\right) \frac{c - u}{c} \right) \\
&= \exp\left(cD\left(\frac{u}{c}\right) - u \right) \left(D\left(\frac{u}{c}\right) + D'\left(\frac{u}{c}\right) \frac{c - u}{c} \right)
\end{aligned}
$$

if $u > c$ and,

$$P(U + V \leq c \mid U = u) = 1 \qquad \text{if } u \leq c.$$

Hence we obtain

$$
\begin{aligned}
&P(U + V \leq c) \\
&= \int_{(-\infty, 0]} P(V \leq c - u | U = u) \exp(u) \, du \\
&= \int_{[c, 0]} \exp\left(cD\left(\frac{u}{c}\right) \right) \left(D\left(\frac{u}{c}\right) + D'\left(\frac{u}{c}\right) \frac{c - u}{c} \right) du + \exp(c) \\
&= \exp(c) - c \int_{[0,1]} \exp(cD(u)) \left(D(u) + D'(u)(1 - u) \right) du. \qquad \square
\end{aligned}
$$

The following result extends Lemma 6.3.1 to a df H, which satisfies condition (6.4). Recall that $H_1(c) = H(c(1,0))$, $c \leq 0$, is the first marginal distribution of H.

Lemma 6.3.2. *Suppose that the df $H(u,v)$, $u, v \leq 0$, of (U, V) satisfies condition (6.4). Let again $\tilde{h}_z(c) = \frac{\partial}{\partial z} H_z(c)$. Then we have, for c close to 0,*

$$P(U + V \leq c) = H_1(c) - \int_0^1 ch_z(c) + \tilde{h}_z(c)(1 - z)\, dz.$$

Proof. Repeating the arguments in the proof of Lemma 6.3.1, we obtain, for $0 > u > c$,

$$P(U + V \leq c \mid U = u)$$
$$= \frac{1}{h_1(u)} \lim_{\varepsilon \downarrow 0} \frac{H(u + \varepsilon, c - u) - H(u, c - u)}{\varepsilon}$$
$$= \frac{1}{h_1(u)} \lim_{\varepsilon \downarrow 0} \frac{H_{\frac{u+\varepsilon}{c+\varepsilon}}(c + \varepsilon) - H_{\frac{u}{c}}(c)}{\varepsilon}$$
$$= \frac{1}{h_1(u)} \left(h_{\frac{u}{c}}(c) + \tilde{h}_{\frac{u}{c}}(c) \frac{c - u}{c^2} \right)$$

by making use of Taylor's formula and the continuity of the partial derivatives of $H_z(c)$.

Since $P(U + V \leq c \mid U = u) = 1$ if $u \leq c$, we obtain by integration and substitution, for c close to 0,

$$P(U + V \leq c)$$
$$= H_1(c) + \int_c^0 h_{\frac{u}{c}}(c) + \tilde{h}_{\frac{u}{c}}(c) \frac{c - u}{c^2}\, du$$
$$= H_1(c) - \int_0^1 ch_z(c) + \tilde{h}_z(c)(1 - z)\, dz. \qquad \square$$

Assume now that the df $H(u,v)$ of (U, V) coincides for u, v close to 0 with the general GPD $W(u,v) = 1 + (u + v)D(u/(u + v))$. Repeating the arguments in the proof of Lemma 6.3.1 one obtains the following result.

Lemma 6.3.3. *We have, for $c < 0$ close to 0*

$$P(U + V \leq c) = 1 + 2c \left(1 - \int_{[0,1]} D(u)\, du \right).$$

THE PICKANDS TRANSFORM

The conditioning technique in the proof of Lemma 6.3.1 also enables us to obtain the following sharper version of Theorem 5.6.2 on the distribution of the Pickands coordinates $C = U + V$ and $Z = U/(U + V)$ in the case, where (U, V) follows a bivariate GPD. This result is true for an arbitrary dependence function D different from the constant function 1 and requires no higher order differentiability conditions on D. Recall that in the bivariate case, a GPD W can be defined by $W(u, v) = 1 + \log(G(u, v))$ for $\log(G(u, v)) \geq -1$, see Lemma 5.1.1.

Theorem 6.3.4. *Suppose that (U, V) follows a GPD with dependence function D, which is not the constant function 1. Then we have for $c_0 \in [-1, 0)$ the following facts:*

(i) *Conditional on $C = U + V > c_0$, the Pickands coordinates $Z = U/(U + V)$ and $C = U + V$ are independent.*

(ii) *C is on $(-1, 0)$ uniformly distributed, precisely,*

$$P(C > c_0) = |c_0|2 \left(1 - \int_{[0,1]} D(z)\,dz\right)$$

and, thus,

$$P(C \geq uc_0 \mid C > c_0) = u, \qquad 0 \leq u \leq 1.$$

(iii) *Conditional on $C > c_0$, Z has the df*

$$F(z) := \frac{D'(z)z(1 - z) + D(z)(2z - 1) + 1 - 2\int_0^z D(u)\,du}{2\left(1 - \int_{[0,1]} D(u)\,du\right)}, \qquad z \in [0, 1].$$

If (U, V) follows a bivariate EVD, then the statements in the preceding result are asymptotically true for $c_0 \uparrow 0$.

Theorem 6.3.5. *Suppose that (U, V) follows a bivariate EVD G with dependence function D, which is not the constant function 1. Then we have for $c_0 \uparrow 0$:*

(i) *Conditional on $C = U + V > c_0$, the Pickands coordinates $Z = U/(U + V)$ and $C = U + V$ are asymptotically for $c_0 \uparrow 0$ independent:*

$$P(C \geq uc_0, Z \leq z \mid C > c_0)$$
$$= P(C \geq uc_0 \mid C > c_0)P(Z \leq z \mid C > c_0) + O(c_0), \qquad 0 \leq u \leq 1.$$

(ii) *We have $P(C > c_0) = |c_0|2 \left(1 - \int_{[0,1]} D(z)\,dz\right)(1 + O(c_0))$ and, thus,*

$$P(C \geq uc_0 \mid C > c_0) = u(1 + O(c_0)), \qquad 0 \leq u \leq 1.$$

(iii) *Conditional on $C > c_0$, Z has for $c_0 \uparrow 0$ the df*

$$P(Z \leq z \mid C > c_0) = F(z) + O(c_0), \qquad 0 \leq z \leq 1,$$

where the df F is defined in Theorem 6.3.4.

MARSHALL-OLKIN GPD RANDOM VECTORS

The predominant example of an EVD with a non-smooth dependence function D is the Marshall-Olkin df ([323])

$$G_\lambda(x,y) = \exp\left(x + y - \lambda \max(x,y)\right), \quad x,y \le 0, \quad \lambda \in [0,1]$$

with the dependence function

$$D_\lambda(z) = 1 - \lambda \min(z, 1-z),$$

see Example 4.3.4. The pertaining GPD is

$$W_\lambda(x,y) = 1 + x + y - \lambda \max(x,y).$$

G_λ is the df of

$$(U,V) := \left(\max\left(\frac{Z_1}{1-\lambda}, \frac{Z_0}{\lambda}\right), \max\left(\frac{Z_2}{1-\lambda}, \frac{Z_0}{\lambda}\right)\right),$$

where Z_0, Z_1, Z_2 are independent standard reverse exponential rv, with the convention

$$(U,V) = (Z_1, Z_2) \quad \text{if} \quad \lambda = 0$$

and

$$(U,V) = (Z_0, Z_0) \quad \text{if} \quad \lambda = 1.$$

It is well known that $W_0(x,y) = 1 + x + y$ is the distribution of $(-\eta, -1 + \eta)$, where η is uniformly distributed on $(0,1)$. The following result is an extension to arbitrary $\lambda \in [0,1]$. It can be verified by elementary computations.

Proposition 6.3.6. *The Marshall-Olkin GPD W_λ is the df of the rv*

$$(U,V) := \frac{1}{\lambda - 2}(\eta, \eta) \cdot 1_{\{0\}}(\varepsilon)$$

$$+ \frac{1}{\lambda - 2}\left((\lambda - 1)\eta - (\lambda - 2), \eta\right) \cdot 1_{\{1\}}(\varepsilon)$$

$$+ \frac{1}{\lambda - 2}\left(\eta, (\lambda - 1)\eta - (\lambda - 2)\right) \cdot 1_{\{2\}}(\varepsilon),$$

where η, ε are independent rv, η is uniformly distributed on $(0,1)$ and ε takes the value $0, 1, 2$ with probabilities

$$P(\varepsilon = 0) = \frac{\lambda}{2 - \lambda}, \quad P(\varepsilon = 1) = P(\varepsilon = 2) = \frac{1 - \lambda}{2 - \lambda}.$$

The rv (U, V) with df $W_\lambda(x, y) = 1 + x + y - \lambda \max(x, y)$ is, thus, generated in a two-step procedure. First, the rv ε takes one of the values $0, 1, 2$ with probability $\lambda/(2 - \lambda), (1 - \lambda)/(2 - \lambda)$ and $(1 - \lambda)/(2 - \lambda)$, respectively. If $\varepsilon = 0$, then (U, V) is set to

$$(U, V) = (U, U),$$

where U is uniformly distributed on $(1/(\lambda - 2), 0)$. If $\varepsilon = 1$, then

$$(U, V) = \left(U, \frac{-1 - U}{1 - \lambda} \right),$$

where U is uniformly distributed on $(-1, 1/(\lambda - 2))$. And if $\varepsilon = 2$, then (U, V) is set to

$$(U, V) = \big(U, (\lambda - 1) U - 1 \big),$$

where U is uniformly distributed on $(1/(\lambda - 2), 0)$. The distribution of (U, V) is, thus, concentrated on the three lines

$$
\begin{aligned}
y &= x, & 1/(\lambda - 2) &\le x \le 0, \\
y &= (-1 - x)/(1 - \lambda), & -1 &\le x \le 1/(\lambda - 2), \\
y &= (\lambda - 1)x - 1, & 1/(\lambda - 2) &\le x \le 0.
\end{aligned}
$$

The distribution of the Pickands transform in case of an underlying GPD W_λ is now an immediate consequence.

Corollary 6.3.7. *If (U, V) has the df $W_\lambda(x, y) = 1 + x + y - \lambda \max(x, y)$, then the Pickands transform $(Z, C) = (U/(U + V), U + V)$ satisfies*

$$
\begin{aligned}
(Z, C) =_D & \left(\frac{1}{2}, \frac{2}{\lambda - 2}\eta \right) 1_{\{0\}}(\varepsilon) \\
& + \left(\frac{(\lambda - 1)\eta - (\lambda - 2)}{\lambda\eta - (\lambda - 2)}, \frac{\lambda\eta - (\lambda - 2)}{\lambda - 2} \right) 1_{\{1\}}(\varepsilon) \\
& + \left(\frac{\eta}{\lambda\eta - (\lambda - 2)}, \frac{\lambda\eta - (\lambda - 2)}{\lambda - 2} \right) 1_{\{2\}}(\varepsilon),
\end{aligned}
$$

where η, ε are defined in Proposition 6.3.6 and $=_D$ denotes equality in distribution.

The df of Z is, thus, not continuous in case of the Marshall-Olkin GPD W_λ with $\lambda > 0$. It has positive mass $P(Z = 1/2) = P(\varepsilon = 0) = \lambda/(2 - \lambda)$ at $1/2$, which is the probability that (U, V) takes values on the diagonal $y = x$. The condition $C > c_0 > 1/(\lambda - 2)$ can only be satisfied if $\varepsilon = 0$ and, thus, we have $Z = \frac{1}{2}$, conditioned on $C > c_0 > 1/(\lambda - 2)$. Note that $1/(\lambda - 2) \le -1/2$ for $\lambda \in [0, 1]$.

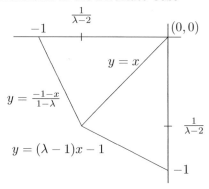

FIGURE 6.3.1. Support lines of Marshall-Olkin GPD
$W_\lambda(x,y) = 1 + x + y - \lambda \max(x,y)$.

Marshall-Olkin EVD Random Vectors

Suppose next that (U,V) follows the Marshall-Olkin df $G_\lambda(x,y) = \exp(x + y - \lambda \max(x,y)), x, y \le 0, \lambda \in [0,1]$. Then

$$g_\lambda(x,y) := \frac{\partial^2}{\partial x \partial y} G_\lambda(x,y) = (1-\lambda) G_\lambda(x,y)$$

exists for all (x,y) with $x \ne y$ and, hence, g_λ is a density of G on the set $\{(x,y) \in (-\infty, 0]^2 : x \ne y\}$. We obtain that the Pickands transform $(Z, C) = (U/(U + V)), U + V)$ corresponding to (U, V) has the density

$$f_\lambda(z,c) = (1-\lambda)|c| \exp\left(c\, D_\lambda(z)\right)$$
$$= (1-\lambda)|c| \exp\left(c\left(1 - \lambda \min(z, 1-z)\right)\right)$$

on $([0,1]\backslash\{1/2\}) \times (-\infty, 0)$.

Lemma 6.3.8. *If (U, V) has the df $G_\lambda(x,y) = \exp(x + y - \lambda \max(x,y)), 0 < \lambda \le 1$, then the Pickands transform (Z, C) satisfies*

$$
\begin{aligned}
P(C > c) &= -\lambda c/2 + O(c^2), \\
P(Z = 1/2 \,|\, C > c) &= 1 + O(c), \\
P(C > u\,c \,|\, C > c) &= u\left(1 + O(c)\right),
\end{aligned}
$$

as $c \uparrow 0$ uniformly for $u \in [0,1]$.

Proof. First we consider the case $\lambda < 1$. We have for $c < 0$,

$$
\begin{aligned}
&P(U + V > c) \\
&= P(U + V > c, U \ne V) + P(2U > c, U = V).
\end{aligned}
$$

The first probability on the right-hand side equals

$$\int_{(c,0)} \int_{[0,1/2)} f_\lambda(z,b)\, dz\, db + \int_{(c,0)} \int_{(1/2,1]} f_\lambda(z,b)\, dz\, db$$

$$= \int_{(c,0)} \int_{[0,1/2)} (1-\lambda)|b| \exp\big(b(1-\lambda z)\big)\, dz\, db$$

$$+ \int_{(c,0)} \int_{(1/2,1]} (1-\lambda)|b| \exp\big(b(1-\lambda(1-z))\big)\, dz\, db$$

$$= O(c^2).$$

From the representation

$$(U,V) = \left(\max\left(\frac{Z_0}{\lambda}, \frac{Z_1}{1-\lambda}\right), \max\left(\frac{Z_0}{\lambda}, \frac{Z_2}{1-\lambda}\right) \right)$$

we obtain that

$$P(2U > c, U = V)$$

$$= P\left(Z_0 > \frac{\lambda c}{2}, Z_0 \geq \frac{\lambda}{1-\lambda} \max(Z_1, Z_2) \right)$$

$$= \int_{(-\infty,0)} \int_{\left(\max\left(\frac{\lambda c}{2}, \frac{\lambda}{1-\lambda} t\right), 0\right)} 2\exp(x)\exp(2t)\, dx\, dt,$$

since $\max(Z_1, Z_2)$ has the density $2\exp(2t)$, $t \leq 0$. The above integral equals

$$\int_{\left(-\infty, \frac{1-\lambda}{2}c\right]} \int_{\left(\frac{\lambda c}{2}, 0\right)} 2\exp(x)\exp(2t)\, dx\, dt$$

$$+ \int_{\left(\frac{1-\lambda}{2}c, 0\right)} \int_{\left(\frac{\lambda}{1-\lambda}t, 0\right)} 2\exp(x)\exp(2t)\, dx\, dt$$

$$= \int_{\left(-\infty, \frac{1-\lambda}{2}c\right]} 2\exp(2t)\left(1 - \exp\left(\frac{\lambda c}{2}\right)\right)\, dt$$

$$+ \int_{\left(\frac{1-\lambda}{2}c, 0\right)} 2\exp(2t)\left(1 - \exp\left(\frac{\lambda}{1-\lambda}t\right)\right)\, dt$$

$$= \left(1 - \exp\left(\frac{\lambda c}{2}\right)\right)\exp\big((1-\lambda)c\big)$$

$$+ 1 - \exp\big((1-\lambda)c\big) - 2\frac{1-\lambda}{2-\lambda}\left(1 - \exp\left(\frac{2-\lambda}{2}c\right)\right)$$

$$= -\frac{\lambda c}{2} + O(c^2).$$

As a consequence, we obtain that

$$P(Z \neq 1/2 \mid C > c)$$

$$= \frac{P(U < V, U + V > c) + P(U > V, U + V > c)}{P(U + V > c)}$$

$$= O(c)$$

and, moreover, that

$$P(C > uc \mid C > c) = P(U + V > uc \mid U + V > c)$$
$$= u(1 + O(c)), \qquad 0 \leq u \leq 1.$$

This proves the assertion in case $\lambda < 1$. If $\lambda = 1$, i.e., $G_1(x, y) = \exp(\min(x, y))$, we have $U = V$ and, hence, $Z = 1/2, C = 2U$. Then we obtain

$$P(C > c) = P\left(U > \frac{c}{2}\right) = 1 - \exp\left(\frac{c}{2}\right) = -\frac{c}{2} + O(c^2),$$

$$P\left(Z = \frac{1}{2} \mid C > c\right) = 1,$$

$$P(C > u, c \mid C > c) = u(1 + O(c)),$$

which completes the proof. $\qquad\qquad\qquad\qquad\qquad\qquad\qquad$ □

In the case $\lambda = 0$, i.e., $G_0(x, y) = \exp(x+y)$, the components of the Pickands transform $Z = U/(U+V)$ and $C = U+V$ are independent, with Z being uniformly distributed on $(0, 1)$ and C having the density $|x| \exp(x)$, $x \leq 0$; see, e.g. Lemma 1.6.6 in Reiss [385]. In this case we obtain

$$
\begin{aligned}
P(C > c) &= O(c^2), \\
P(Z = 1/2) &= 0, \\
P(C \geq uc \mid C > c) &= u^2(1 + O(c)).
\end{aligned}
$$

A notable difference in this case $\lambda = 0$ of independence of U and V from the case $\lambda > 0$ is the fact that C/c, conditioned on $C > c$, does *not* approach for $c \uparrow 0$ the uniform distribution. Instead it has by the last expansion above the limiting df $F(u) = u^2$, $0 \leq u \leq 1$. We will see in Lemma 6.5.2 that this distinct behavior is true for an EVD G with arbitrary dependence function D. This result is extended to a more general framework in Corollary 6.5.1, which is then used to test for the case of dependence.

For any underlying $\lambda \in [0, 1]$ we have $P(C > c) = \lambda |c|/2 + O(c^2)$. Suppose we have a sample $(U_1, V_1), \ldots, (U_n, V_n)$ of independent copies of (U, V). Then the estimator

$$\hat{\lambda}_n := \frac{2}{n|c|} \sum_{i=1}^n 1(C_i > c) =: \frac{2}{n|c|} K(n)$$

of λ suggests itself, where $C_i = U_i + V_i$. The Moivre-Laplace theorem implies that, with underlying $\lambda > 0$,

$$(n|c|)^{1/2}(\hat{\lambda}_n - \lambda) = 2(n|c|)^{-1/2}\big(K(n) - \lambda|c|/2\big) \longrightarrow_D N(0, 2\lambda),$$

provided that the threshold $c = c(n)$ satisfies $n|c| \to \infty$, $nc^3 \to 0$ as $n \to \infty$. The estimator $\hat{\lambda}_n$ is, however, outperformed by those estimators defined below, see Theorem 6.4.5 and the subsequent discussion.

6.4　The Tail Dependence Function

The tail dependence parameter of X and Y with joint bivariate EVD with dependence function D is $\chi = 2(1 - D(1/2))$, see Section 6.1 for details. In the sequel we will investigate a generalization to arbitrary $D(z)$, $z \in [0, 1]$.

THE TAIL DEPENDENCE FUNCTION

Denote by

$$\vartheta := \vartheta(z) := \frac{1 - D(z)}{\min(z, 1 - z)} \in [0, 1], \qquad z \in [0, 1],$$

the *canonical dependence function* or *tail dependence function* with the convention $\vartheta(0) := \lim_{z \downarrow 0} \vartheta(z)$, $\vartheta(1) := \lim_{z \uparrow 1} \vartheta(z)$. Reiss and Thomas [389], (10.12), introduce the canonical dependence function, but they use a different standardization of D. The particular value

$$\vartheta(1/2) = 2(1 - D(1/2))$$

is the *canonical parameter* or *tail dependence parameter*. We refer to Falk and Reiss [151] for its significance in bivariate EVD models. The cases $\vartheta(z) = 0$ and $\vartheta(z) = 1$, $z \in [0, 1]$, now characterize independence and complete dependence in the EVD model with underlying df G.

The canonical dependence function corresponding to the Marshall-Olkin dependence function $D(z) = 1 - \lambda \min(z, 1 - z)$ with parameter $\lambda \in [0, 1]$ is, for example, the constant function $\vartheta(z) = \lambda$. The logistic dependence function $D(z) = (z^\lambda + (1 - z)^\lambda)^{1/\lambda}$ with parameter $\lambda \geq 1$ has the canonical dependence function

$$\vartheta(z) = \begin{cases} \frac{1}{z} - \left(1 + \left(\frac{1-z}{z}\right)^\lambda\right)^{1/\lambda} & \text{if } z \leq \frac{1}{2}, \\[2mm] \frac{1}{1-z} - \left(1 + \left(\frac{z}{1-z}\right)^\lambda\right)^{1/\lambda} & \text{if } z \geq \frac{1}{2}. \end{cases}$$

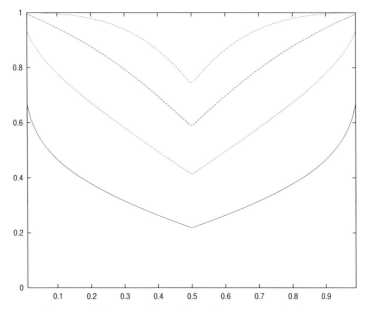

FIGURE 6.4.1. Canonical logistic dependence function with $\lambda = 1.2$, 1.5, 2 and 3, from bottom to top.

If the df H of the rv (U, V) is a GPD, then we have, for $u \in (-1, 0)$ and $z \in (0, 1)$,

$$\vartheta(z) = \begin{cases} P(U > (1 - z)u \mid V > zu), & \text{if } z \in (0, 1/2], \\ P(V > zu \mid U > (1 - z)u), & \text{if } z \in [1/2, 1). \end{cases}$$

If the df H of (U, V) is in the δ-neighborhood of a GPD, then we have, for $u \in (-1, 0)$ and $z \in (0, 1)$,

$$\left. \begin{array}{ll} P(U > (1 - z)u \mid V > zu), & \text{if } z \in (0, 1/2], \\ P(V > zu \mid U > (1 - z)u), & \text{if } z \in [1/2, 1), \end{array} \right\} = \vartheta(z)\Big(1 + O(|u|^\delta)\Big).$$

As a consequence, $\vartheta(1/2)$ coincides with the tail dependence parameter discussed, for example, in Reiss and Thomas [389], (2.57), Coles et al. [74] and Coles [71], Section 8.4, thus, continuing work that was started by Geffroy [168], [169] and Sibuya [415]. A directory of coefficients of tail dependence is given in Heffernan [215].

Note that the canonical dependence function can be extended to arbitrary dimensions by putting

$$\vartheta(\mathbf{z}) := \frac{1 - D(\mathbf{z})}{1 - \max(z_1, \ldots, z_{d-1}, 1 - \sum_{i \le d-1} z_i)},$$

where $D(\mathbf{z})$ is a dependence function defined in (4.32), $\mathbf{z} = (z_1, \ldots, z_{d-1})$. Then we have $0 \le \vartheta(\mathbf{z}) \le 1$. The constant dependence function $D(\mathbf{z}) = 1$, which characterizes independence of the margins of the corresponding EVD G, is then mapped onto $\vartheta_1(\mathbf{z}) = 0$, while the case of complete dependence $D(\mathbf{z}) = \max(z_1, \ldots, z_{d-1}, 1 - \sum_{i \le d-1} z_i)$ is mapped onto $\vartheta_2(\mathbf{z}) = 1$. The Marshall-Olkin dependence function with parameter λ, defined in Example 4.3.4, has, for example, the constant canonical dependence function $\vartheta(\mathbf{z}) = \lambda$ and is just the convex combination of the extremal points $\vartheta_1(\mathbf{z}) = 0$ and $\vartheta_2(\mathbf{z}) = 1$ of the convex set of all canonical dependence functions $\vartheta(\mathbf{z})$. In the sequel we discuss, however, only the bivariate case with the univariate $\vartheta(z)$ with $z \in [0, 1]$.

From Lemma 6.2.1 we obtain for a tail dependence function the representation

$$
\begin{aligned}
\vartheta(z) &= \frac{1 - D(z)}{\min(z, 1 - z)} \\
&= \frac{1}{\min(z, 1 - z)} \int_{[0,z]} 1 - M(x)\, dx \\
&= \begin{cases} 1 - \frac{1}{z} \int_{[0,z]} M(x)\, dx, & \text{if } z \le \frac{1}{2}, \\ \frac{1}{1-z} \int_{[z,1]} M(x)\, dx - 1, & \text{if } z > \frac{1}{2}. \end{cases}
\end{aligned}
$$

Recall that $\int_{[0,1]} M(x)\, dx = D(1) = 1$. This implies that

$$
\vartheta(0) = \lim_{z \downarrow 0} \vartheta(z) = 1 - M(0)
$$

and

$$
\vartheta(1) = \lim_{z \uparrow 1} \vartheta(z) = \lim_{z \uparrow 1} M(z) - 1.
$$

Lemma 6.4.1. *The tail dependence function $\vartheta(z)$ is continuous, monotone decreasing for $z \in (0, 1/2]$ and monotone increasing for $z \in [1/2, 1)$. Its minimum value is the tail dependence parameter $\vartheta(1/2) \ge 0$, and its maximum is $\max(\vartheta(0), \vartheta(1)) \le 1$.*

Proof. The convexity of D implies, for $\lambda, x, y \in [0, 1]$,

$$
D\Big((1 - \lambda)x + \lambda y\Big) \le (1 - \lambda)D(x) + \lambda D(y)
$$

$$
\implies 1 - D\Big((1 - \lambda)x + \lambda y\Big) \ge (1 - \lambda)\Big(1 - D(x)\Big) + \lambda\Big(1 - D(y)\Big)
$$

$$
\implies 1 - D(z_1) \ge \frac{z_1}{z_2}\Big(1 - D(z_2)\Big), \qquad 0 < z_1 < z_2 < 1,
$$

by putting $x := 0, y := z_2, \lambda := z_1/z_2$, and recalling that $D(0) = 1$. This yields

$$
\frac{1 - D(z_1)}{z_1} \ge \frac{1 - D(z_2)}{z_2}, \qquad 0 < z_1 < z_2 < 1.
$$

From the general inequality $D(z) \geq \max(z, 1-z)$, $z \in [0,1]$, we obtain further

$$1 - D(z) \leq 1 - \max(z, 1-z) = \min(z, 1-z)$$
$$\implies 1 - D(z) \leq z, \qquad z \in [0, 1/2].$$

The function $\vartheta(z) = (1 - D(z))/z$, $z \in (0, 1/2]$, is, therefore, bounded by 1 from above and it is monotone decreasing. The same arguments with $\lambda := (z_2 - z_1)/(1 - z_1)$ imply that $\vartheta(z) = (1 - D(z))/(1 - z)$, $z \in [1/2, 1)$, is bounded by 1 as well and monotone increasing. □

AN ESTIMATOR OF THE TAIL DEPENDENCE FUNCTION

Let $(U_1, V_1), \ldots, (U_n, V_n)$ be independent copies of a rv (U, V), whose df H is in the δ-neighborhood of the GPD with dependence function D,

An obvious estimator of the pertaining tail dependence function is

$$\hat{\vartheta}_{n,c}(z) := \frac{1 - \hat{D}_{n,c}(z)}{\min(z, 1-z)},$$

where

$$\hat{D}_{n,c}(z) = \frac{1}{nc} \sum_{i=1}^{n} 1(U_i > cz \text{ or } V_i > c(1-z))$$

$$= \frac{1}{c} \left(1 - \frac{1}{n} \sum_{i=1}^{n} 1(U_i \leq cz, V_i \leq c(1-z)) \right)$$

was defined in Section 4.2. The asymptotic normality of $\hat{\vartheta}_{n,c}(z)$ is now an immediate consequence of Lemma 5.5.6.

Lemma 6.4.2. *If $c = c_n < 0$ satisfies $n|c| \to \infty$, $n|c|^{1+2\delta} \to 0$ as $n \to \infty$, then we have, for $z \in (0,1)$,*

$$(n|c|)^{1/2}(\hat{\vartheta}_{n,c}(z) - \vartheta(z)) \longrightarrow_D N\left(0, \frac{D(z)}{(\min(z, 1-z))^2} \right).$$

The estimator $\hat{\vartheta}_{n,c}(z)$ is, consequently, not efficient, see Theorem 6.4.5 below and the subsequent discussion. It can, however, immediately be used for defining a goodness-of-fit test for the Marshall-Olkin distribution. In this case $\vartheta(z)$ is a constant function and, thus,

$$\hat{\vartheta}_{n,c}(z_2) - \hat{\vartheta}_{n,c}(z_1) = (\hat{\vartheta}_{n,c}(z_2) - \vartheta(z_2)) - (\hat{\vartheta}_{n,c}(z_1) - \vartheta(z_1))$$

is automatically centered, where $0 < z_1 < z_2 < 1$. The following result is a consequence of the central limit theorem for triangular arrays.

Lemma 6.4.3. *Suppose that the df H of (U,V) is in the δ-neighborhood of the Marshall-Olkin GPD. If $c = c_n < 0$ satisfies $n|c| \to \infty$, $n|c|^{1+2\delta} \to 0$ as $n \to \infty$, then we have, for $0 < z_1 < z_2 < 1$,*

$$(n|c|)^{1/2}(\hat{\vartheta}_{n,c}(z_2) - \hat{\vartheta}_{n,c}(z_1)) \longrightarrow_D N(0, \sigma^2_{z_1,z_2,\lambda}),$$

where

$$\sigma^2_{z_1,z_2,\lambda} = 2\frac{z_2 - z_1}{z_1^* z_2^*} - \left(\frac{z_2^* - z_1^*}{z_1^* z_2^*}\right)^2 + \frac{\lambda}{z_1^* z_2^*}\left(2\min(z_1, 1-z_2) - z_1^* - z_2^*\right)$$

with $z_i^* = \min(z_i, 1 - z_i)$, $i = 1, 2$.

A CHARACTERIZATION OF THE MARSHALL-OLKIN DEPENDENCE FUNCTION

The following result characterizes the canonical Marshall-Olkin dependence function. It states, precisely, that the canonical Marshall-Olkin dependence function is under some weak symmetry conditions characterized by the fact that its right derivative at $1/2$ is 0. This fact can be used, for example, to define a goodness-of-fit test for the Marshall-Olkin dependence function; we refer to Falk and Reiss [154] for details.

Theorem 6.4.4. *Suppose that the dependence function D is symmetric about $1/2$, i.e., $D(z) = D(1 - z)$, $0 \le z \le 1$. Then the derivative from above $\vartheta'(1/2) := \lim_{h \downarrow 0}(\vartheta(1/2 + h) - \vartheta(1/2))/h$ exists in $\mathbb{R}$, and we have*

$$\vartheta'(1/2) = 0$$

iff ϑ is the canonical Marshall-Olkin dependence function $\vartheta(z) = \lambda \in [0, 1]$, $z \in [0, 1]$.

Proof. Lemma 6.2.1 implies that $D(z)$ is differentiable from above for $z \in [0, 1)$ with derivative $D'(z) = M(z) - 1$ and, thus,

$$\vartheta'(z) = \frac{-D'(z)(1 - z) + 1 - D(z)}{(1 - z)^2}.$$

This yields

$$\vartheta'(1/2) = 0 \iff D'(1/2) = 2(1 - D(1/2)).$$

The assertion is now immediate from Theorem 3.3 in Falk and Reiss [154]. □

For the canonical logistic dependence function we obtain, for example, that $\vartheta'(1/2) = 4(1 - 2^{1/\lambda-1})$, which is different from 0 unless $\lambda = 1$. But the logistic dependence function with parameter $\lambda = 1$ coincides with the Marshall-Olkin dependence function with parameter $\lambda = 0$.

The following example shows that the symmetry condition on D in Theorem 6.4.4 cannot be dropped. Choose a number $p \in (1, 2)$ and put

$$x := \frac{p-1}{p} \in (0, 1/2).$$

Let ν be that measure on $[0, 1]$ which puts mass p on x and mass $2 - p$ on 1. Then we have

$$\nu([0,1]) = 2, \ \int_{[0,1]} u \, d\nu(u) = px + 2 - p = 1$$

and

$$D(z) = z \int_{[0,z]} (1-u) \, d\nu(u) + (1-z) \int_{(z,1]} u \, d\nu(u)$$

$$= \begin{cases} 1 - z, & \text{if } 0 \le z < x, \\ 2 - p + z(p-1), & \text{if } x \le z \le 1. \end{cases} \tag{6.16}$$

The tail dependence function

$$\vartheta(z) = \frac{1 - D(z)}{\min(z, 1 - z)}$$

satisfies in this case

$$\vartheta(z) = \begin{cases} 1, & \text{if } 0 \le z < x, \\ p - 1, & \text{if } 1/2 \le z \le 1. \end{cases} \cdot$$

Hence we have $\vartheta'(1/2) = 0$, but ϑ is not the canonical Marshall-Olkin dependence function. This is an example of a dependence function, which is neither symmetric about $1/2$, i.e., it does not satisfy $D(z) = D(1 - z)$, $z \in [0, 1]$, nor does $D(z)$, $z \in [0, 1]$, attain its minimum at $z = 1/2$, see Lemma 6.1.2.

LAN and Efficient Estimation of $\vartheta(z)$

Let again $(U_1, V_1), \ldots, (U_n, V_n)$ be independent copies of the rv (U, V), whose df H is in the δ-neighborhood of a GPD W.

We will establish in the sequel LAN of the loglikelihood function in a multinomial model.

Fix $z \in (0, 1)$, suppose that $\vartheta = \vartheta(z) \in (0, 1)$, and divide the quadrant $\{(x, y) : x, y < 0\}$ into the four subsets

$$S_{11} = S_{11}(t) = \left\{ (x, y) : t \frac{z}{\min(z, 1-z)} < x < 0, \ t \frac{1-z}{\min(z, 1-z)} < y < 0 \right\},$$

$$S_{12} = S_{12}(t) = \left\{ (x, y) : x \le t \frac{z}{\min(z, 1-z)}, \ t \frac{1-z}{\min(z, 1-z)} < y < 0 \right\},$$

$$S_{21} = S_{21}(t) = \left\{ (x, y) : t \frac{z}{\min(z, 1-z)} < x < 0, \ y \le t \frac{1-z}{\min(z, 1-z)} \right\},$$

$$S_{22} = S_{22}(t) = \left\{ (x, y) : x \le t \frac{z}{\min(z, 1-z)}, \ y \le t \frac{1-z}{\min(z, 1-z)} \right\},$$

where $t = t(n) \uparrow 0$ as $n \to \infty$. Denote by

$$n_{ij} = \sum_{m=1}^{n} 1\Big((U_m, V_m) \in S_{ij}\Big)$$

the number of observations in S_{ij}. This gives a 2×2 Table of the observations.

The vector (n_{ij}) is multinomial $B(n, (p_{ij}))$ distributed with parameters n and

$$p_{11} = P\Big(U > t\tfrac{z}{\min(z,1-z)}, V > t\tfrac{1-z}{\min(z,1-z)}\Big) = |t|\vartheta\Big(1 + O(|t|^\delta)\Big),$$

$$p_{12} = P\Big(U \le t\tfrac{z}{\min(z,1-z)}, V > t\tfrac{1-z}{\min(z,1-z)}\Big) = |t|\Big(\tfrac{1-z}{\min(z,1-z)} - \vartheta\Big)\Big(1 + O(|t|^\delta)\Big),$$

$$p_{21} = P\Big(U > t\tfrac{z}{\min(z,1-z)}, V \le t\tfrac{1-z}{\min(z,1-z)}\Big) = |t|\Big(\tfrac{z}{\min(z,1-z)} - \vartheta\Big)\Big(1 + O(|t|^\delta)\Big),$$

$$p_{22} = P\Big(U \le t\tfrac{z}{\min(z,1-z)}, V \le t\tfrac{1-z}{\min(z,1-z)}\Big) = 1 + |t|\Big(\vartheta - \tfrac{1}{\min(z,1-z)}\Big) + O(|t|^{1+\delta}).$$

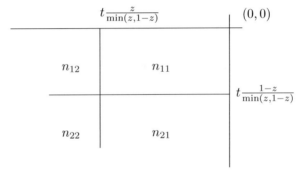

FIGURE 6.4.2. 2×2 table of the observations.

Define the alternative

$$\vartheta(n) := \vartheta + c\delta_n,$$

where $c \in \mathbb{R}$ is arbitrary and

$$\delta_n := \frac{1}{(n|t|)^{1/2}}.$$

Denote the corresponding cell probabilities under $\vartheta(n)$ by p_{ijn}. We require in the following that

$$\vartheta < \frac{\max(z, 1-z)}{\min(z, 1-z)}.$$

This is a mild condition, since we have

$$\vartheta \le \frac{\max(z, 1-z)}{\min(z, 1-z)}$$

by the general inequality $D(z) \geq \max(z, 1 - z)$ anyway. In the following result we establish LAN of the multinomial distributed vector (n_{ij}). For a proof we refer to Falk and Reiss [153].

Theorem 6.4.5. *Suppose that $t = t_n \uparrow 0$ and that $nt_n \to \infty$, $nt_n^{1+\delta} \to 0$. Then we have, for $0 < z < 1$ and $0 < \vartheta = \vartheta(z) < 1$ for the loglikelihood ratio under ϑ, the expansion*

$$L_n = \log\left(\frac{B(n, (p_{ijn}))}{B(n, (p_{ij}))} (n_{ij}) \right)$$

$$= cZ_n - \frac{c^2}{2} \frac{\frac{\max(z,1-z)}{\min(z,1-z)} - \vartheta^2}{\vartheta(1-\vartheta)\left(\frac{\max(z,1-z)}{\min(z,1-z)} - \vartheta\right)} + o_p(1)$$

with the central sequence

$$Z_n = \delta_n \left(\frac{n_{11}}{\vartheta} - \frac{n_{12}}{\frac{1-z}{\min(z,1-z)} - \vartheta} - \frac{n_{21}}{\frac{z}{\min(z,1-z)} - \vartheta} + nt \right)$$

$$\longrightarrow_D N\left(0, \frac{\frac{\max(z,1-z)}{\min(z,1-z)} - \vartheta^2}{\vartheta(1-\vartheta)\left(\frac{\max(z,1-z)}{\min(z,1-z)} - \vartheta\right)} \right).$$

The above result together with the Hájek-LeCam convolution theorem provides us with the asymptotically minimum variance within the class of regular estimators of ϑ. This class of estimators $\tilde{\vartheta}_n$ is defined by the property that they are asymptotically unbiased under $\vartheta_n = \vartheta + c\delta_n$ for any $c \in \mathbb{R}$, precisely

$$\delta_n^{-1}(\tilde{\vartheta}_n - \vartheta_n) \longrightarrow_{D_{\vartheta_n}} Q_\vartheta,$$

where the limit distribution Q_ϑ does not depend on c; see Section 8.4 and 8.5 in Pfanzagl [367] for details.

An efficient estimator of ϑ within the class of regular estimators of ϑ has necessarily the minimum limiting variance

$$\sigma^2_{\text{minimum}} := \vartheta(1-\vartheta) \frac{\frac{\max(z,1-z)}{\min(z,1-z)} - \vartheta}{\frac{\max(z,1-z)}{\min(z,1-z)} - \vartheta^2},$$

which is the inverse of the limiting variance of the central sequence.

Put

$$\hat{\vartheta}_n(z) := \hat{\vartheta}_n := \begin{cases} \frac{n_{11}}{n_{11}+n_{21}}, & \text{if } z \leq \frac{1}{2}, \\[2mm] \frac{n_{11}}{n_{11}+n_{12}}, & \text{if } z > \frac{1}{2}. \end{cases}$$

For $z = 1/2$, the estimator $\hat{\vartheta}_n(1/2)$ coincides with the estimator of the tail dependence parameter $\vartheta(1/2) = 2(1 - D(1/2))$ investigated in Falk and Reiss [151].

Lemma 6.4.6. *We have, under the conditions of Theorem 6.4.5,*

$$(n|t_n|)^{1/2}(\hat\vartheta_n - \vartheta) \longrightarrow_D N\Big(0, \vartheta(1-\vartheta)\Big).$$

Proof. We have, for $z \le 1/2$,

$$(n|t_n|)^{1/2}(\hat\vartheta_n - \vartheta)$$
$$= (n|t_n|)^{1/2}\frac{n_{11} - \vartheta(n_{11}+n_{21})}{n_{11}+n_{21}}$$
$$= (n|t_n|)^{1/2}\frac{(n_{11}-\vartheta n|t_n|)(1-\vartheta) - (n_{21}-(1-\vartheta)n|t_n|)\vartheta}{n_{11}+n_{21}}$$
$$= (n|t_n|)^{-1/2}\Big((n_{11}-\vartheta n|t_n|)(1-\vartheta) - (n_{21}-(1-\vartheta)n|t_n|)\vartheta\Big) + o_p(1)$$
$$\to_D N\Big(0, \vartheta(1-\vartheta)\Big)$$

by the central limit theorem. The case $z > 1/2$ is shown in complete analogy. $\square$

The estimator $\hat\vartheta_n$ does not have, therefore, asymptotically minimum variance. The modification

$$\hat\vartheta_{n,\text{eff}}(z) := \hat\vartheta_{n,\text{eff}}$$
$$:= \hat\vartheta_n + \frac{\hat\vartheta_n(1-\hat\vartheta_n)}{\frac{\max(z,1-z)}{\min(z,1-z)} - \hat\vartheta_n^2} \times \begin{cases} \frac{1-z}{z} - \hat\vartheta_n - \frac{n_{12}}{n|t|}, & \text{if } z \le \frac12, \\ \frac{z}{1-z} - \hat\vartheta_n - \frac{n_{21}}{n|t|}, & \text{if } z > \frac12, \end{cases}$$

however, satisfies

$$\delta_n^{-1}\big(\hat\vartheta_{n,\text{eff}} - \vartheta\big) = \sigma^2_{\text{minimum}} Z_n + o_p(1) \longrightarrow_D N\big(0, \sigma^2_{\text{minimum}}\big),$$

which follows from elementary computations. Note that $\hat\vartheta_{n,\text{eff}}(1/2)$ is the efficient estimator of the tail dependence parameter defined in Falk and Reiss [151]. The modified estimator is by the Hájek-LeCam convolution theorem an efficient estimator in the class of regular estimators with the rate δ_n, see Sections 8.4 and 8.5 in the book by Pfanzagl [367].

6.5 Testing Tail Dependence against Residual Tail Dependence

Effects of mis-specification such as modelling dependent data by independent random variables are described by Dupuis and Tawn [116]. Following Ledford and Tawn [311], they come up with the conclusion

> *It would seem appropriate to test for significant evidence against asymptotic dependence before proceeding with an asymptotic independent model.*

Testing whether tail independence holds is, therefore, mandatory in a data analysis of extreme values. More precisely, we will test tail dependence against residual tail dependence where the latter concept may be regarded as a refinement of tail independence, cf. p. 262.

Testing the Tail Dependence under Differentiable Spectral Expansions

Let (X, Y) be a rv with values in $(-\infty, 0]^2$ whose df $H(x, y)$ satisfies a differentiable spectral expansion of length $k + 1$, i.e.,

$$h_z(c) = D(z) + \sum_{j=1}^{k} B_j(c)A_j(z) + o(B_k(c)), \quad c \uparrow 0, \qquad (6.17)$$

uniformly for $z \in [0, 1]$, cf. (6.5). Then according to Section 6.1, $H(x, y)$ is in the bivariate domain of attraction of the EVD G with Pickands dependence function D. Furthermore, X and Y are tail independent iff $D = 1$.

Let now $(X_1, Y_1), \ldots, (X_n, Y_n)$ be independent copies of (X, Y). If diagnostic checks of $(X_1, Y_1), \ldots, (X_n, Y_n)$ suggest X, Y to be independent in their upper tail region, modelling with dependencies leads to the overestimation of probabilities of extreme joint events. Some inference problems caused by model mis-specification are, for example, exploited by Dupuis and Tawn [116].

Because we want to prove that tail independence holds, we put tail dependence into the null hypothesis—thereby following the advice by J. Pfanzagl (translated from German), see [365], p. 95.

As null hypothesis select the opposite of what you want to prove and try to reject the null hypothesis.

Thus, our first aim is to test

$$\mathrm{H}_0: \ D(z) \neq 1 \quad \text{against} \quad \mathrm{H}_1: \ D(z) = 1.$$

It is an obvious idea to test the tail dependence by estimating the dependence function D and to test for $D \neq 1$ or $D = 1$. The latter approach was carried out by Deheuvels and Martynov [105]. A similar approach was suggested by Capéraà et al. [59]; see also the discussion in Section 3.2.1 of Kotz and Nadarajah [293].

In the sequel the testing will be based on the random distance $C = X + Y$. We will establish the fact that the conditional distribution of $(X + Y)/c$, given $X + Y > c$, has limiting df $F(t) = t^{1+\beta}$, $t \in [0, 1]$, as $c \uparrow 0$ for a certain parameter $\beta > 0$ iff $D(z) = 1$, i.e., iff X and Y are tail independent. If D is not the constant function 1, then the limiting df is that of the uniform distribution on $[0, 1]$, namely, $F(t) = t$, $t \in [0, 1]$. Therefore the testing is reduced to

$$\mathrm{H}_0: F_0(t) = t \quad \text{against} \quad \mathrm{H}_1: F_\beta(t) = t^{1+\beta}, \ \beta > 0$$

which means that we are testing a simple null hypothesis against a composite alternative, cf. Frick et al. [161]. The null hypothesis where $D \neq 1$ or $\beta = 0$ represents tail dependence while the alternative with $D = 1$ and $\beta > 0$ stands for tail independence. The parameter $\beta > 0$ is a particular exponent of variation of the underlying spectral expansion (6.17). If condition (6.8) is satisfied by the function A_1, then β can be chosen as β_1 belonging to the first regularly varying function B_1 in the expansion, cf. (6.18) in Corollary 6.5.1.

Since $\beta_1 > 0$ measures the residual tail dependence, cf. (6.10), one can also speak of testing tail dependence against residual tail dependence. The case $\beta = 1$, which was presented in the second edition of this book, still holds for EVDs and can be regarded as a special case.

The above result will be utilized to define a test on tail dependence of X and Y which is suggested by the Neyman-Pearson lemma. As the Neyman-Pearson test at the level α does not depend on the parameter β, we will get a uniformly most powerful test. The test will be applied to the exceedances $X_i + Y_i > c$ among the sample $(X_1, Y_1), \ldots, (X_n, Y_n)$. To make the test procedure applicable, the data have to be transformed to the left lower quadrant first. This transformation is achieved by means of the marginal empirical dfs, cf. Ledford and Tawn [311], see also Coles et al. [74] and Reiss and Thomas [390], p. 331. The type I and II error rates will be investigated by various simulations.

Concerning goodness-of-fit tests for the case $\beta = 1$ that are based on Fisher's κ, on the Kolmogorov-Smirnov test as well as on the chi-square goodness-of-fit test we again refer to the second edition of this book.

A Conditional Distribution of the Distance $C = X + Y$

The following auxiliary result is actually one of the main results of the present considerations.

Corollary 6.5.1. *Assume that (X, Y) is a random vector with df H which satisfies the conditions (6.4)-(6.7), hence satisfying a differentiable spectral expansion*

$$h_z(c) = D(z) + \sum_{j=1}^{k} B_j(c) A_j(z) + o(B_k(c)), \quad c \uparrow 0,$$

uniformly for $z \in [0, 1]$ and some $k \in \mathbb{N}$ with Pickands dependence function D.

  (i) *(Tail Dependence) If $D(z) \neq 1$, we have*

$$P(X + Y > ct \mid X + Y > c) \longrightarrow t, \quad c \uparrow 0,$$

  uniformly for $t \in [0, 1]$.

 (ii) *(Residual Tail Dependence) If $D(z) = 1$, we have*

$$P(X + Y > ct \mid X + Y > c) \to t^{1 + \beta_\kappa}, \quad c \uparrow 0,$$

  uniformly for $t \in [0, 1]$ provided that

$$\kappa = \min\left\{ j \in \{1, \ldots, k\} : (2 + \beta_j) \int_0^1 A_j(z)\, dz - A_j(0) - A_j(1) \neq 0 \right\}$$
$$(6.18)$$

  exists.

Proof. From Lemma 6.3.2 we obtain, for c close to 0,

$$P(X + Y > ct \mid X + Y > c)$$

$$= \frac{\int_{ct}^{0} h_1(x)\, dx + \int_0^1 cth_z(ct) + \tilde{h}_z(ct)(1-z)\, dz}{\int_c^0 h_1(x)\, dx + \int_0^1 ch_z(c) + \tilde{h}_z(c)(1-z)\, dz} := \frac{I}{II}, \qquad (6.19)$$

where

$$I := \int_{ct}^{0} 1 + \sum_{j=1}^{k} B_j(x)A_j(1) + o(B_k(x))\, dx$$

$$+ \int_0^1 ct\left(D(z) + \sum_{j=1}^{k} B_j(ct)A_j(z) + o(B_k(ct)) \right) dz + \int_0^1 \tilde{h}_z(ct)(1-z)\, dz,$$

$$II := \int_c^0 1 + \sum_{j=1}^{k} B_j(x)A_j(1) + o(B_k(x))\, dx$$

$$+ \int_0^1 c\left(D(z) + \sum_{j=1}^{k} B_j(c)A_j(z) + o(B_k(c)) \right) dz + \int_0^1 \tilde{h}_z(c)(1-z)\, dz.$$

Using partial integration we obtain

$$\int_0^1 \tilde{h}_z(c)(1-z)\, dz$$

$$= \int_c^0 h_0(x)\, dx - \int_0^1 \int_c^0 h_z(x)\, dx\, dz$$

$$= \int_c^0 1 + \sum_{j=1}^{k} B_j(x)A_j(0) + o(B_k(x))\, dx$$

$$- \int_0^1 \int_c^0 D(z) + \sum_{j=1}^{k} B_j(x)A_j(z) + o(B_k(x))\, dx\, dz$$

$$= -c + \sum_{j=1}^{k} A_j(0) \int_c^0 B_j(x)\, dx + o\left(\int_c^0 B_k(x)\, dx \right)$$

$$- \int_0^1 -cD(z) + \sum_{j=1}^{k} \int_c^0 B_j(x)\, dx\, A_j(z) + o\left(\int_c^0 B_k(x)\, dx \right) dz$$

$$= c\left(\int_0^1 D(z)\, dz - 1 \right) - \sum_{j=1}^{k} \int_c^0 B_j(x)\, dx \left(\int_0^1 A_j(z)\, dz - A_j(0) \right)$$

$$+ o\left(\int_c^0 B_k(x)\, dx \right)$$

as well as

$$\int_c^0 1 + \sum_{j=1}^k B_j(x) A_j(1) + o(B_k(x)) \, dx$$

$$= -c + \sum_{j=1}^k \int_c^0 B_j(x) \, dx \, A_j(1) + o\left(\int_c^0 B_k(x) \, dx\right)$$

and

$$\int_0^1 c(D(z) + \sum_{j=1}^k B_j(c) A_j(z) + o(B_k(c)) \, dz$$

$$= c \int_0^1 D(z) \, dz + \sum_{j=1}^k c B_j(c) \int_0^1 A_j(z) \, dz + o(c B_k(c)).$$

The same goes for ct instead of c. Substituting the above expansions in equation (6.19), we obtain with $L := \int_0^1 D(z) \, dz - 1$, $N_j := \int_0^1 A_j(z) \, dz - A_j(0) - A_j(1)$, $M_j := \int_0^1 A_j(z) \, dz$,

$$P(X + Y > ct \mid X + Y > c)$$

$$= \frac{2ctL - \sum_{j=1}^k \left(\int_{ct}^0 B_j(x) \, dx \, N_j - ct B_j(ct) M_j\right) + o\left(ct B_k(ct) + \int_{ct}^0 B_k(x) \, dx\right)}{2cL - \sum_{j=1}^k \left(\int_c^0 B_j(x) \, dx \, N_j - c B_j(c) M_j\right) + o\left(c B_k(c) + \int_c^0 B_k(x) \, dx\right)}$$

$$= \frac{2ctL + \frac{ct B_\kappa(ct)}{1+\rho}\left((2+\rho) \int_0^1 A_\kappa(z) \, dz - A_\kappa(0) - A_\kappa(1)\right) + o(ct B_\kappa(ct))}{2cL + \frac{c B_\kappa(c)}{1+\rho}\left((2+\rho) \int_0^1 A_\kappa(z) \, dz - A_\kappa(0) - A_\kappa(1)\right) + o(c B_\kappa(c))}, \quad (6.20)$$

with κ as defined in (6.18). Representation (6.20) is due to Karamata's theorem about regularly varying functions, which implies $\int_c^0 B_j(t) \, dt \sim -\frac{1}{1+\beta_j} B_j(c) c$ as $c \uparrow 0$ for $j = 1, \ldots, k$. Finally, applying the conditions (6.6) and (6.7) to the regulary varying function B_κ yields the desired assertions.   □

We remark that the existence of κ in (6.18) is actually not a strong condition. Provided that B_1 in the spectral expansion is absolutely continuous and has a monotone derivative we have in general

$$A_1(z) \geq A_1(1) z^{1+\beta_1} + A_1(0)(1-z)^{1+\beta_1}, \quad z \in [0,1],$$

and, hence, $(2+\beta_1) \int_0^1 A_1(z) \, dz - A_1(0) - A_1(1) \geq 0$ anyway: If $D(z) = 1$, $z \in [0,1]$, we have

$$0 \leq \lim_{c \uparrow 0} \frac{P(Y > c \mid X > tc)}{B_1(c)} \tag{6.21}$$

$$= \lim_{c \uparrow 0} \frac{1 - H_1(tc) - H_0(c) + H_{t/(t+1)}(c(t+1))}{B(c)(1 - H_1(tc))}$$

$$= \frac{A_1\left(\frac{t}{t+1}\right)(t+1)^{1+\beta_1} - A_1(0) - A_1(1)t^{1+\beta_1}}{t(1+\beta_1)}$$

for arbitrary $t \in (0,\infty)$, and, hence,

$$A\left(\frac{t}{t+1}\right) \geq A(0)\frac{1}{(t+1)^{1+\beta_1}} + A(1)\left(\frac{t}{t+1}\right)^{1+\beta_1}.$$

Putting $z = t/(t+1)$, we obtain $(2+\beta_1)\int_0^1 A_1(z)\,dz - A_1(0) - A_1(1) \geq 0$, $z \in [0,1]$, cf. Theorem 2 in Frick at al. [161].

From (6.21) we conclude that $\kappa > 1$ if $P(Y > c \mid X > tc)$ converges faster to 0 than $B_1(c)$. With regard to the definition (6.2) of the tail dependence parameter, which equals 0 if $D = 1$, one can say that the tail independence is rather strong in this case.

The parameter κ does not exist for a GPD with $A_j(z) = 0$, $j \geq 1$. In that case we have $P(U + V \geq c) = 0$ for c close to 0 iff $D(z) = 1$, i.e., iff U and V are tail independent. If they are not tail independent, then Corollary 6.5.1 (i) becomes applicable. Testing for tail dependence of U, V in case of an upper GPD tail is, therefore, equivalent to testing for $P(U + V \geq c) > 0$ for some $c < 0$.

If the df H in Corollary 6.5.1 is an EVD with Pickands dependence function D, the pertaining spectral density satisfies the expansion

$$h_z(c) = D(z) + cD(z)^2 + o(c), \quad c \uparrow 0,$$

cf. (6.11). Obviously, $A(z) = A_1(z) = D(z)^2$ and $\beta = \beta_1 = 1$. If $D(z) = 1$, we have $(2+\beta_1)\int_0^1 A_1(z)\,dz - A_1(0) - A_1(1) = 1$. Therefore the parameter β_κ in Corollary 6.5.1(ii) is given by $\beta_\kappa = 1$ and the conditional limiting df of the radial component is $F(t) = t^2$.

This result for EVDs can also be proved directly as shown by the following lemma. See also the special result at the end of Section 6.3.

Lemma 6.5.2. We have, uniformly for $t \in [0,1]$ as $c \uparrow 0$,

$$P(X+Y > ct \mid X+Y > c) = \begin{cases} t^2(1+O(c)), & \text{if } D(z) = 1, z \in [0,1], \\ t(1+O(c)), & \text{otherwise.} \end{cases}$$

Proof. From Lemma 6.3.1 and the Taylor expansion of exp we obtain, uniformly for $t \in [0,1]$ and c close to 0,

$$P(X+Y > ct \mid X+Y > c)$$
$$= \frac{1 - \exp(ct) + ct\int_{[0,1]}\exp(ctD(u))(D(u) + D'(u)(1-u))\,du}{1 - \exp(c) + c\int_{[0,1]}\exp(cD(u))(D(u) + D'(u)(1-u))\,du}$$
$$= \frac{-ct + ct\int_{[0,1]} D(u) + D'(u)(1-u)\,du + O((ct)^2)}{-c + c\int_{[0,1]} D(u) + D'(u)(1-u)\,du + O(c^2)}$$

$$= t(1 + O(c))$$

if D is not the constant function 1. This follows from partial integration:

$$\int_{[0,1]} D(u) + D'(u)(1-u)\, du = 2 \int_{[0,1]} D(u)\, du - 1 \quad \in (0,1]$$

and the facts that $D(z) \in [1/2, 1]$, $D(0) = 1$.

If $D(z)$ is the constant function 1, then we obtain, uniformly for $t \in [0,1]$ and c close to 0,

$$
\begin{aligned}
P(X + Y > ct \mid X + Y > c) &= \frac{1 - \exp(ct) + ct\exp(ct)}{1 - \exp(c) + c\exp(c)} \\
&= \frac{-ct - (ct)^2/2 + ct(1+ct) + O((ct)^3)}{-c - c^2/2 + c(1+c) + O(c^3)} \\
&= t^2(1 + O(c)).
\end{aligned}
$$
$\qquad\square$

TEST STATISTIC BASED ON THE DISTANCE C

Suppose that we have n independent copies $(X_1, Y_1), \ldots, (X_n, Y_n)$ of (X, Y). Fix $c < 0$ and consider only those observations $X_i + Y_i$ among the sample with $X_i + Y_i > c$. Denote these by $C_1, C_2, \ldots, C_{K(n)}$ in the order of their outcome. Then C_i/c, $i = 1, 2, \ldots$ are iid with common df $F_c(t) := P(X + Y > ct \mid X + Y > c)$, $t \in [0, 1]$, and they are independent of $K(n)$, which is binomial $B(n, q)$-distributed with $q = 1 - (1 - c)\exp(c)$ if c is close to zero and D is the constant function 1. This is a consequence of Theorem 1.3.1. We will now consider the Neyman-Pearson test.

We have to decide, roughly, whether the df of $V_i := C_i/c$, $i = 1, 2, \ldots$ is either equal to the null hypothesis $F_0(t) = t$ or the alternative $F_\beta(t) = t^{1+\beta}$, $0 \le t \le 1$. Assuming that these approximations of the df of $V_i := C_i/c$ were exact and that $K(n) = m > 0$, we choose the test statistic $\sum_{i=1}^m \log V_i$. Under F_0 it is distributed according to the gamma df

$$H_m(t) = \exp(t) \sum_{i=0}^{m-1} \frac{(-t)^i}{i!}, \quad t \le 0,$$

on the negative half-line with parameter m. The Neyman-Pearson test at the level α is then given by

$$C_{m,\alpha} = \left\{ \sum_{i=1}^m \log V_i > H_m^{-1}(1-\alpha) \right\}, \tag{6.22}$$

i.e., the null hypothesis H_0 is rejected if the test statistic exceeds the $(1 - \alpha)$-quantile of the gamma df. The power function for the level-α test is

$$\psi_{m,\alpha}(\beta) = 1 - H_m\Big((1+\beta) H_m^{-1}(1-\beta) \Big), \quad \beta \ge 0,$$

and can be approximated by

$$\psi_{m,\alpha}(\beta) \approx 1 - \Phi((1+\beta)\Phi^{-1}(1-\alpha) - \beta m^{1/2}), \quad \beta \geq 0,$$

for large m by the central limit theorem, where Φ is the standard normal df.
Similarly, the p-value of the optimal test is given by

$$p = 1 - \exp\left(\sum_{i=1}^{m} \log V_i\right) \sum_{j=0}^{m-1} \frac{(-\sum_{i=1}^{m} \log V_i)^j}{j!} \approx \Phi\left(-\frac{\sum_{i=1}^{m} \log V_i + m}{m^{1/2}}\right).$$

For a discussion of the previous testing problem see also Frick et al. [161].

SIMULATIONS OF p-VALUES

The following figures exemplify numerous simulations that we did to evaluate the
performance of the Neyman-Pearson test for tail dependence against tail independence defined above. Figure 6.5.1 shows quantile plots of 100 independent realizations of the p-value, based on $K(n) = m = 25$ exceedances over the particular
threshold under the hypothesis H_0 of tail dependence of X and Y.

The 100 p-values were ordered, i.e., $p_{1:100} \leq \cdots \leq p_{100:100}$, and the points
$(i/101, p_{i:100})$, $1 \leq i \leq 100$, were plotted.

The underlying df is the logistic df with parameter $\lambda = 1.5$ and $\lambda = 2.5$ (see
Example 4.3.5).

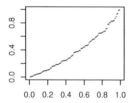 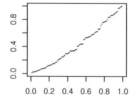 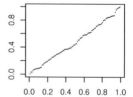

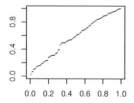 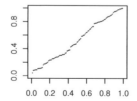 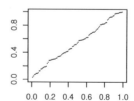

FIGURE 6.5.1. Quantile plots of 100 values with underlying logistic df with
$\lambda = 1.5$ (above) and $\lambda = 2.5$ (below) and 25 exceedances over the thresholds
$c = -0.45$ (left), $c = -0.35$ (middle), and $c = -0.1$ (right).

The three almost straight lines formed by the quantile plots in the lower part of Figure 6.5.1 (where $\lambda = 2.5$) visualize that the correct type I error rate is achieved for each of the chosen thresholds. Therefore, the distribution of the Neyman-Pearson test is not affected by the threshold if the tail dependence is sufficiently strong. In case of a weaker tail dependence ($\lambda = 1.5$) the upper part of Figure 6.5.1 shows that the distribution is slightly affected by too small thresholds, cf. also Frick et al. [161].

Next we simulate deviations from the tail dependence and consider (X, Y) having a standard normal df with correlation $\rho \in (0, 1)$. The plots in Figure 6.5.2 and 6.5.3 were generated in the same way as in Figure 6.5.1. Figure 6.5.2 visualizes tests of tail dependence against residual tail dependence with underlying standard normal df with correlation $\rho = 0.5$.

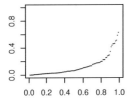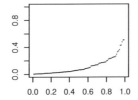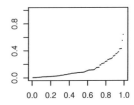

FIGURE 6.5.2. Quantile plots of 100 p-values with underlying standard normal df with $\varrho = 0.5$ and 25 exceedances over the thresholds $c = -0.45$ (left), $c = -0.35$ (middle), $c = -0.1$ (right).

It turns out that the distribution of the p-value is now shifted to the left under tail independence, i.e., the type II error rate is quite small.

Figure 6.5.3 shows how the type II error rate is influenced by the size of the correlation coefficient ρ. If ρ is close to 1, the parameter β of the underlying differentiable spectral expansion is small, cf. (6.12), meaning that we are close to the null hypothesis. In this case the quantile plot of the p-values almost reaches a straight line, i.e., the type II error rate is rather high and the test is likely to fail, see also Frick et al. [161].

Testing Tail Dependence in Arbitrary Dimension

Next we extend the previous results for bivariate dfs to those in arbitrary dimension d. In the bivariate case we have seen that the conditional distribution of $(X + Y)/c$, given $X + Y > c$, has limiting df $F(t) = t$ or $F(t) = t^{1+\beta}$, if $D \neq 1$ or $D = 1$, respectively, provided that the df H of the rv (X, Y) satisfies a spectral expansions with leading term $D(z)$. In the multivariate case spectral expansions are no longer suitable, therefore we use expansions of Pickands densities where the leading term

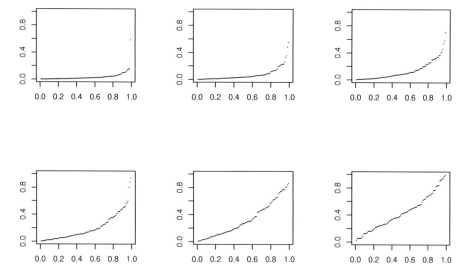

FIGURE 6.5.3. Quantile plots of 100 p-values with underlying standard normal df with $\rho = 0.2$ (above left), $\rho = 0.35$ (above middle), $\rho = 0.5$ (above right), $\rho = 0.65$ (below left), $\rho = 0.8$ (below middle), $\rho = 0.95$ (below right) and 25 exceedances over the threshold $c = -0.4$.

is the Pickands density φ of d-variate GPD W with Pickands dependence function D. According to (5.40) the above conditions are equivalent to $\int_0^1 \varphi(z)\, dz > 0$ and $\int_0^1 \varphi(z)\, dz = 0$, respectively, in the bivariate case. This result will now be generalized to arbitrary dimensions. For a proof of the following lemma see Frick and Reiss [162].

Lemma 6.5.3. *Assume that the random vector* $\mathbf{X} = (X_1, \ldots, X_d)$ *has a Pickands density which satisfies the conditions* (5.42)-(5.44), *where* φ *is the Pickands density of a* d-*variate GPD with Pickands dependence function* D.

(i) *If* $\int_R \varphi(\mathbf{z})\, d\mathbf{z} > 0$, *then*

$$P\left(\sum_{i \leq d} X_i > ct \mid \sum_{i \leq d} X_i > c\right) \to t, \quad c \uparrow 0,$$

uniformly for $t \in [0, 1]$.

(ii) *If* $\int_R \varphi(\mathbf{z})\, d\mathbf{z} = 0$ *and* (5.44) *holds with* $0 < \beta_1 < \beta_2 < \cdots < \beta_k$, *then*

$$P\left(\sum_{i \leq d} X_i > ct \mid \sum_{i \leq d} X_i > c\right) \to t^{1+\beta_\kappa}, \quad c \uparrow 0,$$

uniformly for $t \in [0,1]$ provided that

$$\kappa = \min \left\{ j \in \{1, \ldots, k\} : \int_R \tilde{A}_j(\mathbf{z})\, d\mathbf{z} \neq 0 \right\}$$

exists.

Note that in the bivariate case the parameter κ is the same as in (6.18) if the functions A_j of the spectral expansion are twice continuously differentiable and

$$\tilde{A}_j(z) = -\beta_j A_j(z) - \frac{\beta_j}{1+\beta_j} A'_j(z)(1-2z) + \frac{1}{1+\beta_j} A''_j(z)z(1-z),$$

$j = 1, \ldots, k$. For in this case we have

$$\int_0^1 \tilde{A}_j(z)\, dz = -(2+\beta_j) \int_0^1 A_j(z)\, dz + A_j(0) + A_j(1),$$

$j = 1, \ldots, k$, cf. Lemma 6.1.3(i).

From Lemma 5.6.3 we know that $\int_R \varphi(\mathbf{z})\, d\mathbf{z} > 0$ implies $D \neq 1$. Therefore, part (i) of Lemma 6.5.3 stands for multivariate tail dependence. On the other hand, $\int_R \varphi(\mathbf{z})\, d\mathbf{z} = 0$ is equivalent to $D_{rs} = 1$ for at least one pair $r, s \in \{1, \ldots, d\}$, where D_{rs} is the bivariate pairwise Pickands dependence function, defined in (5.39). Hence, Lemma 6.5.3 (ii) represents tail independence in at least one bivariate marginal distribution. For a Pickands dependence function D satisfying the symmetry condition (4.34) we have that $\int_R \varphi(\mathbf{z})\, d\mathbf{z} = 0$ is equivalent to $D = 1$, i.e., to multivariate tail independence. In this case we can directly compute the Pickands density of an EVD G with reverse exponential margins:

$$\begin{aligned}
f(\mathbf{z}, c) &= |c|^{d-1} \left(\frac{\partial^d}{\partial x_1 \ldots \partial x_d} \exp(x_1 + \cdots + x_d) \right) (T^{-1}(\mathbf{z}, c)) \\
&= |c|^{d-1} \exp(c) \\
&= |c|^{d-1} + o(|c|^{d-1}), \quad c \uparrow 0.
\end{aligned}$$

Therefore, we have $\rho_\kappa = d - 1$ and the conditional limiting distribution of the radial component $\sum_{i \le d} X_i$ is given by $F(t) = t^d$, $t \in [0,1]$, in conformity with Lemma 6.5.3 of the second edition of this book, cf. also Example 2 in Frick and Reiss [162].

The result of Lemma 6.5.3 leads to the same testing problem as before. By analogy with the bivariate case we consider the observations $C_i = \sum_{k \le d} X_k^{(i)}$, $1 \le i \le K(n)$, from a sample $(X_1^{(j)}, \ldots, X_d^{(j)})$, $1 \le j \le n$, where $\sum_{k \le d} X_k^{(i)} > c$. Conditional on $K(n) = m$, the optimal test suggested by the Neyman-Pearson lemma for testing

$$H_0 : F_0(t) = t \quad \text{against} \quad H_1 : F_\beta(t) = t^{1+\beta}, \, \beta > 0$$

based on $V_i = C_i/c$, $1 \le i \le m$, is again given by (6.22) and the power function as well as the p-value remain the same.

Notice, though, that the alternative of the testing problem in the multivariate framework has to be interpreted differently unless D satisfies the symmetry condition (4.34). As we have seen above, the null hypothesis stands for multivariate tail dependence and a rejection means that there is significance for tail independence in at least one bivariate marginal distribution. In this case one can proceed with an intersection-union test by testing each bivariate marginal distribution on tail dependence to find out whether there is significance for multivariate tail independence, see Frick and Reiss [162].

Finally, we consider the situation where we have univariate margins that are not necessarily reverse exponential. Let $G_{\alpha_1,\dots,\alpha_d}$ be an EVD whose i-th marginal G_i is an arbitrary standard EVD,

$$G_i(x) = \exp(\psi_{\alpha_i}(x)), \qquad 1 \le i \le d,$$

where

$$\psi_\alpha(x) := \begin{cases} -(-x)^\alpha, & x < 0, \text{ if } \alpha > 0, \\ -x^\alpha, & x > 0, \text{ if } \alpha < 0, \\ -\exp(-x), & x \in \mathbb{R}, \text{ if } \alpha = 0, \end{cases}$$

defining, thus, the family of (reverse) Weibull, Fréchet and the Gumbel distribution $\exp(\psi_\alpha(x))$. Up to a location or scale shift, $G_{\alpha_1,\dots,\alpha_d}$ is the family of possible d-dimensional EVD.

Note that

$$G_{\alpha_1,\dots,\alpha_d}\left(\psi_{\alpha_1}^{-1}(x_1),\dots,\psi_{\alpha_d}^{-1}(x_d)\right) = G_{1,\dots,1}(x_1,\dots,x_d), \quad x_i < 0, \ 1 \le i \le d,$$

where $G_{1,\dots,1} = G$ has reverse exponential margins.

If the df of the rv $(X_1,\dots,X_d)$ coincides in its upper tail with $G_{\alpha_1,\dots,\alpha_d}$, then the df of $(\psi_{\alpha_1}(X_1),\dots,\psi_{\alpha_d}(X_d))$ coincides ultimately with G. We can test, therefore, for tail dependence of $(X_1,\dots,X_d)$ by applying the preceding results to $\sum_{i\le d} \psi_{\alpha_i}(X_i)$ in place of $\sum_{i\le d} X_i$.

6.6 Estimation of the Angular Density in Bivariate Generalized Pareto Models

We will now investigate a non-parametric estimation method in bivariate GP models. In many applications it is of great importance to have a good insight into the tail dependence structure of a given data set. Estimating the angular density for that purpose is also popular in extreme value models, see for example Coles and Tawn [72], [73] or Coles et al. [74].

We mainly focus on the bivariate case in this section, since it has special properties which make the estimation easier. However we will also give a brief discussion of the general multivariate case at the end of this section.

The Bivariate Angular Density

Recall that the bivariate Pickands dependence function is defined by

$$D(t) = \int_{[0,1]} \max\left(ut, (1-u)(1-t)\right) \nu(du), \tag{6.23}$$

where ν is a measure on $[0,1]$ with

$$\nu\left([0,1]\right) = d \quad \text{and} \quad \int_{[0,1]} u\nu(du) = 1, \tag{6.24}$$

see Section 4.3. ν is called the *angular measure*. As we have seen in Theorem 4.3.1, the characterization of this measure by (6.24) is necessary and sufficient to define a Pickands dependence function,

Also recall from the beginning of Section 5.9 that the df $L(z) = \nu\left([0,z]\right)$ of the measure ν is called *angular distribution*. If the measure ν, restricted to $(0,1)$, possesses a density we denote it by l and call it the *angular density*. The restriction to the interior of $[0,1]$ helps us here to avoid certain special cases. We will see in this section that under certain regularity conditions, the angular component of the Pickands coordinates of GPD distributed rv follow a suitably scaled angular distribution, thus our choice of the name. In the literature it is also common to call the angular measure/distribution/density the *spectral* measure/distribution/density.

We have seen in Theorem 5.6.2 that Pickands coordinates are important in GPD models, since they decompose GP distributed rv into two conditionally independent components, given that the radial component exceeds some high value. The distribution of the radial component is then the uniform distribution, the angular component has the density $f(\mathbf{z}) := \varphi(\mathbf{z})/\zeta$, where φ is the Pickands density and its integral is again denoted by $\zeta := \int_{(0,1)} \varphi(\mathbf{z})\,dz$.

In the next theorem we compute the angular density for smooth GPD. This result will be crucial in what follows. A proof is given in Section 2 of Michel [334]. For a general multivariate version of this result we refer to Michel [330], Theorem 2.2.4.

Theorem 6.6.1. *Let the GPD W have continuous partial derivatives of order 2. Then the corresponding angular density l is given by*

$$l\left(\frac{\frac{1}{x_1}}{\frac{1}{x_1} + \frac{1}{x_2}}\right) = \frac{x_1^2 x_2^2}{\left(-\frac{1}{x_1} - \frac{1}{x_2}\right)^{-3}} \frac{\partial^d}{\partial x_1 \cdots \partial x_d} W(x_1, x_2).$$

We briefly illustrate with the logistic family why we estimate the angular density for the investigation of the tail dependence structure. The logistic family has, according to Section 3.5.1 in Kotz and Nadarajah [293], for $1 \le \lambda < \infty$ the angular density

$$l_\lambda(z) = (\lambda - 1)z^{-\lambda-1}(1-z)^{-\lambda-1}\left(z^{-\lambda} + (1-z)^{-\lambda}\right)^{1/\lambda-2}$$

and according to Theorem 2.4 of Michel [331] the Pickands density

$$\varphi_\lambda(z) = (\lambda - 1)z^{\lambda - 1}(1 - z)^{\lambda - 1}\left(z^\lambda + (1 - z)^\lambda\right)^{1/\lambda - 2}.$$

Both reduce to 0 for $\lambda = 1$. The angular density has for miscellaneous λ the graphs from Figure 6.6.1.

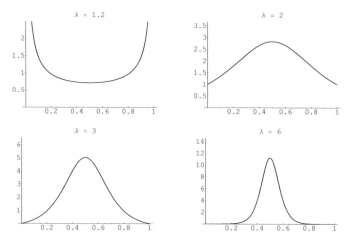

FIGURE 6.6.1. Logistic angular densities for different λ.

Graphs with the same parameters are plotted for the Pickands density in Figure 6.6.2.

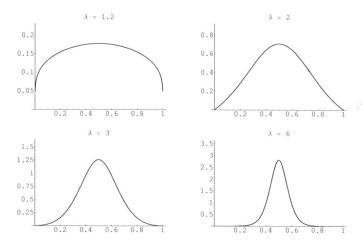

FIGURE 6.6.2. Logistic Pickands densities for different λ.

For the investigation of the tail dependence structure, the behavior of $l(z)$ especially at the boundary of $(0, 1)$, is of interest. A convergence of the angular density to ∞ at the boundary indicates a high degree of independence, convergence to 0 indicates a high degree of dependence. This follows from the fact that in the case of independence the angular measure ν has all its mass at the points 0 and 1, and in the case of complete dependence it has its complete mass at the point $1/2$, see Section 3 of Michel [332].

The tail dependence structure is not distinctly visualized in such a way by the Pickands density, since all functions have a maximum in the interior and converge to 0 at the boundary in the logistic case. However, we will see that the Pickands density will play a crucial role in these considerations.

AN ALTERNATIVE REPRESENTATION OF THE ANGULAR DENSITY

The angular and the Pickands density of a bivariate GPD have a close connection. This is the content of the following theorem.

Theorem 6.6.2. *Let (X_1, X_2) follow a bivariate GPD W, which has partial derivatives of order 2. Suppose that $\zeta = \int_{(0,1)} \varphi(\mathbf{z}) \, dz > 0$. Then we have for the angular density*

$$l(z) = \frac{\varphi(1-z)}{z(1-z)} = \frac{\zeta f(1-z)}{z(1-z)},$$

where φ is the Pickands density and f is defined as in Theorem 5.6.2.

Proof. Using Theorem 6.6.1 and inserting the inverse Pickands transformation (5.36) we get

$$l\left(\frac{\frac{1}{cz}}{\frac{1}{cz} + \frac{1}{c(1-z)}}\right) = -\frac{(cz)^2(c(1-z))^2}{\left(\frac{1}{cz} + \frac{1}{c(1-z)}\right)^{-3}} \frac{\partial^2}{\partial x_1 \partial x_2} W\left(T_P^{-1}(z, c)\right)$$

$$\Longleftrightarrow l(1-z) = (-c)\frac{z^2(1-z)^2}{(z(1-z))^3} \frac{\partial^2}{\partial x_1 \partial x_2} W\left(T_P^{-1}(z, c)\right)$$

$$\Longleftrightarrow l(1-z) = \frac{|c|}{z(1-z)} \frac{\partial^2}{\partial x_1 \partial x_2} W\left(T_P^{-1}(z, c)\right).$$

With the definition of the Pickands density φ from Theorem 5.6.2 and replacing z by $1 - z$ we finally obtain the equation

$$l(z) = \frac{\varphi(1-z)}{z(1-z)} = \frac{\zeta f(1-z)}{z(1-z)}, \qquad 0 < z < 1. \qquad \square$$

If the rv X_1, X_2 are exchangeable, i.e., if the distributions of (X_1, X_2) and (X_2, X_1) coincide, then the assertion of Theorem 6.6.2 reduces to

$$l(z) = \frac{\varphi(z)}{z(1-z)} = \frac{\zeta f(z)}{z(1-z)}.$$

ESTIMATION OF THE PICKANDS DENSITY

We have above a connection between the angular and the Pickands density. We will now first estimate the Pickands density and then use this estimator to get to our goal of estimating the angular density.

Let $(\tilde{X}_1, \tilde{X}_2) < 0$ be a bivariate rv following a GPD W. Suppose that we have n independent copies $(\tilde{X}_1^{(i)}, \tilde{X}_2^{(i)})$ of $(\tilde{X}_1, \tilde{X}_2)$, and denote by $\tilde{Z}_i := \tilde{X}_1^{(i)}/(\tilde{X}_1^{(i)} + \tilde{X}_2^{(i)})$ and $\tilde{C}_i := \tilde{X}_1^{(i)} + \tilde{X}_2^{(i)}$ the corresponding Pickands coordinates, $i = 1, \ldots, n$. Fix a threshold c close to 0, and consider only those observations $(\tilde{X}_1^{(i)}, \tilde{X}_2^{(i)})$ with $\tilde{C}_i > c$. Denote these by $(X_1^{(1)}, X_2^{(1)}), \ldots, (X_1^{(m)}, X_2^{(m)})$, where $m = K(n)$ is the random number of observations with $\tilde{C}_i > c$. From Theorem 1.3.1 we know that $K(n)$ and the $(X_1^{(j)}, X_2^{(j)})$ are all independent rv, that $K(n)$ is binomial $B(n, p)$ distributed with $p = P(\tilde{C} > c)$, and that the Z_j have the density $f(z)$ from Theorem 5.6.2.

A natural estimator of f is the kernel density estimator with kernel function k and bandwidth $h > 0$,

$$\hat{f}_m(z) := \frac{1}{mh} \sum_{i=1}^{m} k\left(\frac{z - Z_i}{h}\right), \tag{6.25}$$

where $Z_i := X_1^{(i)}/(X_1^{(i)} + X_2^{(i)})$, $i = 1, \ldots, m$.

As is known from the standard literature on kernel density estimators, the choice of a suitable bandwidth h is a crucial problem. This bandwidth is highly dependent on the density to be estimated itself. So there is the need for an automatic bandwidth selection.

Michel [334] recommends using the bandwidth

$$h = S_m \left(\frac{4}{3m}\right)^{1/5}, \tag{6.26}$$

with

$$S_m := \left(\frac{1}{m-1} \sum_{i=1}^{m} (Z_i - \bar{Z}_m)^2\right)^{1/2}, \quad \bar{Z}_m := \frac{1}{m} \sum_{i=1}^{m} Z_i,$$

the empirical standard deviation and the arithmetic mean of the Z_j. Using the empirical standard deviation for the definition of the bandwidth as done here is also known as *data sphering*.

To further improve the estimation, another recommendation is to use reflection techniques, which is done for all following estimators. Details are described in Section 4 of Michel [334].

Example 6.6.3. Taking k to be the normal kernel, we did simulations of estimator (6.25) using Algorithm 5.7.6 for the generation of rv following a logistic GPD. In Figure 6.6.3 we present some results. In each case 50 observations were simulated, which exceed the threshold $c = -0.1$. We did this for different λ; the corresponding bandwidth h, chosen according to (6.26), is also given.

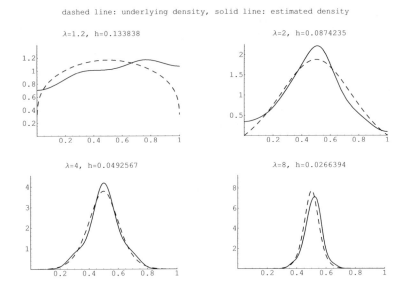

FIGURE 6.6.3. Simulations of the Pickands density estimator with different logistic parameters λ and $m = 50$, $c = -0.1$.

ESTIMATION OF THE ANGULAR DENSITY

By estimating f we are now able to estimate (a constant multiple of) the angular density and derive a graphical tool for the investigation of the tail dependence structure.

We obtain from Theorem 6.6.2 that

$$g(z) := \frac{f(1-z)}{z(1-z)} = \frac{l(z)}{\zeta}, \quad 0 < z < 1.$$

The function g, which is a constant multiple of the angular density l, determines if the underlying distribution of (X_1, X_2) is closer to the case of independence or the case of dependence. A peak of $g(z)$ near 0 and 1 indicates that our observations come from a distribution which is closer to the independence case, whereas a peak in the interior of the unit interval determines that we are closer to the dependence case.

With the ability to estimate $f(z)$ by $\hat{f}_m(z)$, we have also gained the ability to estimate $g(z)$ by

$$\hat{g}_m(z) := \frac{\hat{f}_m(1-z)}{z(1-z)}. \tag{6.27}$$

Example 6.6.4. The estimator (6.27) was simulated in Figure 6.6.4 for the logistic case with $m = 50$, $c = -0.1$ and different λ. Once again automatic bandwidth selection, data sphering and reflection techniques were included for practical purposes.

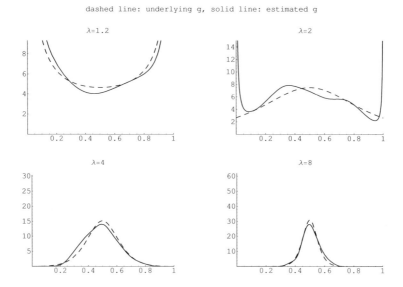

FIGURE 6.6.4. Simulations of the angular density estimator with different logistic parameters λ and $m = 50$, $c = -0.1$.

For λ noticeably smaller or larger than 2, the functions g and $\hat{g}_m$ have the same behavior at the boundary. For λ close to 2, g and its estimator $\hat{g}_m$ seem to behave differently when approaching the boundary. This is due to numerical effects coming from the division by $z(1-z)$. Convergence of $\hat{g}_m$ to 0 when approaching the boundary is a clear sign of dependence. In contrast to this, one has to be careful when $\hat{g}_m$ tends to ∞ at the boundary.

Under suitable regularity conditions the asymptotic normality of estimator (6.27) can be shown. More precisely, if $h = o\left(m^{-1/5}\right)$, then

$$(mh)^{1/2}\left(\hat{g}_m(z) - \frac{l(z)}{\zeta}\right) \to_D N\left(0, \frac{f(1-z)}{z^2(1-z)^2}\int k^2(u)\,du\right).$$

We refer to Theorem 5.2 of Michel [334] for details on the assumptions and a proof.

Note that the distribution of the angular components of the Pickands coordinates is independent of the threshold c. This could also be a tool for the graphical verification of the GPD model assumption: One could use different thresholds and compare the resulting estimators of the angular density. If the data actually follow a GPD, then all estimators should basically give the same graphic. If the graphics differ heavily, then one can have doubts about the GPD model assumption. Such considerations are also used to check the EVD approximation in threshold models by Joe et al. [276] and Coles and Tawn [73].

The Problem of Generalization to the Trivariate Case

We have seen previously that we could find a multiplicative decomposition of the angular density l into

$$l(z) = \kappa \frac{\varphi(z)}{z(1-z)},$$

in the case of exchangeability, where κ is a constant depending only on l but not on z. One can wonder if such a decomposition is also possible for the trivariate case. The natural generalization would be that

$$l(z_1, z_2) = \kappa \frac{\varphi(z_1, z_2)}{z_1 z_2 (1 - z_1 - z_2)}$$

for exchangeable models where again κ is a constant only depending on l. We will, however, see that this equation does not hold. In fact, we will see that there is no multiplicative decomposition of a differentiable l such that

$$l(z_1, z_2) = \kappa b(z_1, z_2)\varphi(z_1, z_2),$$

with κ depending only on l, and b differentiable of order 1, depending only on z_1 and z_2 but not on l.

To show this, we will use the trivariate logistic case, where the angular and the Pickands density have the representations

$$l_\lambda(z_1, z_2) = (\lambda - 1)(2\lambda - 1) (z_1 z_2)^{-\lambda-1} (1 - z_1 - z_2)^{-\lambda-1}$$
$$\times \left(z_1^{-\lambda} + z_2^{-\lambda} + (1 - z_1 - z_2)^{-\lambda} \right)^{1/\lambda - 3}$$

and

$$\varphi_\lambda(z_1, z_2) = (\lambda - 1)(2\lambda - 1) (z_1 z_2)^{\lambda-1} (1 - z_1 - z_2)^{\lambda-1}$$
$$\times \left(z_1^{\lambda} + z_2^{\lambda} + (1 - z_1 - z_2)^{\lambda} \right)^{1/\lambda - 3}$$

on $S := \{(z_1, z_2) \in (0,1)^2 : z_1 + z_2 < 1\}$ being the open unit simplex in $\mathbb{R}^2$, see Section 3.5.1 in Kotz and Nadarajah [293] and Theorem 2.4 of Michel [331].

Our goal is to disprove the equation

$$l_\lambda(z_1, z_2) = \kappa(\lambda)b(z_1, z_2)\varphi_\lambda(z_1, z_2).$$

Theorem 6.6.5. *There does not exist a function $\kappa : (1, \infty) \to \mathbb{R}$ and a differentiable function $b(z_1, z_2) : S \to \mathbb{R}$ such that for $\lambda > 1$ the decomposition*

$$l_\lambda(z_1, z_2) = \kappa(\lambda)b(z_1, z_2)\varphi_\lambda(z_1, z_2)$$

holds, where l_λ is the angular density and φ_λ is the Pickands density of a trivariate logistic GPD $W(x) = 1 - ||x||_\lambda = 1 - \left(|x_1|^\lambda + |x_2|^\lambda + |x_3|^\lambda\right)^{1/\lambda}$ with parameter λ.

Proof. Suppose that there exist functions κ and b such that

$$l_\lambda(z_1, z_2) = \kappa(\lambda)b(z_1, z_2)\varphi_\lambda(z_1, z_2), \quad (z_1, z_2) \in S.$$

We know that $\varphi_\lambda > 0$ for $\lambda > 1$. Dividing by φ_λ we get

$$\frac{l_\lambda(z_1, z_2)}{\varphi_\lambda(z_1, z_2)} = \kappa(\lambda)b(z_1, z_2).$$

Since also $l_\lambda(z_1, z_2) > 0$ for any $(z_1, z_2) \in S$, we can assume without loss of generality that $\kappa(\lambda) > 0$ and $b(z_1, z_2) > 0$. Therefore, we get

$$\log\left(\frac{l_\lambda(z_1, z_2)}{\varphi_\lambda(z_1, z_2)}\right) = \log(\kappa(\lambda)) + \log(b(z_1, z_2))$$

by taking the logarithm on both sides. Computing the partial derivative with respect to z_1 (or with respect to z_2 leading to the same results due to the exchangeability of z_1 and z_2 in the logistic case), we have

$$a(\lambda, z_1, z_2) := \frac{\partial}{\partial z_1} \log\left(\frac{l_\lambda(z_1, z_2)}{\varphi_\lambda(z_1, z_2)}\right) = \frac{\frac{\partial}{\partial z_1} b(z_1, z_2)}{b(z_1, z_2)},$$

which is constant with respect to λ. This will be our contradiction since we will show that there exist $\lambda_1, \lambda_2 > 1$ and $(z_1, z_2) \in S$ such that

$$a(\lambda_1, z_1, z_2) \neq a(\lambda_2, z_1, z_2).$$

With the above representations of l_λ and φ_λ we compute (details are left to the reader)

$$a(\lambda, z_1, z_2)$$

$$= 2\lambda\left(\frac{1}{1 - z_1 - z_2} - \frac{1}{z_1}\right) + (3\lambda - 1)\frac{z_1^{-\lambda-1} - (1 - z_1 - z_2)^{-\lambda-1}}{z_1^{-\lambda} + z_2^{-\lambda} + (1 - z_1 - z_2)^{-\lambda}}$$

$$+ (3\lambda - 1)\frac{z_1^{\lambda-1} - (1 - z_1 - z_2)^{\lambda-1}}{z_1^\lambda + z_2^\lambda + (1 - z_1 - z_2)^\lambda}.$$

By choosing $z_1 = z_2 = 1/4$ we get (details again left to the reader)

$$a\left(\lambda, \frac{1}{4}, \frac{1}{4}\right) = -4\lambda + 2(3\lambda - 1)\left(\frac{2^{\lambda+1} - 1}{2^{\lambda+1} + 1} + \frac{1 - 2^{\lambda-1}}{2^{\lambda-1} + 1}\right).$$

But since

$$a\left(2, \frac{1}{4}, \frac{1}{4}\right) = -\frac{32}{9} \neq -\frac{636}{85} = a\left(3, \frac{1}{4}, \frac{1}{4}\right),$$

we have completed the proof.   □

Therefore, the estimator of the angular density presented in this section cannot be analogously transferred to the multivariate case. The bivariate case seems to be a special case.

AN ALTERNATIVE ESTIMATION FOR THE MULTIVARIATE CASE

An alternative method of estimating the angular density in multivariate GP models is shown in Michel [330], Chapter 5. It uses the Pickands coordinates with respect to Fréchet margins, which we have introduced in (5.61).

As already mentioned in Section 5.9, it can be shown that

$$P\left(\mathbf{Z} \in B \mid C = -r, \mathbf{Z} \in Q_{r,s}\right) = \frac{1}{\chi(r,s)} \int_{B \cap Q_{r,s}} l(\mathbf{z})\, d\mathbf{z}$$

holds with $\mathbf{Z}$ and C being the random Pickands coordinates with regard to Fréchet margins, B is some Borel set in $R = \left\{\mathbf{x} \in (0,1)^{d-1} : \sum_{i \leq d-1} x_i < 1\right\}$, $Q_{r,s}$ is defined in (5.63) and $\chi(r,s)$ is defined in (5.64), which is close to a constant if r is large.

Based on this result, the intuitive approach is to use the angular component of the Pickands coordinates with regard to Fréchet margins to estimate the angular density.

Assume that we have n independent copies $\tilde{\mathbf{X}}^{(1)}, \ldots, \tilde{\mathbf{X}}^{(n)}$ of a rv $\mathbf{X}$, which follows a GPD on K_s from (5.62) with some $s > 0$. Denote by $\tilde{\mathbf{Z}}^{(i)}$ and $\tilde{C}^{(i)}$ the corresponding Pickands coordinates with regard to Fréchet margins, $i = 1, \ldots, n$. Choose a large threshold $r > 0$, and consider only those observations $\tilde{\mathbf{X}}^{(i)}$ with $\tilde{\mathbf{X}}^{(i)} \in A_{r,s}$ as defined in (5.67), i.e., $\tilde{\mathbf{X}}^{(i)} \in K_s$ and $\tilde{C}^{(i)} < -r$. We denote these by $\mathbf{X}^{(1)}, \ldots, \mathbf{X}^{(m)}$. They are independent of the binomially distributed rv $m = K(n)$, independent from each other and identically distributed, see Theorem 1.3.1. A natural estimator for the angular density l is, thus, a kernel density estimator with kernel k, bandwidth $h > 0$ and data sphering

$$\hat{l}_{m,r}(\mathbf{z}) = d\frac{1}{(\det \mathbf{S}_m)^{1/2}\, m h^{d-1}} \sum_{i=1}^{m} k\left(\frac{\mathbf{S}_m^{-1/2}\left(\mathbf{z} - \mathbf{Z}^{(i)}\right)}{h}\right). \qquad (6.28)$$

Here $\mathbf{S}_m$ denotes the covariance matrix of the $\mathbf{Z}^{(i)}$ and $\mathbf{S}_m^{-1/2}$ its inverse symmetric root. Data sphering is a concept from multivariate analysis, where the data are first multiplied by the inverse symmetric root of their covariance matrix, then the density is estimated, and afterwards transformed back. In the univariate case, this reduces to dividing the data by their empirical standard deviation as we have done for the bivariate estimator in the previous section. For more information on data sphering, see for example, Falk et al. [145], Chapter 6.

Corollary 5.6.5 of Michel [330] shows that under certain regularity conditions the estimator (6.28) is asymptotically normal with mean $(d/d^*)l(\mathbf{z})$ and variance $(d^2/d^*)l(\mathbf{z}) \int k^2(\mathbf{u}) \, d\mathbf{u}$ for r and $m \to \infty$, where d^* is as in Section 5.9 the mass of the angular measure ν in the interior of the unit simplex. Thus, the factor d in the estimator (6.28) is included to get an asymptotically unbiased estimator of $l(\mathbf{z})$ in the case $d^* = d$.

For practical applications, Michel [330], Section 5.2 recommends using the normal kernel, as bandwidth

$$h = \left(\frac{4}{m(d+1)} \right)^{1/(d+3)} \tag{6.29}$$

and reflection techniques, e.g. presented in Sections 2.1 and 8.2 of Reiss and Thomas [389].

Detailed results of numerous simulations are described in Section 5.2 of Michel [330]. A major finding is that $\hat{l}_{m,r}(\mathbf{z})$ has the same problems as the parametric estimators based on the angular density presented in Section 5.9, when we are close to the case of independence, producing a non-negligible bias. This does not come as a surprise, since both use the same approximation of l, see (5.65).

As a consequence, the estimation presented in the first half of this section should be used in the bivariate case, since it does not suffer from this bias and is more reliable. In higher dimensions, however, the estimator (6.28) is presently, to the best of our knowledge, the only existing option.

Chapter 7

Multivariate Extremes: Supplementary Concepts and Results

In this chapter we will deal with exceedances and upper order statistics (besides maxima), with the point process approach being central for these investigations. Extremes will be asymptotically represented by means of Poisson processes with intensity measures given by max-Lévy measures as introduced in Section 4.3.

In Sections 7.1 and 7.2, the approximations are formulated in terms of the variational distance defined by

$$d(\nu_0, \nu_1) := \sup_B |\nu_0(B) - \nu_1(B)|$$

for finite measures ν_0 and ν_1, where the sup ranges over the measurable sets B. We also write $d(X, Y)$ to express the variational distance between the distributions of rv X and Y, see page 7.

Thinned empirical processes, which generalize truncated processes by allowing random truncating sets, are introduced in Section 7.3. Local asymptotic normality of these processes is established under the condition that we deal with rare events, that is, the probability of thinning converges to zero.

7.1 Strong Approximation of Exceedances

We introduce different concepts of multivariate exceedances and provide Poisson approximations which hold with respect to the variational distance. We particularly continue our discussion in Section 5.1 about generalized Pareto distributions (GPD). This section is based on two articles by Kaufmann and Reiss [285], [286].

M. Falk et al., *Laws of Small Numbers: Extremes and Rare Events*, 3rd ed.,
DOI 10.1007/978-3-0348-0009-9_7, © Springer Basel AG 2011

Approximation Technique

Approximations of functionals $T(N_n)$ of the empirical process N_n (cf. (7.3)) may be carried out by means of the Poisson approximation in conjunction with the coupling inequality $d(X, Y) \leq P\{X \neq Y\}$ for rv X and Y.

We indicate in which way the limiting distribution of $T(N_n)$ can be established by using this approach. The limiting distribution $\mathcal{L}(T(N^*))$, where N^* is a Poisson process, may be computed in the following manner:

(a) **Coupling**: replace $T(N_n)$ by $T(N_{nt})$, where N_{nt} is the empirical process N_n truncated outside of a "rare event";

(b) **first-order Poisson approximation:** replace $T(N_{nt})$ by $T(N_{nt}^*)$, where N_{nt}^* is a Poisson process which has the same intensity measure as N_{nt};

(c) **second-order Poisson approximation:** replace $T(N_{nt}^*)$ by $T(N_t^*)$, where N_t^* is the truncation of a Poisson process N^*;

(d) **coupling:** replace $T(N_t^*)$ by $T(N^*)$.

As a special case let us consider the maximum functional $T = \max$. Notice that $T(N_n)$ is the usual maximum of a sample of iid rv. The limiting df G of $T(N_n)$ is the df of $\max(N^*)$, where N^* is a Poisson process with mean value function $\log(G)$. Likewise one may prove that the k-th largest value of the points of N^* has the df

$$G \sum_{0 \leq j \leq k-1} (-\log(G))^j / j!, \tag{7.1}$$

which is the limiting df of the k-th largest order statistic of n iid rv. For $T = \max$ we may as well utilize the following identity instead of the coupling, namely,

$$P\left(\max(N_t^*) \leq y\right) = P(\max(N^*) \leq y) = G(y) \tag{7.2}$$

for $y > t$. This result can easily be generalized to the k-th largest point of N_t^* and to the multivariate framework.

First-Order Poisson Approximation

Let $\boldsymbol{X}_{ni} = (X_{ni1}, \ldots, X_{nid})$, $i \leq n$, be a sample of iid d-variate rv. Let

$$N_n = \sum_{i \leq n} \varepsilon \boldsymbol{X}_{ni} \tag{7.3}$$

be the empirical process. The pertaining intensity measure is

$$\nu_n = nP(\boldsymbol{X}_{n1} \in \cdot).$$

Let N_n^* be a Poisson process with the same intensity measure ν_n as N_n and denote by $N_{n\mathbf{t}}$ and $N_{n\mathbf{t}}^*$ the truncations of N_n and N_n^* outside of $[-\infty, \mathbf{t}]^{\complement}$. Then

$$d(N_{n\mathbf{t}}, N_{n\mathbf{t}}^*) \leq \nu_n[-\infty, \mathbf{t}]^{\complement}/n.$$

A replacement of $N_{n\mathbf{t}}$ by $N_{\mathbf{t}}^*$ (second-order Poisson approximation) is generally not possible because the variational distance $d\big(\nu_n(\cdot \cap [-\infty, \mathbf{t}]^{\complement}), \nu(\cdot \cap [-\infty, \mathbf{t}]^{\complement})\big)$ can be very large; e.g. these measures have disjoint supports in the case of rv with independent components (cf. (4.12)).

A first-order Poisson approximation of the empirical process truncated outside of a set $\{(x, y) \in [0, \infty)^2 : (x^2 + y^2)^{1/2} \geq r\}$ might be of interest in view of the results by Einmahl et al. [120].

PATHWISE EXCEEDANCES

We are going to consider the following situation: A random mechanism generates the vector $\boldsymbol{x} = (x_1, \ldots, x_d)$, yet we are merely able to observe those x_j above the level t_j. We note x_j if $x_j > t_j$ and, otherwise t_j, if at least one of the components x_j of $\boldsymbol{x}$ exceeds t_j. Consequently, given $\boldsymbol{x} \in [-\infty, \mathbf{t}]^{\complement}$ put

$$m_{\mathbf{t}}(\boldsymbol{x}) := (\max(x_j, t_j))_{j \leq d},$$

and $M_{\mathbf{t}}\nu =: m_{\mathbf{t}}\nu(\cdot \cap [-\infty, \mathbf{t}]^{\complement})$ for every measure ν. As a special case of the latter definition one obtains

$$M_{\mathbf{t}}(\mu) = \sum_i 1_{[-\infty, \mathbf{t}]^{\complement}}(\boldsymbol{x}_i) \varepsilon_{m_{\mathbf{t}}(\mathbf{x}_i)} \tag{7.4}$$

for point measures $\mu = \sum_i \varepsilon_{\mathbf{x}_i}$. Thus, the missing components of $\boldsymbol{x}_i$ are replaced by the corresponding thresholds.

For non-void $K \subset \{1, \ldots, d\}$ define the projections $\pi_K(\boldsymbol{x}) := (x_k)_{k \in K}$ and the sets

$$C(K, \mathbf{t}) := \{\boldsymbol{x} \in [-\infty, \infty) : x_k > t_k, \ k \in K\}. \tag{7.5}$$

Define the map $\Pi_{K, \mathbf{t}}$ on the space of measures ν by

$$\Pi_{K, \mathbf{t}}\nu := \pi_K \nu_{C(K, \mathbf{t})}, \tag{7.6}$$

where the right-hand side is the measure induced by π_K and the measure $\nu_{C(K, \mathbf{t})}$ which is ν truncated outside of $C(K, \mathbf{t})$.

In the following theorem, the pathwise exceedance process $M_{\mathbf{t}}N_n$ will be approximated by a Poisson process $M_{\mathbf{t}}N^*$.

Theorem 7.1.1. *Let N_n be the empirical process in (7.3) and N^* a Poisson process with intensity measure ν which is the max-Lévy measure of a max-id df G.*

Then

$$d\big(M_\mathbf{t} N_n, M_\mathbf{t} N^*\big) \leq \frac{-\log(G(\mathbf{t}))}{n} + \frac{5}{2} d\big(M_\mathbf{t}\nu_n, M_\mathbf{t}\nu\big)$$

$$\leq \frac{-\log(G(\mathbf{t}))}{n} + c(d)\, A_{n\mathbf{t}},$$

where c(d) is a constant depending only on the dimension d, and

$$A_{n\mathbf{t}} = \sum_K d\big(\Pi_{K,\mathbf{t}}\nu_n, \Pi_{K,\mathbf{t}}\nu\big) \tag{7.7}$$

with K ranging over all non-void subsets of $\{1,\dots,d\}$.

Proof. By means of the first-order Poisson approximation and the monotonicity theorem we obtain

$$d\big(M_\mathbf{t} N_n, M_\mathbf{t} N_n^*\big) \leq \nu_n[-\infty,\mathbf{t}]^\complement/n,$$

where N_n^* is a Poisson process having the intensity measure ν_n of N_n. Moreover, applying formula (3.8) in Reiss [387] (which holds also for the constant 3/2 in place of 3) and the triangle inequality, we obtain

$$d\big(M_\mathbf{t} N_n, M_\mathbf{t} N^*\big) \leq \nu_n[-\infty,\mathbf{t}]^\complement/n + \frac{3}{2} d\big(M_\mathbf{t}\nu_n, M_\mathbf{t}\nu\big)$$

$$\leq \frac{-\log(G(\mathbf{t}))}{n} + \frac{5}{2} d\big(M_\mathbf{t}\nu_n, M_\mathbf{t}\nu\big),$$

because $\nu[-\infty,\mathbf{t}]^\complement = -\log(G(\mathbf{t}))$. The proof of the first inequality is complete.

The proof of the second inequality is based on a counterpart of the representation of $1-F$ in (4.7). For details see Kaufmann and Reiss [286]. □

We see that there is a direct relationship between the preceding term $A_{n\mathbf{t}}$ and the upper bound in Lemma 4.1.3(ii). The functions $nS_{n,K}$ and L_K generate the measures $\Pi_{K,\mathbf{t}}\nu_n$ and $\Pi_{K,\mathbf{t}}\nu$, respectively.

Generalized Pareto Distributions

The Poisson process N^* truncated outside of $[-\infty,\mathbf{t}]^\complement$ possesses a finite intensity measure and can, therefore, be written as $\sum_{i=1}^\tau \varepsilon_{X_i}$ with rv X_i distributed according to the GPD $Q_\mathbf{t}$ as introduced in (5.3). In addition, $M_\mathbf{t} N^*$ is a Poisson process based on rv with GPD $m_\mathbf{t} Q_\mathbf{t}$. Apparently, this may serve as another version of a GPD. This version was also studied by Tajvidi [436]; its relationship to the bivariate GPD $W = 1 + \log(G)$ was discussed in [389], 2nd ed., Section 10.1.

Vectors of Exceedances

In the following, the map $\Pi_{K,\mathbf{t}}$ defined in (7.6) will also be applied to point processes. Applying the monotonicity theorem one obtains

$$d\big(\big(\Pi_{K,\mathbf{t}}N_n\big)_K,\big(\Pi_{K,\mathbf{t}}N^*\big)_K\big) = d\big(M_{\mathbf{t}}N_n, M_{\mathbf{t}}N^*\big). \qquad (7.8)$$

Define the projection-truncation map $\Pi_{\mathbf{t}}$ by

$$\Pi_{\mathbf{t}}\mu = \big(\Pi_{\{j\},\mathbf{t}}\,\mu\big)_{j\le d}. \qquad (7.9)$$

Note that

$$\Pi_{\mathbf{t}}N_n = (N_{n1t_1},\dots,N_{ndt_d}), \qquad (7.10)$$

where

$$N_{njt_j} = \sum_{i\le n} 1_{(t_j,\infty)}(X_{nij})\varepsilon_{X_{nij}}, \qquad j\le d, \qquad (7.11)$$

are the univariate marginals of the empirical process N_n truncated left of t_j.

If N^* is a Poisson process on $[-\boldsymbol{\infty},\boldsymbol{\infty}) := [-\infty,\infty)^d$ with intensity measure ν, then

$$(N_{1t_1}^*,\dots,N_{dt_d}^*) := \Pi_{\mathbf{t}}N^*$$

is a vector of Poisson processes. As an immediate consequence of Theorem 7.1.1, (5.9) and (5.10) we obtain the following result.

Corollary 7.1.2. *Let N_n be the empirical process in (7.3) and N^* a Poisson process with intensity measure ν, which is the max-Lévy measure of a max-id df G. Then*

$$d\big(\Pi_{\mathbf{t}}N_n, \Pi_{\mathbf{t}}N^*\big) \le \frac{-\log(G(\mathbf{t}))}{n} + c(d)\,A_{n\mathbf{t}}, \qquad (7.12)$$

where $c(d)$ is a constant depending only on d.

Under the mild condition that the marginals G_j of G are continuous at $\alpha(G_j)$, it is proven in Theorem 2.2 of Kaufmann and Reiss [285] that $A_{n,\mathbf{t}} \to 0$ for every $\mathbf{t} > \alpha(G)$ if the left-hand side of (5.13) goes to zero for every $\mathbf{t} > \alpha(G)$ as $n \to \infty$.

Define the Hellinger distance $H(\nu_1,\nu_2)$ between measures ν_1 and ν_2 by

$$H(\nu_1,\nu_2) = \left(\int (f_1^{1/2} - f_2^{1/2})^2\, d\nu_0\right)^{1/2},$$

where f_i is a ν_0-density of ν_i. As on page 7 let $H(X,Y)$ define the Hellinger distance between the distributions of X and Y.

It is an open question whether an inequality of the form

$$H\big(\Pi_{\mathbf{t}}N_n, \Pi_{\mathbf{t}}N^*\big)$$
$$= O\bigg(-\frac{\log(G(\mathbf{t}))}{n} + \sum_K H\big(\Pi_{K,\mathbf{t}}\nu_n, \Pi_{K,\mathbf{t}}\nu\big)\bigg)$$

holds within the framework of (5.13). This inequality holds in the special case of asymptotic independence (see Reiss [386]). Our present method of proof fails in the general framework.

RANDOM THRESHOLDS

Another ingredient of our theory is the notion of an admissible threshold. In the univariate case, a $[-\infty, \infty)$-valued, measurable map T on the space of point measures is an admissible threshold for the df G if

$$\max(x, T(\mu)) = \max(x, T(\mu_x)) \qquad (7.13)$$

for every point measure μ and real x with μ_x denoting the truncation of μ left of x, and

$$P\{T(N^*) \le x\} \to 0, \qquad x \downarrow \alpha(G), \qquad (7.14)$$

where N^* is a Poisson process with mean value function $\log(G)$.

In the d-variate case, a threshold $T = (T_1, \ldots, T_d)$ is admissible (for G) if T_j is admissible for the j-th marginals G_j of G for $j \le d$.

Example 7.1.3. (i) The constant threshold $T(\mu) = t$ is admissible if $t > \alpha(G)$. (ii) If G is continuous at $\alpha(G)$, then

$$T(\mu) = \begin{cases} k\text{-th largest point of } \mu & \mu(\mathbb{R}) \ge k, \\ & \text{if} \\ -\infty & \mu(\mathbb{R}) < k, \end{cases}$$

is admissible. To see this notice that

$$P(T(N^*) \le x) = P(N^*(x, \infty) \le k - 1)$$
$$= G(x) \sum_{0 \le i \le k-1} \frac{(-\log(G(x)))^i}{i!},$$

where the last expression converges to zero as $x \downarrow \alpha(G)$ due to the continuity of G at $\alpha(G)$. If $k = 1$ we will write $T(\mu) = \max(\mu)$.

Our main result unifies and extends several results known in the literature. In view of Example 7.1.3, the following Theorem 7.1.4 deals with the joint distribution of the k largest order statistics as well as with point processes of exceedances of non-random thresholds.

Theorem 7.1.4. *Let G be a max-id df with max-Lévy measure ν. If $A_{nt} \to 0$, $n \to \infty$, for every $t > \alpha(G)$, then for every admissible threshold T,*

$$d\Big(\big(N_{n1T_1(N_{n1})}, \ldots, N_{ndT_d(N_{nd})}\big), \big(N^*_{1T_1(N^*_1)}, \ldots, N^*_{dT_d(N^*_d)}\big)\Big) \to 0 \qquad (7.15)$$

as $n \to \infty$.

To prove this result recall that, according to Corollary 7.1.2,

$$d\big(\boldsymbol{\Pi_t} N_n, \boldsymbol{\Pi_t} N^*\big) \to 0, \qquad n \to \infty, \tag{7.16}$$

for every $\boldsymbol{t} > \boldsymbol{\alpha}(G)$ if $A_{n\boldsymbol{t}} \to 0$ as $n \to \infty$ for every $\boldsymbol{t} > \boldsymbol{\alpha}(G)$.

MULTIVARIATE MAXIMA

As already indicated in the univariate case, we may deduce a limit theorem for maxima from the corresponding result for processes of exceedances. In the following we do not distinguish between a df and the pertaining probability measure in our notation.

Corollary 7.1.5. *If the marginal df G_j of G are continuous at $\alpha(G_j)$, then under the conditions of Theorem 7.1.4,*

$$d\Big(\mathcal{L}\big(\max_{i \leq n} \boldsymbol{X}_i\big), G\Big) \to 0, \qquad n \to \infty. \tag{7.17}$$

Proof. Apply Theorem 7.1.4 to thresholds $T_j(\mu) = \max(\mu)$ as defined in Example 7.1.3(ii). Identify $N_{nj\,\max(N_{nj})}$ with $\max_{i \leq n} X_{i,j}$ and $N^*_{j\,\max(N^*_j)}$ with $\max(N^*_j)$. Moreover, using again avoidance probabilities we have

$$P((\max(N^*_1), \ldots, \max(N^*_d)) \leq \boldsymbol{x}) = P(N^*([-\boldsymbol{\infty}, \boldsymbol{x}]^{\complement}) = 0)$$
$$= G(\boldsymbol{x}).$$

The proof is complete. □

Note that the preceding continuity condition holds for max-stable df G. A corresponding result holds for the largest order statistics taken in each of the components.

7.2 Further Concepts of Extremes

Many attempts have been made to extend the concept of order statistics from the univariate to the multivariate framework. We refer to Barnett [31] and, in addition, to Reiss [385] for supplementary remarks.

Subsequently, we define certain subsets $\mathcal{P}\{\boldsymbol{x}_1, \ldots, \boldsymbol{x}_n\} \subset \{\boldsymbol{x}_1, \ldots, \boldsymbol{x}_n\}$ such as the vertices of the convex hull, the greatest convex minorant or the set of Pareto points. A common characteristic of these subsets is that, within a stochastic framework, $\mathcal{P}\{\boldsymbol{X}_1, \ldots, \boldsymbol{X}_n\} \subset B_n$ with high probability, where B_n is a rare event. One is interested in quantities such as the number of points in $\mathcal{P}\{\boldsymbol{X}_1, \ldots, \boldsymbol{X}_n\}$. Our aim is to show in which way such a question can be shifted from the empirical framework to that of Poisson processes. This is achieved by the coupling argument in conjunction with a first-order Poisson approximation.

VERTICES OF THE CONVEX HULL

One of the celebrated objects in stochastic geometry is the convex hull of rv $\boldsymbol{X}_1, \ldots, \boldsymbol{X}_n$, which is defined by the smallest convex set containing $\{\boldsymbol{X}_1, \ldots, \boldsymbol{X}_n\}$ (initiated by Rényi and Sulanke [391] and further elaborated by Eddy and Gale [119], Brozius and de Haan [56] and Groeneboom [180] among others).

Denote by $\mathcal{P}_v\{\boldsymbol{X}_1, \ldots, \boldsymbol{X}_n\}$ the set of vertices of the boundary of the convex hull. For iid rv $\boldsymbol{X}_i$, which are uniformly distributed on the unit square, it was proven by Groeneboom [180], Corollary 2.4, that

$$\left(|\mathcal{P}_v\{\boldsymbol{X}_1, \ldots, \boldsymbol{X}_n\}| - \frac{8}{3}\log(n) \right) \bigg/ \left(\frac{40}{27}\log(n) \right)^{1/2} \tag{7.18}$$

is asymptotically standard normal. This result can be verified by a Poisson approximation in conjunction with tedious calculations of the number of vertices of the boundary of the convex hull of the Poisson points. Note that the expected number of vertices is of order $\log(n)$, a result already proven by Rényi and Sulanke [391]. Because the expected number is not fixed as $n \to \infty$, a weak convergence result for the point process pertaining to $\mathcal{P}_v\{\boldsymbol{X}_1, \ldots, \boldsymbol{X}_n\}$ is not sufficient to verify (5.19) (see Groeneboom [180], page 328, Lemma 2.2, Corollary 2.2 and, in addition, Reiss [387], pages 215 and 216, for a general discussion of this question).

The computations by Groeneboom [180] first concern the vertices of the left-lower boundary of the convex hull (greatest convex minorant). Then the proof can be completed by introducing the corresponding processes for the other corners of the unit square; these four processes are asymptotically independent. The decisive first step will be discussed within the framework of Pareto points in the following subsection.

PARETO POINTS

Another example concerns the set of Pareto points $\mathcal{P}_a(A)$ of a given set $A = \{\boldsymbol{x}_1, \ldots, \boldsymbol{x}_n\} \subset [0,1]^2$, where

$$\mathcal{P}_a(A) := \left\{ \boldsymbol{x} \in A : \min_{i \le d}(x_i - y_i) \le 0, \, \boldsymbol{y} \in A \right\}. \tag{7.19}$$

Thus, $\boldsymbol{x} \in A$ is a Pareto point if for each $\boldsymbol{y} \in A$ at least one of the components of $\boldsymbol{x}$ is smaller than or equal to the corresponding component of $\boldsymbol{y}$. Given a simple point measure $\mu = \sum_{\mathbf{x} \in A} \varepsilon_{\mathbf{x}}$, let $\mathcal{P}_a(\mu) := \sum_{\mathbf{x} \in \mathcal{P}_a(A)} \varepsilon_{\mathbf{x}}$ be the point measure built by the Pareto points of A.

We give some details, whereby our excursion to Pareto points follows the line of research by Witte [462]. In the following lines we merely deal with the special case of the empirical process

$$N_n = \sum_{i \le n} \varepsilon_{\mathbf{X}_i}$$

of iid rv $\boldsymbol{X}_i$ with common uniform distribution Q on the unit square $[0, 1]^2$. For early results concerning Pareto points we refer to Barndorff-Nielsen and Sobel [30]. Applications in conjunction with multicriteria optimization are mentioned in Berezovskiy and Gnedin [33] and Bruss and Rogers [57].

Let

$$B_n = \{(x, y) \in [0, 1]^2 : xy \le \alpha(n)\}.$$

Fubini's theorem yields

$$Q(B_n) = \alpha(n)(1 - \log(\alpha(n))), \tag{7.20}$$

showing that B_n is a rare event when $\alpha(n)$ is small. The crucial step in our computations is Lemma 5.1.1 in Witte [462], where it is proved that

$$P\{\mathcal{P}_a\{\boldsymbol{X}_1, \ldots, \boldsymbol{X}_n\} \subset B_n\} \ge 1 - n(1 - \alpha(n))^{n-1}. \tag{7.21}$$

Therefore, $Q(B_n) \to 0$ and the left-hand side of (5.22) tends to 1 as $n \to \infty$ for suitably chosen $\alpha(n)$. Thus, the Pareto points form a subset of B_n with high probability.

Applying (5.21) and (5.22), one obtains a bound on the variational distance between the distributions of $\mathcal{P}_a N_n$ and $\mathcal{P}_a N_n^*$, where N_n^* is a Poisson process with the same intensity measure $\nu_n = nQ$ as N_n.

Lemma 7.2.1. *Let $\tau(n)$ be a Poisson rv with parameter n. We have*

$$d\big(\mathcal{P}_a N_n, \mathcal{P}_a N_n^*\big) \le n\,(1 - \alpha(n))^{n-1} + n \int (1 - \alpha(k + 1))^k\, d\mathcal{L}(\tau(n))(k)$$
$$+ \alpha(n)\,(1 - \log(\alpha(n)))\,.$$

Proof. Because of the special structure of the set B_n and the fact that $\boldsymbol{x}$ is a Pareto point of A if, and only if,

$$(-\infty, \boldsymbol{x}] \cap A = \{\boldsymbol{x}\}, \tag{7.22}$$

one obtains $\mathcal{P}_a(A) \cap B_n = \mathcal{P}_a(A \cap B_n)$. This implies

$$(\mathcal{P}_a N)_{B_n} = \mathcal{P}_a (N_{B_n})$$

for point processes N, where N_{B_n} is the truncation of N outside of B_n.

Therefore, using the coupling argument and a first-order Poisson approximation one obtains from (5.21), (5.22) and the monotonicity theorem that

$$d\big(\mathcal{P}_a N_n, (\mathcal{P}_a N_n^*)_{B_n}\big)$$
$$\le d\big(\mathcal{P}_a N_n, \mathcal{P}_a(N_{n, B_n})\big) + d\big(N_{n, B_n}, N_{n, B_n}^*\big)$$
$$\le n(1 - \alpha(n))^{n-1} + \alpha(n)(1 - \log(\alpha(n))).$$

Applying the convexity theorem and (7.21) again one obtains

$$d\big(((\mathcal{P}_a N_n^*)_{B_n}, \mathcal{P}_a N_n^*\big) \leq \int d\big((\mathcal{P}_a N_k)_{B_n}, \mathcal{P}_a N_k\big)\, d\mathcal{L}(\tau(n))(k)$$

$$\leq \int k(1 - \alpha(k))^{k-1}\, d\mathcal{L}(\tau(n))(k)$$

$$= n \int (1 - \alpha(k+1))^k\, d\mathcal{L}(\tau(n))(k).$$

Combining the preceding inequalities we obtain the assertion.   □

Taking $\alpha(n) = (2\log(n+1))/(n-1)$ and using the inequality $1 + x \leq \exp(x)$ one obtains

$$d\big(\mathcal{P}_a N_n, \mathcal{P}_a N_n^*\big) \leq (4\log(n))/n. \tag{7.23}$$

To prove the asymptotic normality of the number $|\mathcal{P}_a\{\boldsymbol{X}_1, \ldots, \boldsymbol{X}_n\}|$ of Pareto points one may utilize another characterization of Pareto points. These are the points of discontinuity of the greatest piecewise constant, decreasing and right-continuous minorant of the points of the empirical process (and likewise of the Poisson process N_n^*). Replacing the point measures $\sum_i \varepsilon_{(y_i, z_i)}$ by $\sum_i \varepsilon_{(ny_i, z_i)}$ we obtain processes with the Lebesgue measure truncated outside of $[0, n] \times [0, 1]$ as intensity measures. This indicates that the question of evaluating the number of Pareto points can be handled via the point process of jump points of the min-extremal-F process, where $F(x) = 1 - e^{-x}$ (cf. (7.36)).

We may also speak of a Pareto point $\boldsymbol{x}$ of order k within $A = \{\boldsymbol{x}_1, \ldots, \boldsymbol{x}_n\}$ if for each $\boldsymbol{y} \in A$ at least k of the components of $\boldsymbol{x}$ are smaller than or equal to the corresponding components of $\boldsymbol{y}$. The previous Pareto points are of order $k = 1$. Moreover, the multivariate minimum is a Pareto point of order n if the minimum is an element of the original sample.

7.3   Thinned Empirical Processes

In this section we introduce thinned empirical processes, which generalize truncated processes by allowing random truncating sets. Local asymptotic normality is established under the condition that the probability of thinning converges to zero. An application to density estimation leads to a fuzzy set density estimator, which is efficient in a parametric model. This section is based on a paper by Falk and Liese [140].

INTRODUCTION

Let $X_1, \ldots, X_n$ be independent copies of a random element (re) X, which is realized in some arbitrary sample space S, equipped with a σ-field $\mathcal{D}$. Suppose that an observation, if falling on an point $x \in S$, is merely counted with probability $\varphi(x) \in [0, 1]$. We assume throughout that the *thinning function* $\varphi : (S, \mathcal{D}) \longrightarrow [0, 1]$ is measurable.

With the particular choice $\varphi(x) = 1_D(x)$ for some $D \in \mathcal{D}$, this setup underlies for example the peaks-over-threshold approach (POT) in extreme value analysis or nonlinear regression analysis (see e.g. Section 1.3); in right-censoring models for observations in $S = \mathbb{R}$, the function $\varphi(x) = P(Y > x) = 1 - P(Y \le x)$ is the conditional probability that the outcome $X = x$ is *not* censored by an independent censoring variable Y (see e.g. Section II.1 of Andersen et al. [13]). In general this setup can be utilized for modelling missing observations with $\varphi(x)$ being the conditional probability that the outcome $X = x$ enters the data set.

A unified approach for the above models is offered by the concept of thinned empirical point processes . First, we identify each point $x \in S$ with the pertaining Dirac measure $\varepsilon_x(B) = 1_B(x)$, $B \in \mathcal{D}$. Thus, we can identify the re X with the random Dirac measure ε_X. The re $U\varepsilon_X(\cdot)$ then models the outcome of X, which is counted merely with probability $\varphi(x)$, given $X = x$. Here U is a rv (rv) with values in $\{0, 1\}$, such that

$$P(U = 1 | X = x) = \varphi(x), \quad x \in S.$$

Assume now that (X_i, U_i), $i = 1, \ldots, n$, are independent copies of (X, U). Then,

$$N_n^\varphi(\cdot) := \sum_{i=1}^n U_i \varepsilon_{X_i}(\cdot),$$

is a *thinned empirical point process* with underlying empirical process

$$N_n(\cdot) := \sum_{i=1}^n \varepsilon_{X_i}(\cdot)$$

(Reiss [387], Section 2.4, Daley and Vere-Jones [90], Example 8.2). The thinned process N_n^φ now models the sampling scheme that an observation X_i, if falling on $x \in S$, enters the data set only with probability $\varphi(x)$.

The process N_n^φ is a re in the set $\mathbb{M} := \{\mu = \sum_{j=1}^n \varepsilon_{x_j} : x_1, \ldots, x_n \in S, n = 0, 1, 2, \ldots\}$ of finite point measures on $(S, \mathcal{D})$, equipped with the smallest σ-field $\mathcal{M}$ such that for any $B \in \mathcal{D}$ the projection

$$\pi_B : \mathbb{M} \longrightarrow \{0, 1, 2, \ldots\}, \quad \pi_B(\mu) := \mu(B)$$

is measurable (cf. Section 1.1 of [387]).

TRUNCATED EMPIRICAL PROCESSES

In the particular case $\varphi(x) = 1_D(x)$, $x \in S$, where D is a given measurable subset of S, the thinned process N_n^φ equals the truncated empirical process

$$N_n^D(\cdot) := \sum_{i=1}^n 1_D(X_i)\varepsilon_{X_i}(\cdot) = \sum_{i=1}^n \varepsilon_{X_i}(\cdot \cap D).$$

If the subset D is a *rare event*, i.e., $D = D(n)$ depends on the sample size n and satisfies $P(X \in D) \longrightarrow_{n \to \infty} 0$, the process N_n^D has been investigated in a series of papers: Various Poisson approximations are derived in the preceding chapters, efficient statistical procedures based on N_n^D in certain parametric models are established by Höpfner [219], Höpfner and Jacod [220] and Marohn [320], bounds for the loss of information due to truncation are computed in Falk and Marohn [144], local asymptotic normality (LAN) of the loglikelihood ratio of N_n^D in quite general parametric models was established in Falk [136] and Marohn [321].

In particular in Falk [136] a characterization of the central sequence in the LAN-expansion of N_n^D was established, which provides an *if and only if* condition on an underlying parametric family of distributions P_ϑ such that just the (random) number $K(n) := N_n^D(S) = N_n(D)$ of actually observed data contains asymptotically the complete statistically relevant information about the true parameter ϑ_0 that is contained in N_n^D. This paper explained the observation that this phenomenon typically occurs in the POT approach in extreme value theory (Falk [135], [136]) by giving a precise description of the mathematical structures yielding this effect; see also the discussion after Theorem 2.4.4.

We will extend these characterizations for truncated empirical processes to thinned processes by considering sequences $\varphi = \varphi_n$ with $\alpha_n := E(\varphi_n(X)) \to_{n \to \infty} 0$.

A Very Brief Sketch of Some LAN Theory

For easier access we give in the following a very brief and informal introduction to the concept of the powerful LAN theory , which is mainly due to LeCam [308] (cf. also Strasser [430] and LeCam and Yang [309] for the general theory; for applications in estimation problems we refer to the books by Ibragimov and Has'minskii [268] and Pfanzagl [367]. A very readable introduction to both, estimation and testing is in Chapter 8 of Andersen et al. [13]) as well as in Chapter 7 of van der Vaart [449].

Suppose that the distribution $\mathcal{L}(N_n^\varphi)$ of N_n^φ is governed by some parameter $\vartheta \in \Theta \subset \mathbb{R}$, i.e., $\mathcal{L}(N_n^\varphi) = \mathcal{L}_\vartheta(N_n^\varphi)$, where we assume just for the sake of simplicity that the parameter space Θ is one-dimensional. Fix $\vartheta_0 \in \Theta$ and choose $\delta_n \longrightarrow_{n \to \infty} 0$. Introduce a local parametrization by setting $\vartheta_n = \vartheta_0 + \delta_n \xi$. The loglikelihood ratio

$$L_n(\vartheta_n|\vartheta_0) := \log\left\{\frac{d\mathcal{L}_{\vartheta_n}(N_n^\varphi)}{d\mathcal{L}_{\vartheta_0}(N_n^\varphi)}\right\}(N_n^\varphi)$$

can be expanded as

$$L_n(\vartheta_n|\vartheta_0) = \xi Z_{(n)} - \frac{\xi^2}{2} + o_{P_{\vartheta_0}}(1).$$

The rv $Z_{(n)}$ is supposed to be asymptotically standard normal

$$Z_{(n)} \longrightarrow_{D_{\vartheta_0}} N(0,1),$$

where $\longrightarrow_{D_\vartheta}$ denotes convergence in distribution under the parameter ϑ and thus,

$$L_n(\vartheta_n|\vartheta_0) \longrightarrow_{D_{\vartheta_0}} N(-\frac{\xi^2}{2}, \xi^2). \qquad \text{(LAN)}$$

LeCam's First Lemma then implies that the distributions $\mathcal{L}_{\vartheta_n}(N_n^\varphi)$ and $\mathcal{L}_{\vartheta_0}(N_n^\varphi)$ are mutually contiguous with $Z_{(n)} \to_{D_{\vartheta_n}} N(\xi,1)$, yielding

$$L_n(\vartheta_n|\vartheta_0) \longrightarrow_{D_{\vartheta_n}} N\left(\frac{\xi^2}{2}, \xi^2\right).$$

Results by LeCam [308] and Hájek [196] imply further that asymptotically optimal tests and estimates can be based on $Z_{(n)}$, $n \in \mathbb{N}$, which is therefore called the *central sequence*.

THE BASIC REPRESENTATION LEMMA

A characterization of the central sequence in the special case $\varphi = 1_D$ with $D = D(n)$ satisfying $P(X \in D) \longrightarrow_{n \to \infty} 0$, $nP(X \in D) \longrightarrow_{n \to \infty} \infty$ was established in Falk [136] and Marohn [321]. In particular an iff condition was established in Falk [136] on the family of underlying distributions such that, with $K(n) = N_n(D)$ and $\alpha_n = P(X \in D)$,

$$Z_{(n)} = \frac{K(n) - n\alpha_n}{\left(n\alpha_n(1-\alpha_n)\right)^{1/2}}, \qquad n \in \mathbb{N},$$

is the central sequence. Note that $Z_{(n)} \longrightarrow_{D_{\vartheta_n}} N(0,1)$ by the Moivre-Laplace theorem.

The derivation of the preceding result was eased by the fact that the truncated process N_n^D is a binomial process. Denote by $Y_1, Y_2, \ldots$ those observations among $X_1, \ldots, X_n$, which actually fall into the subset D. Then we can write

$$N_n^D = \sum_{i=1}^n \varepsilon_{X_i}(\cdot \cap D) = \sum_{j=1}^{K(n)} \varepsilon_{Y_j},$$

where $Y_1, Y_2, \ldots$ behave like independent copies of a re Y, whose distribution $P(Y \in \cdot) = P(X \in \cdot \mid X \in D)$ is the conditional distribution of X, given $X \in D$, and $Y_1, Y_2, \ldots$ are independent of their total number $K(n)$, which is binomial $B(n, P(X \in D))$-distributed (see Theorem 1.3.1).

The fact that a thinned empirical process N_n^φ is *in general* a binomial process, enables us to establish in the following LAN of $N_n^{\varphi_n}$ for an arbitrary sequence φ_n, $n \in \mathbb{N}$, of thinning functions satisfying $\alpha_n = E(\varphi_n(X)) \longrightarrow_{n \to \infty} 0$, $n\alpha_n \longrightarrow_{n \to \infty} \infty$.

Denote by $Y_1, Y_2, \ldots, Y_{K(n)}$ those observations among $X_1, \ldots, X_n$ that are actually observed, i.e., for which $U_i = 1$, and let again $K(n) = N_n^\varphi(S) = \sum_{i=1}^n U_i$ be their total number. By $=_D$ we denote equality in distribution.

The following result is an immediate consequence of the fact that a thinned process can be represented as a projection of a truncated process; see Section 2.4 of Reiss [387].

Lemma 7.3.1. *Let $\varphi : S \to [0,1]$ be an arbitrary measurable thinning function and put $\alpha := E(U) = P(U = 1) = E(\varphi(X))$. If $0 < \alpha < 1$, we have*

$$N_n^\varphi = \sum_{j=1}^{K(n)} \varepsilon_{Y_j} =_D \sum_{j=1}^{K(n)} \varepsilon_{W_j},$$

where $K(n)$ is $B(n, \alpha)$-distributed, $W_1, W_2, \ldots, W_n$ are iid res with common distribution

$$P_W(\cdot) = P(X \in \cdot \mid U = 1)$$

and $K(n)$ and the vector $(W_1, W_2, \ldots, W_n)$ are independent.

This result shows that for a general thinned process the actually observed res Y_j can be handled like iid res, whose common distribution is $P_W(\cdot)$, and they are independent of their total number $K(n)$, which is $B(n, \alpha)$-distributed. The choice $\varphi = 1_D$ yields again the well-known fact for truncated processes mentioned above.

The Model Assumptions

Our statistical model, underlying the thinned process $N_n^\varphi = \sum_{i=1}^n U_i \varepsilon_{X_i}$ for the sample size n, is the assumption that the measure $P(X \in \cdot, U = 1) = \mathcal{L}(X, U)(\cdot \times \{1\})$ is a member of a parametric family

$$P(X \in \cdot, U = 1) = P_\vartheta(X \in \cdot, U = 1) =: Q_\vartheta(\cdot), \qquad \vartheta \in \Theta \subset \mathbb{R}^d,$$

where $Q_\vartheta(S) = P_\vartheta(U = 1) =: \alpha_\vartheta \in [0,1]$. Note that in our model the parameter space $\Theta \subset \mathbb{R}^d$ is fixed and *does not depend on the sample size n*, the measures Q_ϑ however may depend on n, possibly due to a variation of the thinning function $\varphi = \varphi_{n,\vartheta}$. In particular the probability α_ϑ that an observation X_i enters the data set, then depends on n, i.e., $\alpha_\vartheta = \alpha_{n,\vartheta}$. Set

$$Q_{n,\vartheta}(B) = \int_B \varphi_{n,\vartheta}(x) \, P_\vartheta(dx).$$

Suppose that ϑ_0 is an inner point of Θ. We assume that P_ϑ, $\vartheta \in \Theta$, is dominated by the σ-finite measure μ and denote by $f_{n,\vartheta} = dQ_{n,\vartheta}/d\mu$ the density. To calculate the likelihood ratio for the thinned point process we use Example 3.1.2 in Reiss [387] and Lemma 7.3.1 to get

$$L_n(\vartheta|\vartheta_0) =_{D_{\vartheta_0}} \sum_{j=1}^{K(n)} \log\left\{ \frac{f_{n,\vartheta}(Y_j)}{f_{n,\vartheta_0}(Y_j)} \frac{\alpha_{n,\vartheta_0}}{\alpha_{n,\vartheta}} \right\} + K(n) \log\left\{ 1 + \frac{\alpha_{n,\vartheta} - \alpha_{n,\vartheta_0}}{\alpha_{n,\vartheta_0}} \right\}$$

$$+ (n - K(n)) \log\left\{ 1 + \frac{\alpha_{n,\vartheta_0} - \alpha_{n,\vartheta}}{1 - \alpha_{n,\vartheta_0}} \right\}.$$

Note that absolute continuity $Q_{n,\vartheta} \ll Q_{n,\vartheta_0}$ for ϑ close to ϑ_0 is a consequence of condition (7.24) below. Now we localize our model by setting $\vartheta_{n,i} = \vartheta_{0,i} + \xi_{n,i} = \vartheta_{0,i} + \xi_i \delta_{n,i}$, where $\delta_{n,i} \longrightarrow_{n\to\infty} 0$, $1 \leq i \leq d$. Taking formally the derivative with respect to ϑ in $L_n(\vartheta, \vartheta_0)$, we get the formal expansion

$$L(\vartheta_n|\vartheta_0) =_{D_{\vartheta_0}} \sum_{i=1}^{d} \sum_{j=n}^{K(n)} \left\{ \left(\frac{\partial}{\partial \vartheta_i} \log(f_{n,\vartheta_0}) \right)(Y_j) - \frac{\partial}{\partial \vartheta_i} \log(\alpha_{n,\vartheta_0}) \right\} \delta_{n,i}\, \xi_i$$

$$+ \left(\sum_{i=1}^{d} \frac{\partial}{\partial \vartheta_i} \log(\alpha_{n,\vartheta_0})\, \delta_{n,i}\, \xi_i \right) K(n)$$

$$+ \left(\sum_{i=1}^{d} \frac{\partial}{\partial \vartheta_i} \log(1 - \alpha_{n,\vartheta_0})\, \delta_{n,i}\, \xi_i \right) (n - K(n)).$$

We assume that the sequence $f_{n,\vartheta}$ admits the expansion

$$f_{n,\vartheta} = f_{n,\vartheta_0}\left(1 + \langle \vartheta - \vartheta_0, g_n \rangle + \langle \vartheta - \vartheta_0, h_{n,\vartheta} \rangle \right), \tag{7.24}$$

where $g_n = (g_{n,1}, \ldots, g_{n,d})$, $h_{n,\vartheta} = (h_{n,\vartheta,1}, \ldots, h_{n,\vartheta,d})$ are Borel measurable functions and the remainder term satisfies

$$\frac{\int h_{n,\vartheta_n,i}^2\, dQ_{n,\vartheta_0}}{\int g_{n,i}^2\, dQ_{n,\vartheta_0}} \longrightarrow_{n\to\infty} 0, \tag{7.25}$$

where

$$\vartheta_{n,i} = \vartheta_{0,i} + \xi_i \delta_{n,i}, \quad \delta_{n,i} = \left(n \int g_{n,i}^2\, dQ_{n,\vartheta_0} \right)^{-1/2}$$

and the ξ_i are arbitrary.

A Crucial Condition

Denote by W_n a rv with $\mathcal{L}(W_n) = \alpha_{n,\vartheta_0}^{-1} Q_{n,\vartheta_0}$. We index expectations E_ϑ etc. with the underlying parameter. We suppose that $||\delta_n|| \longrightarrow_{n\to\infty} 0$ and that

$$\alpha_{n,\vartheta_0} \longrightarrow_{n\to\infty} 0, \qquad n\alpha_{n,\vartheta_0} \longrightarrow_{n\to\infty} \infty. \tag{7.26}$$

Besides further regularity conditions listed below, we suppose the following crucial condition on the tangent functions $g_{n,i}$, $1 \leq i \leq d$:

$$c_{ni} := \frac{E_{\vartheta_0}\left(g_{n,i}(W_n)\right)}{\left(E_{\vartheta_0}\left(g_{n,i}^2(W_n)\right)\right)^{1/2}}$$

$$= \frac{\int g_{n,i}\, dQ_{n,\vartheta_0}}{\alpha_{n,\vartheta_0}^{1/2}\left(\int g_{n,i}^2\, dQ_{n,\vartheta_0}\right)^{1/2}} \qquad \longrightarrow_{n\to\infty} c_i \in [-1,1], \qquad (7.27)$$

note that $c_{ni} \in [-1,1]$ anyway. Theorem 7.3.2 below shows that the central sequence in the LAN expansion of L_n consists of the total number $K(n)$ of actually observed variables $Y_1, \ldots, Y_{K(n)}$ and of $g_{n,i}(Y_j)$, $1 \le i \le d$, $1 \le j \le n$. It turns out that the number $1 - c_i^2 \in [0,1]$ reflects the part that $g_{n,i}(Y_1), \ldots, g_{n,i}(Y_{K(n)})$ contribute to the central sequence. In the particular case, where $|c_i| = 1$, $1 \le i \le d$, which typically occurs in EVD models (Falk [135], [136]), Marohn [321]), the variables $g_{n,i}(Y_j)$ do not contribute to the central sequence, which consequently consists *only* of the *number* $K(n)$ of actually observed data.

This crucial condition (7.27) was introduced in [136] in the form $1/c_{ni}^2 - 1 \longrightarrow_{n\to\infty} \tilde{c}_i \in [0,\infty)$. But this formulation excludes the case $\tilde{c}_i = \infty$, i.e., $c_i = 0$, which typically occurs in regression analysis (Falk [136], Example 2.4, Marohn [321], Example 3 (c), Falk and Marohn [142]).

Further Regularity Conditions

If $|c_i| < 1$ we need a Lindeberg type condition . To be more precise, we set

$$A_{n,i,\varepsilon} := \left\{ |g_{n,i}| > \varepsilon \left(n \int g_{n,i}^2\, dQ_{n,\vartheta_0} \right)^{1/2} \right\}$$

and require that

$$L_n(\varepsilon) = \frac{\int_{A_{n,i,\varepsilon}} g_{n,i}^2\, dQ_{n,\vartheta_0}}{\int g_{n,i}^2\, dQ_{n,\vartheta_0}} \longrightarrow_{n\to\infty} 0 \qquad (7.28)$$

for every $\varepsilon > 0$. Note that this simple form of the Lindeberg condition is due to the fact that we have iid rv for a fixed sample size n. If for some $\varepsilon > 0$,

$$\frac{\int |g_{n,i}|^{2+\varepsilon}\, dQ_{n,\vartheta_0}}{\left(\int g_{n,i}^2\, dQ_{n,\vartheta_0}\right)^{1+\varepsilon/2}} \longrightarrow_{n\to\infty} 0, \qquad (7.29)$$

then we obtain from Hölder's inequality

$$\frac{\int_{A_{n,i,\varepsilon}} g_{n,i}^2\, dQ_{n,\vartheta_0}}{\int g_{n,i}^2\, dQ_{n,\vartheta_0}} \le \left(\frac{\int |g_{n,i}|^{2+\varepsilon}\, dQ_{n,\vartheta_0}}{\left(\int g_{n,i}^2\, dQ_{n,\vartheta_0}\right)^{1+\varepsilon/2}} \right)^{2/(2+\varepsilon)} \longrightarrow_{n\to\infty} 0,$$

so that (7.28) is satisfied. The assumption (7.29) is a Ljapunov type condition.

We suppose that $g_{n,i_1}(W_n)$ and $g_{n,i_2}(W_n)$ are asymptotically uncorrelated for $i_1 \neq i_2$, i.e.,

$$
\frac{E_{\vartheta_0}\Big(\big(g_{n,i_1}(W_n) - E_{\vartheta_0}(g_{n,i_1}(W_n))\big)\big(g_{n,i_2}(W_n) - E_{\vartheta_0}(g_{n,i_2}(W_n))\big)\Big)}{E_{\vartheta_0}(g_{n,i_1}^2(W_n))^{1/2} E_{\vartheta_0}(g_{n,i_2}^2(W_n))^{1/2}}
$$

$$
= \frac{\int g_{n,i_1} g_{n,i_2}\, dQ_{n,\vartheta_0}}{\Big(\int g_{n,i_1}^2\, dQ_{n,\vartheta_0}\Big)^{1/2}\Big(\int g_{n,i_2}^2\, dQ_{n,\vartheta_0}\Big)^{1/2}} - c_{ni_1} c_{ni_2}
$$

$$
\longrightarrow_{n\to\infty} 0 \qquad \text{if } i_1 \neq i_2. \tag{7.30}
$$

The Main Result

For every fixed n denote by Y_j, $j = 1, \ldots, n$, iid rv with common distribution $(\alpha_{n,\vartheta_0})^{-1} Q_{n,\vartheta_0}$. Now we are ready to state our main result, which provides LAN of thinned processes.

Theorem 7.3.2. *If the conditions (7.24)-(7.27) and (7.28), (7.30) are satisfied, then we have*

$$
L_n\Big(\vartheta_n|\vartheta_0\Big)
$$

$$
=_{D_{\vartheta_0}} \Big(\sum_{i=1}^d \xi_i c_i\Big) \frac{K(n) - n\alpha_{n,\vartheta_0}}{(n\alpha_{n,\vartheta_0}(1 - \alpha_{n,\vartheta_0}))^{1/2}} - \frac{1}{2}\Big(\sum_{i=1}^d \xi_i c_i\Big)^2
$$

$$
+ \frac{1}{(n\alpha_{n,\vartheta_0})^{1/2}} \sum_{j=1}^{K(n)} \sum_{i:|c_i|<1} \xi_i \frac{g_{n,i}(Y_j) - E_{\vartheta_0}\Big(g_{n,i}(Y_j)\Big)}{E_{\vartheta_0}\Big(g_{n,i}^2(Y_j)\Big)^{1/2}}
$$

$$
- \frac{1}{2} \sum_{i=1}^d \xi_i^2(1 - c_i^2) + o_{P_{\vartheta_0}}(1)
$$

$$
\longrightarrow_{D_{\vartheta_0}} N\Bigg(-\frac{1}{2}\Big(\big(\sum_{i=1}^d \xi_i c_i\big)^2 + \sum_{i=1}^d \xi_i^2(1 - c_i^2)\Big), \big(\sum_{i=1}^d \xi_i c_i\big)^2 + \sum_{i=1}^d \xi_i^2(1 - c_i^2)\Bigg),
$$

where $K(n)$ and $Y_1, Y_2, \ldots$, are independent and $K(n)$ has a binomial distribution with parameters n and α_{n,ϑ_0}.

With the particular thinning function $\varphi(x) = 1_D(x)$, the preceding result implies Theorem 1.1 in Falk [136] and the main result in Marohn [321].

The Moivre-Laplace theorem implies that

$$
Z_{n1} := \frac{K(n) - n\alpha_{n,\vartheta_0}}{\big(n\alpha_{n,\vartheta_0}(1 - \alpha_{n,\vartheta_0})\big)^{1/2}} \longrightarrow_{D_{\vartheta_0}} N(0,1);
$$

recall that $K(n)$ is binomial $B(n, \alpha_n)$-distributed under parameter ϑ_0, with $n\alpha_{n,\vartheta_0}$ $\longrightarrow_{n \to \infty} \infty$. The proof of Theorem 7.3.2 shows that

$$Z_{n2} := \frac{1}{(n\alpha_{n,\vartheta_0})^{1/2}} \sum_{j=1}^{K(n)} \sum_{i:|c_i|<1} \xi_i \frac{g_{n,i}(Y_j) - E_{\vartheta_0}(g_{n,i}(Y))}{E_{\vartheta_0}(g_{n,i}^2(Y))^{1/2}}$$

$$\longrightarrow_{D_{\vartheta_0}} N\left(0, \sum_{i=1}^{d} \sigma_i^2 (1 - c_i^2)\right);$$

the independence of $K(n)$ and $Y_1, Y_2, \ldots$ implies that Z_{n1} and Z_{n2} are asymptotically independent. The limiting normal distribution in Theorem 7.3.2 is therefore a consequence of the convolution theorem for normal distributions.

EXAMPLE: RIGHT-CENSORED DATA

Suppose that we observe right-censored data $\sum_{i=1}^{n} U_i \varepsilon_{X_i}$, where $U_i = 1(Z_i > X_i)$ and Z_i is independent of X_i. Assume that the censoring distribution $\mathcal{L}(Z)$ is an exponential one with parameter λ_n and that X follows an exponential distribution with parameter $\vartheta > 0$, i.e., we have, for the sample size n and $1 \le i \le n$,

$$P(Z_i > x) = P(Z > x) = \exp(-\lambda_n x), \qquad x \ge 0,$$
$$P(X_i > x) = P_\vartheta(X > x) = \exp(-\vartheta x), \qquad x \ge 0.$$

In this case we have, for a Borel set $B \subset [0, \infty)$,

$$Q_\vartheta(B) = P_\vartheta(X \in B, U = 1) = P_\vartheta(X \in B, Z > X)$$

$$= \int_B P(Z > x)\vartheta \exp(-\vartheta x)\, dx$$

$$= \int_B \vartheta \exp(-(\lambda_n + \vartheta)x)\, dx$$

and thus,

$$f_\vartheta(x) := \vartheta \exp(-(\lambda_n + \vartheta)x), \quad x \ge 0,$$

is a Lebesgue density of Q_ϑ. We, consequently, obtain

$$\alpha_{n,\vartheta} = Q_{n,\vartheta}\left([0, \infty)\right) = \int_0^\infty f_\vartheta(x)\, dx = \frac{\vartheta}{\lambda_n + \vartheta}.$$

Fix $\vartheta_0 > 0$. Iterated Taylor expansion implies, for $x \ge 0$ and ϑ close to ϑ_0,

$$\frac{f_\vartheta(x)}{f_{\vartheta_0}(x)} = \frac{\vartheta}{\vartheta_0} \exp\left((\vartheta_0 - \vartheta)x\right)$$

$$= \exp\left(\log\left(1 + \frac{\vartheta - \vartheta_0}{\vartheta_0}\right) + (\vartheta_0 - \vartheta)x\right)$$

$$= 1 + (\vartheta_0 - \vartheta)\left(x - \frac{1}{\vartheta_0}\right)$$
$$+ O\left((\vartheta_0 - \vartheta)^2 (\vartheta_0 - \vartheta)^2 (x^2 + 1) \exp(|\vartheta_0 - \vartheta|x)\right)$$
$$=: 1 + \left(\vartheta_0 - \vartheta\right)g(x) + \left(\vartheta_0 - \vartheta\right)h_{n,\vartheta}(x).$$

Hence, condition (7.24) is satisfied with $d = 1$ and

$$\int g^2 \, dQ_{n,\vartheta_0} = \int_0^\infty (x - \frac{1}{\vartheta_0})^2 \vartheta_0 \exp\left(-(\lambda_n + \vartheta_0)x\right) dx$$
$$= \frac{1}{\lambda_n} \frac{\frac{1}{\vartheta_0} + \frac{\vartheta_0}{\lambda_n^2}}{\left(1 + \frac{\vartheta_0}{\lambda_n}\right)^3} = \frac{1}{\lambda_n \vartheta_0}(1 + o(1))$$

if $\lambda_n \longrightarrow_{n\to\infty} \infty$. We require in addition that $\lambda_n/n \longrightarrow_{n\to\infty} 0$, which implies $\alpha_{n,\vartheta} \longrightarrow_{n\to\infty} 0$, $n\alpha_{n,\vartheta} \longrightarrow_{n\to\infty} \infty$. Since, moreover,

$$\int g \, Q_{\vartheta_0} = -\frac{\lambda_n}{(\lambda_n + \vartheta_0)^2} = -\frac{1}{\lambda_n}(1 + o(1)),$$

condition (7.27) is also satisfied with $c = -1$:

$$c_n = \frac{\int g \, dQ_{n,\vartheta_0}}{\alpha_{n,\vartheta_0}^{1/2}(\int g^2 \, dQ_{n,\vartheta_0})^{1/2}} = -\frac{\frac{1}{\lambda_n}(1 + o(1))}{\left(\frac{\vartheta_0}{\lambda_n}\right)^{1/2}(1 + o(1))\left(\frac{1}{\lambda_n \vartheta_0}\right)^{1/2}(1 + o(1))}$$
$$\longrightarrow_{n\to\infty} -1.$$

The central sequence in the LAN expansion of Theorem 7.3.2 will consist in this example therefore *only* of the number $K(n) = \sum_{i=1}^n U_i \varepsilon_{X_i}([0,\infty))$ of uncensored observations. It remains to verify condition (7.25) with $\vartheta_n = \vartheta_0 + \xi \delta_n(1 + o(1)) = \vartheta_0 + \xi \vartheta_0^{1/2}(\lambda_n/n)^{1/2}(1 + o(1))$:

$$\frac{\int h_{n,\vartheta}^2 \, dQ_{n,\vartheta_0}}{\int g^2 \, dQ_{n,\vartheta_0}} = O\left(\lambda_n\left(\vartheta_0 - \vartheta_n\right)^2 \int_0^\infty (x^2 + 1)^2 \exp\left(2|\vartheta_0 - \vartheta_n|x\right) f_{\vartheta_0}(x) \, dx\right)$$
$$= O\left(\frac{\lambda_n}{n}\right) = o(1).$$

Hence we obtain from Theorem 7.3.2 the LAN expansion

$$L_n\left(\vartheta_n|\vartheta_0\right) = \xi \frac{n\alpha_n - K(n)}{(n\alpha_n(1 - \alpha_n))^{1/2}} - \frac{\xi^2}{n} + o_{P_{\vartheta_0}}(1)$$
$$\longrightarrow_{D_{\vartheta_0}} N\left(-\xi^2/2, \xi^2\right).$$

The complete statistical information about the underlying parameter, which is contained in the thinned process $\sum_{i=1}^{n} U_i \varepsilon_{X_i}$, is in this example already contained in $K(n) = \sum_{i=1}^{n} U_i \varepsilon_{X_i}([0, \infty)) = U_1 + \cdots + U_n$, but *not* in the actually observed non-censored data $Y_1, \ldots, Y_{K(n)}$. This phenomenon typically occurs in the peaks-over-threshold approach (POT) in extreme value theory (Falk [135], [136]), whereas the converse case $c = 0$ typically occurs in regression analysis (Falk [136], Example 2.4, Marohn [321], Example 3 (c)).

AN EFFICIENT ESTIMATOR

Proposition 7.3.3. *We consider the particular case* $d = 1$ *and* $c \in \{1, -1\}$. *Suppose that the following regularity conditions are satisfied:*

(a) $\delta_\varepsilon := \inf_{n, \vartheta: |\vartheta - \vartheta_0| > \varepsilon} |\alpha_{n,\vartheta} - \alpha_{n,\vartheta_0}| > 0$ *for any* $\varepsilon > 0$.

(b) $\alpha_{n,\vartheta}$ *is differentiable near* $\vartheta = \vartheta_0$ *for any* $n \in \mathbb{N}$ *with* $\inf_{n, \vartheta: |\vartheta - \vartheta_0| \leq \varepsilon_0} |\alpha'_{n,\vartheta}| \geq C > 0$ *for some* $\varepsilon_0 > 0$ *and some* $C > 0$.

(c) $\sup_{\vartheta: |\vartheta - \vartheta_0| \leq K n^{-1/2}} |\int h_{n,\vartheta} \, dQ_{n,\vartheta_0}| \longrightarrow_{n \to \infty} 0$ *for any* $K > 0$.

Then an asymptotically efficient estimator of the underlying parameter ϑ_0 *based on the thinned process is given by the solution* $\hat{\vartheta}_n$ *of the equation*

$$\alpha_{n,\hat{\vartheta}_n} = \frac{K(n)}{n}.$$

Proof. Put $\delta_n := 1/(n \int g^2 \, dQ_{n,\vartheta_0})^{1/2}$, $c_n := \int g \, dQ_{n,\vartheta_0} / (\alpha_{n,\vartheta_0} \int g^2 \, dQ_{n,\vartheta_0})^{1/2}$. The expansion

$$
\begin{aligned}
\frac{K(n)}{n} = \alpha_{n,\hat{\vartheta}_n} = Q_{n,\hat{\vartheta}_n}(S) &= \int f_{\hat{\vartheta}_n} \, d\mu \\
&= \int 1 + \left(\hat{\vartheta}_n - \vartheta_0\right) g + \left(\hat{\vartheta}_n - \vartheta_0\right) h_{n,\hat{\vartheta}} \, dQ_{n,\vartheta_0} \\
&= \alpha_{n,\vartheta_0} + \left(\hat{\vartheta}_n - \vartheta_0\right) \int g \, dQ_{n,\vartheta_0} + r_n
\end{aligned}
$$

implies that

$$\frac{K(n) - n\alpha_{n,\vartheta_0}}{(n\alpha_{n,\vartheta_0}(1 - \alpha_{n,\vartheta_0}))^{1/2}} = \frac{c_n}{(1 - \alpha_{n,\vartheta_0})^{1/2}} \delta_n^{-1} \left(\hat{\vartheta}_n - \vartheta_0\right) + \frac{n^{1/2} r_n}{\alpha_{n,\vartheta_0}^{1/2}(1 - \alpha_{n,\vartheta_0})^{1/2}}.$$

Note that

$$\{|\hat{\vartheta}_n - \vartheta_0| > \varepsilon\} \subset \{|\alpha_{n,\hat{\vartheta}_n} - \alpha_{n,\vartheta_0}| \geq \delta_\varepsilon\} \subset \{|K(n)/n - \alpha_{n,\vartheta_0}| \geq \delta_\varepsilon\}$$

for arbitrary $\varepsilon > 0$. Since $\delta_\varepsilon > 0$ by condition (a) we have consistency of $\hat{\vartheta}_n$. Condition (b) now implies

$$\alpha_{n,\hat{\vartheta}_n} = \alpha_{n,\vartheta_0} + \alpha'_{n,\tilde{\vartheta}_n}(\hat{\vartheta}_n - \vartheta_0),$$

where $\tilde{\vartheta}_n$ is between ϑ_0 and $\hat{\vartheta}_n$. We, consequently, obtain

$$|n^{1/2}(\hat{\vartheta}_n - \vartheta_0)| \leq C^{-1} n^{1/2} |\alpha_{n,\hat{\vartheta}_n} - \alpha_{n,\vartheta_0}| = C^{-1} n^{1/2} \left| \frac{K(n)}{n} - \alpha_{n,\vartheta_0} \right| = O_P(\alpha_{n,\vartheta_0}^{1/2}).$$

Condition (c) now implies that $n^{1/2} r_n / \alpha_{n,\vartheta_0}^{1/2} = o_{P_{\vartheta_0}}(1)$ and, hence, we obtain

$$\delta_n^{-1}\left(\hat{\vartheta}_n - \vartheta_0\right) = c \frac{K(n) - n\alpha_{n,\vartheta_0}}{(n\alpha_{n,\vartheta_0}(1 - \alpha_{n,\vartheta_0}))^{1/2}} + o_{P_{\vartheta_0}}(1).$$

The expansion of Theorem 7.3.2 is by LeCam's First Lemma also valid under the alternative $\vartheta_n = \vartheta_0 + \xi\delta_n$,

$$L_n(\vartheta_n|\vartheta_0) = c\xi \frac{K(n) - n\alpha_{n,\vartheta_0}}{(n\alpha_{n,\vartheta_0}(1 - \alpha_{n,\vartheta_0}))^{1/2}} - \frac{\xi^2}{2} + o_{P_{\vartheta_0}}(1),$$

where now

$$\frac{K(n) - n\alpha_{n,\vartheta_0}}{(n\alpha_{n,\vartheta_0}(1 - \alpha_{n,\vartheta_0}))^{1/2}} \longrightarrow_{D_{\vartheta_n}} N(c\xi, 1).$$

Consequently,

$$\delta_n^{-1}\left(\hat{\vartheta}_n - \vartheta_0\right) \longrightarrow_{D_{\vartheta_0}} N(0,1), \qquad \delta_n^{-1}\left(\hat{\vartheta}_n - \vartheta_n\right) \longrightarrow_{D_{\vartheta_n}} N(0,1),$$

i.e., $\hat{\vartheta}_n$ is a *regular* estimator, asymptotically unbiased under ϑ_0 as well as under ϑ_n. Its limiting variance 1 coincides with that of the central sequence $(K(n) - n\alpha_{n,\vartheta_0})/(n\alpha_{n,\vartheta_0}(1 - \alpha_{n,\vartheta_0}))^{1/2}$ and thus Hájeks [196] Convolution Theorem now implies that $\hat{\vartheta}_n$ actually has minimum limiting variance among all regular estimates that are based on the thinned empirical process. □

APPLICATION TO FUZZY SET DENSITY ESTIMATION

Consider independent copies $X_1, \ldots, X_n$ of a rv X in $\mathbb{R}^d$, whose distribution $\mathcal{L}(X)$ has a Lebesgue density p near some fixed point $x_0 \in \mathbb{R}^d$. The problem is the estimation of $p(x_0)$. We will establish in the following a parametric model for this non-parametric problem and we will show, how the preceding results can be utilized to prove efficiency of a fuzzy set density estimator within this model. Though seemingly quite similar to a usual kernel density estimator, the fuzzy set estimator has surprising advantages over the latter. Just for notational simplicity we assume in the following that $d = 1$; all subsequent considerations can be generalized to the case $d > 1$ in a straightforward manner.

A common estimator of $p(x_0)$ is the kernel density estimator

$$\hat{p}_n(x_0) := \frac{1}{n b_n} \sum_{i=1}^{n} k\left(\frac{x_0 - X_i}{b_n}\right),$$

where the kernel function k satisfies $\int k(x)\,dx = 1$, $\int x\,k(x)\,dx = 0$ and $b_n > 0$ is a bandwidth.

In contrast to the kernel estimator, which assigns weight to all points of the sample, we now select points from the sample with different probabilities. As we have to evaluate the local behavior of the distribution of X, it is obvious that only observations X_i in a neighborhood of x_0 can reasonably contribute to the estimation of $p(x_0)$. Our set of observations in a neighborhood of x_0 can now be described by the thinned process

$$N_n^{\varphi_n} = \sum_{i=1}^{n} U_i\, \varepsilon_{X_i},$$

where U_i decides, whether X_i belongs to the neighborhood of x_0 or not.

Precisely,

$$\varphi_n(x) := P(U_i = 1 \mid X_i = x)$$

is the probability that the observation $X_i = x$ belongs to the neighborhood of x_0. Note that this neighborhood is not explicitly defined, but it is actually a *fuzzy set* in the sense of Zadeh [468], given by its *membership function* φ_n. The thinned process $N_n^{\varphi_n}$ is, therefore, a fuzzy set representation of the data, where we assume that $(X_1, U_1), \ldots, (X_n, U_n)$ are iid copies of (X, U). For a review of fuzzy set theory and its applications we refer to the monograph by Zimmermann [474].

It is plausible to let $\varphi_n(x)$ depend on the distance $|x - x_0|$ and to put

$$\varphi_n(x) := \varphi\left(\frac{x - x_0}{b_n}\right), \qquad x \in \mathbb{R},$$

where the function φ has values in $[0, 1]$ and $b_n > 0$ is a scaling factor or bandwidth. We assume that φ is continuous at 0 with $\varphi(0) > 0$. To keep the conditions on p as general as possible, we require that $\varphi(x) = 0 = k(x)$ if $|x| > K$, where $K > 0$ is some fixed constant.

Put now

$$b_n := \frac{a_n}{\int \varphi(x)\,dx},$$

where $a_n > 0$ satisfies $na_n \longrightarrow_{n\to\infty} \infty$, $na_n^5 \longrightarrow_{n\to\infty} 0$. Elementary computations imply that under the above conditions

$$(n\,a_n)^{1/2}\,(\widehat{p}_n(x_0) - p(x_0)) \longrightarrow_{D_{\vartheta_0}} N\left(0, p(x_0) \int k^2(x)\,dx \int \varphi(x)\,dx\right).$$

A simple analysis shows, moreover, that the fuzzy set density estimator

$$\widehat{\vartheta}_n := \frac{K(n)}{na_n} = \frac{N_n^{\varphi_n}(\mathbb{R})}{na_n} = \frac{\sum_{i=1}^{n} U_i}{na_n}$$

satisfies

$$(na_n)^{1/2}\left(\widehat{\vartheta}_n - p(x_0)\right) \longrightarrow_D N\left(0, p(x_0)\right),$$

provided $\int \varphi(x) x\, dx = 0$ and that the density p is twice differentiable near x_0 with bounded second derivative. Note that $\widehat{\vartheta}_n$ depends only on the number $K(n)$ of non-thinned observations and that its limiting normal distribution is independent of φ.

For the particular choice $\varphi^* := 1_{[-K,K]}$ we obtain from the Cauchy-Schwarz inequality

$$1 = \int_{-K}^{K} k(x)\, dx \leq \left(\int_{-K}^{K} k^2(x)\, dx \right)^{1/2} (2K)^{1/2}$$

$$= \left(\int_{-K}^{K} k^2(x)\, dx \right)^{1/2} \left(\int \varphi^*(x)\, dx \right)^{1/2}$$

and, thus, the limiting normal distribution $N\left(0, p(x_0) \int k^2(x)\, dx \int \varphi^*(x)\, dx\right)$ of the kernel density estimator $\widehat{p}_n(x_0)$ is more spread out than that of $\widehat{\vartheta}_n$ with the thinning function φ^*. Note that $\widehat{p}_n(x_0)$ and $\widehat{\vartheta}_n$ use the same bandwidth sequence b_n and have the same rate of convergence $(na_n)^{-1/2}$.

THE ESTIMATOR $\widehat{\vartheta}_n$ IS ACTUALLY EFFICIENT

Now we study the efficiency of $\widehat{\vartheta}_n$ within the class of all estimators based on randomly selected points from the sample. To this end we use the LAN-approach and apply Theorem 7.3.2 to special parametric submodels.

Precisely we require

$$p_\vartheta(x) = \vartheta + r(\vartheta, x),$$

where $\vartheta \in \Theta \subset (0, \infty)$ and r satisfies

$$r(\vartheta, x_0) = 0, \qquad \vartheta \in \Theta. \tag{7.31}$$

The parameter ϑ resembles, therefore, the possible value of the unknown density p at x_0, with $p(x_0) = \vartheta_0$ being the actual one. We assume in addition that

$r(\vartheta, x)$ is continuous near (ϑ_0, x_0),

$\dfrac{\partial}{\partial \vartheta} r(\vartheta, x)$ exists in a neighborhood of (ϑ_0, x_0) and is continuous at (ϑ_0, x_0),

$\dfrac{\partial^2}{\partial x^2} r(\vartheta_0, x)$ exists for x near x_0 and is bounded. $\tag{7.32}$

Note that (7.31) implies

$$\frac{\partial}{\partial \vartheta} r(\vartheta, x_0) = 0, \qquad \vartheta \in \Theta. \tag{7.33}$$

We apply Theorem 7.3.2 and obtain that our model leads to an LAN expansion of $L_n(\vartheta_n|\vartheta_0)$, where condition (7.27) is satisfied with $c = 1$ and, hence, the central sequence is given by

$$Z_{n1} := \frac{K(n) - n\alpha_{n,\vartheta_0}}{(n\alpha_{n,\vartheta_0}(1 - \alpha_{n,\vartheta_0}))^{1/2}}.$$

Proposition 7.3.4. *If in addition to the above assumptions on φ and b_n, conditions 7.31-7.33 are satisfied, then*

$$\widehat{\vartheta}_n = \frac{K(n)}{n\,a_n}$$

is an efficient estimator of ϑ_0 in the set of all regular estimates that are based on the sequence of thinned processes $N_n^{\varphi_n}$, $n \in \mathbb{N}$.

ASYMPTOTICALLY BIASED DENSITY ESTIMATORS

It is further interesting to compare $\widehat{\vartheta}_n$ as a non-parametric density estimator with a kernel density estimator also in the case where both are asymptotically biased, i.e., where $n\,b_n^5 \not\to_{n\to\infty} 0$. Suppose to this end that the density $p(x)$ is twice differentiable near x_0 and that p'' is continuous at x_0. A simple analysis shows that the mean squared error of the non-parametric fuzzy set density estimator

$$\widehat{\vartheta}_n = \frac{1}{n\,a_n} \sum_{i=1}^{n} U_i$$

with

$$P(U_i = 1 \mid X_i = x) = \varphi\left(\frac{x - x_0}{b_n}\right)$$

and $b_n = a_n / \int \varphi(x)dx$ can be expanded as

$$E\left((\widehat{\vartheta}_n - p(x_0))^2\right) = \frac{p(x_0)}{n\,a_n} + b_n^4 \left(\frac{p''(x_0)}{2}\int \varphi(x)x^2\,dx / \int \varphi(x)\,dx\right)^2$$
$$+ o\left(\frac{1}{n\,a_n} + a_n^4\right).$$

The usual kernel density estimator $\widehat{p}_n(x_0) = (nb_n)^{-1}\sum_{i=1}^{n} k\left((x_0 - X_i)/b_n\right)$ has, on the other hand, the mean squared error

$$E\left((\widehat{p}_n(x_0) - p(x_0))^2\right)$$

$$= \frac{p(x_0)}{n\,a_n} \int k^2(x)\,dx \int \varphi(x)\,dx + b_n^4 \left(\frac{p''(x_0)}{2}\int k(x)x^2\,dx\right)^2 + o\left(\frac{1}{n\,a_n} + a_n^4\right).$$

If $nb_n^5 \not\to 0$, then the second terms in both asymptotic expansions cannot be neglected compared with the first one. While with the particular choice $\varphi^* = 1_{[-K,K]}$ the first term in the above expansion of the mean squared error of $\widehat{p}_n(x_0)$ is greater than that of $\widehat{\vartheta}_n$, this is in general not true for the second term. Take, for example, the popular Epanechnikov kernel

$$k_E(x) = (3/4)(1 - x^2)1_{[-1,1]}(x),$$

i.e., $K = 1$. Then we have

$$\int k_E(x)x^2 \, dx = 1/5$$

but

$$\int_{-1}^{1} \varphi^*(x)x^2 \, dx \Big/ \int_{-1}^{1} \varphi^*(x) \, dx = 1/3.$$

We have, on the other hand,

$$\int_{-1}^{1} \varphi^*(x) \, dx \int k_E^2(x) \, dx = 6/5$$

and, thus, the mean squared error of the Epanechnikov kernel with bandwidth b_n of order $n^{-1/5}$ can be larger as well as smaller than that of $\widehat{\vartheta}_n$, depending on $p(x_0)$ and $p''(x_0)$.

7.4 Max-Stable Stochastic Processes

In the following we study stochastic processes $\boldsymbol{X} = (X(t))_{t \in T}$. Corresponding to the multivariate case, arithmetic operations and relations are meant componentwise. Thus, e.g.,

$$\boldsymbol{a}^{-1}(\boldsymbol{X} - \boldsymbol{b}) := \big(a(t)^{-1}(X(t) - b(t))\big)_{t \in T},$$

where $\boldsymbol{a} = (a(t))_{t \in T} > 0$ and $\boldsymbol{b} = (b(t))_{t \in T}$.

MAX-STABILITY

The notion of max-stability can be generalized to the infinite-dimensional setting. Let $\boldsymbol{X} = (X(t))_{t \in T}$ be a stochastic process and let $\boldsymbol{X}^{(1)}, \ldots, \boldsymbol{X}^{(n)}$ be independent copies of $\boldsymbol{X}$. The maximum is again taken componentwise, that is,

$$\max_{i \leq n} \boldsymbol{X}^{(i)} := \Big(\max_{i \leq n} X^{(i)}(t)\Big)_{t \in T}.$$

Now, $\boldsymbol{X}$ is called max-stable, if for every $n \in \mathbb{N}$ there are normalizing functions $\boldsymbol{a}_n > 0$ and $\boldsymbol{b}_n$ such that

$$\max_{i \leq n} \Big(\boldsymbol{a}_n^{-1}\big(\boldsymbol{X}_n^{(i)} - \boldsymbol{b}_n\big)\Big) =_D \boldsymbol{X}$$

in the sense of having the same finite-dimensional marginal distributions. To prove the max-stability, one has to verify that

$$P\left(a_n(t)^{-1}\left(X(t)-b_n(t)\right) \le y_t, \, t \in T_0\right)^n = P\left(X(t) \le y_t, \, t \in T_0\right) \qquad (7.34)$$

for every finite $T_0 \subset T$ and $n \in \mathbb{N}$.

Likewise, one may introduce the notion of a max-infinitely divisible process (cf. Vatan [452] and Balkema et al. [23]).

We proceed by discussing an important example of max-stable processes which will later be reconsidered in the light of theoretical results.

MAX-STABLE EXTREMAL PROCESSES

First let $T = \mathbb{N}$. Let Y_i, $i \in \mathbb{N}$, be a sequence of iid rv with common df F. Put $X(i) = Y_{i:i}$. Verify that for every $m \in \mathbb{N}$ and $1 \le n_1 < n_2 < \cdots < n_m$,

$$P(X(n_1) \le x_1, X(n_2) \le x_2, \ldots, X(n_m) \le x_m)$$

$$= F\left(\min_{1 \le i \le m} x_i\right)^{n_1} F\left(\min_{2 \le i \le m} x_i\right)^{n_2 - n_1} \cdots F(x_m)^{n_m - n_{m-1}}.$$

If F is max-stable with $F^n(b_n + a_n x) = F(x)$, then it is a simple exercise to show that $\boldsymbol{X} = (X(i)_{i \in \mathbb{N}}$ is max-stable with $\boldsymbol{a}_n$ and $\boldsymbol{b}_n$ being equal to the constants a_n and b_n, respectively. This concept can be extended to the continuous time domain $T = (0, \infty)$.

An extremal-F process $\boldsymbol{X} = (X(t))_{t>0}$ pertaining to a df F has the following property: For every $m \in \mathbb{N}$ and $0 < t_1 < t_2 < \cdots < t_m$,

$$P\left(X(t_1) \le x_1, X(t_2) \le x_2, \ldots, X(t_m) \le x_m\right)$$

$$= F\left(\min_{1 \le i \le m} x_i\right)^{t_1} F\left(\min_{2 \le i \le m} x_i\right)^{t_2 - t_1} \cdots F(x_m)^{t_m - t_{m-1}}.$$

The preceding remark about max-stability is also valid for the continuous time version of the extremal process.

Let $\sum_i \varepsilon_{(Y_i, Z_i)}$ be a Poisson process with intensity measure $\lambda_0 \times \nu$, where λ_0 is the Lebesgue measure restricted to $(0, \infty)$ and ν has the measure generating function $\log(F)$. Then

$$X(t) := \sup\{Z_i : Y_i \le t, \, i \in \mathbb{N}\}, \qquad t > 0, \qquad (7.35)$$

defines an extremal-F process (cf. Pickands [370], Resnick [393]). It is the smallest piecewise constant, increasing and right-continuous majorant of the Poisson points. This extremal process takes values in the space $D(0, \infty)$ equipped with the Borel-σ-field of the Skorohod topology if F is continuous.

Likewise one may define a min-extremal-F process which is related to minima instead of maxima. Then take

$$X(t) = \inf\{Z_i : Y_i \le t, \, i \in \mathbb{N}\}, \qquad t > 0, \qquad (7.36)$$

and the measure ν with generating function $-\log(1 - F)$.

GENERATION OF MAX-STABLE PROCESSES BY CONVOLUTIONS

In the following we merely deal with Poisson processes related to the Gumbel df G_3 in view of our main example which concerns Brownian motions. Let N_3^* be a Poisson process with mean value function

$$\Psi_3(x) = \log(G_3)(x) = -e^{-x}.$$

Denote by ν_3 the pertaining intensity measure. Recall that $\max(N_3^*)$ has the df G_3, where the maximum of a point process is again the maximum of its points. Let $\boldsymbol{X} = (X(t))_{t \in T}$ be a stochastic process, where the finite-dimensional case is included if T is finite. Let $h(u, (x(t))_{t \in T}) = (u + x(t))_{t \in T}$. Let N^* be the Poisson process with intensity measure $h(\nu_3 \times \mathcal{L}(\boldsymbol{X}))$, which is the measure induced by h and the product $\nu_3 \times \mathcal{L}(\boldsymbol{X})$. Hence, a copy of $\boldsymbol{X}$ is added independently to every point of N_3^*. Let $\max(N^*)$ be defined by the componentwise maximum of the points of N^* and put

$$(M(t))_{t \in T} := \max(N^*). \tag{7.37}$$

In the following we assume that

$$b(t) := \int e^r \, d\mathcal{L}(X(t))(r) \in (0, \infty), \qquad t \in T. \tag{7.38}$$

Theorem 7.4.1. *If condition* (7.38) *holds, then*

(i) $P(M(t) \le x) = \exp(-\exp(-z + \log(b(t))))$;

(ii) $(M(t))_{t \in T}$ *is max-stable.*

Proof. Put $Q = \mathcal{L}(X(t)_{t \in T})$. Using again avoidance probabilities, we obtain

$$P(M(t) \le z) = P\big(N^*\{x : x(t) > z\} = 0\big)$$
$$= \exp\big(-h(\nu \times Q)\{x : x(t) > z\}\big).$$

Moreover,

$$g(\nu \times Q)\{x : x(t) > z\} = (\nu \times Q)\big\{(u, y) : u + y(t) > z\big\}$$
$$= \int \nu\left((z - r, \infty)\right) \, d\mathcal{L}(X(t))(r)$$
$$= e^{-z} \int e^r d\mathcal{L}(X(t))(r).$$

Hence, (i) holds. We verify (7.34) to prove the max-stability:

$$P(M(t) - \log(n) \le z_t, \, t \in T_0)^n$$
$$= P(N^*\{x : x(t) > z_t + \log(n) \text{ for some } t \in T_0\} = 0)^n$$

$$= \exp\left(-\int n\nu\left(\min_{t\in T_0}(z_t - x(t)), \infty\right) d\mathcal{L}\left((X(t))_{t\in T_0}\right)\right)$$

$$= \exp\left(-\int \nu\left(\min_{t\in T_0}(z_t - x(t)), \infty\right) d\mathcal{L}\left((X(t))_{t\in T_0}\right)\right)$$

$$= P(M(t) \le z_t, \, t \in T_0).$$

Thus, (7.34) holds. $\qquad\qquad\qquad\qquad\qquad\qquad\qquad\qquad\qquad\qquad$ □

A Max-Stable Process
Corresponding to Brownian Motion

Let C be the space of continuous functions on $[0, \infty)$ equipped with the topology of uniform convergence on bounded intervals. Let $\mathcal{C}$ be the Borel-σ-field on C. Put

$$C_0 = \{x \in C : x(0) = 0\}$$

and denote by $\mathcal{C}_0$ the trace of $\mathcal{C}$ in C_0.

Recall that a stochastic process $B = (B(t))_{t\ge 0}$ with values in C_0, equipped with the σ-field $\mathcal{C}_0$, is a standard Brownian motion if

(i) the increments $B(t_1) - B(t_0), B(t_2) - B(t_1), \ldots, B(t_{m+1}) - B(t_m)$ are independent for $m \in \mathbb{N}$ and $t_0 < t_1 < \cdots < t_{m+1}$;

(ii) $\mathcal{L}(B(t) - B(s)) = N_{(0, t-s)}, \quad 0 \le s < t.$

Moreover, B is characterized by the following properties: B is a Gaussian process (that is, the finite-dimensional margins are normal rv) with mean function $E(B(t)) = 0$ and covariance function $K(s,t) = E(B(s)B(t)) = \min(s,t)$. Let

$$(X(t))_{t\ge 0} = (B(t) - t/2)_{t\ge 0}$$

and set

$$(M(t))_{t\ge 0} := \max(N^*) \tag{7.39}$$

as in (7.37). Condition (7.38) holds, because

$$\int e^r d\mathcal{L}(B(t) - t/2)(r) = \int e^r dN_{(-t/2, t)}(r) = \int dN_{(t/2, t)} = 1, \tag{7.40}$$

and, hence, we know from Theorem 7.4.1 that $(M(t))_{t\ge 0}$ is max-stable. This special max-stable process was dealt with by Brown and Resnick [55]. In the following lemma we compute the univariate and bivariate margins. Let H_λ be again the bivariate df in Example 4.1.4.

Lemma 7.4.2. *For every $t > 0$:*

(i) $P\{M(t) \le z\} = \exp(-e^{-z})$;

(ii) $P(M(0) \leq z_1, M(t) \leq z_2) = H_{t^{1/2}/2}(z_1, z_2)$.

Proof. (i) follows from Theorem 7.4.1(i) and (7.40). Moreover,

$$P(M(0) \leq z_1, M(t) \leq z_2)$$
$$= \exp\left(-g(\nu \times Q)\{x \in C : x(0) > z_1 \text{ or } x(t) > z_2\}\right)$$

and

$$g(\nu \times Q)\{x \in C : x(0) > z_1 \text{ or } x(t) > z_2\}$$
$$= \int \nu\{u : u > z_1 \text{ or } u > z_2 - r\} \, d\mathcal{L}(B(t) - t/2)(r)$$
$$= \int_{-\infty}^{z_2 - z_1} e^{-z_1} \, dN_{(-t/2,t)}(r) + \int_{z_2 - z_1}^{\infty} e^{-z_2 + r} \, dN_{(-t/2,t)}(r)$$
$$= \Phi\left(\lambda + \frac{z_2 - z_1}{2\lambda}\right) e^{-z_1} + \Phi\left(\lambda + \frac{z_1 - z_2}{2\lambda}\right) e^{-z_2}$$

for $\lambda = t^{1/2}/2$. The proof is complete. $\square$

Generally, the finite-dimensional margins of $(M(t))_{t>0}$ are special cases of the rv in Example 4.1.4.

MAXIMA OF INDEPENDENT BROWNIAN MOTIONS

Let B, B_i, $i \in \mathbb{N}$, be independent Brownian motions. Consider

$$X_{ni}(t) = b_n\{B_i(1 + tb_n^{-2}) - b_n(1 + t/(2b_n^2))\},$$

where b_n is again defined by $b_n = n\varphi(b_n)$ with φ denoting the standard normal density.

The following result is due to Brown and Resnick [55], who used a slightly different normalization.

Theorem 7.4.3. *We have*

$$\left(\max_{i \leq n} X_{ni}(t)\right)_{t \geq 0} \to_D \max(N^*), \qquad n \to \infty,$$

where N^* is the Poisson process defined in (7.39).

The basic idea in the proof of Theorem 7.4.3 is the decomposition

$$B_i(1 + tb_n^{-2}) = B_i(1) + b_n^{-1} B_i^*(t),$$

where B_i^* are iid standard Brownian motions which are independent of $B_i(1)$, $i \leq n$. We have

$$X_{ni}(t) = b_n(B_i(1) - b_n) + (B_i^*(t) - t/2).$$

From this representation we see that one is dealing with a question related to that in Theorem 7.4.1.

It was proven by H. Drees (personal communication) that the convergence also holds in the variational distance, if the domain of t is restricted to a finite interval (cf. also Reiss [387], E.6.8).

Since Theorem 7.4.3 implies the weak convergence of the finite-dimensional margins we know that $\max_{i \leq n} X_{ni}(t)$ is asymptotically distributed as $M(t)$, where again $(M(t))_{t \geq 0} = \max(N^*)$. From this identity we may deduce again that $M(t)$ is a standard Gumbel rv.

THEORETICAL RESULTS

Because a process $\boldsymbol{X} = (X(t))_{t \in T}$ is max-stable if all finite-dimensional margins are max-stable, these margins have representations as given in (4.23). If the univariate margins are of Fréchet-form and $N^* = \sum_i \varepsilon_{(Y_i, Z_i)}$, one finds f_t, $t \in T$, such that the process

$$\left(\max_i \left(Z_i f_t(Y_i) \right) \right)_{t \in T}$$

has the same finite-dimensional distributions as $\boldsymbol{X}$. Notice that this is also the construction of extremal processes in (7.35) with $f_t = 1_{[0,t]}$.

The constructions around Theorem 7.4.1 and Lemma 7.4.2 can easily be described within the general framework of max-stable processes with univariate margins of Gumbel-form. We have $\rho = \mathcal{L}(\boldsymbol{X})$ and $f_t(x) = \exp(x_t)$. Notice that condition (7.38) corresponds to $0 < \int f_t \, d\rho < \infty$. We refer to de Haan [187], [188] and Vatan [452] for theoretical, and to Coles [70] and further literature cited therein for applications.

PART III

Non-IID Observations

Chapter 8

Introduction to the Non-IID Case

We present in the following some examples to motivate the extension of the classical extreme value theory for iid sequences to a theory for non-iid sequences. We introduce different classes of non-iid sequences together with the main ideas. The examples show that suitable restrictions for each class are needed to find limit results which are useful for applications.

8.1 Definitions

By $\{X_i,\ i \geq 1\}$ we denote in the following a sequence of real-valued rv X_i with marginal distributions F_{X_i}. In contrast to the iid case we do not assume that these marginal distributions are identical. Furthermore, the independency assumption may also be dropped.

Such generalizations are required in many applications, where the rv are dependent, as for instance in time series of ecological data, sulfate and ozone concentration values and their exceedances of the threshold set by the government, the daily rainfall amount, the daily maximum or minimum temperature.

Often, these time series exhibit in addition a trend or a seasonal component, sometimes of a periodic nature. Furthermore, the variances of the X_i's are often observed to be non-constant. In general, we have to specify the kind of non-stationarity for such time series. Their extremes can only be treated with a more general theory for the extreme values.

In addition, a more general theory reveals also the limitations of the classical theory of the iid case. We gain a deeper insight into the properties of the classical theory and its relation to the general one.

The iid case can be generalized in various ways by not assuming the independence or the identical distributions $F_{X_i}(\cdot) = F_{X_1}(\cdot)$ of the X_i's.

M. Falk et al., *Laws of Small Numbers: Extremes and Rare Events*, 3rd ed.,
DOI 10.1007/978-3-0348-0009-9_8, © Springer Basel AG 2011

(i) The best known special case concerning dependent sequences is the *station-ary* case. A random sequence is called (strongly) *stationary*, if the finite-dimensional df of the random sequence are such that

$$F_{X_{i_1}, X_{i_2}, \ldots, X_{i_k}}(\cdot) = F_{X_{i_1+m}, X_{i_2+m}, \ldots, X_{i_k+m}}(\cdot), \qquad (8.1)$$

holds for any $\{i_j \in I\!N, j = 1, \ldots, k\}$ and $k, m \in I\!N$. Obviously, this implies $F_{X_i}(\cdot) = F_{X_1}(\cdot)$ for all $i \geq 1$ (by setting $(k = 1)$ in (8.1)). If (8.1) does not hold, then the random sequence is usually called *non-stationary*.

(ii) On the other hand, we may sometimes assume independence of the rv, but dealing with non-identical marginal distributions F_{X_i}. Such sequences are called *independent* random sequences. They are used to model for instance the extremes of some ecological data mentioned above, because one can as-sume approximate independence of the data sets sufficiently separated in time (or space).

The class of non-stationary random sequences is rather large; an extreme value theory for such a general class of non-stationary random sequences does not exist. Since in applications certain special models are considered, it is worthwhile to deal with such particular non-stationary random sequences also. Appropriate subclasses have been introduced and treated, for instance the random sequences $\{X_i, \ i \geq 1\}$ of the form $X_i = \mu_i + s_i Y_i$, with some trend values μ_i, scaling values s_i and a stationary random sequence $\{Y_i, \ i \geq 1\}$. Another subclass consists of non-stationary random sequences which have identical marginal distributions: $F_{X_i}(\cdot) = F_{X_1}(\cdot)$, for all $i \geq 1$; for instance a sequence of standardized Gaussian rv is such a random sequence.

We find that in the stationary or the independent case the behavior of ex-tremes and exceedances of a level u may deviate substantially from that in the iid case. In the following this will be illustrated by some simple examples which will also imply the kind of restrictions needed to develop an extreme value theory for a rather large class of non-stationary random sequences. The exact mathematical formulation of such necessary conditions are stated in the following chapters which present the general results.

8.2 Stationary Random Sequences

In this section we assume that the random sequence $\{X_i, \ i \geq 1\}$ satisfies (8.1). Let $F_{X_i}(\cdot) = F(\cdot)$ for all $i \geq 1$. In the iid case the exceedances of a high level u can be considered as rare events. If the level u tends to the upper endpoint $\omega(F)$ of the distribution $F(\cdot)$, the number of exceedances (up to time n) can asymptotically be approximated by a Poisson rv.

- The Poisson-approximation holds in the iid case, iff the level $u = u_n$ is such that $u_n \to \omega(F)$ and $n(1 - F(u_n)) \to \tau < \infty$ as $n \to \infty$ (which means that if $\omega(F) < \infty$, $\omega(F)$ is a continuity point of the distribution $F(\cdot)$).

Under certain conditions the Poisson approximation still holds for stationary random sequences. It is obvious that in many of these cases the level $u = u_n$ has necessarily to tend to $\omega(F)$, being a continuity point of $F(\cdot)$ if $\omega(F) < \infty$.

Example 8.2.1. Let $X_i = W + Y_i$, $i \geq 1$, where W, Y_i, $i \geq 1$, are independent rv with $W \sim F_W$, $Y_i \sim F_Y$. The sequence X_i is stationary and each X_i depends on the rv W in the same way. If there exists a sequence $\{b_n, \ n \geq 1\}$ such that

$$P\Big(|\max_{i \leq n} Y_i - b_n| > \epsilon\Big) \to 0 \qquad \text{for any} \quad \epsilon > 0,$$

then as $n \to \infty$

$$P\Big(\max_{i \leq n} X_i \leq w + b_n\Big) = P\Big(W + \max_{i \leq n} Y_i \leq w + b_n\Big)$$

$$\to_D P(W \leq w) = F_W(w).$$

This shows that any df F_W could occur as limit distribution of extreme values of a stationary random sequence. Note that in this case $u_n \not\to \omega(F)$. In this example, X_i and X_j have the same dependence structure, for every pair i, j, $j \neq i$, even if i and j are rather far apart, (i.e., if $|i - j|$ is large or $|i - j| \to \infty$). The Poisson approximation does not make sense in this case, since events of the Poisson process, which are separated in time, are independent. If second moments of the rv X_i exist, the random sequence $\{X_i, \ i \geq 1\}$ is called *equally correlated*, since $\text{Corr}(X_i, X_j) = \rho$, $i \neq j$.

Example 8.2.2. If we take the maximum instead of the sum of the rv W and Y_i, we gain further insight into the theory on stationary sequences. Again, let Y_i and W be as in Example 8.2.1 and define

$$X_i = \max(W, Y_i).$$

Assume now that $\{a_n, \ n \geq 1\}$ and $\{b_n, \ n \geq 1\}$ are such that as $n \to \infty$

$$F_Y^n(a_n x + b_n) \to_D G(x),$$

thus $a_n x + b_n \to \omega(F_Y)$, for all x with $G(x) > 0$. This means that F_Y belongs to the domain of attraction of the extreme value distribution (EVD) G, denoted by $F_Y \in D(G)$ (cf. Section 2.1 and Leadbetter et al. [303], Galambos [167], Resnick [393]). Then

$$P\Big(\max_{i \leq n} X_i \leq a_n x + b_n\Big) = P\Big(W \leq a_n x + b_n, \ \max_{i \leq n} Y_i \leq a_n x + b_n\Big)$$

$$= F_W(a_n x + b_n)F_Y^n(a_n x + b_n) \to_D G(x)$$

if and only if $1 - F_W(a_n x + b_n) \to 0$, or equivalently $\omega(F_Y) \geq \omega(F_W)$. (If $\omega(F_Y) = \omega(F_W) < \infty$, then $\omega(F_W)$ has to be a continuity point of F_W.)

If $\omega(F_Y) < \omega(F_W)$, we get

$$P\Big(\max_{i \leq n} X_i \leq w \Big) \to_D \begin{cases} F_W(w) & \text{if} \quad w \geq \omega(F_Y), \\ 0 & \text{if} \quad w < \omega(F_Y). \end{cases}$$

Concerning the dependence between X_i and X_j, $i \neq j$, the following is observed.

$$\begin{aligned} P(X_i \leq u, \ X_j \leq u) &= P(W \leq u, \ Y_i \leq u, \ Y_j \leq u) \\ &= F_W(u) F_Y^2(u) \\ &= P(X_i \leq u) P(X_j \leq u) \\ &\quad + F_W(u)(1 - F_W(u)) F_Y^2(u). \end{aligned} \tag{8.2}$$

If now $u \to \omega(F_Y)$, then the second term of (8.2) is asymptotically negligible, if and only if $1 - F_W(u) \to 0$ which means $\omega(F_Y) \geq \omega(F_W)$. This implies that the events $\{X_i \leq u\}$ and $\{X_j \leq u\}$ are asymptotically independent. This is equivalent to the asymptotic independence of $\{X_i > u\}$ and $\{X_j > u\}$. We shall show that a Poisson approximation is possible in this case.

For the case $\omega(F_Y) < \omega(F_W)$, neither the asymptotic independence nor the Poisson approximation hold. Furthermore, if $F_W(a_n x + b_n) \to 1$, then

$$P(X_i > a_n x + b_n) = 1 - F_W(a_n x + b_n) F_{Y_i}(a_n x + b_n) \to 0.$$

This convergence holds even uniformly in i. This property of the random sequence $\{X_i, i \geq 1\}$ is called *uniform asymptotic negligibility* (*uan*) which means that none of the rv X_i has a significant influence on the extremes; each X_i could be deleted without losing important information on the extremes. The uan definition is given in (8.3) for the general case.

The idea of the asymptotic independence of the exceedances occurring in separated time intervals has to be formulated mathematically in an appropriate way, such that the Poisson approximation can be proved for a general class of stationary random sequences.

8.3   Independent Random Sequences

We consider now independent rv X_i with df $F_{X_i}(\cdot)$, which are in general non-identical. Thus the distribution of the maximum $M_n = \max_{i \leq n} X_i$ is simply given by

$$P(M_n \leq u) = P\Big(\max_{i \leq n} X_i \leq u \Big) = \prod_{i \leq n} F_{X_i}(u).$$

From the point of view of applications, this representation is often useless, since the $F_{X_i}(\cdot)$ are not known in general. Therefore one might try to approximate the distribution for the maximum asymptotically.

Necessary for the Poisson approximation is the uan condition

$$\sup_{i \le n} p_{i,n} = \sup_{i \le n} P(X_i > u_n) = \sup_{i \le n} \left[1 - F_{X_i}(u_n) \right] \to 0 \tag{8.3}$$

as $n \to \infty$. This means that the level $u_n \to u_\infty \ge \omega(F_{X_i})$, for all $i \ge 1$, in other words, it is assumed that a single exceedance at a time point i is negligible. Under the uan condition the number of exceedances

$$N_n = \sum_{i \le n} 1(X_i > u_n)$$

is asymptotically a Poisson rv with parameter $\lambda \in [0, \infty)$, iff

$$\sum_{i \le n} \left[1 - F_{X_i}(u_n) \right] \to \lambda \tag{8.4}$$

as $n \to \infty$. Condition (8.4) generalizes the condition $n(1 - F_X(u_n)) \to \lambda = -\log(G(x))$ with $u_n = a_n x + b_n$ in the case of iid sequences. The error of approximation can be computed; using a result of Barbour and Holst [28], we get that the variational distance

$$d(N, N_n) = \sup_{A \subset \mathbb{N}} |P(N \in A) - P(N_n \in A)|,$$

where $N \sim \text{Poisson}(\lambda)$, is bounded from above by

$$(1 \wedge \lambda^{-1}) \sum_{i \le n} \left[1 - F_{X_i}(u_n) \right]^2.$$

By (8.3) and (8.4), this is $O(\sup_{i \le n} p_{i,n})$, converging to 0 as $n \to \infty$, thus the variational distance converges to 0. This proves the Poisson approximation.

To find possible asymptotic distributions of $M_n = \max_{i \le n} X_i$, further conditions are necessary. The following well-known example shows that if further conditions are not posed, every df can occur as limit distribution (compare Example 8.2.1).

Example 8.3.1. Let $G(\cdot)$ be any df. Then also $G^\gamma(\cdot)$ is a df for any $\gamma \in (0, 1]$ and we can define a sequence of independent rv $\{X_i, i \ge 1\}$ such that

$$X_i \sim G^{\gamma_i}, \quad \text{where} \quad \gamma_i \in (0,1) \quad \text{with} \quad \sum_{i \le n} \gamma_i \to 1 \quad \text{as } n \to \infty.$$

Then

$$P\left(\max_{i \le n} X_i \le u \right) = \left(G(u) \right)^{\sum_{i \le n} \gamma_i} \to_D G(u).$$

Note that the uan condition (8.3) with $u_n \equiv u$ does not hold in this example.

However, in the case of independent random sequences property (8.3) is not sufficiently restrictive in order to obtain only the EVD of the iid case as possible limit distributions. The class of limit distributions for the extremes in the non-iid case is much larger. To find a reasonable subclass of limit distributions for applications, further conditions have to be posed or special models of random sequences have to be considered.

On the other hand, in many cases the distribution of the maxima of non-identically distributed rv can be approximated by the distribution of the maxima of a suitable iid sequence which implies that the limit distribution of this maxima is one of the EVD.

Example 8.3.2. Let $U_i \sim \mathrm{Exp}(1)$, $i \geq 1$, be iid rv and set $X_i = U_i + \log(c_i)$ with $c_i > 0$ for all $i \geq 1$. Then with

$$u_n(x) = \log\Big(\sum_{i\leq n} c_i\Big) + x$$

and

$$\min_{i\leq n}\big(u_n(x) - \log(c_i)\big) \to \infty, \qquad n \to \infty,$$

the probability of no exceedance of $u_n(x)$ at i is

$$P(X_i \leq u_n(x)) = P(U_i \leq u_n(x) - \log(c_i))$$
$$= 1 - \exp\big(-u_n(x) + \log(c_i)\big).$$

Thus the distribution of the normalized maxima converges

$$P\Big(\max_{i\leq n} X_i \leq u_n(x)\Big) = \prod_{i\leq n} F_{X_i}(u_n(x))$$
$$= \exp\Big\{-\sum_{i\leq n}[1 - F_{X_i}(u_n(x))](1 + o(1))\Big\}$$
$$= \exp\Big\{-\Big(\sum_{i\leq n} c_i\Big)^{-1} e^{-x}\sum_{i\leq n} c_i(1 + o(1))\Big\}$$
$$\to_D \exp(-e^{-x}) = G_3(x).$$

Note that the iid case ($X_i = U_i$, i.e., $c_i \equiv 1$) belongs to this class and it is well known that

$$P\Big(\max_{i\leq n} U_i \leq a_n x + b_n\Big) \to_D G_3(x)$$

with $a_n \equiv 1, b_n = \log(n)$ as suggested by the definition of $u_n(x)$ given above. Even if $c_i \not\equiv 1$, but such that

$$\log\Big(\sum_{i\leq n} c_i\Big) = \log(n) + o(1)$$

or equivalently

$$\frac{1}{n} \sum_{i \leq n} c_i \to 1,$$

we get the same limit distribution G_3 with the same normalizing constants $a_n = 1$ and $b_n = \log(n)$ as in the iid case. It is easy to see that the uan condition (8.3) holds.

In this example the 'non-stationarity' has no influence on the asymptotic distribution of the maxima. It is questionable whether the non-stationarity has an influence on the extreme order statistics. We shall note in Section 8.3 that in some cases the limit point process of exceedances is still the same as in the iid case. The following example shows that this statement is not always true.

Example 8.3.3. Daley and Hall [89] discussed special models of independent sequences with monotone trends and variance inhomogeneities. Let $\{X_i, \ i \geq 1\}$ be an iid random sequence and

$$1 = w_1(\gamma) \geq w_i(\gamma) \geq w_{i+1}(\gamma) \to 0 \quad \text{as} \quad i \to \infty \quad \text{and} \quad w_i(\gamma) \uparrow 1 \ \text{as} \ \gamma \to 1,$$
$$0 = v_1(\beta) \leq v_i(\beta) \leq v_{i+1}(\beta) \to \infty \quad \text{as} \quad i \to \infty \quad \text{and} \quad v_i(\beta) \downarrow 0 \ \text{as} \ \beta \to 0.$$

Now define $M(\gamma, \beta) = \sup_{i \geq 1}\{w_i(\gamma)X_i - v_i(\beta)\}$, the supremum of the weighted and shifted sequence, with shifts $v_i(\beta)$ and weights $w_i(\gamma)$. Daley and Hall analyzed the class of limit distributions of $M(\gamma, \beta)$ as $\gamma \to 1$ and $\beta \to 0$. Obviously, the EVD $G_{1,\alpha}, G_{2,\alpha}$ and G_3 belong to this class. Two special cases are the geometrically weighted iid sequences and linearly shifted iid sequences with suprema

$$W(\gamma) = \sup_{i \geq 1}\{\gamma^{i-1} X_i\} \quad \text{and} \quad S(\beta) = \sup_{i \geq 1}\{X_i - (i-1)\beta\},$$

respectively.

1) Suppose that $F = F_X \in D(G_{1,\alpha})$. Then also $W(\gamma)$, suitably normalized, converges in distribution to $G_{1,\alpha}$. The following more general result holds for $M(\gamma, 0)$, the supremum of a weighted iid sequence. If the weights $w_i(\gamma)$ are such that

$$\infty > \sum_{i \geq 1} \left[1 - F(1/w_i(\gamma))\right] \to \infty \qquad \text{as} \ \gamma \to 1,$$

then

$$P(a(\gamma)M(\gamma, 0) \leq x) \to_D G_{1,\alpha}(x) \qquad \text{as} \ \gamma \to 1,$$

where

$$a(\gamma) = \sup \left\{ a : \sum_{i \geq 1} \left[1 - F(\frac{1}{aw_i(\gamma)})\right] \leq 1 \right\}.$$

Note that since $a(\gamma)$ is monotone decreasing in γ and tends to 0 as $\gamma \to 1$ by the assumptions, we have

$$P(a(\gamma)M(\gamma,0) \le x) = \prod_{i \ge 1} F\Big(\frac{x}{a(\gamma)w_i(\gamma)}\Big)$$

$$\sim \exp\Big(-x^{-\alpha}\sum_{i \ge 1}\{1 - F(1/a(\gamma)w_i(\gamma))\}\Big)$$

$$\to \exp(-x^{-\alpha}) = G_{1,\alpha}(x), \qquad x > 0.$$

For instance if $w_i(\gamma) = \gamma^{i-1}$ with $1 - F(x) \sim cx^{-\alpha}$, as $x \to \infty$, $c > 0$, then the result holds with $a(\gamma) = [(1 - \gamma^\alpha)/c]^{1/\alpha}$.

The linearly shifted iid random sequence $\{X_i - (i-1)\beta, i \ge 1\}$ is now analyzed. It is necessary to assume $\alpha > 1$ to guarantee that $S(\beta)$ is well defined (i.e., $S(\beta) < \infty$ a.s.). Then define, for sufficiently small β,

$$a(\beta) = \sup\{a \in (0,1) : (1 - F(1/a))/a \le (\alpha - 1)\beta\}.$$

Note that $a(\beta) \to 0$ as $\beta \to 0$. Then

$$P(a(\beta)S(\beta) \le x) = \prod_{i \ge 1} P(X_i \le (i-1)\beta + x/a(\beta))$$

$$\sim \exp\{-\sum_{i \ge 0}[1 - F(i\beta + x/a(\beta))]\}$$

$$\sim \exp\{-\beta^{-1}\int_{x/a(\beta)}^{\infty}(1 - F(z))dz\}$$

$$\sim \exp\{-\beta^{-1}(\alpha - 1)^{-1}(x/a(\beta))[1 - F(x/a(\beta))]\}$$

$$\to \exp\{-x^{-\alpha+1}\} = G_{1,\alpha-1}(x), \qquad x > 0,$$

since $1 - F(\cdot)$ is regularly varying and

$$[1 - F(1/a(\beta))]\,[a(\beta)\beta(\alpha - 1)]^{-1} \to 1$$

as $\beta \to 0$. Therefore $G_{1,\alpha-1}$ is the limit distribution of $S(\beta)$.

2) Daley and Hall showed also that if $F \in D(G_3)$ then both $W(\gamma)$ and $S(\beta)$ converge in distribution to $G_3(\cdot)$ (compare with Example 8.3.2). If $F \in D(G_{2,\alpha})$, a similar statement holds.

By replacing u_n in (8.3) by $x/a(\gamma)$ and $x/a(\beta)$, respectively, it can be shown that the uan condition is satisfied by the sequences analyzed in this example, for any $x > 0$.

The rv $S(\beta)$ of the shifted iid model is related to a rv in a somewhat different problem. Define $\tau(\beta) = \sup\{i \ge 1 : X_i > i\beta\}$ which is the last exit time of the random sequence $\{X_i,\ i \geq 1\}$ from the region $\{(t,x)\ :\ x\ \leq\ t\beta\}$. Hüsler [235] analysed the limit distributions of $\tau(\beta)$ as $\beta \to 0$, assuming that the X_i 's are iid rv. (For extensions to stationary sequences see Hüsler [236]).

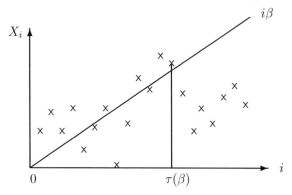

FIGURE 8.3.1. Last exit time $\tau(\beta)$ of the random sequence X_i.

The following equations reveal the equivalence of the asymptotic results for $S(\beta)$ and $\tau(\beta)$. We have, for every $l \in \mathbb{N}$,

$$
\begin{aligned}
1\{\tau(\beta) < l\} =\ & 1\{X_i \leq i\beta,\ i \geq l\} \\
=_D\ & 1\{X_k - (k-1)\beta \leq l\beta,\ k \geq 1\} \\
=\ & 1\{S(\beta) \leq l\beta\} \\
=\ & 1\{S(\beta)/\beta \leq l\}.
\end{aligned}
$$

Hence, $a(\beta)\beta\tau(\beta)$ has the same limit distribution as $a(\beta)S(\beta)$.

8.4   Non-stationary Random Sequences

From the above discussion of special extensions of the iid case we deduce ideas on how to develop a reasonably general, applicable theory for non-stationary random sequences.

We observed in Example 8.3.2 that maxima of special non-stationary sequences behave asymptotically as those of iid (or stationary) sequences. As it will be shown, this is true for a fairly large class of non-stationary sequences. This implies that in such cases only the three types of EVD occur as limit distribution for the maxima and therefore in some applications the problem reduces to one which can be handled using standard statistical procedures.

However, deviations from the iid or stationary case are possible and will be observed if the Poisson approximation for exceedances is considered instead of the weak limit for the maxima. A question of interest is which conditions have to be

posed on the non-stationary sequences so that the classical theory of extremes is
still applicable.

Example 8.4.1. Let $\{X_i,\ i \geq 1\}$ be a Gaussian sequence with non-constant mean
values $E(X_i) = \mu_i$ and variances $\mathrm{Var}(X_i) = \sigma_i^2$. The sequences $\{\mu_i,\ i \geq 1\}$ and
$\{\sigma_i,\ i \geq 1\}$ are sometimes periodic, for example in time series with a seasonal
component.

A particular subclass of non-stationary Gaussian sequences is defined by $X_i =
\mu_i + \sigma_i Y_i,\ i \geq 1$, where $\{Y_i,\ i \geq 1\}$ is a stationary standardized Gaussian sequence.
We have for the maximum $M_n = \sup\{X_i,\ i \leq n\}$ the simple relationship

$$\{M_n \leq u\} = \{Y_i \leq (u - \mu_i)/\sigma_i,\ i \leq n\}.$$

This means that instead of dealing with the maxima of the non-stationary sequence
$\{X_i,\ i \geq 1\}$ one has to deal now with the exceedances of a non-constant boundary
by a stationary Gaussian sequence.

Example 8.4.2. Example 8.4.1 can easily be extended by replacing the standardized Gaussian sequence with any stationary random sequence $\{Y_i,\ i \geq 1\}$. Note
that the sequences in Example 8.3.2 and 8.3.3 are of this type with $\{Y_i\}$ being iid
rv. Ballerini and McCormick [25] used such a model to analyze rainfall data. They
supposed that mean values μ_i and variances σ_i^2 are periodic.

Remark 8.4.3. Finally, we note that the theory is developed mainly for random
sequences. Stochastic processes in continuous time (parameter) and with continuous paths give rise to further complications. The following simple idea can be
applied in many cases. By partitioning the interval $[0, T]$ we can rewrite the maximum

$$M_T = \sup\{X(t),\ 0 \leq t \leq T\}$$

in terms of the maximum of a finite number of rv Z_j defined by

$$Z_j = \sup\Big\{X(t),\ t \in [(j-1)h, jh)\Big\}, \qquad j \leq [T/h]$$

and

$$Z^* = \sup\Big\{X(t),\ t \in [h[T/h], T]\Big\},$$

where $h > 0$. In this way the suprema of stochastic processes in continuous time
can be treated in the context of the theory for the maxima of random sequences.
The suprema of Gaussian processes are analyzed in Chapter 10.

8.5 Triangular Arrays of Discrete Random Variables

We introduced the uniform asymptotic negligibility condition as a necessary assumption to show the Poisson approximation in Section 8.2. We assumed that

$$\sup_{i \le n} P(X_i > u_n) =: \sup_{i \le n} p_{i,n} = o(1)$$

as $n \to \infty$, where u_n can be the linear normalization $u_n = a_n x + b_n$ or some other sequence tending to the right endpoint of the distributions of the X_i's.

In the derivation of the limit distributions of the maxima of independent, but not necessarily identically distributed rv, we investigate the product

$$\prod_{i \le n} P(X_i \le u_n) \sim \exp\left(-\sum_{i \le n} p_{i,n} \right).$$

More generally, we can consider a triangular array of independent rv $X_{i,n}$ which means that for each fixed n the rv $X_{i,n}, i \le n$, are independent. Then defining the factors $p_{i,n} = P(X_{i,n} > u_n)$ with u_n any suitable (large) boundary value or normalization, the distribution of the maximum of the $X_{i,n}, i \le n$, is given by the same product as above being approximated also by $\exp(-\sum_{i \le n} p_{i,n})$.

The connection to the iid case is obvious by setting $X_{i,n} = (X_i - b_n)/a_n$ and $u_n = x$. Hence, this scheme of triangular arrays of rv includes also the models introduced in Section 8.3. This scheme is analytically not much different and allows derivation with the same effort and method laws for maxima as well as for rare events.

Let us consider Poisson rv. It is known that the distribution of the linearly normalized maximum of n iid Poisson rv does not converge weakly to a limit distribution because the jumps of the discrete df not satisfy the condition

$$\frac{P(X = k)}{P(X \ge k)} \to 0 \quad \text{as } k \to \infty.$$

See e.g. Leadbetter et al. [303]. Here X is Poisson distributed, with $P(X = k)/P(X \ge k) \to 1$ as $k \to \infty$.

Anderson [14], [15] showed for Poisson variables X_i with parameter λ that there is a sequence of integers I_n for which

$$\lim_{n \to \infty} P\left(\max_{1 \le i \le n} X_i = I_n \text{ or } I_n + 1 \right) = 1.$$

Obviously no normalizing function $u_n(x)$ can be found which leads to a non-degenerate limit distribution. Thus the distribution of $\max_{1 \le i \le n} X_i$ concentrates increasingly on a pair of consecutive integers as $n \to \infty$. The asymptotic properties of the sequence of integers I_n have been characterized by Kimber [288]. This holds for any fixed λ.

However, if λ is large, the concentration on the two values I_n and $I_n + 1$ happens slowly with n. If we would let λ tend to ∞ with n, can we expect another behavior of the distribution of M_n? Because the Poisson distribution converges to a normal distribution as $\lambda \to \infty$, could the limit distribution of M_n be as the maximum of (approximately) normal rv, hence the Gumbel distribution? However, this is expected to depend on the rate of $\lambda = \lambda_n \to \infty$. Therefore, let $\{X_{i,n}, i \leq n\}$, $n \geq 1$, denote a triangular array of Poisson rv, which are iid with parameter λ_n for fixed n. Then the following Gumbel limit was derived in Anderson et al. [16].

Theorem 8.5.1. *Suppose that λ_n grows with n in such a way that for some integer $r \geq 0$,*

$$\log(n) = o(\lambda_n^{(r+1)/(r+3)}).$$

Then there is a linear normalization

$$u_n(x) = \lambda_n + \lambda_n^{1/2}(\beta_n^{(r)} + \alpha_n x)$$

such that

$$\lim_{n\to\infty} P\left(\max_{1\leq i\leq n} X_{i,n} \leq u_n(x) \right) = \exp(-e^{-x}).$$

It was shown that $\beta_n^{(r)}$ is the solution of the equation

$$h_n(x) = \frac{x^2}{2} + \log(x) + \frac{1}{2}\log(2\pi) - x^2 \sum_{j=1}^{r} c_j \left(\frac{x}{\lambda_n^{1/2}} \right)^j = \log(n)$$

and

$$\alpha_n = (2\log(n))^{-1/2}.$$

The constants c_j depend on the moments of the Poisson r.v. In general $\beta_n^{(r)} \sim (2\log(n))^{1/2}$. More explicitly, for $r = 0, 1, 2$, we use

$$\beta_n^{(0)} = (2\log(n))^{1/2} - \frac{\log\log(n) + \log(4\pi)}{2(2\log(n))^{1/2}},$$

$$\beta_n^{(1)} = (2\log(n))^{1/2} - \frac{\log\log(n) + \log(4\pi)}{2(2\log(n))^{1/2}} + \frac{1}{6}\frac{2\log(n)}{\lambda_n^{1/2}},$$

and

$$\beta_n^{(2)} = \beta_n^{(0)} + (2\log(n))^{1/2} \left(\frac{1}{6}\frac{(2\log(n))^{1/2}}{\lambda_n^{1/2}} - \frac{1}{24}\frac{(2\log(n))}{\lambda_n} \right).$$

The case $r = 0$, i.e., $\log(n) = o(\lambda_n^{1/3})$, or equivalently $(\log(n))^3/\lambda_n = o(1)$, covers the rather fast-growing λ_n, hence a rather good approximation of the Poisson to the normal rv. So the discreteness of the Poisson distribution has no effect on the limiting distribution since we use the same normalization as in the case of iid normal rv. If $r \geq 1$, then this discreteness has a limited influence on the convergence of M_n, since the normalization has only to be adapted to the speed of growth of λ_n by chosen $\beta_n^{(r)}$ appropriately. As λ_n grows less fast, the adaption gets more involved.

The arguments of Anderson et al. [16] for a Gumbel limit do not depend critically on the rv being Poisson. They extended the above result to row-wise maxima of certain triangular arrays of independent variables, each converging in distribution to a normal rv at a certain speed. Then similar results can be derived which depend again on the growth of the parameters or the speed of convergence to the normal distribution.

Example 8.5.2. Let us consider another simple example. For fixed n let $X_{i,n}$ be a discrete uniform rv on the values $1, \ldots, N_n$ where $N_n \to \infty$ as $n \to \infty$. Then for $k = k_n$ such that $k_n/N_n \to 0$ we get

$$P(M_n \leq N_n - k_n) = (1 - k_n/N_n)^n \sim \exp(-k_n n/N_n).$$

Hence the convergence depends on the behavior of n/N_n and suitably chosen k_n. If $n/N_n \to c \in (0, \infty)$, then $k_n = k$ are the normalization values and the limit is e^{-kc}, the discrete geometric type distribution. If $n/N_n \to c = 0$, then $k_n = [xN_n/n] = xN_n/n + O(1)$ with $x > 0$, and the limit is e^{-x}, hence the limit distribution of the normalized maximum M_n is G_{-1}. If $n/N_n \to c = \infty$, then it is easily seen that $P(M_n = N_n) \to 1$.

Such a result was found earlier by Kolchin [291] where he considered multinomially distributed rv with N equally probable events. These limits depend in a similar way on the behavior of n/N as $n \to \infty$ with $N = N_n \to \infty$. Since he approximated the multinomial distribution by the Poisson distribution, his investigations were rather similar for the particular case of Poisson rv. Related results for order statistics, including expansions, may be found in [385], Section 4.6 (with a discussion on page 150).

Nadarajah and Mitov [344] showed that this behavior holds also for maxima of discrete rv from the uniform, binomial, geometric and negative binomial distribution, with varying parameters. They derived the suitable normalizations for the convergence of the normalized maximum to a limit distribution. This limit is the Gumbel distribution in case of binomial, geometric and negative binomial distribution.

Finally we mention that such a triangular scheme can be based also on rv. Coles and Pauli [75] extended the univariate problem of Anderson et al. [16] to the bivariate problem with Poisson distributed rv. They considered the case that $(X_{i,n}, Y_{i,n}), i \leq n$, is a triangular array of independent Poisson rv, defined by $X_{i,n} = U_{i,n} + W_{i,n}$ and $Y_{i,n} = U_{i,n} + W_{i,n}$ with independent Poisson rv $U_{i,n}, V_{i,n}, W_{i,n}$ with parameters $\lambda_n - d_n, \lambda_n - d_n$, and d_n, respectively. Hence $X_{i,n}$ and $Y_{i,n}$ are dependent Poisson rv, each with parameter λ_n and covariance d_n. If d_n is the dominating term, meaning that d_n/λ_n tends to 1, we can expect an asymptotic dependence of the extremes of the two components. More precisely, they showed that if

$$(1 - d_n/\lambda_n) \log(n) \to \lambda^2 \in (0, \infty)$$

and

$$\log(n) = o(\lambda_n^{(r+1)/(r+3)}) \quad \text{for some integer } r \geq 0,$$

then there exists a sequence $u_n(x)$ (defined above as the normalization in the univariate result of Anderson et al. [16]) such that

$$\lim_{n \to \infty} P\left(\max_{i \leq n} X_{i,n} \leq u_n(x), \max_{i \leq n} Y_{i,n} \leq u_n(y) \right) = H_\lambda(x, y)$$

where H_λ denotes the bivariate EVD defined in Example 4.1.4 having a copula function which was derived from the bivariate Gaussian distribution.

This result holds also for the particular cases where (i) $\lambda = 0$ assuming in addition that $\lambda_n - d_n \to \infty$, and where (ii) $\lambda = \infty$ assuming that $d_n/\lambda_n \to 1$. If $\lambda = 0$, the row-wise maxima $\max_{i \leq n} X_{i,n}$ and $\max_{i \leq n} Y_{i,n}$ are asymptotically completely dependent, and if $\lambda = \infty$, then they are asymptotically independent.

Chapter 9

Extremes of Random Sequences

We develop the general theory of extremes and exceedances of high boundaries by non-stationary random sequences. Of main interest is the asymptotic convergence of the point processes of exceedances or of clusters of exceedances. These results are then applied for special cases, as stationary, independent and particular non-stationary random sequences.

9.1 Introduction and General Theory

In this section we consider general non-stationary random sequences $\{X_i, \ i \geq 1\}$. The rv X_i are real-valued with marginal distributions $F_{X_i}(\cdot) = F_i(\cdot)$; extensions to rv are possible with some additional effort. (See Section 11.5 and Chapter 4). The aim of this section is to present a rather general and unified theory to derive the most important results. In doing this we will pose conditions which are slightly more restrictive than essentially needed. The more general results can be found in the literature.

We deal with rare events, in this context with the exceedances of a boundary by a random sequence. Of interest in this section are the occurrence times of such rare events and not the excesses above the boundary. The boundary $\{u_{ni}, \ i \leq n, \ n \geq 1\}$ for a given n is non-constant in general (see Figure 9.1.1). In Chapter 8 we showed that such an extension is natural and needed. As already mentioned, the Poisson approximation of the sequence of exceedances is one of the topics we are interested in. We begin by discussing non-stationary random sequences. In the following sections we deal with certain cases of non-stationary random sequences and apply the results to stationary and to independent ones.

M. Falk et al., *Laws of Small Numbers: Extremes and Rare Events*, 3rd ed.,
DOI 10.1007/978-3-0348-0009-9_9, © Springer Basel AG 2011

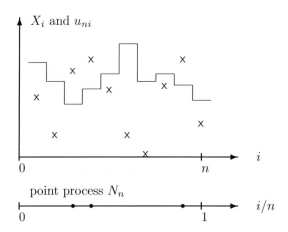

FIGURE 9.1.1. Exceedances of boundary values u_{ni} (as step function) by a random sequence X_i (symbol: x) and the related point process N_n of exceedances.

We mentioned in Section 8.2 that for the general case the uan condition is necessary in order to prove the Poisson approximation. Thus from now on we assume that

$$\sup_{i \leq n} P(X_i > u_{ni}) = \sup_{i \leq n}(1 - F_i(u_{ni})) \to 0, \qquad n \to \infty.$$

Of main interest is the point process N_n of the exceedances, counting the number of points i/n at which the rv X_i exceeds the boundary u_{ni}:

$$N_n(\cdot) = \sum_{i \leq n} \delta_{i/n}(\cdot)\, 1(X_i > u_{ni})$$

where $\delta_x(\cdot)$ denotes the point measure at x: $\delta_x(B) = 1$ if $x \in B$, and 0, else. The point process N_n is defined on the unit interval $(0, 1]$ (see Figure 9.1.1). For any Borel set $B \subset (0, 1]$, $N_n(B)$ is the rv counting the number of exceedances $X_i > u_{ni}$ for $i \in nB$. Letting $B = (0, 1]$ and $u_{ni} = u_n$, $i \leq n$, we get the following relation between N_n and the maximum M_n:

$$\{N_n((0, 1]) = 0\} = \{X_i \leq u_n, \, i \leq n\} = \{M_n \leq u_n\}$$

for any n. However, the point process N_n contains more information about the behavior of the exceedances and the extreme order statistics. We shall show for instance that exceedances in pairwise disjoint time intervals $B_i \in (0, 1]$, $i = 1, \ldots, k$, are, under certain restrictions, asymptotically independent.

N_n is said to *converge in distribution* to N ($N_n \to_D N$) if the rv $(N_n(B_j))_{j \leq k}$ converges in distribution to $(N(B_j))_{j \leq k}$ for arbitrary Borel sets $B_j \subset (0, 1]$, $j \leq k$, $k \in \mathbb{N}$, on the boundaries of which N has no point, with probability 1.

In the following the discussion will be confined to the case where the limiting point process N is a Poisson process. This simplifies the proof of the convergence $N_n \to_D N$, since we can make use of the property that the Poisson point process is *simple*, (multiple points do not occur with probability 1), iff the intensity measure is atomless. Furthermore, we only consider point processes where the corresponding intensity measure λ is a Radon measure, which means that $\lambda(B) = E(N(B)) < \infty$, for any Borel set $B \subset (0, 1]$. This is equivalent to $\lambda((0, 1]) < \infty$. The following theorem is an application of a result of Kallenberg [282], (see Leadbetter et al. [303], Resnick [393]).

Theorem 9.1.1. *Let N, N_n, $n \geq 1$, be point processes on $(0, 1]$. Assume that N is simple, with $\lambda((0, 1]) < \infty$. If*

a) $E(N_n((c, d])) \to E(N((c, d]))$ *for every $(c, d] \subset (0, 1]$, and*

b) $P(N_n(B) = 0) \to P(N(B) = 0)$ *for every B, which is a finite union of intervals $B_j = (c_j, d_j] \subset (0, 1]$,*

then $N_n \to_D N$.

Remark 9.1.2. This result is also valid for processes on $\mathbb{R}_+^d$ or $(0, 1]^d$, $(d \geq 1)$, if the semiclosed intervals are replaced by the semiclosed rectangles in $\mathbb{R}_+^d$ or $(0, 1]^d$, $(d \geq 1)$, respectively.

We now check whether the conditions of Theorem 9.1.1 are satisfied by the exceedance point process N_n. We easily derive that

$$E\Big(N_n\big((c, d]\big) \Big) = E\Big(\sum_{i \leq n} \delta_{i/n}\big((c, d]\big) 1(X_i > u_{ni}) \Big)$$

$$= \sum_{i \leq n} E\Big(\delta_{i/n}\big((c, d]\big) 1(X_i > u_{ni}) \Big)$$

$$= \sum_{i \in (nc, nd]} E(1(X_i > u_{ni})) = \sum_{i \in (nc, nd]} P(X_i > u_{ni}).$$

Therefore condition a) holds if the sum $\sum_{i \in nB}(1 - F_i(u_{ni}))$ converges to $\lambda(B) = E(N(B))$ for any interval B or equivalently

$$\sum_{i \leq nt} P(X_i > u_{ni}) = \sum_{i \leq nt}(1 - F_i(u_{ni})) \to \lambda(t) = \lambda((0, t]), \qquad (9.1)$$

as $n \to \infty$ for every $t \in (0, 1]$ with $\lambda(1) < \infty$.

Note that we use the same notation for the function $\lambda(\cdot)$ and the intensity measure $\lambda(\cdot)$. Each defines the other in a unique way. Thus the measure $\lambda(\cdot)$ is atomless iff the function $\lambda(\cdot)$ is continuous. The question whether condition a) of Theorem 9.1.1 is satisfied, is thus reduced to the question whether one can find a suitable normalization $\{u_{ni}, i \le n\}$ which only depends on the marginal distributions F_i. If the random sequence is stationary and the boundaries are constant for each n, i.e., $u_{ni} = u_n$, then (9.1) is equivalent to

$$n(1 - F(u_n)) \to \lambda < \infty, \tag{9.2}$$

Thus $\lambda(t) = \lambda t = \lambda m((0, t])$, m denoting the Lebesgue measure.

Long Range Dependence

In condition a) the dependence among the rv X_i does not play a role but it becomes important in condition b) of Theorem 9.1.1. Since the limit point process N is a Poisson process, we have

$$P(N(B) = 0) = \prod_j P(N(B_j) = 0)$$

for finitely many disjoint intervals $B_j \subset (0, 1]$. This also holds asymptotically for point processes N_n if the numbers $N_n(B_j)$ of exceedances of the boundary u_{ni} by X_i, occurring in disjoint (separated) intervals nB_j, become approximately independent for large n. The importance of this property was already shown in Example 8.2.1. This property is called *mixing*. There are several mixing conditions in the literature. For maxima and exceedances the mixing property has to be formulated with respect to the events $\{X_i \le u_{ni}\}$ only, given in Hüsler [237], applying ideas of Leadbetter (c.f. Leadbetter [303]). Let $\alpha_{n,m}$ be such that

$$|P(X_{i_1} \le u_{ni_1}, \ldots, X_{i_k} \le u_{ni_k}, X_{j_1} \le u_{nj_1}, \ldots, X_{j_l} \le u_{nj_l})$$
$$- P(X_{i_1} \le u_{ni_1}, \ldots, X_{i_k} \le u_{ni_k}) P(X_{j_1} \le u_{nj_1}, \ldots, X_{j_l} \le u_{nj_l})|$$
$$\le \alpha_{n,m}$$

for any integers $0 < i_1 < i_2 < \cdots < i_k < j_1 < \cdots < j_l \le n$ for which $j_1 - i_k \ge m$.

Definition 9.1.3. $D(u_{ni})$ holds for the random sequence $\{X_i\}$ with respect to the boundary $\{u_{ni}, i \le n, n \ge 1\}$, if there exists a sequence $\{m_n\}$ such that $\alpha_{n,m_n} \to 0$ and $m_n \bar{F}_{n,max} \to 0$ as $n \to \infty$, where $\bar{F}_{n,max} = \sup_{i \le n}(1 - F_i(u_{ni}))$.

In the following cases we have $\lambda((0, 1]) > 0$ implying $\liminf_n n\bar{F}_{n,max} > 0$. Hence the assumption $m_n \bar{F}_{n,max} \to 0$ implies $m_n = o(n)$.

We can always choose a sequence $\{k_n\}$ of integers such that

$$\lim_{n \to \infty} k_n m_n \bar{F}_{n,max} = 0 \quad \text{and} \quad \lim_{n \to \infty} k_n \alpha_{n,m_n} = 0. \tag{9.3}$$

For instance, $k_n = [\min(m_n \bar{F}_{n,max}, \alpha_{n,m_n})]^{-1/2}$ is such a sequence. Note that $\{k_n\}$ can be bounded or can tend to ∞, but by the same reasons $k_n m_n = o(n)$. In the following $\{k_n\}$ denotes always a sequence satisfying (9.3).

Lemma 9.1.4. *Suppose that $D(u_{ni})$ holds for the random sequence $\{X_i,\ i \geq 1\}$ with respect to the boundary $\{u_{ni}\}$. Let B_j ($= B_{j,n}$), $j \leq k_n$, be disjoint intervals of $(0,1]$, where (9.3) holds for k_n. Then, as $n \to \infty$,*

$$P\left(N_n\left(\bigcup_{j \leq k_n} B_j\right) = 0\right) - \prod_{j \leq k_n} P(N_n(B_j) = 0) \to 0.$$

This lemma is proved in the usual way, using the mixing property $k_n - 1$ times to approximate for each l, $2 \leq l \leq k_n$, $P(N_n(\bigcup_{j \leq l-1} B_j \cup B_l) = 0)$ by $P(N_n(\bigcup_{j \leq l-1} B_j) = 0)P(N_n(B_l) = 0)$. If the nB_j's are separated by m_n, the statement follows, since $\{k_n\}$ is chosen such that $k_n \alpha_{n,m_n} \to 0$ as $n \to \infty$. If they are not separated by m_n, then the B_j's are approximated by B_j^*'s constructed from B_j by deleting a small interval of length m_n/n at the right end of each B_j. These nB_j^*'s are separated by m_n and the approximation error tends to 0 since $k_n m_n \bar{F}_{n,max} \to 0$ (cf. Leadbetter et al. [303], Hüsler [237], [240] and Leadbetter and Nandagopalan [304]). By this lemma, the verification of condition b) in Theorem 9.1.1 is reduced to the verification of the convergence

$$P(N_n(B) = 0) \to P(N(B) = 0)$$

for any $B = (c,d] \subset (0,1]$. B may be split up into k_n suitably chosen disjoint intervals $B_{j,n}$, $j \leq k_n$, such that $m(B_{j,n}) \to 0$. Applying Lemma 9.1.4 once more

$$P(N_n(B) = 0) - \prod_{j \leq k_n} P(N_n(B_{j,n}) = 0) \to 0.$$

LOCAL DEPENDENCE

The problem is now reduced to the consideration of $P(N_n(B_{j,n}) = 0) = P(X_i \leq u_{ni}, i \in nB_{j,n})$. We note that only the local dependence behavior of the random sequence $\{X_i\}$ is involved in this probability. However, the local dependence behavior of this sequence is not restricted by the mixing condition $D(u_{ni})$ and we need a further condition $D^*(u_{ni})$, which restricts the local dependence. Let $\{k_n\}$ satisfy (9.3) and let α_n^* be such that

$$\sum_{i<j<j+1\in I} P(X_i > u_{ni},\ X_j \leq u_{nj},\ X_{j+1} > u_{n,j+1}) \leq \alpha_n^*$$

for any intervals $I = \{i_1 \leq i \leq i_2 \leq n\} \subset \mathbb{N}$ with

$$\sum_{i \in I} P(X_i > u_{ni}) \leq \sum_{i \leq n} P(X_i > u_{ni})/k_n.$$

Definition 9.1.5. $D^*(u_{ni})$ holds for the random sequence $\{X_i\}$ with respect to the boundary $\{u_{ni}\}$, if $k_n \alpha_n^* \to 0$ as $n \to \infty$.

Note that Condition $D^*(u_{ni})$ excludes the possibility of clustering of upcrossings in a small interval I, because it excludes cases where the random sequence $\{X_i\}$ oscillates rapidly around the boundary $\{u_{ni}\}$.

Lemma 9.1.6. *Suppose that condition $D(u_{ni})$ and $D^*(u_{ni})$ hold for $\{X_i\}$ with respect to the boundary $\{u_{ni}\}$. Then, as $n \to \infty$,*

$$P(X_i \le u_{ni},\ i \le nt) \to \exp(-\mu(t))$$

for $t \le 1$, for some function $\mu(t)$, if and only if

$$\sum_{i \le nt-1} P(X_i \le u_{ni},\ X_{i+1} > u_{n,i+1}) \to \mu(t), \tag{9.4}$$

for $t \le 1$, where $\mu(\cdot)$ is a bounded function.

Obviously, $\mu(\cdot)$ is a non-decreasing positive function with $\mu(1) < \infty$. Note that the sum in (9.4) can be taken also over all terms $i \le nt$ for $t < 1$ which will give asymptotically the same result. This holds also for $t = 1$, by letting $u_{n,n+1}$ be some value such that $P(X_{n+1} > u_{n,n+1})$ tends to 0.

Lemma 9.1.6 is not formulated in terms of exceedances as in statement (9.1), but in terms of upcrossings (condition (9.4)). In general,

$$\mu(t) \le \lambda(t), \qquad t \le 1.$$

It is easy to verify that if $\mu(1) = \lambda(1)$ then $\mu(\cdot) \equiv \lambda(\cdot)$, since for any $t \le 1$,

$$0 \le \sum_{i \le nt-1} [P(X_{i+1} > u_{n,i+1}) - P(X_i \le u_{ni},\ X_{i+1} > u_{n,i+1})]$$

$$\le \sum_{i \le n-1} [P(X_{i+1} > u_{n,i+1}) - P(X_i \le u_{ni},\ X_{i+1} > u_{n,i+1})]$$

$$\to \lambda(1) - \mu(1) = 0,$$

as $n \to \infty$.

Instead of $D^*(u_{ni})$ one might use weaker conditions as shown by Chernick et al. [67] for stationary sequences. They used the condition $D^{(k)}(u_n)$, where $k \ge 2$ is fixed and the boundary u_n is constant. This condition $D^{(k)}(u_n)$ is said to hold if

$$\lim_{n \to \infty} nP\left(X_1 > u_n > \max_{2 \le i \le k} X_i,\ \max_{k+1 \le j \le r_n} X_j > u_n \right) = 0,$$

where $r_n = [n/k_n], k_n = o(n)$ with k_n satisfying (9.3) (i.e., $k_n \alpha_{n,m_n} \to 0$ and $k_n m_n/n \to 0$ for the stationary case). Obviously, $D^{(k)}(u_n)$ implies $D^{(k+1)}(u_n)$ and $D^{(2)}(u_n)$ corresponds to $D^*(u_n)$ in the stationary case, which was proposed by Leadbetter and Nandagopalan [304]. Weaker conditions like $D^{(k)}(u_n)$ are useful in dealing with special random sequences (see Chernick et al. [67] and Section 11.5).

Point Process of Exceedances

We assume in the following always that $\mu(\cdot)$ and $\lambda(\cdot)$ are continuous functions such that the corresponding Poisson point processes are simple. However, the results hold also if these functions are discontinuous at some points $t \leq 1$ (see Section 11.5).

If now $\lambda(\cdot) \equiv \mu(\cdot)$, then upcrossings and exceedances occur with the same intensity and the corresponding limit point processes are the same.

Theorem 9.1.7. *Suppose that conditions $D(u_{ni})$ and $D^*(u_{ni})$ hold for the random sequence $\{X_i,\ i \geq 1\}$ with respect to the boundary $\{u_{ni},\ i \leq n,\ n \geq 1\}$. If (9.1) and (9.4) hold with $\mu(\cdot) \equiv \lambda(\cdot)$ being continuous, then*

$$N_n \to_D N \qquad as \quad n \to \infty,$$

N being a Poisson process on $(0,1]$ with intensity measure $\lambda(\cdot)$.

Proof. To prove the statement we apply the mentioned lemmas showing the two conditions of Theorem 9.1.1. The first statement follows by the convergence of the mean numbers of exceedances and upcrossings to $\mu(t) = \lambda(t)$ for any t. The second statement of Theorem 9.1.1 needs long range dependence. Let B_j be finitely many disjoint intervals of $(0,1]$, $j \leq J$, with $B = \bigcup_j B_j$. Then condition $D(u_{ni})$ implies by Lemma 9.1.4 that

$$P(N_n(B) = 0) = \prod_{j \leq J} P(N_n(B_j) = 0) + o(1).$$

We can partition $(0,1]$ into $B_l^*, l \leq k_n$, with k_n as in (9.3) and such that the mean number of exceedances in any of the B_l^* is bounded by $\sum_{i \leq n} P(X_i > u_{ni})/k_n$. These subintervals partition each B_j into disjoint $B_{j,l} = B_j \cap B_l^*, l \leq k_n$. Note that some of these intervals $B_{j,l}$ are empty and thus can be deleted. Again Lemma 9.1.4 implies by long range dependence that

$$P(N_n(B) = 0) = \prod_{j \leq J, l \leq k_n} P(N_n(B_{j,l}) = 0) + o(1).$$

Because of the uniform bound for $P(N_n(B_{j,l}) > 0)$ which tends to 0 as $n \to \infty$, we can approximate the product by

$$\exp\left(-(1 + o(1)) \sum_{j,l} P(N_n(B_{j,l}) > 0)\right).$$

Using the local dependence condition $D^*(u_{ni})$ and Bonferroni's inequality, each term of the last sum can be bounded by

$$P(N_n(B_{j,l}) > 0) \leq \sum_{i \in nB_{j,l}} P(X_i \leq u_{ni}, X_{i+1} > u_{n,i+1}) + \bar{F}_{n,max}$$

and

$$P(N_n(B_{j,l}) > 0)$$

$$\geq \sum_{i \in nB_{j,l}} P(X_i \leq u_{ni}, X_{i+1} > u_{n,i+1})$$

$$- \sum_{i < i', i, i' \in nB_{j,l}} P(X_i \leq u_{ni}, X_{i+1} > u_{ni}, X_{i'} \leq u_{ni'}, X_{i'+1} > u_{n,i'+1}).$$

The double sum is bounded by α_n^* for each j, l. Since $k_n \alpha_n^* \to 0$, the sum (on j, l) of all the double sums tends also to 0 as $n \to \infty$. Summing the first sums in the Bonferroni inequality we get $\sum_j \mu(B_j)$ in the limit by (9.4) and (9.3). Hence combining terms it shows that $P(N_n(B) = 0) \to \exp(-\sum_j \mu(B_j)) = \exp(-\mu(B)) = \exp(-\lambda(B)) = P(N(B) = 0)$ which is the second statement of Theorem 9.1.1 and implies the stated convergence result.   □

For a converse statement see Section 11.1. In general, the Poisson process is non-homogeneous, since $\lambda(t) \neq \lambda t$. Theorem 9.1.7 states that if $\lambda(\cdot) = \mu(\cdot)$ then asymptotically each exceedance is an upcrossing, i.e., if there is an upcrossing at i, then $i+1, i+2, \ldots$ are asymptotically not exceedance points. The fact that N is a Poisson process suggests that upcrossings occur separated in time, i.e., they do not cluster.

The result of Theorem 9.1.7 follows also by assuming the so-called condition $D'(u_{ni})$ (see Leadbetter et al. [303] for the stationary case with constant boundary and Hüsler [237] for the non-stationary case):

$$\lim_{n \to \infty} k_n \sum_{i < j \in I} P(X_i > u_{ni}, X_j > u_{nj}) = 0,$$

for the same sets I as in the condition $D^*(u_{ni})$. $D'(u_{ni})$ implies that the probability of exceedances occurring at neighboring points is asymptotically 0. Hence, if $D'(u_{ni})$ and (9.1) hold, then $D^*(u_{ni})$ and (9.4) hold with $\mu(1) = \lambda(1)$ and $N_n \to N$, the Poisson process with intensity measure λ.

POINT PROCESS OF UPCROSSINGS

In some applications one observes a different behavior of the upcrossings and exceedances. If there is an upcrossing at a given point then the sequence remains above the boundary for the next few time points. It is then obvious that $\mu(t) < \lambda(t)$ for some $t \leq 1$. Since some exceedances occur in clusters, the limit point process cannot be a simple Poisson process. It is in general possible to show that N_n converges to an (extended) compound Poisson process $\sum \beta_i \delta_{\tau_i}$ where the β_i's are independent rv in $I\!N$ (cf. Hsing et al. [231] for the stationary case and Nandagopalan et al. [347] for the non-stationary case). The β_i's (the cluster sizes) are independent of the τ_i's (the occurrence times of the clusters), but not identically distributed in general. In the stationary case, $\beta_i =_D \beta_1$, and N is a compound Poisson process (see Section 11.2).

However, the point process $\tilde{N}_n$, consisting of the upcrossings only, can be approximated by a Poisson process. An upcrossing at the point i/n is given by the event $\{X_{i-1} \le u_{n,i-1}, X_i > u_{ni}\}, i \le n$, and the corresponding point process $\tilde{N}_n$ of upcrossings by

$$\tilde{N}_n(\cdot) = \sum_{1 < i \le n} \delta_{i/n}(\cdot)\, 1(X_{i-1} \le u_{n,i-1}, X_i > u_{ni}).$$

Theorem 9.1.8. *Suppose that the conditions $D(u_{ni})$ and $D^*(u_{ni})$ hold for the random sequence $\{X_i,\ i \ge 1\}$ with respect to the boundary $\{u_{ni},\ i \le n,\ n \ge 1\}$. Then (9.4) implies*

$$\tilde{N}_n \to_D \tilde{N} \qquad as \quad n \to \infty,$$

where $\tilde{N}$ is a Poisson process on $(0,1]$ with intensity measure $\mu(\cdot)$.

A similar result can be shown for the point process of downcrossings given by the events $\{X_{i-1} > u_{n,i-1},\ X_i \le u_{ni}\},\ i \le n$. The points τ_i's of the compound Poisson process $\sum \beta_i \delta_{\tau_i}$, mentioned above, are the points of the (underlying) Poisson process $\tilde{N}$ of occurrences of clusters of exceedances. The whole cluster is thinned here or replaced by the first (or last) exceedance. But we might also consider the whole cluster as an event.

Point Process of Clusters

We can define the point process N_n^* of cluster positions. Let B_j, $j \le k_n$, be small intervals which form a partition of $(0,1]$. Then the events $\{N_n(B_j) \ne 0\}$ define the point process

$$N_n^*(\cdot) = \sum_{j \le k_n} \delta_{t_j}(\cdot)\, 1(N_n(B_j) \ne 0),$$

with some $t_j = t_j(n) \in B_j$, representing the position of the cluster. Thus, for any Borel set $B \subset (0,1]$,

$$N_n^*(B) = \sum_{j \le k_n, t_j \in B} 1(N_n(B_j) \ne 0)$$

counts the number of clusters of exceedances with cluster position in B. We might choose for instance the first point, the center or the last point of B_j as t_j. Note that by this approach or definition there might be runs of exceedances which are separated by B_j into two clusters or there might be runs within B_j which are joint to form one cluster. However, Theorem 9.1.10 states that both, the run and the block definition of clusters, lead asymptotically to the same Poisson process, if nB_j are suitably growing sets. Another way to define the process of cluster positions of exceedances is given by $\sum_{j \leq k_n} 1(N_n(B_j \cap B) \neq 0)$, which may differ from N_n^* for finite n, but they are asymptotically equivalent. N_n^* can be approximated by a simple Poisson process, since the multiplicities representing the cluster sizes are not accounted for.

In choosing B_j one has to take into consideration the non-stationarity of the sequence X_i and the non-constant boundaries $\{u_{ni}\}$. Thus, for a given n we choose successively $0 = i_0 < i_1 < i_2 < \cdots < i_{k_n} \leq n$ such that

$$\sum_{i_{j-1} < i \leq i_j} (1 - F_i(u_{ni})) \leq \sum_{i \leq n} (1 - F_i(u_{ni}))/k_n$$

and

$$\sum_{i_{j-1} < i \leq i_j + 1} (1 - F_i(u_{ni})) \geq \sum_{i \leq n} (1 - F_i(u_{ni}))/k_n,$$

are satisfied. Define $B_j = (i_{j-1}/n, i_j/n]$ for $j \leq k_n$. Note that the time domain $\{i : i \leq n\}$ is split up with respect to the probabilities of the exceedances and the i_j's are chosen maximally. By this choice the possible exceedances at the points $i_{k_n} + 1, \ldots, n$ are not considered. Because of the maximally chosen i_j's it follows easily that $\sum_{i_{k_n} < i \leq n} (1 - F_i(u_{ni})) \leq k_n \bar{F}_{n,max}$. Thus this marginal effect of exceedances in these last points is asymptotically negligible. We need also that the $P(N_n(B_j) \neq 0)$ are uniformly converging to 0 which is simply implied if the term $\sum_{i \leq n} (1 - F_i(u_{ni}))/k_n$ tends to 0 which is assumed in the following. This condition is obviously true if the sum is bounded, which holds if $\lambda(1) < \infty$. Further, we should fix the cluster position. A reasonable choice for the t_j's is given for instance by the right endpoints of B_j: $t_j = i_j/n$.

Theorem 9.1.9. *Suppose that the conditions $D(u_{ni})$ and $D^*(u_{ni})$ hold for the random sequence $\{X_i, i \geq 1\}$ with respect to the boundary $\{u_{ni}, i \leq n, n \geq 1\}$. If*

$$\sum_{j: t_j \leq t} P(N_n(B_j) \neq 0) \to \mu^*(t), \qquad t \leq 1,$$

with $\mu^*(\cdot)$ continuous and $\mu^*(1) < \infty$, with the B_j's constructed as above, then

$$N_n^* \to_D N^* \qquad as \quad n \to \infty,$$

where N^* is a Poisson process on $(0, 1]$ with intensity measure $\mu^*(\cdot)$.

The two point processes, the point process of upcrossings $\tilde{N}_n$ and the point process of clusters N_n^* are asymptotically related. Assuming that the limits $\mu(t)$ and $\mu^*(t)$ hold, we show that (9.3) implies $\mu^*(t) \leq \mu(t) (\leq \lambda(t))$ for all $t \in (0, 1]$. Furthermore, (9.3) together with $D^*(u_{ni})$ implies $\mu^* = \mu$ and therefore N_n^* and $\tilde{N}_n$ converge to the same Poisson process $\tilde{N}$ with intensity measure μ.

$$\mu^*(t) = \lim_{n\to\infty} \sum_{j:i_j \leq nt} P(N_n(B_j) \neq 0)$$

$$\leq \lim_{n\to\infty} \left(\sum_{j:i_j \leq nt} \left(\sum_{i \in nB_j} P(X_i \leq u_{ni}, X_{i+1} > u_{n,i+1}) \right. \right.$$

$$\left. \left. + P(X_{i_{j-1}+1} > u_{n,i_{j-1}+1}) \right) \right)$$

$$\leq \lim_{n\to\infty} \left(\sum_{i \leq i_{j(t)}} P(X_i \leq u_{ni}, X_{i+1} > u_{n,i+1}) + O(k_n \bar{F}_{n,max}) \right)$$

$$\leq \lim_{n\to\infty} \sum_{i \leq nt} P(X_i \leq u_{ni}, X_{i+1} > u_{n,i+1}) = \mu(t),$$

where $j(t)$ is the largest j such that $i_j \leq nt$ for fixed n. Conversely,

$$\mu^*(t) = \lim_{n\to\infty} \sum_{j:i_j \leq nt} P(N_n(B_j) \neq 0) = \lim_{n\to\infty} \sum_{j \leq j(t)} P(N_n(B_j) \neq 0)$$

$$\geq \lim_{n\to\infty} \left(\sum_{j \leq j(t)} \left(\sum_{i \in nB_j} P(X_{i-1} \leq u_{n,i-1}, X_i > u_{ni}) \right. \right.$$

$$\left. \left. - \sum_{i<l \in nB_j} P(X_{i-1} \leq u_{n,i-1}, X_i > u_{ni}, X_{l-1} \leq u_{n,l-1}, X_l > u_{nl}) \right) \right)$$

$$\geq \lim_{n\to\infty} \left(\sum_{i \leq i_{j(t)}} P(X_{i-1} \leq u_{n,i-1}, X_i > u_{ni}) + O(k_n \alpha_n^*) \right)$$

$$= \lim_{n\to\infty} \sum_{i \leq nt} P(X_{i-1} \leq u_{n,i-1}, X_i > u_{ni}) = \mu(t),$$

by condition $D^*(u_{ni})$ and the uniformity assumption. This result holds for all choices of $t_j \in B_j$. Therefore Theorem 9.1.9 implies

Theorem 9.1.10. *Suppose that the conditions $D(u_{ni})$ and $D^*(u_{ni})$ hold for the random sequence $\{X_i,\ i \geq 1\}$ with respect to the boundary $\{u_{ni},\ i \leq n,\ n \geq 1\}$. Define N_n^* as above, where $t_j \in B_j$, $j \leq k_n$, are chosen arbitrarily. Then (9.4) implies*

$$N_n^* \to_D \tilde{N} \qquad as \quad n \to \infty$$

where $\tilde{N}$ is a Poisson process on $(0,1]$ with intensity measure $\mu(\cdot)$.

The theory for exceedances can easily be generalized to other rare events $\{X_i \in A_{ni}\}$, where the A_{ni}'s form a triangular array of Borel sets of $\mathbb{R}$. This is dealt with in Section 11.1.

9.2   Applications: Stationary Sequences

Let us now assume that the random sequence $\{X_i,\ i \geq 1\}$ is stationary and that the boundaries are constant $u_{ni} = u_n, i \leq n$, for each $n \geq 1$. With these restrictions the presented conditions in the previous section are simplified. Instead of (9.1) we assume now (9.2), i.e., $n(1 - F(u_n)) \to \lambda$. In $D^*(u_n)$ define $r_n^* = [n/k_n]$, k_n as in (9.3) (i.e., $k_n m_n = o(n)$ and $k_n \alpha_{nm_n} \to 0$ since $\bar{F}_{n,max} = O(1/n)$), and

$$\alpha_n^* = r_n^* \sum_{2 \leq j \leq r_n^* - 1} P(X_1 > u_n,\ X_j \leq u_n,\ X_{j+1} > u_n).$$

Moreover instead of (9.4) we assume

$$\lim_{n \to \infty} nP(X_1 \leq u_n,\ X_2 > u_n) = \mu < \infty. \tag{9.5}$$

Obviously, $\mu(t) = t\mu$ and $\lambda(t) = t\lambda$ are continuous. The general results derived in Section 9.1 can be reformulated for stationary sequences. For this purpose we define the intervals B_j (of fixed length r_n^*/n) as follows:

$$B_j = \left(\frac{(j-1)r_n^*}{n},\ \frac{jr_n^*}{n} \right].$$

Note that r_n^* denotes the number of rv in a block B_j. We can choose any of the points of B_j as t_j, (or even j/k_n, since the right endpoint of B_j is $jr_n^*/n = (1 + O(k_n/n))j/k_n$). From now on let $t_j = jr_n^*/n$. We have again $\mu^*(\cdot) \equiv \mu(\cdot)$.

Corollary 9.2.1. *Suppose that the conditions $D(u_n)$ and $D^*(u_n)$ hold for the stationary random sequence $\{X_i,\ i \geq 1\}$ with respect to the constant boundary $\{u_n,\ n \geq 1\}$. Let $\hat{N}$ be a homogeneous Poisson process with intensity measure $\mu(\cdot)$ defined by $\mu(t) = \mu t$.*

  (i) *(9.2) and (9.5) with $\lambda = \mu$ imply*

$$N_n \to_D \hat{N} \qquad as \quad n \to \infty.$$

  (ii) *(9.5) implies*

$$\tilde{N}_n \to_D \hat{N} \qquad as \quad n \to \infty$$

  and

$$N_n^* \to_D \hat{N} \qquad as \quad n \to \infty.$$

  (iii) *(9.2) and (9.5) imply*

$$P(M_n \leq u_n) \to \exp(-\mu) = \exp(-\theta\lambda),$$

  where $\theta = \mu/\lambda \leq 1$.

These results were given by Leadbetter and Nandagopalan [304]. θ is called the *extremal index* (Leadbetter [300]) and has the following property: Let $F \in D(G)$, G an extreme value distribution (EVD), and $u_n(x) = a_n x + b_n$ be a suitable normalization. If the condition $D(u_n(x))$ holds for all x and $P(M_n \leq u_n(x))$ converges to some $H(x)$, then $H(x) = (G(x))^\theta$, with $0 \leq \theta \leq 1$. Under some regularity assumptions the expected value of the cluster size of exceedances is equal to $1/\theta$ (if $\theta > 0$), giving a nice interpretation of the extremal index. Corollary 9.2.1 states that the limit distribution of M_n depends on θ only. However, Hsing [225] shows that the limit distribution of the k-th maximum ($k > 1$) depends also on the cluster size distribution, if clustering occurs, i.e., if $\theta < 1$. If $\theta = 1$, then there is no clustering of the exceedances and from the limiting Poisson process, we easily deduce the limiting distribution of the k-th largest order statistics $X_{n-k+1;n}$: $\lim_{n\to\infty} P(X_{n-k+1;n} \leq b_n + a_n x) = P(N((0,1]) < k) = \exp(-\lambda) \sum_{l<k} \lambda^l/l!$ for any $k \geq 1$ fixed. Even some joint events of extreme order statistics are implied in the same way from the limiting Poisson process result. For example, consider $\lim_{n\to\infty} P(X_{n-k+1;n} \leq b_n + a_n x < X_{n-m+1;n}) = P(m \leq N((0,1]) < k) = \exp(-\lambda) \sum_{m\leq l<k} \lambda^l/l!$ with $1 \leq m < k$. For the joint distributions of extreme order statistics one has to introduce multiple boundaries and point processes in $\mathbb{R}^2$, cf. Leadbetter et al. [303].

Example 9.2.2. Let $\{X_i, \ i \geq 1\}$ be a Gaussian sequence, with mean values 0, variances 1 and autocorrelations $r_n = E(X_1 X_{n+1})$. Berman [36] showed that M_n converges in distribution to the EVD G_3 if $r_n \log(n) \to 0$ as $n \to \infty$. It is known that this condition, called Berman's condition, implies the conditions $D(u_n)$ and $D'(u_n)$ (cf. Leadbetter, Lindgren and Rootzen [303]), where $u_n = \sqrt{2\log(n)} + (\frac{1}{2}\log(4\pi\log(n)) - \log(\lambda))/\sqrt{2\log(n)}$.

As mentioned above condition $D'(u_n)$ is more restrictive than $D^*(u_n)$. $D'(u_n)$ can be formulated much easier in the stationary case, redefining α'_n:

$$\alpha'_n = r_n^* \sum_{2 \leq k \leq r_n^*} P(X_1 > u_n, X_j > u_n).$$

$D'(u_n)$ holds if $k_n \alpha'_n \to 0$ as $n \to \infty$. Therefore the point processes $N_n, \tilde{N}_n$ and N_n^* converge in distribution to the same Poisson process N and $\theta = 1$, i.e., asymptotically there is no clustering of exceedances. Note that if $\lambda = \mu$ and $D^*(u_n)$ holds, then $D'(u_n)$ holds also. For, $\lambda = \mu$ implies $P(X_1 > u_n) - P(X_1 \leq u_n, X_2 > u_n) = P(X_1 > u_n, X_2 > u_n) = o(1/n)$, hence together with $D^*(u_n)$ the condition $D'(u_n)$ holds.

Condition $D'(u_n)$ is verified in the Gaussian case by using that

$$P(X_i > u_n, X_j > u_n) \leq (1-\Phi(u_n))^2 + O\big(|r(i-j)|\exp\{-u_n^2/(1+|r(i-j)|)\}\big) \quad (9.6)$$

for $i \neq j$ (see Leadbetter et al. [303]). Then a straightforward calculation shows that

$$\alpha_n^* \leq \alpha'_n = O(1/k_n^2) \quad \text{as} \quad k_n \to \infty.$$

We get also by (9.6),

$$\mu = \lim_{n\to\infty} nP(X_1 \le u_n, X_2 > u_n)$$
$$= \lim_{n\to\infty} \Big(nP(X_2 > u_n) - nP(X_1 > u_n, X_2 > u_n) \Big)$$
$$= \lambda.$$

Example 9.2.3. Let $\{X_i,\ i \ge 1\}$ be a max-autoregressive random sequence, i.e.,

$$X_i = c\max(X_{i-1},\ Y_i) \qquad \text{for}\quad i \ge 2,$$

where $c \in (0,1)$, X_1 is a rv with an arbitrary distribution $F_1(\cdot)$, and $\{Y_i,\ i \ge 2\}$ is an iid random sequence, independent of X_1. Let $F_i(\cdot)$ and $H(\cdot)$ denote the distribution of X_i and Y_i, respectively. Then

$$F_i(x) = F_{i-1}(x/c)H(x/c)$$

for $i \ge 2$. The sequence $\{X_i\}$ is stationary iff $F_i(\cdot) \equiv F(\cdot)$ for $i \ge 1$ and

$$F(x) = F(x/c)H(x/c) \tag{9.7}$$

holds (cf. Alpuim [7]). Such a simulated sequence is shown in Fig. 9.2.1 with $c = 0.85$.

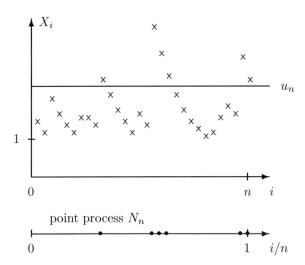

FIGURE 9.2.1. Max-autoregressive sequence $\{X_i\}$ (symbol: x) and the related point process N_n of exceedances.

We assume that $F \in D(G_{1,\alpha})$, i.e., there exists a sequence $\{a_n,\ n \ge 1\}$ of positive constants such that $F^n(a_n x) \to \exp(-x^{-\alpha})$, $x > 0$, $\alpha > 0$. This is equivalent to $H \in D(G_{1,\alpha})$ by (9.7). Using (9.7) we get

$$P(M_n \le a_n x) = P(X_1 \le a_n x, \ Y_i \le a_n x/c, \ 2 \le i \le n)$$
$$= F(a_n x) H^{n-1}(a_n x/c)$$
$$= \Big(F(a_n x)\Big)^n \Big/ \Big(F(a_n x/c)\Big)^{n-1}$$
$$\to G_{1,\alpha}(x)/G_{1,\alpha}(x/c)$$
$$= \exp\Big(-x^{-\alpha}(1 - c^\alpha)\Big).$$

Furthermore, $\lambda = \lim_{n\to\infty} n\Big(1 - F(a_n x)\Big) = x^{-\alpha}$ and

$$\mu = \lim_{n\to\infty} nP(X_1 \le a_n x, \ X_2 > a_n x)$$
$$= \lim_{n\to\infty} nP(X_1 \le a_n x, \ Y_2 > a_n x/c)$$
$$= \lim_{n\to\infty} n\Big[1 - H(a_n x/c)\Big]$$
$$= \lim_{n\to\infty} n\Big[1 - F(a_n x) - \big(1 - F(a_n x/c)\big)\Big]$$
$$= x^{-\alpha}(1 - c^\alpha).$$

Therefore $\theta = \mu/\lambda = 1 - c^\alpha$. It is also possible to verify $D(a_n x)$ and $D^*(a_n x)$ for $x > 0$. For instance for any sequence $\{k_n\}$ with $k_n \to \infty$ (satisfying (9.3)),

$$k_n \alpha_n^* = k_n r_n^* \sum_{j \le r_n^*} P(X_1 > a_n x, \ X_j \le a_n x, \ X_{j+1} > a_n x)$$
$$= k_n r_n^* \sum_{j \le r_n^*} P(X_1 > a_n x, \ X_j \le a_n x, \ Y_{j+1} > a_n x/c)$$
$$\le n\Big[1 - H(a_n x/c)\Big] \sum_{j \le r_n^*} P(X_1 > a_n x)$$
$$\le n\frac{1 - F(a_n x) - (1 - F(a_n x/c))}{F(a_n x/c)} r_n^* \Big(1 - F(a_n x)\Big)$$
$$= O(r_n^*/n) = O(1/k_n) \to 0 \qquad \text{as} \quad k_n \to \infty.$$

In the same way one can show that $\alpha_{n,m} \to 0$ for any $m \ge 1$. It was proved that the cluster sizes K have asymptotically a geometric distribution: $P(K = i) = (1 - c^\alpha)c^{\alpha(i-1)}$, and $E(K) = (1 - c^\alpha)^{-1} = 1/\theta$ (Alpuim [7]). Extensions to non-stationary max-AR(1)-sequences are discussed by Alpuim et al. [10], Catkan [62]. See also Chapter 10 for applications of max-AR(1) with respect to general rare events.

Other stationary sequences such as moving average processes, queues and Markov chains, are discussed for instance in Chernick et al. [67], O'Brien [357], Perfekt [362], [363], Rootzén ([399], [400]), Serfozo ([410], [409]) and Smith [422].

9.3   Applications: Independent Sequences

The independent random sequences, which have a trend or a non-constant variance function, represent a typical case which deviates from the iid case (compare with Examples 8.3.2 and 8.3.3). Often such sequences are periodic as in the following example. Clustering is not possible in the sense as discussed. Condition D holds, and also D^*, D' assuming that the boundary satisfies $\lambda((0,1]) < \infty$.

Example 9.3.1. Let $\{Y_i,\ i \geq 1\}$ be an iid sequence with marginal distribution $H \in D(G_3)$, i.e., there exist $\{a_n\}$ and $\{b_n\}$ such that $H^n(a_n x + b_n) \to G_3(x)$. Let

$$X_i = c_i Y_i,$$

where the real sequence $\{c_i,\ i \geq 1\}$ is periodic with period p. Suppose that $c_i > 0$ for $i \geq 1$. We define $c^* = \max_{i \leq p} c_i$, $q = \sharp\{i \leq p : c_i = c^*\}$, $u_n(x) = a_n x + b_n$, and $u_n^*(x) = c^* u_n(x)$. Then

$$
\begin{aligned}
\lambda(t) &= \lim_{n \to \infty} \sum_{i \leq nt} P(X_i > u_n^*(x)) \\
&= \lim_{n \to \infty} \sum_{i \leq p} \Big(1 - H(u_n^*(x)/c_i)\Big) nt/p \\
&= \lim_{n \to \infty} q\Big(1 - H(u_n(x))\Big) nt/p \\
&= \log\Big(1/G_3(x)\Big) tq/p
\end{aligned}
$$

for, if i is such that $c_i < c^*$, then $1 - H(u_n(x)c^*/c_i) = o(1/n)$ (see de Haan [184], Corollary 2.4.2)). Because of the independence of the X_i's, we get $\mu(\cdot) = \lambda(\cdot)$, $k_n \alpha_n^* = O(1/k_n) \to 0$ for any sequence $\{k_n\}$ with $k_n \to \infty$ and $D(u_n^*(x))$ and $D^*(u_n^*(x))$ hold. The results of Section 9.2 imply that the three point processes N_n, $\tilde{N}_n$ and N_n^* converge in distribution to the same homogeneous Poisson process N with intensity $\lambda = \log(1/G_3(x))q/p$. Since the period p is fixed and since asymptotically only the points i with $c_i = c^*$ contribute to the extreme values and to exceedances of the level $u_n^*(x)$ by $\{X_i\}$, the influence of the non-stationarity consists of a thinning of the point process of exceedances of $u_n(x)$ by $\{Y_i\}$. As special case let Y_i be standard normal rv, with c_i being the standard deviations of X_i. Hence the important points for the possible exceedances are the points of maximal variances. The Poisson convergence of the point process of exceedances imply also the limiting distribution of the k-th extremes $X_{n-k+1;n}$, the k-th largest of $\{X_i, i \leq n\}$.

Example 9.3.2. We consider again the shifted iid sequence analyzed by Daley and Hall [89] (see Example 8.3.3). Let $S_\beta = \sup\{X_i - (i-1)\beta,\ i \geq 1\}$ where the X_i's are iid rv with distribution $F \in D(G_{1,\alpha})$, $\alpha > 1$. We deal here with the convergence of S_β with respect to the continuous parameter β, which extends in an obvious way the convergence results obtained with respect to the discrete parameter n. For $x > 0$ we define the point process N_β on $[0, \infty)$ such that, for any Borel set B,

$$N_\beta(B) = \sum_{i \in (\beta a(\beta))^{-1} B} 1(X_i - (i-1)\beta > x/a(\beta)),$$

where $a(\beta) \to 0$ as $\beta \to 0$ such that $\bar{F}(1/a(\beta)) = 1 - F(1/a(\beta)) \sim a(\beta)\beta(\alpha - 1)$, (see Example 8.3.3). We prove that $N_\beta \to_D N$ as $\beta \to 0$, where N is a non-homogeneous Poisson process with intensity measure $\nu(\cdot)$:

$$\nu([s, \infty)) = (x + s)^{-(\alpha - 1)}.$$

Because of Remark 9.1.2 we can apply Theorem 9.1.1. Thus in proving $N_\beta \to_D N$ we have to verify conditions a) and b) of Theorem 9.1.1:

a) Let $B = [s, t)$. Then

$$E(N_\beta(B)) = \sum_{i \in \left(\beta a(\beta)\right)^{-1} B} P(X_i > (i-1)\beta + x/a(\beta))$$

$$= \sum_{i \in \left[s/\left(\beta a(\beta)\right), t/\left(\beta a(\beta)\right)\right)} \bar{F}\left(i\beta + x/a(\beta)\right) + o(1)$$

$$\sim \beta^{-1} \int_{(x+s)/a(\beta)}^{(x+t)/a(\beta)} \bar{F}(u) du$$

$$\to (x + s)^{-(\alpha - 1)} - (x + t)^{-(\alpha - 1)}$$

$$= \nu(B)$$

since $F \in D(G_{1,\alpha})$ means that $\bar{F}$ is regularly varying.

b) By the independence of the rv X_i,

$$P\left(N_\beta\left(\bigcup_{j \le k} B_j\right) = 0\right) = \prod_{j \le k} P(N_\beta(B_j) = 0)$$

$$\to \prod_{j \le k} \exp\left(-\nu(B_j)\right)$$

$$= \exp\left(-\nu\left(\bigcup_{j \le k} B_j\right)\right)$$

if $B_j = [s_j, t_j)$ are disjoint intervals.

This statement implies the convergence result obtained in Example 8.3.3.

$$P\Big(a(\beta)\sup_{i\geq 1}\{X_i - (i-1)\beta\} \leq x\Big)$$
$$= P\big(N_\beta([0,\infty)) = 0\big)$$
$$\to \exp\Big(-\nu\big([0,\infty)\big)\Big)$$
$$= \exp\big(-x^{-(\alpha-1)}\big)$$
$$= G_{1,\alpha-1}(x).$$

For any time interval $[s,t]$ further information about the exceedances of the boundary $x/a(\beta)$ can be obtained:

$$P\Big(\sup_{i\in[s/(\beta a(\beta)),t/(\beta a(\beta)))}\{X_i - (i-1)\beta\} \leq x/a(\beta)\Big)$$
$$= P\big(N_\beta([s,t)) = 0\big)$$
$$\to \exp\{-\nu([s,t))\}$$
$$= \exp\{-(x+s)^{-(\alpha-1)} + (x+t)^{-(\alpha-1)}\}$$
$$= G_{1,\alpha-1}(x+s)/G_{1,\alpha-1}(x+t).$$

This suggests that we redefine N_β by a finite number of rv X_i, $i \leq I(\beta)$:

$$N_\beta = \sum_{i\leq I(\beta)} \delta_{i\beta a(\beta)}1(X_i - (i-1)\beta > x/a(\beta)),$$

where $I(\beta)$ is suitably increasing with $\beta \to 0$. Both definitions of N_β lead to the same limit point process, since

$$\sum_{i\geq I(\beta)} 1(X_i - (i-1)\beta > x/a(\beta))$$

converges in probability to 0, if $I(\beta)\beta a(\beta) \to \infty$.

Note that also the results of Section 9.2 can be used by defining $u_{\beta i} = (i-1)\beta + x/a(\beta)$ with $(\beta a(\beta))^{-1}$ replacing n for the time rescaling. Then obviously $D(u_{\beta i})$ and $D^*(u_{\beta i})$ hold. The derivation above shows that $\lambda(t) = x^{-(\alpha-1)} - (x+t)^{-(\alpha-1)} = \mu(t)$.

In Resnick [393] the last exit time problem, which is related to the shifted iid sequence (cf. Example 8.3.3), is treated in a different way using point processes methods and the continuous mapping theorem.

9.4   Applications: Non-stationary Sequences

Example 9.4.1. Let $\{X_i,\ i \geq 1\}$ be now a non-stationary Gaussian sequence. Without loss of generality we suppose that $E(X_i) = 0$ and $E(X_i^2) = 1$. We consider a generalization of Example 9.2.2. Suppose that

$$\sup_{i \neq j} |r(i,j)| < 1$$

and

$$\sup_{|i-j| \geq n} |r(i,j)| \log(n) \to 0, \qquad n \to \infty$$

and that the boundaries $\{u_{ni}, i \leq n, n \geq 1\}$ are such that

$$\lambda(t) = \lim_{n \to \infty} \sum_{i \leq nt} (1 - \Phi(u_{ni})),$$

is continuous. The point processes N_n, $\tilde{N}_n$ and N_n^* converge to the same Poisson process N with intensity measure $\lambda(\cdot)$ if the mixing conditions $D(u_{ni})$ and $D^*(u_{ni})$ hold and $\lambda(\cdot) \equiv \mu(\cdot)$. The verification of all these conditions is still possible but rather tedious because of the non-stationarity and the non-constant boundaries (Hüsler [237], [240] and [241]). To overcome the problem of non-stationarity, a special comparison method was developed, which relies on grouping the rv in an appropriate way (Hüsler [237]). The grouping is defined by grouping first similar values of the boundary together and then splitting each of these groups into appropriate 'neighboring' and 'far distant' time points applying the long range dependence condition. The first grouping is not needed in the stationary case with a constant boundary.

CONVERGENCE RATE OF THE POISSON APPROXIMATION

The point process of exceedances of a stationary process converges under the mentioned conditions to a homogenous Poisson process, see Section 9.2. In the case of a non-stationary sequence the point process of exceedance converges to a non-homogeneous Poisson process discussed in Section 9.1. Of interest is also the rate of these convergences. Under additional conditions this rate can be characterized. We are discussing in this section the convergence rate in the case of a Gaussian sequence. We assume again that the non-stationary Gaussian sequence is such that the correlation values r_{ij} satisfy $|r_{ij}| \leq \rho_{|i-j|}$ for all $i \neq j$ where $\rho_n < 1$ for $n \geq 1$ and $\rho_n \log(n) \to 0$ (Berman's condition). The boundary values u_{ni} for the exceedances are such that

$$\limsup_{n \to \infty} \lambda_n = \limsup_{n \to \infty} \sum_{i \leq n} P(X_i < u_{ni}) < \infty$$

with $u_{n,\min} = \min_{i \leq n} u_{ni} \to \infty$ as $n \to \infty$. Then the number of exceedances $N_n = \sum_{i \leq n} 1(X_i < u_{ni})$ can be approximated by a Poisson rv $P_0(\lambda_n)$ with mean λ_n.

We investigate the approximation error of N_n and $P_0(\lambda_n)$. The deviation is measured again by the variational distance $d(N_n, P_0(\lambda_n))$. Applying a result of Barbour et al [29] we get the following bound for the variational distance:

$$d(N_n, P_0(\lambda_n)) \leq \frac{1 - e^{-\lambda_n}}{\lambda_n} \left(\sum_{i \leq n} \pi_{ni}^2 + \sum_{i \neq j \leq n} |\mathrm{Cov}(1(X_i > u_{ni}), 1(X_j > u_{nj}))| \right)$$

where $\pi_{ni} = P(X_i > u_{ni}) = \bar{\Phi}(u_{ni})$ assuming that the random sequences X_i is standardized. The first sum is asymptotically negligible. The second sum depends on the correlations of the rv X_i and on the boundary values u_{ni}. Obviously the minimum boundary value $u_{n,\min}$ plays an important role. We do not restrict in any way the index of the point where this minimum of the boundary occurs. Also the largest correlation less than 1 is important. Thus let $\rho = \sup\{0, r_{ij}, i \neq j\}$ which is smaller than 1. For this general situation it is shown (see Hüsler and Kratz [256]) that the second sum, hence the variational distance, can be bounded:

if $\rho > 0$ by

$$d(N_n, P_0(\lambda_n)) = O\left(\frac{1}{u_{n,\min}} \exp\left(-\frac{1 - \rho}{2(1 + \rho)} u_{n,\min}^2 \right) + \Delta\left(\exp(u_{n,\min}^2/\alpha) \right) \right)$$

where $\Delta(s) = \sup_{k \geq s}\{\rho_k \log(k)\}$,

if $\rho = 0$ by

$$d(N_n, P_0(\lambda_n)) = O\left(u_{n,\min} \exp\left(-\frac{1}{2} u_{n,\min}^2 \right) \sum_{j \leq n} \rho_j \right).$$

The first term of the bound in case $\rho > 0$ dominates the second one if $\Delta(s)$ tends sufficiently fast to 0. In this case the given rate of convergence depends only on the maximal correlation and the minimum boundary value. If the maximal correlation is not occurring with indices i, where the minimum boundary value is observed, then the bound for the variational distance could be improved. However, assuming the boundary is constant and the sequence is stationary, then the bound simplifies to the known result of the stationary Gaussian case, given in Rootzén [398], Smith [419] and Holst and Janson [218]. Note that the rate in the given bound is for this particular stationary case the best possible one, given by Rootzén [398]. Some other cases and examples are discussed in Hüsler and Kratz [256], too. Similar results are derived for Gaussian processes which are stationary and mean square differentiable, see Kratz and Rootzén [296]. The convergence rate of the distribution of M_n of a stationary infinite moving average Gaussian sequence was also analyzed by Barbe and McCormick [26] generalizing the result of Rootzén [398]. Their results fit well to the above result on the variational distance showing that the largest correlation ρ (smaller than 1) plays the most crucial role for the convergence rate.

Example 9.4.2. In a similar way we can drop the stationarity condition (9.7) in Example 9.2.3. We suppose that $H \in D(G_{1,\alpha})$ and $X_1 \sim F_1$ is such that

$$\sum_{0 \leq i \leq n-1} \bar{F}_1(a_n x/c^i) \to 0, \qquad n \to \infty$$

for all $x > 0$. This insures that X_1 is not dominating the sequence $\{X_i\}$. Here $\{a_n\}$ is the normalizing sequence, i.e., $H^n(a_n x) \to G_{1,\alpha}(x)$. These assumptions imply

$$
\begin{aligned}
P(M_n \leq a_n x) &= P(X_1 \leq a_n x, \; Y_i \leq a_n x/c, \; 2 \leq i \leq n) \\
&= F_1(a_n x) H^{n-1}(a_n x/c) \\
&\to G_{1,\alpha}(x/c).
\end{aligned}
$$

We show that $\lambda(t)$ exists. We use the fact that, for $i \geq 2$,

$$1 - F_i(a_n x) = 1 - F_1(a_n x/c^{i-1}) \prod_{1 \leq j \leq i-1} H(a_n x/c^j).$$

Since $H(\cdot)$ is regularly varying, we have, for n large,

$$1 - F_i(a_n x) = 1 - F_1(a_n x/c^{i-1}) \exp\left(-(1 + o(1))\bar{H}(a_n x)\frac{c^\alpha(1 - c^{\alpha(i-1)})}{1 - c^\alpha}\right)$$

$$= (1 + o(1))x^{-\alpha}n^{-1}\frac{c^\alpha(1 - c^{\alpha(i-1)})}{1 - c^\alpha} + \bar{F}_1(a_n x/c^{i-1}),$$

uniformly in $i \leq n$, where $\bar{H}(x) = 1 - H(x)$. Hence

$$\sum_{i \leq nt}(1 - F_i(a_n x)) = (1 + o(1))x^{-\alpha}t\left(\frac{c^\alpha}{1 - c^\alpha} + o(1)\right) + o(1)$$

$$\to tx^{-\alpha}\frac{c^\alpha}{(1 - c^\alpha)} = \lambda(t).$$

Using similar arguments we verify that $\mu(t) = tx^{-\alpha}c^\alpha$. Therefore

$$\theta(t) = \mu(t)/\lambda(t) = 1 - c^\alpha = \theta$$

for all $t \leq 1$. Note that $\theta < 1$ since $c > 0$. It is also possible to verify the mixing conditions $D(u_n)$ and $D^*(u_n)$ with $u_n = a_n x$, for all $x > 0$.

We observe that the non-stationarity is asymptotically negligible. We obtain the same limits as in the stationary case. It is even possible to replace the constant c by positive rv C_i, which are independent, but not identically distributed rv on $(0, 1]$. Under certain conditions such as $F_{C_i} \to_w F_{C_\infty}$, similar results can be derived for the more general case (Catkan [62]).

In the same way, Example 9.3.1 can be generalized by taking any stationary random sequence Y_i which satisfies the mixing conditions.

Remark 9.4.3. In the stationary case $\theta = \mu/\lambda = \mu(t)/\lambda(t)$, $t \leq 1$. In the non-stationary case we define analogously $\theta(t) = \mu(t)/\lambda(t) \leq 1$, $t \leq 1$, as in Example 9.4.2. However, in general $\theta(t)$ is not constant in t. Moreover, θ may depend on the boundary $u_{ni}(x)$ (cf. Hüsler [240]), shown with the following example. This is the reason why the extremal index does not play the same role in the general non-stationary case as in the stationary case.

Example 9.4.4. Consider an iid sequence Y_i with continuous marginal distribution F. Define the normalization $u_n(\tau) = \bar{F}^{-1}(\tau/n)$, for any $\tau > 0$, and the dependent non-stationary sequence

$$X_i = Y_{[(i+1)/2]}, \quad i \geq 1.$$

It implies simply that $P(X_i \leq u_n(\tau)) \to \exp(-\tau/2)$ and further that $\theta = 1/2$ for the constant boundary. For, if there is an exceedance by the iid Y_i, then there is a cluster of two exceedances by X_i's. The cluster distribution is 2 with probability 1. Now define for some constant $d \leq 1$ a non-constant boundary

$$u_{ni}(\tau) = \begin{cases} u_n(d\tau) & \text{for } i \text{ odd}, \\ u_n((2-d)\tau) & \text{for } i \text{ even}. \end{cases}$$

If $d = 0$, then let $u_n(0) = \omega(F)$. Then the mean number of exceedances by X_i's is asymptotically τ, and the cluster sizes are either 2 with probability $d/(2-d)$ (occurring if some $Y_i > u_n(d\tau)$), or 1 with probability $2(1-d)/(2-d)$ (occurring if some $Y_i \in (u_n((2-d)\tau), u_n(d\tau))$. The extremal index turns out to be $\theta = (2-d)/2 \in [1/2, 1]$. Because of the regularity of the boundary values we have also $\theta(t) = \theta$ for any $t \leq 1$, since $\lambda(t) = \lambda\tau$ and $\mu(t) = t\tau(2-d)/2$ for any $t \leq 1$.

Let us change now further the boundary. Define for some $s \in (0,1)$

$$u_{ni}(\tau) = \begin{cases} u_n(d\tau) & \text{for } i \leq ns, i \text{ odd}, \\ u_n((2-d)\tau) & \text{for } i \leq ns, i \text{ even}, \\ u_n(\tau) & \text{for } i > ns. \end{cases}$$

The mean number of exceedances up to t is still $\lambda(t) = t\tau$. The mean number of upcrossings up to time t is $\mu(t) = (2-d)t\tau/2$ for $t \leq s$, and $= (t-s+(2-d)s)\tau/2$ for $t \geq s$. It implies that $\theta(t) = \mu(t)/\lambda(t) = (2-d)/2$ for $t \leq s$, and $= (1+(1-d)s/t)/2$. Setting $t = 1$, we note that the extremal index $\theta(1) = (1+(1-d)s)/2 \in [0.5, 1]$, and depends now on s. It is not only dependent on the dependence type of the sequence X_i, but also on the boundary type.

In general, the limiting behavior of the point process of exceedances, stated in Section 11.2, reveals a distinct insight into the behavior of the exceedances and provides more information than the extremal index alone.

Remark 9.4.5. Often data in applications show an almost periodic behavior. Such applications are modeled as $X_i = h(i)Y_i + g(i)$ where $g(i)$, $h(i)$ (> 0) are periodic functions and $\{Y_i\}$ is a stationary random sequence satisfying the mixing conditions (cf. Ballerini and McCormick [25], Oliveira and Turkman and Oliveira [448], Nandagopalan et al. [347]).

A particular case was discussed in Example 9.3.1 with $g(i) \equiv 0$. Depending on the underlying distribution F_Y, the extreme values mostly occur at time points j where $h(j)$ takes its maximum value, since these values of the random sequence dominate the other values X_i. This induces a thinning of the point process of exceedances of $\{Y_i\}$, as mentioned.

If $g(i) \not\equiv 0$, then by usual standardization of $X(i)$ the events $\{X(i) \leq u\}$ are transformed to $\{Y(i) \leq (u - g(i))/h(i)\}$. Often the maxima is expected to occur at points where the boundary $(u - g(i))/h(i)$ is smallest. If these values dominate the behavior of the maximum, then the point process of exceedances by $X(i)$ is also a thinning of the point process of exceedances by $Y(i)$.

9.5 Extensions: Random Fields

The approach of this chapter for non-stationary random sequences which satisfy the mixing type conditions, can be generalized to random fields. The main question is the extension of the long range mixing condition $D(u_n)$ or $\Delta(u_n)$. This was done by Leadbetter and Rootzén [306] via a coordinate-wise extension for stationary random fields and by Perreira and Ferreira [364] for non-stationary random fields.

Let us consider, for notational reasons, only the case of a two-dimensional time parameter $\mathbf{t} = (t_1, t_2)' \in \mathbb{R}_+^2$. Leadbetter and Rootzén [306] introduced weak mixing conditions for stationary random fields by following the idea of the distributional condition $D(u_n)$, instead of assuming a strong mixing condition. In the condition, basically, one wants to consider as few as possible events of the type $\{M(A) \leq u_n\}$, where $A \subset \mathbb{R}_+^2$. Since there is no natural past and future in higher dimensions, they formulate the condition in a sequential manner, restricting first the mixing with respect to the first coordinate t_1 and then for the the second coordinate t_2. Obviously, one might interchange the coordinates and would define a somewhat different mixing condition. By this way, the number of events to consider for verification of the mixing type condition, is substantially reduced.

Let us mention this condition in detail. For the maximum

$$M(\mathbf{T}) = \max\{X(t_1, t_2), 0 < t_j \leq T_j, j = 1, 2\},$$

with $\mathbf{T} = (T_1, T_2)'$, we define blocks of rv of two-dimensional size $r_1 r_2$ for some $r_j = o(T_j), j = 1, 2$, where $T_1 T_2$ is the number of rv considered in $M(\mathbf{T})$. The rectangle $(0, T_1] \times (0, T_2]$ is now split into blocks $B_{ij} = ((i-1)r_1, ir_1] \times ((j-1)r_2, jr_2]$ with $i \leq k_1 = [T_1/r_1]$ and $j \leq k_2 = [T_2/r_2]$. There are $k_1 k_2$ such blocks, and at most $(k_1 + k_2 - 1)$ additional incomplete blocks which have asymptotically no impact, similar to the one-dimensional time case.

Then we assume mixing with respect to the first coordinate, if for any rectangles $B_1 = (0, t_1] \times (0, h]$, with $0 \leq h \leq T_2$ and $0 < t_1 < T_1$, and $B_2 = (t, s] \times (0, h]$ with $s - t \leq r_1, s < T_1$, which are at least separated by a distance ℓ_1 w.r.t. the first coordinate, i.e., $t \geq t_1 + \ell_1$, we have, for some $\alpha_1(\cdot, \cdot)$,

$$|P(M(B_1) \leq u, M(B_2) \leq u) - P(M(B_1) \leq u)P(M(B_2) \leq u)| \leq \alpha_1(r_1, \ell_1)$$

where $\alpha_1(\cdot, \cdot)$ denotes the mixing function for the first coordinate. We assume as usual that $k_1\alpha_1(r_1, \ell_1) \to 0$, with $u = u_\mathbf{T} \to x_F$ as $\mathbf{T} = (T_1, T_2)' \to \infty$, with $T_i \to \infty$, $i = 1, 2$, where x_F denotes the endpoint of the marginal distribution F of the stationary random field X. The behavior of $\mathbf{T} = (T_1, T_2)' \to \infty$ is fixed by a continuous, strictly increasing positive function ψ with $\psi(T) \to \infty$ as $T \to \infty$, which defines $\mathbf{T} = (T_1, \psi(T_1))' = (T_1, T_2)'$.

Now for the second coordinate we assume that for any rectangles $B_1 = (0, r_1] \times (0, h]$ with $0 < h \leq T_2$, and $B_2 = (0, r_1] \times (t, s]$ with $s - t \leq r_2, s \leq T_2$, which are separated by at least ℓ_2, i.e., $t \geq h + \ell_2$, we have, for some $\alpha_2(\cdot, \cdot, \cdot)$,

$$|P(M(B_1) \leq u, M(B_2) \leq u) - P(M(B_1) \leq u)P(M(B_2) \leq u)| \leq \alpha_2(r_1, r_2, \ell_2)$$

where $\alpha_2(\cdot, \cdot, \cdot)$ denotes the mixing function for the second coordinate. Note that α_2 also depends on r_1 from the mixing condition of the first coordinate. We assume here similarly that $k_1 k_2 \alpha_2(r_1, r_2, \ell_2) \to 0$, with $u = u_\mathbf{T} \to x_F$ as $\mathbf{T} \to \infty$.

If these two coordinate-wise weak mixing conditions hold, then one can show that, for $B_\mathbf{T} = (0, k_1 r_1] \times (0, k_2 r_2]$,

$$P(M(B_\mathbf{T}) \leq u_\mathbf{T}) = P^{k_1 k_2}(M(J) \leq u_\mathbf{T}) + o(1)$$

as $\mathbf{T} \to \infty$ with $J = (0, r_1] \times (0, r_2]$.

With this weak long-range dependence restriction, one has to investigate only the local behavior of the maximum on a small rectangle J to derive as in the one-dimensional 'time' parameter case, the limiting distribution of $M(B_\mathbf{T})$ and to discuss a possible cluster index, see Leadbetter and Rootzén [306].

This concept was used and extended by Perreira and Ferreira [364] for non-stationary random fields and non-constant boundaries $u_{\mathbf{T}, \mathbf{t}}$. They discuss also conditions which restrict the local behavior of the random field and which permit us to derive the cluster index for random fields, and mention some examples of random fields, in particular 1-dependent ones.

Chapter 10

Extremes of Gaussian Processes

In this chapter continuous Gaussian processes and their extremes, exceedances and sojourns above a boundary are treated. Results are derived for stationary and locally stationary Gaussian processes. The asymptotic results are then applied to a statistical problem related to empirical characteristic functions. In addition, some results on other non-stationary Gaussian processes are discussed. The relation between the continuous process and its discrete approximation on a certain fine grid is a rather interesting issue, in particular for simulations or approximations.

10.1 Introduction

In this chapter we consider Gaussian processes $\{X(t),\ t \geq 0\}$ in continuous time with (a.s.) continuous paths. Denote by

$$m(t) = E(X(t)) \qquad \text{and} \qquad \sigma(t, s) = \text{Cov}(X(t), X(s)),$$

the mean value and the covariance function, respectively. Let $\Phi(\cdot)$ and $\varphi(\cdot)$ be the df and the density function of the unit normal law, respectively.

In the following we mainly deal with the extremes of a Gaussian process, in particular with

$$M_T = \sup\{X(t),\ 0 \leq t \leq T\}.$$

The main topics of interest are

1) the distribution of M_T for fixed T, especially the behavior of $P(M_T > u)$ as $u \to \infty$, and

2) the convergence of the distribution of M_T, as $T \to \infty$.

M. Falk et al., *Laws of Small Numbers: Extremes and Rare Events*, 3rd ed.,
DOI 10.1007/978-3-0348-0009-9_10, © Springer Basel AG 2011

Related to these topics is the time spent above the boundary $u_T(\cdot)$ by the process $X(\cdot)$, which means the sojourn time of $\{X(t), t \leq T\}$, in $\{(t, x) : 0 \leq t \leq T, x \in (u_T(t), \infty)\}$:

$$\int_0^T 1(X(t) > u_T(t))\, dt.$$

The boundary $u_T(t)$ is usually not constant for fixed T. This integral can be rewritten as a sum of subintegrals

$$\int_{I_j} 1(X(t) > u_T(t))\, dt$$

where I_j denotes the j-th sojourn of the process $X(\cdot)$ above the boundary $u_T(t)$, $t \leq T$. The exceedances are usually clustering, thus it is reasonable to approximate the point process of exceedances by a compound Poisson process. The cluster size, i.e., the sojourn time above the boundary in our case, could be defined as the mark of a marked point process.

It is also of interest to analyze the point process of upcrossings and the point process of cluster positions as in Chapter 9. If the path behavior of $X(\cdot)$ is restricted in a suitable way, we may approximate both point processes by a simple Poisson process as for the processes $\tilde{N}_n$ and N_n^* defined in Chapter 9.

The problems 1) and 2) are closely related to each other. If the upper tail behavior of the distribution of M_T for fixed T is known, then the asymptotic behavior of the distribution of M_T $(T \to \infty)$ can be analyzed using the long range mixing condition. As in the discrete time case we choose an 'extreme' boundary $u_T(t)$, which converges to ∞, the upper endpoint of $\Phi(\cdot)$, as $T \to \infty$:

$$\inf\{u_T(t),\ t \leq T\} \to \infty.$$

In analyzing stationary processes we often use a constant boundary, $u_T(t) \equiv u_T = a_T x + b_T$. In a non-stationary Gaussian process a non-constant boundary is rather typical, as mentioned earlier. Often the point of minimal boundary value plays an important role. This case will be analyzed. A major part of the investigation of maximum of a continuous process $X(t)$ is also based on the relation of the maximum of the process $X(t)$ to the maximum of the rv $X(t_i)$, by sampling $X(t)$ at some discrete time points t_i which is investigated extensively in the following.

10.2 Stationary Gaussian Processes

We deal first with stationary Gaussian processes and assume without loss of generality that $m(t) = 0$, $\sigma(t, t) = 1$ and $\sigma(t, s) = \rho(t - s)$, $t, s \geq 0$. M_T can be rewritten as the maximum of finitely many rv:

$$\begin{aligned} M_T &= \max\{M([(j-1)h, jh)),\ j = 1, \ldots, n = [T/h],\ M([nh, T])\} \\ &\approx \max\{M([(j-1)h, jh)),\ j = 1, \ldots, n\}, \qquad \text{for}\quad h > 0, \end{aligned}$$

where $M(I) = \sup\{X(t),\ t \in I\}$ for any interval $I \subset [0,\infty)$. If we define $Z_j = M([(j-1)h, jh))$, then M_T is approximated by the maxima of these identically distributed rv Z_j, $j \leq n$. This leads us to the case of random sequences.

LOCAL BEHAVIOR OF EXTREMES

But we have to restrict $\rho(\cdot)$ in a suitable way in order to make use of the results of Section 9.2. We have to investigate the behavior of

$$P(Z_1 > u) = P(M_h > u) \quad \text{as} \quad u \to \infty,$$

which also provides a solution to problem 1). Although the inequality

$$P(M_h > u) \geq P(X(0) > u) = 1 - \Phi(u) \sim \varphi(u)/u = \exp(-u^2/2)/(u\sqrt{2\pi})$$

gives a sharp bound for some special processes, e.g., $\rho(s) = 1$, $s \leq h$, i.e., $X(t) = X(0)$ a.s., for $t \leq h$, it is of less use in general.

We can bound $P(M_h > u)$ from above:

$$P(M_h > u) \leq P(X(0) > u) + P(X(0) \leq u,\ \exists t \leq h : X(t) > u).$$

The second term usually dominates the first term. It is clear that the second term depends on the local path behavior of the Gaussian process, which suggests that $P(M_h > u)$ does also. The local path behavior of $X(\cdot)$ is determined by the local behavior of the correlation function $\rho(t)$ as $t \to 0$. Therefore we restrict $\rho(\cdot)$ in the following way, for t in a neighborhood of 0:

$$1 - \rho(t) \sim C|t|^\alpha \quad \text{as} \quad t \to 0, \quad 0 < \alpha \leq 2, \quad C > 0. \tag{10.1}$$

This condition also implies that the paths of the Gaussian process are continuous with probability 1 (Fernique [157]). Furthermore, the paths are a.s. differentiable if $\alpha = 2$ with additional conditions, but a.s. not differentiable for $\alpha < 2$. A well-known example of a stationary Gaussian process satisfying this condition is the Ornstein-Uhlenbeck process with $\rho(t) = \exp(-C|t|)$, hence $\alpha = 1$.

Theorem 10.2.1. *Let $\{X(t),\ t \geq 0\}$ be a standardized, stationary Gaussian process such that (10.1) holds with $0 < \alpha \leq 2$. Then for any $h > 0$ and $u \to \infty$:*

$$P(M_h > u) \sim hu^{2/\alpha - 1}\varphi(u)C^{1/\alpha}H_\alpha, \tag{10.2}$$

where H_α is a constant depending on α only.

A detailed proof of this result, due to Pickands [369], and the definition of H_α can be found for instance in Leadbetter et al. [303]. It is known that $H_1 = 1$ and $H_2 = 1/\sqrt{\pi}$.

The so-called *Pickands constant* H_α is defined with respect to a fractional Brownian motion $Y(t)$ with parameter α. In the literature the so-called Hurst parameter H is often used to define the fractional Brownian motion instead of the parameter α. The Hurst parameter H is simply $H = \alpha/2$. The fractional Brownian motion is a centered Gaussian process with covariance given by

$$Cov(Y(t), Y(s)) = |t|^\alpha + |s|^\alpha - |t - s|^\alpha, \quad t, s \in \mathbb{R}.$$

Note the difference of definition of the covariance, because sometimes the fractional Brownian motion is defined to be standardized such that the variance of $Y(t)$ is equal to $|t|^\alpha$ whereas the variance is in our definition $2|t|^\alpha$. The Brownian motion is the particular case with $\alpha = 1$ where $Cov(Y(t), Y(s)) = 2 \min(t, s)$. The case with $\alpha = 2$ is very particular and not of much interest, since this fractional Brownian motion has the representation $Y(t)) = tY(1)$. Note also that the correlation of increments

$$\mathrm{Corr}(Y(t) - Y(s), Y(t') - Y(s')) = \frac{|t - s'|^\alpha + |t' - s|^\alpha - |t - t'|^\alpha - |s - s'|^\alpha}{2(|t - s'|^\alpha |t - s'|^\alpha)^{1/2}}.$$

It shows the different dependence structure of the fractional Brownian motions since for non-overlapping time intervals $[s, t]$ and $[s', t']$ these correlations are positive if $\alpha > 1$, negative for $\alpha < 1$, and 0 if $\alpha = 1$. Now H_α is given by

$$H_\alpha = \lim_{T \to \infty} \frac{1}{T} \int_{-\infty}^{0} e^{-x} P\left(\sup_{0 \le t \le T} Y(t) > -x\right) dx.$$

The evaluation of this expression is rather difficult. Some analytical bounds are given by Shao [411]. Simply simulating the fractional Brownian motion and deriving the distribution of $\sup_{0 \le t \le T} Y(t)$ does not give reasonable accurate estimates H_α even with very fast computers.

Another definition of H_α was found by Hüsler [251] analyzing the relation between the discrete and continuous Gaussian processes. It was shown (for more details see also Section 11.5) that H_α is a limit of a cluster index and is equal to

$$H_\alpha = \lim_{\gamma \to 0} \int_{0}^{\infty} e^{-x} P(\max_{k \ge 1} V_k \le -x) \, dx / \gamma$$

where V_k are normal variables with means and covariances

$$E(V_k) = -(\gamma k)^\alpha / 2, \quad \mathrm{Cov}(V_k, V_j) = \gamma^\alpha (k^\alpha + j^\alpha - |k - j|^\alpha).$$

This definition is slightly simpler for numerical derivation.

The constants H_α were derived by sophisticated simulation techniques for the subinterval $\alpha \in (0.8, 1.8)$ (Piterbarg and Romanova, [378]). The whole domain $0 < \alpha \le 2$ was not possible because of the applied technique for simulation. In addition, the simulation has to stabilize sufficiently well for each α to believe in the derived simulated constants. Earlier calculations of Michna [335] were not correct. His simulations for some $\alpha > 1$ indicated a discontinuity at $\alpha = 1$ which corresponds to the Brownian motion. Such a discontinuity does not exist. Earlier, Breitung conjectured that $H_\alpha = 1/\Gamma(1/\alpha)$. The simulation results of Piterbarg and Romanova show that this conjecture is not correct.

Now let us return to the discussion of the main problem. An important step of the proof of Theorem 10.2.1 consists in approximating the maxima and the process $X(t)$ with continuous points $t \in [0, h]$ by that with discrete time points $iq \le h$, $q > 0$, $i \in \mathbb{N}$. Choosing $q = q(u)$ sufficiently small, the exceedances of u by the process $\{X(t),\ t \le h\}$ can be well approximated by the exceedances of u by the sequence $\{X(iq),\ iq \le h\}$. It turns out that q should be chosen such that $qu^{2/\alpha} \to 0$ as $u \to \infty$. Finally,

$$P(\exists i : iq \le h,\ X(iq) > u)$$

can be analyzed using the methods worked out for stationary random sequences.

LIMIT BEHAVIOR OF EXTREMES

Using Theorem 10.2.1 we can now solve the second problem. (10.2) shows that $P(M_h > u)$ converges exponentially fast to 0 as $u \to \infty$. Therefore we expect the limit distribution of M_T to be Gumbel $G_3(\cdot)$. The Berman condition introduced for Gaussian sequences in Section 9.2 can be reformulated for Gaussian processes: $\rho(t)\log(t) \to 0$ as $t \to \infty$. This condition implies that

$$P(M_T \le u_T) \sim \prod_{j \le n} P(M(I_j) \le u_T) = [P(M_h \le u_T)]^n,$$

where again $I_j = [(j-1)h, jh)$, $j \le n$, $n = [T/h]$ for some fixed $h > 0$. With (10.2), the problem is reduced to finding a suitable normalization $u_T = u_T(x) = a_T x + b_T$ such that

$$\frac{T}{h}\left(h u_T^{2/\alpha - 1} \varphi(u_T) C^{1/\alpha} H_\alpha\right) \to e^{-x}$$

as $T \to \infty$. This leads to

Theorem 10.2.2. *Let $\{X(t),\ t \ge 0\}$ be a standardized stationary Gaussian process such that (10.1) and the Berman condition*

$$\rho(t)\log(t) \to 0 \qquad as \quad t \to \infty \tag{10.3}$$

hold. Then

$$P((M_T - b_T)/a_T \le x) \to_D G_3(x) = \exp(-e^{-x}), \tag{10.4}$$

where $a_T = 1/\sqrt{2\log(T)}$,

$$b_T = \sqrt{2\log(T)} + a_T\left(\frac{2-\alpha}{2\alpha}\log\log(T) + \log\left\{C^{1/\alpha} H_\alpha 2^{(2-\alpha)/(2\alpha)}/\sqrt{2\pi}\right\}\right).$$

This result was first proved by Pickands [369] and then extended and completed by Berman [37], Qualls and Watanabe [380] and Lindgren, de Maré and Rootzén [314].

Maxima of Discrete and Continuous Processes

In the proof of the above Theorem 10.2.2 we need to approximate the continuous maxima M_T by the maxima M_n in discrete time related to the sequence $X(iq), i \leq n = [T/q(u)]$ with sufficiently small $q = q(u)$. The approximation is based on a sufficiently dense grid with points iq. As mentioned it is necessary that $q = q(u)$ is chosen such that $qu^{2/\alpha} \to 0$ as $u \to \infty$. If this relation does not hold, what is the relation between M_T and M_n? Does it still mean that $M_T = M_n + o(1)$? Since M_n depends on q, it is better to denote this dependence by setting $M_T^q = M_n$. Piterbarg [377] analyzes this relation and shows that three different cases can occur, namely that M_T and M_T^q can be asymptotically completely dependent, dependent or independent. Obviously, if the grid points are rather dense, then the two maxima differ very little and they are completely dependent. Conversely, if the grid points are rather sparse, with a $q < q_0$, small, we may expect that the two maxima differ substantially, such that they are independent in the limit. This is shown in the next theorem.

Theorem 10.2.3. *Assume that the Gaussian process $X(t), t \geq 0$ is stationary and standardized, satisfying condition (10.1) (with $C = 1$) and Berman's condition (10.3). Let*

$$qu_T^{2/\alpha} \to Q, \quad \text{with } 0 \leq Q \leq \infty.$$

i) *If $Q = 0$, then*

$$P(M_T \leq u_T(x), M_T^q \leq u_T(y)) \to \min(G_3(x), G_3(y))$$
$$= \exp(-\exp(-\min(x, y)))$$

with the usual normalization $u_T = a_T x + b_T$ given in Theorem 10.2.2.

ii) *If $Q \in (0, \infty)$, then*

$$P(M_T \leq u_T(x), M_T^q \leq u_T^q(y)) \to \exp(-e^{-x} - e^{-y} + D(x, y)))$$

where $D(x, y)$ depends on α and Q, and $u_T^q(y) = a_T y + b_T^q$ with $b_T^q = \sqrt{2 \log(T)} - \log(H_{a,\alpha}^{-2} 2\pi (\log(T))^{1-2/\alpha})/(2a_T)$.

iii) *If $Q = \infty$, then*

$$P(M_T \leq u_T(x), M_T^q \leq u_T^q(y)) \to G_3(x)G_3(y)$$
$$= \exp(-e^{-x} - e^{-y})).$$

Besides the dependence of M_T and M_T^q, the difference between the two rv was investigated. More relevant for simulations and numerical approximations is the difference between M_T and the maxima M_T^Y where Y is the approximating Gaussian process which is based on the points $X(i/n)$ and (e.g.) linear interpolations:

$$Y(t) = (i + 1 - nt)X(i/n) + (nt - i)X((i+1)/n)$$

for $t \in [i/n, (i+1)/n]$. The difference process $X(t) - Y(t)$ is again a Gaussian process, but not a stationary one. Therefore we will discuss this approach in Section 10.3.

CROSSINGS OR LEVEL SETS OF SMOOTH GAUSSIAN PROCESSES

If the Gaussian process $X(t)$ is rather regular, with differentiable paths, one may define level crossing sets and the number of such crossings $N_u(X, I)$, as well as the number of upcrossings $U_u(X, I)$ and downcrossings $D_u(X, I)$ of u in the interval I where

$$N_u(X, I) = |\{t \in I : X(t) = u\}|,$$

$$U_u(X, I) = |\{t \in I : X(t) = u, X'(t) > 0\}|$$

and

$$D_u(X, I) = |\{t \in I : X(t) = u, X'(t) < 0\}|.$$

The well-known Rice formula gives the expectation of the number of crossings for a standard Gaussian process $X(t)$ in case the second spectral moment λ_2 exists. We have

$$E(N_u(X, I)) = \sqrt{\lambda_2} \exp(-u^2/2)|I|/\pi$$

(see Rice [394], [395], or e.g. Cramér and Leadbetter [79], Azaïs and Wschebor [20]). Under additional conditions, higher moments of the number of crossings can be derived, as well as bounds for the maximum term $M_I = \max_{t \in I} X(t)$. For some recent results see Mercadier [328].

For instance, under additional conditions and if the sixth spectral moment λ_6 exists, the tail of the distribution of M_T of a standard, stationary Gaussian process $X(t)$ is derived by Azaïs et al. [19]. They showed that for $I = [0, T]$ and $u \to \infty$,

$$P(M_T > u) = 1 - \Phi(u) + \sqrt{\frac{\lambda_2}{2\pi}} T \phi(u)$$

$$- (1 + o(1)) \frac{3\sqrt{3}(\lambda_4 - \lambda_2^2)^{9/2}}{2\pi \lambda_2^{9/2}(\lambda_2 \lambda_6 - \lambda_4^2)} \frac{T}{u^5} \phi\left(\sqrt{\frac{\lambda_4}{\lambda_4 - \lambda_2^2}} u\right).$$

This result was shown first by Piterbarg [374] for sufficiently small T. The concepts of crossings, upcrossings and downcrossings can be extended for regular random fields, in particular for Gaussian ones. Of interest are then the level sets $\{t \in I : X(t) = u\}$ where $t \in \mathbb{R}^d$ and $I \subset \mathbb{R}^d$ is compact. Such results on crossings and level sets are discussed in Azaïs and Wschebor [20] with several applications in different topics.

Instead of a fixed level u, one may also consider a smooth level function $u(t)$ and investigate the number of crossings $\{t \in I : X(t) = u(t)\}$ of such a level function, see Kratz and León [294], [295].

The curve crossings can be considered also for a discrete time process $X_i, i \in \mathbb{Z}$, and a general curve u_i. The number C_n of curve crossings is defined as

$$C_n = \sum_{1 \le i \le n} 1((X_{i-1} - u_{i-1})(X_i - u_i) < 0)$$

which adds the up- and down-crossings together. Zhao and Wu [473] derived a central limit result for C_n as $n \to \infty$ for short-range dependent stationary sequences X_i and for certain curves u_i. Such limit results were derived first for particular sequences, e.g. for linear sequences by Wu [465] and [466]. For long-range dependent sequences, the behavior of C_n is more complicated. A particular case of a long range linear sequence is studied in Zhao and Wu [473].

10.3   Non-stationary Gaussian Processes

The theorems of Section 10.2 can be extended to non-stationary Gaussian processes with continuous paths. Let $\tilde{X}(t)$ be any non-stationary Gaussian process with mean $m(\cdot)$ and covariance $\sigma(\cdot, \cdot)$ such that $\mathrm{Corr}(\tilde{X}(t), \tilde{X}(s)) = 1$ if and only if $t = s$. By standardizing $\tilde{X}(\cdot)$, we get a Gaussian process $X(\cdot)$ with zero mean and unit variance. Hence, when dealing with exceedances, we introduce now non-constant boundaries $\{u_T(t), \ t \le T\}$, which are usually continuous functions. This is obvious since even the usual constant boundary $a_T x + b_T$ for $\tilde{X}(\cdot)$ is transformed into a non-constant boundary function for $X(\cdot)$ by the standardization:

$$u_T(t) = \frac{a_T x + b_T - m(t)}{\sigma(t,t)^{1/2}}, \qquad t \le T.$$

One may try to approximate the boundary function u_T by piecewise constant functions with span $h \to 0$ as $T \to \infty$; let u be the approximation of $u_T(\cdot), 0 \le t \le h$. This gives rise to the question of whether the solution to the first problem (Theorem 10.2.1) remains valid for $h \to 0$. If we let $h = h(u)$ tend slowly to 0 so that

$$hu^{2/\alpha} \to \infty \qquad \text{as} \quad u \to \infty, \tag{10.5}$$

then (10.2) still holds (see Hüsler [246]). Note that such a condition is necessary since in the inequality

$$P(M_h > u) \le P(X(0) > u) + P(X(0) \le u, \ \exists t \le h : X(t) > u)$$

the second term of the upper bound dominates the first term only if h is sufficiently large and tends slowly to 0.

- (10.5) implies that the interval $(0, h)$ is sufficiently large such that an up-crossing in $(0, h)$ can occur even if $X(0) < u$, with a positive probability. The probability of this upcrossing or exceedance depends on the local path behavior of the Gaussian process.

- If $h \to 0$ faster (and such that (10.5) does not hold), then we may have

$$P(M_h > u) \sim P(X(0) > u) = 1 - \Phi(u).$$

- Compare the choice of h with the choice of q in the previous section: $qu^{2/\alpha}$ converges slowly to 0 as $u \to \infty$. We observed that $P(M_h > u)$ is approximated by $P(\exists i : iq \leq h,\ X(iq) > u)$, hence the number of rv $X(iq)$ in this event $\{M_h > u\}$ is equal to h/q, still tending to ∞.

Under condition (10.5) we can generalize Theorem 10.2.1 for stationary Gaussian processes, allowing now $h(u) \to 0$.

Theorem 10.3.1. *Let $\{X(t),\ t \geq 0\}$ be a standardized stationary Gaussian process such that (10.1) holds with $0 < \alpha \leq 2$. If $h = h(u) > 0$, $h(u) \to 0$ such that (10.5) holds, then*

$$P(M_h > u)/(hu^{2/\alpha - 1}\varphi(u)) \to C^{1/\alpha}H_\alpha \qquad as \quad u \to \infty. \qquad (10.6)$$

The proof is given in Hüsler [246]. We want to extend this result to non-stationary Gaussian processes. This can be done by approximating non-stationary Gaussian processes by stationary ones. A useful tool for this approximation is the *Slepian inequality* (Slepian [416], cf. Leadbetter et al. [303]).

Theorem 10.3.2 (Slepian). *Let $\{X(t),\ t \geq 0\}$ and $\{Y(t),\ t \geq 0\}$ be two standardized Gaussian processes with continuous paths. If for some $t_0 > 0$,*

$$\mathrm{Cov}(X(t), X(s)) \leq \mathrm{Cov}(Y(t), Y(s)), \qquad t,\ s \leq t_0,$$

then for all $h \leq t_0$ and all u,

$$P\Big(\sup_{t \leq h} X(t) \leq u \Big) \leq P\Big(\sup_{t \leq h} Y(t) \leq u \Big).$$

This inequality is obvious in the special case, where $Y(t) = Y(0)$ a.s., for all t, since then

$$P\Big(\sup_{t \leq h} X(t) \leq u \Big) \leq P(X(0) \leq u) = P\Big(\sup_{t \leq h} Y(t) \leq u \Big).$$

Note that in this case the condition on the covariances is satisfied for any standardized Gaussian process $X(\cdot)$, since $\mathrm{Cov}(Y(t), Y(s)) = \mathrm{Var}(Y(0)) = 1$. In the case where $r(t) \geq 0$ for all $t \leq t_0$, the Slepian inequality implies also

$$P\Big(\sup_{t \in A} X(t) \leq u \Big) \geq \prod_{t \in A} P(X(t) \leq u),$$

where A is a discrete subset of $[0, t_0]$. The inequality above holds even if the constant boundary u is replaced by a non-constant one, $u(t)$, and the events by $\{X(t) \leq u(t),\ t \in A\}$.

LOCALLY STATIONARY GAUSSIAN PROCESSES

Let $X(\cdot)$ be now a non-stationary Gaussian process. If there exist stationary Gaussian processes which have a similar local path behavior as $X(\cdot)$, then we can generalize Theorem 10.3.1, using the Slepian inequality. This is possible if an assumption on the correlation function, similar to (10.1), is satisfied by $X(\cdot)$.

Therefore we introduce the following class of non-stationary Gaussian processes with covariance function $\sigma(t, s)$, given by Berman [38].

Definition 10.3.3. A standardized Gaussian process $X(\cdot)$ is called *locally stationary*, if there exists a continuous function $C(t)$, $t \geq 0$, with

$$0 < \inf\{C(t),\ t \geq 0\} \leq \sup\{C(t),\ t \geq 0\} < \infty$$

such that, for some α, $0 < \alpha \leq 2$,

$$\lim_{s \to 0} \frac{1 - \sigma(t, t + s)}{|s|^\alpha} = C(t) \qquad \text{uniformly in} \quad t \geq 0. \tag{10.7}$$

Note that (10.7) reduces to (10.1) in the stationary case, where $C(t) \equiv C$. This implies that the local path behavior of $X(\cdot)$ is similar to that of a standardized stationary Gaussian process. $C(t)$ determines the variance of the increments of the process near t. Instead of the power function $|s|^\alpha$, one might use a non-negative, monotone increasing function $K(|\cdot|)$ such that $K(0) = 0$ and $K(|s|) > 0$ for $s \neq 0$. For example we can choose a regularly varying function, which fulfills these requirements.

Simple examples of locally stationary Gaussian processes $X(t)$ are time transformations of stationary Gaussian processes $Y(\cdot) : X(t) = Y(g(t))$ for some non-negative differentiable function $g(\cdot)$ with $g'(x) > 0$. Then $C_X(t) = C_Y \cdot (g'(t))^\alpha$ (cf. Hüsler [249]).

If $X(\cdot)$ is a locally stationary Gaussian process, then the asymptotic behavior of an exceedance of u in a small interval h is given by:

Theorem 10.3.4. *Let* $\{X(t),\ t \geq 0\}$ *be a locally stationary Gaussian process with* $0 < \alpha \leq 2$ *and* $C(t)$. *If* (10.5) *holds for* $h = h(u)$ *with* $h(u) \to 0$, *then*

$$\lim_{u \to \infty} P(M(t, t + h) > u)/(hu^{2/\alpha - 1}\varphi(u)) = C^{1/\alpha}(t)H_\alpha$$

uniformly in $t \geq 0$.

Proof. It can be shown that $\rho_C(s) = 1/(1 + C|s|^\alpha)$ with $0 < \alpha \leq 2$ and $C > 0$ is a correlation function of a standardized stationary Gaussian process $Y(\cdot)$. Furthermore, $1 - \rho_C(s) \sim C|s|^\alpha$ as $s \to 0$. For every $\epsilon > 0$ and $t \geq 0$ there exists $C_1 = C_1(t) \leq C(t)$ such that $\sigma(t, t + s) \leq \rho_{C_1}(s)$ for all $s < \epsilon$. Using the Slepian inequality (Theorem 10.3.2) and Theorem 10.3.1 we get

$$C_1^{1/\alpha} H_\alpha \leq \liminf_{u \to \infty} P(M(t, t + h) > u)/(hu^{2/\alpha - 1}\varphi(u)).$$

On the other hand there exists $C_2 = C_2(t) \geq C(t)$ such that $\sigma(t, t+s) \geq \rho_{C_2}(s)$ for all $s < \epsilon$, which leads to an upper bound for the probability of an exceedance

$$C_2^{1/\alpha} H_\alpha \geq \limsup_{u \to \infty} P(M(t, t+h) > u)/(hu^{2/\alpha-1}\varphi(u)).$$

Since this holds for any ϵ and C_i can be chosen such that $C_i \to C(t)$ as $\epsilon \to 0$, $i = 1, 2$, the statement follows, using the uniformity of (10.7). □

As in the stationary case the limit distribution of M_T as $T \to \infty$ can be found, by splitting up the interval $[0, T]$ into subintervals of length $h = h_T \to 0$. Let $n = [T/h]$ and

$$u_{T,min} = \min_{t \leq T} u_T(t) \to \infty. \tag{10.8}$$

We approximate the boundary $u_T(t), t \leq T$ by a step function,

$$\sum_{j \leq n} u_{T,j} 1((j-1)h \leq t < jh) \qquad \text{for} \quad t \leq nh.$$

Then under suitable conditions we get, for T sufficiently large,

$$P(X(t) \leq u_T(t), \, t \leq T) \approx P(X(t) \leq u_{T,j}, \, t \in [(j-1)h, jh), \, j \leq n)$$

$$\approx \prod_{j \leq n} P(X(t) \leq u_{T,j}, \, t \in [(j-1)h, jh))$$

$$\approx \exp\left[-\sum_{j \leq n} P(M([(j-1)h, jh)) > u_{T,j}) \right]$$

$$\approx \exp\left[-\sum_{j \leq n} h H_\alpha C^{1/\alpha}(jh) u_{T,j}^{2/\alpha-1} \varphi(u_{T,j}) \right]$$

$$\approx \exp\left[-H_\alpha \int_0^T C^{1/\alpha}(t) u_T^{2/\alpha-1}(t) \varphi(u_T(t)) \, dt \right]$$

$$=: \exp[-J(T)].$$

The first approximation follows since $u_{T,min} \to \infty$. The second approximation holds by Berman's condition

$$\sup_{|t-s|>\tau} |\sigma(t, s)| \log(\tau) \to 0 \qquad \text{as} \quad \tau \to \infty. \tag{10.9}$$

The third and fourth approximations follow by Theorem 10.3.4, if $h = h_T$ satisfies (10.5) with $u = u_{T,min}$. For the last approximation we need that the integral $J(T)$ can be approximated accurately by Riemann sums with span h_T. This can be done if for some h_T satisfying

$$h_T(u_{T,min}^2/\log(u_{T,min}))^{1/\alpha} \to \infty \qquad \text{as} \quad T \to \infty, \tag{10.10}$$

the following holds:

$$\Delta(T, h_T) = J^+(T) - J^-(T) \to 0 \qquad \text{as} \quad T \to \infty, \tag{10.11}$$

where $J^+(T)$ and $J^-(T)$ denote the upper and lower Riemann sum of $J(T)$, respectively. Note that (10.10) implies (10.5).

Summarizing, the following general result can be proved (Hüsler [246]).

Theorem 10.3.5. *Let $\{X(t), t \geq 0\}$ be a locally stationary Gaussian process with $0 < \alpha \leq 2$ and $C(t)$, such that Berman's condition (10.9) holds. Let $u_T(t)$, $t \leq T$, $T > 0$, be a continuous boundary function, for each T, such that (10.8) holds. If (10.10) and (10.11) hold and if $\limsup_{T \to \infty} J(T) < \infty$, then*

$$P(X(t) \leq u_T(t), \ t \leq T) - \exp(-J(T)) \to 0$$

as $T \to \infty$.

The extremal behavior of Gaussian processes with trend or non-constant variance function can be treated using this result. By analyzing the convergence of $J(T)$ the limiting distribution $G(\cdot)$ and the normalization can be worked out (cf. Hüsler [246]).

CONSTANT BOUNDARIES

In particular, if the boundary function is constant, i.e., $u_T(t) = u_T$, $t \leq T$, then (10.11) is satisfied if

$$C_T = \int_0^T C^{1/\alpha}(t) \, dt$$

can be approximated by Riemann sums C_T^+ and C_T^- with span h_T satisfying (10.10) such that

$$(C_T^+ - C_T^-) u_T^{2/\alpha - 1} \varphi(u_T) \to 0 \tag{10.12}$$

as $T \to \infty$. For this special case we state

Corollary 10.3.6. *Let $\{X(t), t \geq 0\}$ be a locally stationary Gaussian process with $0 < \alpha \leq 2$ and $C(t)$, satisfying (10.9). Let $u_T = a_T x + b_T$ be a constant boundary where*

$$a_T = (2 \log(C_T))^{-1/2}$$

and

$$b_T = \sqrt{2 \log(C_T)} + a_T \left(\frac{2 - \alpha}{2\alpha} \log \log(C_T) + \log \left\{ H_\alpha 2^{(2-\alpha)/(2\alpha)} / \sqrt{2\pi} \right\} \right).$$

If $C(t)$ is such that (10.12) holds, then

$$P(M_T \leq a_T x + b_T) \to_D G_3(x) = \exp(-e^{-x}) \tag{10.13}$$

as $T \to \infty$.

The non-stationarity of the Gaussian process influences the normalization only through C_T. Since in Corollary 10.3.6, $u_T = u_{T,min} \sim \sqrt{2\log(T)}$, (10.10) is satisfied if

$$h_T(\log(T))^{1/\alpha}(\log\log(T))^{-1/\alpha} \to \infty, \qquad \text{as} \quad T \to \infty \tag{10.14}$$

for some $h_T \to 0$.

If $C(t) \equiv C$ is a constant function, then (10.11) holds for any h_T satisfying (10.14) with $C_T = C^{1/\alpha}T$. Hence we get

Corollary 10.3.7. *Let $\{X(t),\ t \geq 0\}$ be a locally stationary Gaussian process with $0 < \alpha \leq 2$ and $C(t)$, satisfying (10.9) and suppose $C(t) \equiv C > 0$. Let $u_T = a_T x + b_T$ be a constant boundary where*

$$a_T = (2\log(T))^{-1/2}$$

and

$$b_T = \sqrt{2\log(T)} + a_T\left(\frac{2-\alpha}{2\alpha}\log\log(T) + \log\left\{C^{1/\alpha}H_\alpha 2^{(2-\alpha)/(2\alpha)}/\sqrt{2\pi}\right\}\right).$$

Then

$$P(M_T \leq a_T x + b_T) \to_D G_3(x) = \exp(-e^{-x}) \tag{10.15}$$

as $T \to \infty$.

Note that we use a normalization which is asymptotically equivalent to the one in Corollary 10.3.6. If the Gaussian process is stationary, this result coincides with that of Theorem 10.2.2. This shows that the expressions for a_T and b_T and the limit distribution $G_3(\cdot)$ do not depend on the stationarity assumption.

MULTIFRACTIONAL PROCESSES

Another generalization of stationary Gaussian processes has been introduced in the discussion of extremes or exceedances. Dębicki and Kisowski [98] defined multifractional Gaussian processes, motivated from multifractional Brownian motions.

The condition on the correlation function, given in (10.1), was weakened by letting the coefficient C depend on t which implied the condition (10.7). We can apply this idea also to the coefficient α, assuming that it may vary with t. From Theorem 10.3.4, it becomes obvious that only the smallest α values contribute to the probability of the maximum value $M([t, t+h])$ above u, asymptotically. Hence, if, for example, $\alpha(t)$ is a step function, we have to consider only the interval $\{s \in [t, t+h] : \alpha(s) = \alpha_{min}\} = I_{\min}$ where the function $\alpha(\cdot)$ has its minimum values $\alpha_{\min} = \min_{s \in [t,t+h]} \alpha(s)$. Then Theorem 10.3.4 holds with this α_{min} instead of the fixed α value, and the length of the interval $I_{\min}$ instead of h. More interesting is the extension to a continuously varying $\alpha(t)$. Again, the minimum value α_{min} is most relevant. We assume for simplicity that $t_{\min} = \arg\min \alpha(s)$ is unique. It could be a boundary point or an inner point of $[t, t+h]$. If t_{min} is an inner point, the probability $P(M([t, t+h]) > u)$ is asymptotically twice the term we present for the case where t_{min} is a boundary point, $t_{min} = t$ or $t_{min} = t + h$.

In addition, we have to investigate how the local behavior of $\alpha(s)$ in $t_{\min}$ influences the probability of interest, since $t_{\min}$ is unique. We simplify the notation by considering the case $t_{\min} = t = 0$. The other cases $t_{\min} = h$ or $t_{\min} \in (0, h)$ follow in the same way. Let us formulate the extended version of Definition 10.3.3.

Definition 10.3.8. A standardized Gaussian process $X(\cdot)$ is called $\alpha(t)$-*locally stationary* on $[0, T]$ with $T \leq \infty$, if there exist continuous functions $C(t)$ and $\alpha(t)$, $0 \leq t \leq T$, with

$$0 < \inf\{C(t),\ t \in [0, T]\} \leq \sup\{C(t),\ t \in [0, T]\} < \infty$$

and $0 < \alpha(t) \leq 2$, such that

$$\lim_{s \to 0} \frac{1 - r(t, t+s)}{|s|^{\alpha(t)}} = C(t) \qquad \text{uniformly for } t \in [0, T]. \tag{10.16}$$

As mentioned, the behavior of $\alpha(t)$ in the vicinity of $t_{min} = 0$ plays a crucial role. It influences the chance of an exceedance in the neighborhood of $t_{min} = 0$. The crucial neigborhood is rather small, with a length tending to 0 as $u \to \infty$. We assume that

$$\alpha_{min} = \min_{t \in [0,h]} \alpha(t) = \alpha(0) > 0$$

and

$$\alpha(t) = \alpha_{min} + Bt^\beta + o(|t|^{\beta+\delta}) \qquad \text{as } t \to 0, \text{ for some positive } B, \beta, \delta. \tag{10.17}$$

Theorem 10.3.9. *Let* $\{X(t),\ 0 \leq t \leq h\}$ *be a* $\alpha(t)$-*locally stationary Gaussian process with* $0 < \alpha(t) < 2$ *and* $C(t)$. *If* (10.16) *and* (10.17) *hold, then for* $h > 0$,

$$\lim_{u \to \infty} \frac{P(M([0, h]) > u)}{u^{2/\alpha_{\min} - 1}(\log u)^{-1/\beta}\, \phi(u)} = (C(0))^{1/\alpha_{\min}} \left(\frac{\alpha_{\min}^2}{2B}\right)^{1/\beta} H_{\alpha_{\min}} \frac{\Gamma(1/\beta)}{\beta}.$$

This is a result of Dębicki and Kisowski [98]. The proof is based on the usual double sum approximation method. We note that the term $(\log u)^{-1/\beta}$ is the impact from the varying $\alpha(t)$ in the neighborhood of $t_{\min} = 0$. Obviously, we can let h depend on u as in Theorem 10.3.4. The result holds also if $h = h(u)$ tends to 0, as long as $h(u) \geq ((\alpha_{\min} \log\log u)/(\beta \log u))^{1/\beta}$. Note that this term is much larger than the one used for the Pickands window with $h(u)u^{2/\alpha} \to \infty$ (cf. (10.5)). So the usual approximation with a Pickands window is not sufficient. The result shows also that the behavior of the function $C(t)$ in the vicinity of $t_{\min}$ has no impact on the asymptotic behavior of the considered probability. The function $C(t)$ might be constant: $C(t) = C(0)$ for $t \in [0, h]$. Also the length h of the interval has no impact on the result.

The authors applied Theorem 10.3.9 to multifractional Brownian motions $B_H(t), t > 0$. These are centered Gaussian processes with

$$E(B_H(t)B_H(s))$$
$$= \frac{1}{2}D(H(s) + H(t)) \left[|s|^{H(s)+H(t)} + |t|^{H(s)+H(t)} - |t - s|^{H(s)+H(t)} \right],$$

where $D(x) = 2\pi/(\Gamma(x+1)\sin(\pi x/2))$ and $H(t)$ is a Hölder function with exponent γ, with $0 < H(t) < \min(1, \gamma)$ for $0 \leq t < \infty$. Then the standardized multifractional Brownian motion is $\alpha(t)$-locally stationary for $t \in [t_0, t_1]$ with $0 < t_0 < t_1 < \infty$, where $\alpha(t) = 2H(t)$ and $C(t) = t^{-2H(t)}/2$.

EXCURSIONS ABOVE VERY HIGH BOUNDARIES

Up to now the boundary functions were related to T in such a way that we could derive a non-degenerate limit for the probability of an upcrossing. If this probability tends to zero we are interested in other quantities such as the rate of convergence. Cuzick [87] derived the convergence rate of

$$P(X(t) > u_n(t) \text{ for some } t \in (0, T)) \to 0 \qquad \text{as} \quad n \to \infty$$

where $X(\cdot)$ is a stationary Gaussian process not necessarily satisfying Berman's condition, and T can be finite or infinite. The extensions to locally stationary Gaussian processes can be found in Bräker [53]. Many further results on sojourns and extreme values of Gaussian and non-Gaussian processes are given in Berman [39]. Some additional results and applications are shown in the following sections.

BOUNDARIES WITH A UNIQUE POINT OF MINIMAL VALUE

Often particular boundaries are such that only a few points play an important role in the approximation of the probability of an exceedance. Such points are often defined to be points with minimal boundary value after eventually standardizing the Gaussian process. This situation occurs for example if the variance of a centered Gaussian process is not constant, having a unique point of maximal variance. Such a case was mentioned in Section 10.2 when a stationary Gaussian process is approximated by a piecewise linear Gaussian process. The asymptotic approximation of the exceedance probability can be derived accurately in such a case, but it is based on a somewhat different approach from the one presented above. Also $J(T)$ would tend to 0.

The well-known standardization of a general non-stationary Gaussian process $X(t)$ with general non-constant variance $\sigma^2(t)$ and non-constant trend $\mu(t)$ links the boundary with the trend and variance. Considering then the supremum of the process $X(t)$ in an interval I corresponds to the event of crossing the boundary after the standardization:

$$\{X(t) \geq x, \text{ for all } t \in I\}$$
$$= \left\{ \frac{X(t) - \mu(t)}{\sigma(t)} \geq \frac{x - \mu(t)}{\sigma(t)} \text{ for all } t \in I \right\}$$
$$= \left\{ Y(t) \geq \frac{x - \mu(t)}{\sigma(t)} =: u_x(t) \text{ for all } t \in I \right\}$$

where $Y(t)$ denotes the standardized Gaussian process with the same correlation function as $X(t)$. Most probable points to cross the boundary are points in the neighborhood of minimal boundary values. Can we state in general that the lowest boundary value determines in such cases the probability that the supremum $M(I)$ is larger than x, for large x? This question will be investigated in the following, for a bounded interval I. We can set $I = [0, 1]$ by time transformation, and $\sup_{t \in I} \sigma(t) = 1$.

Example 10.3.10. (Brownian bridge and motion) Let us consider the well-known case of the Brownian bridge B_0, a centered Gaussian process, with covariance $\sigma(t, s) = \text{Cov}(B_0(t), B_0(s)) = \sigma^2(\min(t, s) - st)$. Assume $\sigma^2 = 1$. Note that the largest value of variance $\sigma(t, t) = t(1 - t)$ is equal to $1/4$, occurring at $t_0 = 0.5$. It is known that
$$P(\sup_{t \in [0,1]} B_0(t) > u) = \exp(-2u^2).$$

We cannot expect that such an exact formula for the df of the supremum can be derived for many Gaussian processes. Hopefully we might approximate accurately the distribution for large values u, not giving only the leading term as with large deviation principles.

The result we are going to derive, holds for rather general Gaussian processes with a unique point of maximal variance. It implies in the case of the Brownian bridge that
$$P\left(\sup_{t \in [0,1]} X(t) > u\right) \sim \exp(-2u^2)$$

as $u \to \infty$.

Consider now the Brownian motion $B(t)$. It is well known that by the reflection principle $P(\sup_{t \in [0,1]} B(t) > u) = 2(1 - \Phi(u))$. The maximal variance occurs at the endpoint $t_0 = 1$, another situation. The exceedance of u by $X(1)$ has the probability $1 - \Phi(u)$ which is not the expression above. The given expression is larger, hence some other values $X(t)$ of the process for t near 1 contribute to the event $\{\sup_{t \in [0,1]} X(t) > u\}$. This problem is also investigated in the following for general Gaussian processes and an accurate expression for $u \to \infty$ is derived.

The general approach considers first the probability that the Gaussian process exceeds the boundary somewhere in the interval $I_{t_0} \subset I$, being a suitable neighborhood of the point t_0 with smallest boundary value. It is assumed that the boundary is minimal in one unique point t_0. Then in a second step it is shown that the probability of an exceedance above the boundary outside of the interval I_{t_0} is asymptotically negligible with respect to the first derived probability.

The probability of an exceedance of a high boundary depends again on the correlation function and the behavior of boundary in the vicinity of the minimal values. We focus in the following on centered processes with mean 0 and a variance function with one unique maximal value at t_0. Typically the following conditions are supposed for general cases.

The behavior of the variance $\sigma^2(t)$ is restricted by

$$1 - \sigma(t) \sim b|t - t_0|^\beta \quad \text{for } |t - t_0| \to 0, \tag{10.18}$$

for some $b > 0, \beta > 0$ and some $t_0 \in I$. t_0 denotes the unique point of maximum variance, hence $\sigma(t) < 1$ for all $t \neq t_0$.

The correlation function $r(t, s) = \text{Corr}(X(t), X(s))$ is restricted by

$$1 - r(t, s) \sim |t - s|^\alpha, \quad \text{for } t \to t_0, s \to t_0 \text{ and } \alpha \in (0, 2]. \tag{10.19}$$

It is related to the behavior of the sample paths of $X(t)$; the larger the parameter α, the smoother the sample paths.

A regularity condition is needed for the remaining points t not in the neighborhood of t_0:

$$E(X(t) - X(s))^2 \leq G|t - s|^\gamma \quad \text{for } t, s \in [0, 1] \text{ and some } G > 0, \gamma > 0. \tag{10.20}$$

Piterbarg [375] proved the following result for a fixed boundary u.

Theorem 10.3.11. *Let $X(t)$ be a Gaussian process with mean 0 satisfying the conditions* (10.18), (10.19) *and* (10.20) *with t_0 the unique point of maximum variance in the interior of $I = [0, 1]$ with $\sigma(t_0) = 1$.*

(i) *If $\beta \geq \alpha$, then*

$$P\left(\sup_{t \in [0,1]} X(t) > u\right) \sim C(\alpha, \beta, b)\, u^{2/\alpha - 2/\beta}\, (1 - \Phi(u)), \quad \text{for } u \to \infty$$

where $C(\alpha, \beta, b) = 2H_\alpha \Gamma(1/\beta)/(\beta b^{1/\beta})$ for $\beta > \alpha$, and $= 2H_\alpha^b$ for $\beta = \alpha$;

(ii) *If $\beta < \alpha$, then*

$$P\left(\sup_{t \in [0,1]} X(t) > u\right) \sim (1 - \Phi(u)), \quad \text{for } u \to \infty.$$

The Pickands constants H_α and H_α^b are discussed below. Intuitively, we can say that if $\beta \geq \alpha$, then the variance or the boundary function $u(t) = u/\sigma(t)$ is smoother than the paths of the process, hence points t in the neighborhood of t_0 also substantially contribute to the probability of the extreme event $\sup X(t) \geq u$. This is similar to the case of a constant variance and constant boundary (hence with $\beta = \infty$). If $\beta < \alpha$, then the paths are now smoother than the variance or boundary function and only $X(t_0)$ asymptotically determines the extreme event, maybe also some values $X(t)$ with $t - t_0$ very small, but because of the strong correlation of $X(t)$ and $X(t_0)$, these values do not influence the probability of the extreme event.

MORE PICKANDS CONSTANTS

The Pickands constants H_α and H_α^b in Theorem 10.3.11 are related to the Gaussian process $\chi(t)$ with continuous paths, with mean function $-|t|^\alpha$ and covariance function $|t|^\alpha + |s|^\alpha - |t - s|^\alpha$, a fractional Brownian motion with trend. They are defined with $T, T_1 \leq T_2$ by

$$H_\alpha^b(T_1, T_2) := E(\exp(\sup_{T_1 \leq t \leq T_2} (\chi(t) - b|t|^\alpha))) < \infty,$$

$$H_\alpha(T) := H_\alpha^0(0, T),$$

$$H_\alpha^b := \lim_{S \to \infty} H_\alpha^b(-S, S) \in (0, \infty)$$

and

$$H_\alpha := \lim_{T \to \infty} H_\alpha(T)/T \in (0, \infty).$$

If the point t_0 is a boundary point of I, say $t_0 = 0$, then Theorem 10.3.11 remains true replacing the constants $C(\alpha, \beta, b)$ by the new constant $C^*(\alpha, \beta, b)$ $= C(\alpha, \beta, b)/2$ if $\beta \geq \alpha$ (Konstant and Piterbarg, [292]). This is obvious because in the case $\beta \geq \alpha$ the interval around $t_0 = 0$ included in $I = [0, 1]$ where exceedances are most probable, is now only half the length as if t_0 is an interior point. If $\beta < \alpha$ still $X(t_0) = X(0)$ exceeds most probably the boundary u.

The length of the interval around t_0 where exceedances are most probable is typically related to the correlation function, which means to the parameter α. This was discussed in Piterbarg and Weber [379]) setting $t_0 = 0$, shifting the interval I. Let $Y(t)$ be a standardized stationary Gaussian process on $[-a, a]$, a small interval around $t_0 = 0$, with correlation function $r(t) = \exp(-t^\alpha)$ with $0 < \alpha \leq 2$ and $a = a(u)$.

Theorem 10.3.12. *Let $Y(t)$ be a standardized stationary Gaussian process with correlation $r(t) = \exp(-t^\alpha)$ with $0 < \alpha \leq 2$. For any $T_1 \leq 0, T_2 > 0, b > 0$ and $\beta > 0$,*

$$P\left(\sup_{t \in [T_1 u^{-2/\alpha}, T_2 u^{-2/\alpha}]} Y(t)(1 - b|t|^\beta) > u\right) \sim H(\alpha, \beta, b, T_1, T_2)(1 - \Phi(u))$$

as $u \to \infty$, where the constant is related to the above Pickands constants:

$$H(\alpha, \beta, b, T_1, T_2) = \begin{cases} 1, & if \quad \beta < \alpha, \\ H_\alpha(T_2 - T_1), & if \quad \beta > \alpha, \\ H_\alpha^b(T_1, T_2), & if \quad \beta = \alpha. \end{cases}$$

The process $Y(t)(1 - b|t|^\beta)$ corresponds to a Gaussian process with non-constant variance $(1 - b|t|^\beta)^2 = 1 - 2b(1 + o(1))|t|^\beta$ for $t \to 0$, with unique point $t_0 = 0$ of maximal variance. The more general case of a correlation function $r(t)$ with $r(t) = 1 - (d + o(1))|t|^\alpha$ ($t \to 0$) where $d > 0$, can be treated by approximating $r(t)$ with $r^*(t) = \exp(-dt^\alpha)$. By transforming the time $t \to td^{1/\alpha}$, we get a process $Y(t)$ with $r_Y(t) = \exp(-t^\alpha)$ which allows us to apply Theorem 10.3.12.

If $t_0 = t_0(u) \in [0, 1]$ depends on u such that $t_0(u) \to 0$, then the speed of convergence of $t_0(u)$ to 0 is relevant for the probability of exceedances of u. If the speed is rather slow, then the probability of an exceedance is as in Theorem 10.3.11. If the speed is very fast, the case is just as if $t_0(u) = 0$. For the intermediate cases we have to apply Theorem 10.3.12 which occurs if the variance function $\sigma_u(t)$ and also the correlation function $\rho_u(t)$ depend on u. This was considered in Hashorva and Hüsler [209] for a family of processes $X_u(t)$. Define for some $\beta > 0$ the interval

$$I_u := [-\delta^*(u, \beta), \delta(u, \beta)],$$

where

$$\delta^*(u, \beta) = \min(\delta(u, \beta), t_0(u))$$

and

$$\delta(u, \beta) := u^{-2/\beta} \log^{2/\beta}(u).$$

The shifted interval $t_0(u) + I_u = [t_0(u) - \delta^*(u, \beta), t_0(u) + \delta(u, \beta)] \subset [0, t_0(u) + \delta(u, \beta)]$ plays a crucial role for the probabilities. The conditions (10.18), (10.19) and (10.20) are reformulated for dependence on u. We assume

$$\sup_{t \in I_u + t_0(u)} \left| \frac{1 - \sigma_u(t)}{b|t - t_0(u)|^\beta} - 1 \right| \to 0 \quad \text{as} \quad u \to \infty \tag{10.21}$$

for some $b > 0$ and $\beta > 0$,

$$\sup_{t, s \in I_u + t_0(u)} \left| \frac{1 - \rho_u(t, s)}{|t - s|^\alpha} - 1 \right| \to 0 \quad \text{as} \quad u \to \infty, \tag{10.22}$$

and for large u,

$$E\left((X_u(t) - X_u(s))^2\right) \le G|t - s|^\gamma, \qquad t, s \in [0, 1], \tag{10.23}$$

with some $G > 0$, $\gamma > 0$.

For the case of the unique point of maximal variance we get now under these conditions the general approximation of the tail of the distribution of the supremum. Let us denote the incomplete Γ-integral by $\Gamma(a, w) = \int_0^w e^{-x} x^{a-1} dx$.

Theorem 10.3.13. *Let $X_u(t), t \in [0,1], u > 0$, be a family of Gaussian processes with means 0, variances $\sigma_u^2(t) \leq \sigma_u^2(t_0(u)) = 1$, and correlations $\rho_u(s,t)$ with $t_0(u)$ being the unique point of maximal variance. Suppose the conditions (10.21), (10.22) and (10.23) hold.*

i) *If $\alpha < \beta$ and $t_0(u)u^{2/\beta} \to C \in [0,\infty]$, then*

$$P\left(\sup_{[0,1]} X_u(t) > u\right) \sim H_\alpha (1 - \Phi(u))\, u^{2/\alpha - 2/\beta}\, \frac{\Gamma(1/\beta) + \Gamma(1/\beta, bC^\beta)}{\beta\, b^{1/\beta}}$$

as $u \to \infty$.

ii) *If $\alpha = \beta$ and $t_0(u)u^{2/\alpha} \to C \in [0,\infty]$, then*

$$P\left(\sup_{[0,1]} X_u(t) > u\right) \sim H_\alpha^b(C)\,(1 - \Phi(u))$$

as $u \to \infty$ where the values $H_\alpha^b(C) = \lim_{S \to \infty} H_\alpha^b(-C, S)$ are given above.

iii) *If $\alpha > \beta$, then as $u \to \infty$,*

$$P\left(\sup_{[0,1]} X_u(t) > u\right) \sim 1 - \Phi(u).$$

These theorems can be transformed for the case where we consider the minimal boundary instead of the maximal variance. Instead of the local behavior of $\sigma(t)$ in the neighborhood of t_0 we have to analyze the behavior of the boundary function in the neighborhood of the minimal boundary value. The condition on the correlation functions remains the same.

Random Variance or Random Boundary

Instead of considering a fixed variance or a non-constant boundary, one may deal with a variance or a boundary which is random, such as in random environment situations. It means that one is interested in the impact of a random variance on the maxima of a Gaussian process. Hence, let us consider a process which is the product $X(t) = Y(t)\eta(t)$ of a standard Gaussian process $Y(t)$ and a random process $\eta(t)$, playing the role of the random standard deviation of $X(t)$. The two processes $Y(t)$ and $\eta(t)$ are assumed to be independent. Note, that the product process $X(t)$ is no longer a Gaussian process. This model is equivalent to the case of the Gaussian process $Y(t)$ with the random non-constant boundary $u/\eta(t)$ in random boundary problems.

If $\eta(t) = \eta$ is a positive rv, not depending on t, we derive the asymptotic result by conditioning on η. We want to investigate the event of an exceedance:

$$\{Y(t)\eta(t) > u \text{ for some } t \in [0,a] \mid \eta\} = \{Y(t) > u/\eta, \text{ for some } t \in [0,a] \mid \eta\}.$$

By our previous results, the probability of this event is influenced mostly by the largest values of η, which makes u/η as small as possible. Hence, it is sensible to restrict the support of η to be bounded. Let s_η denote the finite upper bound of the support of η. It denotes also the largest standard deviation of the process $X(t)$.

Assuming that the Gaussian process is stationary, we get for $u \to \infty$ the asymptotic exact probability by using Theorem 10.3.1, assuming (10.1) and replacing u by the value u/η and integrating on the distribution of η:

$$P_u := P(Y(t)\eta > u, \text{ for some } t \in [0,a])$$

$$= \int P(Y(t) > u/\eta, \text{ for some } t \in [0,a] \mid \eta = y) \, dF_\eta(y)$$

$$\sim aH_\alpha C^{1/\alpha} \int_0^{s_\eta} (u/y)^{2/\alpha-1} \phi(u/y) \, dF_\eta(y).$$

Now it remains to find the asymptotic behavior of the integral, which depends on the behavior of F_η at s_η. If η has a density f_η which is continuous at s_η with $f_\eta(s_\eta) > 0$, then the probability P_u is asymptotically equal to

$$aH_\alpha C^{1/\alpha} s_\eta f_\eta(s_\eta)(u/s_\eta)^{2/\alpha-3} \phi(u/s_\eta).$$

This follows by changing the variable y to $x = u^2(y - s_\eta)$ in the integral, and using simple Taylor expansions.

If the density $f_\eta(s_\eta) = 0$, then its derivatives play an important role in the asymptotic formula. The general result follows again by the same variable transformation and Taylor expansions or alternatively, by the Laplace method. We make use of the following more general approximation which follows by the Laplace method.

Proposition 10.3.14. *Let $g(x)$, $x \in [0,s]$, be a bounded function, which is k-times continuously differentiable in a neighborhood of s, where $g^{(r)}(s) = 0$ for $r = 0, 1, ..., k-1$ and $g^{(k)}(s) \neq 0$. Then for any $\varepsilon \in (0,s)$,*

$$\int_\varepsilon^s g(x)(x/u)\phi(u/x)dx = (-1)^k s^{3k+4} g^{(k)}(s)u^{-3-2k}\phi(u/s)(1+o(1)) \quad (10.24)$$

as $u \to \infty$. If $g(x) = g_1(x)g_2(x)$, $g_1(x)$ is continuous at s with $g_1(s) > 0$, and $g_2(x)$ satisfies the above conditions on g, one replaces $g^{(k)}(s)$ in (10.24) by $g_1(s)g_2^{(k)}(s)$.

For the general case, assume that $f_\eta(x)$ is k-times continuously differentiable in a neighborhood of s_η, with $f_\eta^{(r)}(s_\eta) = 0$ for $r < k$ and $f_\eta^{(k)}(s_\eta) \neq 0$. Then the general result follows by the proposition with $g(x) = x^{-2/\alpha} f_\eta(x)$ and $s = s_\eta$.

Theorem 10.3.15. *Assume that $Y(t)$ is a stationary standard Gaussian process, and that (10.1) holds with $0 < \alpha \leq 2$. Let η be a r.v. with k-times continuously differentiable density $f_\eta(x)$ in the neighborhood of s_η, for some $k \geq 0$. Assume that*

η is independent of $Y(t)$ and that $f_\eta^{(r)}(s_\eta) = 0$ for $r < k$ and $f_\eta^{(k)}(s_\eta) \neq 0$. Then, for $u \to \infty$,

$$P(X(t) = Y(t)\eta > u, \text{ for some } t \in [0, a])$$
$$\sim (-1)^k a H_\alpha C^{1/\alpha} s_\eta^{3k+4-2/\alpha} u^{2/\alpha-3-2k} \phi(u/s_\eta) f_\eta^{(k)}(s_\eta).$$

Since the variance is constant, it means that an exceedance of u/η by $Y(t)$ or an exceedance of u by $X(t)$ could happen anywhere in $[0, a]$. This is not the case if we consider the random variance $\eta(t)$ as non-constant. It is convenient to assume first that the random variance has a unique random maximal value. By conditioning on $\eta(t)$, we may possibly apply the methods of the case with a fixed variance.

Let us consider a simple case first. We investigate the case $\eta(t) = 1 - \zeta|t - t_0|^\beta$ ($\beta > 0$) in a suitable vicinity of t_0 with ζ a rv. For simplicity, let $t_0 = 0$, with the interval $[0, a]$, for some $a > 0$. The other cases are dealt with by the same argumentation; for example, if t_0 is an inner point of the interval $[0, a]$, we have to multiply the resulting probability approximation by 2. Note that a cannot be large, since $1 - \zeta a^\beta$ has to be positive to be the standard deviation of the process $X(t)$. Therefore, we assume that the rv ζ has a finite upper support point s_ζ and $a < (1/s_\zeta)^{1/\beta}$.

Let us first discuss the particular case of a straight line ($\beta = 1$) in detail. By conditioning on ζ, we can use the method of Theorem 10.3.13 to derive, for $\zeta > 0$ and $\alpha < 1$,

$$P\left(\max_{t \in [0,a]} Y(t)(1 - \zeta t) > u \mid \zeta\right) \sim H_\alpha u^{2/\alpha-2}(1 - \Phi(u))/\zeta. \qquad (10.25)$$

We note that the behavior of the distribution of ζ near 0 is now of interest. Either $E(1/\zeta)$ exists, or a more accurate approximation is needed for very small ζ. This approximation is given in Theorem 10.3.16 which is proved in Hüsler et al. [262]. We note that the smoothness parameter α of the paths of $Y(t)$ is quite relevant, implying three different results for $\alpha > 1, = 1$ or < 1.

We assume that in the following theorems the density of ζ is bounded and has a bounded support with upper support point $s_\zeta < \infty$, as mentioned.

Theorem 10.3.16. *Let $\alpha < 1$ and $a \in (0, 1/s_\zeta)$ with $s_\zeta < \infty$. Then*

$$P\left(\max_{t \in [0,a]} Y(t)(1 - \zeta t) > u\right)$$
$$\sim H_\alpha u^{2/\alpha-2}(1 - \Phi(u)) E\left(\frac{1 - \exp(-au^2\zeta)}{\zeta}\right) \qquad (10.26)$$

as $u \to \infty$.

(i) *If in addition $E(1/\zeta) < \infty$, then*

$$P\left(\max_{t \in [0,a]} Y(t)(1 - \zeta t) > u\right) \sim H_\alpha u^{2/\alpha-2}(1 - \Phi(u)) E(1/\zeta)$$

as $u \to \infty$.

(ii) *If in addition the density* $f_\zeta(x)$, $x \geq 0$, *of* ζ *is continuous at* 0, *with* $f_\zeta(0) > 0$, *then*

$$P\left(\max_{t\in[0,a]} Y(t)(1-\zeta t) > u\right) \sim 2H_\alpha f_\zeta(0) u^{2/\alpha-2}(1-\Phi(u))\log(u)$$

as $u \to \infty$.

Statement (i) follows from the result (10.25) and (ii) from the general result (10.26). Quite different is the case $\alpha = 1$, where the paths have the same smoothness as the random function $\eta(t)$.

Theorem 10.3.17. *Let* $\alpha = 1$ *and* $a \in (0, 1/s_\zeta)$ *with* $s_\zeta < \infty$.

(i) *Assume* $E(1/\zeta) < \infty$,

$$P\left(\max_{t\in[0,a]} Y(t)(1-\zeta t) > u\right) \sim H(1-\Phi(u)) \quad \textit{as } u \to \infty,$$

with

$$0 < H = E\left(\exp\left(\max_{[0,\infty]}(\sqrt{2}B(t) - (1+\zeta)t)\right)\right) < \infty,$$

where $B(t)$ *is the standard Brownian motion.*

(ii) *If the density* $f_\zeta(x)$ *of* ζ *is continuous at* 0 *with* $f_\zeta(0) > 0$,

$$P\left(\max_{t\in[0,a]} Y(t)(1-\zeta t) > u\right) \sim 2f_\zeta(0)\,(1-\Phi(u))\,\log(u)$$

as $u \to \infty$.

The case $\alpha > 1$ shows also a different dependence of the probability on the random variance. Since the paths are 'smoother' than the random boundary or the random variance function, it shows that exceedances occur asymptotically only around one point, namely at $t_0 = 0$; other exceedances have no impact on the crossing probability.

Theorem 10.3.18. *Let* $\alpha > 1$ *and* $a \in (0, 1/s_\zeta)$ *with* $s_\zeta < \infty$. *Then*

$$P\left(\max_{t\in[0,a]} Y(t)(1-\zeta t) > u\right) \sim (1-\Phi(u)) \quad \textit{as } u \to \infty.$$

Similar results hold for the more general power functions $1 - \zeta|t|^\beta$ where the three cases need to be distinguished, again depending on $\beta > \alpha, = \alpha$ or $< \alpha$.

Theorem 10.3.19. *Let* $\alpha < \beta \in (0,\infty)$ *and* $a \in (0, s_\zeta^{-1/\beta})$ *with* $s_\zeta < \infty$.

(i) *If $E(\zeta^{-1/\beta}) < \infty$, then*

$$P\left(\max_{t\in[0,a]} Y(t)(1 - \zeta t^\beta) > u\right) \sim \frac{H_\alpha \Gamma(1/\beta)}{\beta} u^{2/\alpha - 2/\beta}(1 - \Phi(u))E(\zeta^{-1/\beta})$$

as $u \to \infty$.

(ii) *If the density $f_\zeta(x)$, $x \geq 0$, is continuous at 0, with $f_\zeta(0) > 0$, then for $0 < \alpha < \beta \in (0,1)$,*

$$P\left(\max_{t\in[0,a]} Y(t)(1 - \zeta t^\beta) > u\right) \sim \frac{H_\alpha a^{1-\beta}}{1 - \beta} f_\zeta(0)u^{2/\alpha - 2}(1 - \Phi(u))$$

as $u \to \infty$.

Note that the second result with $\beta < 1$ is different from the second result of Theorem 10.3.16 with $\beta = 1$. The situation is again different if $\alpha = \beta$. Let $\chi_\alpha(t)$ denote the fractional Brownian motion with mean $-|t|^\alpha$ and covariance $\mathrm{Cov}(\chi_\alpha(t), \chi_\alpha(s)) = |t|^\alpha + |s|^\alpha - |t - s|^\alpha$.

Theorem 10.3.20. *Let $\alpha = \beta \in (0,2]$ with $s_\zeta < \infty$.*

(i) *Assume that $E\zeta^{-1/\beta} < \infty$ and let $a \in (0, s_\zeta^{-1/\beta})$. Then*

$$P\left(\max_{t\in[0,a]} Y(t)(1 - \zeta t^\beta) > u\right) \sim H_\alpha^\zeta(1 - \Phi(u)) \qquad \text{as } u \to \infty,$$

where $0 < H_\alpha^\zeta := E\left(\exp\left[\max_{[0,\infty)}(\chi_\alpha(t) - \zeta t^\alpha)\right]\right) < \infty$.

(ii) *Assume that the density $f_\zeta(x)$, $x \geq 0$, of ζ is positive and continuous at 0, and let $a \in (0, s_\zeta^{-1/\beta})$.*

   (a) *If $\alpha = \beta \in (0,1)$, then*

$$P\left(\max_{t\in[0,a]} Y(t)(1 - \zeta t^\beta) > u\right) \sim \frac{H_\alpha a^{1-\alpha}}{1 - \alpha} f_\zeta(0)u^{2/\alpha - 2}(1 - \Phi(u))$$

   as $u \to \infty$;

   (b) *if $\alpha = \beta \in (1,2]$, then*

$$P\left(\max_{t\in[0,a]} Y(t)(1 - \zeta t^\beta) > u\right) \sim H_\alpha^\zeta(1 - \Phi(u))$$

as $u \to \infty$, where $0 < H_\alpha^\zeta = E\left(\exp\left(\max_{[0,\infty)}(\chi(t) - \zeta t^\alpha)\right)\right) < \infty$.

The result for $\alpha = \beta = 1$ is given in Theorem 10.3.17. Finally, for the case $\alpha > \beta$ we get the approximation as in Theorem 10.3.18, since the paths of the process $X(t)$ are smoother than the random variance. It means that the process crosses the random boundary mostly around $t = 0$, asymptotically.

Theorem 10.3.21. *Let* $2 \geq \alpha > \beta > 0$. *Assume that* ζ *is independent of* $Y(\cdot)$ *with a strictly positive lower bound. Then for any* $a \in (0, s_\zeta^{-1/\beta})$ *with* $s_\zeta < \infty$,

$$P \left(\max_{t \in [0,a]} Y(t)(1 - \zeta t^\beta) > u \right) \sim 1 - \Phi(u)$$

as $u \to \infty$.

The case with the general function $\eta(t) = \eta - \zeta|t|^\beta$ is treated by conditioning on η and then using the above results which now depend on the behavior of the density of ζ/η. In cases $\alpha > \beta$ the rv ζ has no impact. These results are given in two papers. Hüsler et al. [262] dealt first with the cases of $\eta(t)$ being a parabola or a straight line and Zhang [472] (see also Hüsler et al. [263]) with more general functions $\eta(t) = \eta - \zeta|t - t_0|^\beta$ (for any $\beta > 0$) with positive bounded rv η and ζ where t_0 denotes the unique argument with maximal value of $\eta(t)$. By conditioning on η, we can derive an asymptotic expression for the probability of an exceedance. The upper support point of η determines again the asymptotic probability term. We may assume that the two r.v. η and ζ are independent, however, this is not necessary. In the proofs, one conditions first on η and deals then with the conditional distribution of ζ, given η.

APPROXIMATION OF A STATIONARY GAUSSIAN PROCESS

Let $X(t), 0 \leq t \leq 1$, be a stationary Gaussian process which is observed on a grid $t = i/n, 0 \leq i \leq n$. The process $X(t)$ is approximated by a piecewise linear function

$$L(t) = ((i+1) - nt)X(i/n) + (nt - i)X((i+1)/n) \quad \text{for } i/n \leq t \leq (i+1)/n.$$

Of interest is the deviation process $Y(t) = X(t) - L(t)$. It is still a Gaussian process but not stationary as mentioned, because $Y(i/n) = 0$ for all i. Hence, the variance is 0 at the grid points and attains in each subinterval $(i/n, (i+1)/n)$ a maximal value. The variance is cyclo-stationary, being equal for each subinterval $(i/n, (i+1)/n)$. We expect again that the maximal values occur in the neighborhood of the values t with maximal variance. Asymptotically the maxima of the variance function in each subinterval is separated sufficiently, so the maxima of $X(t)$ on the subintervals $(i/n, (i+1)/n)$ are asymptotically independent. Hence a Poisson nature of the cluster of exceedances is observed and the limit distribution of the maximum of $\Delta(t), t \in [0,1]$ can be derived. This is dealt with e.g. in Seleznjev [406], Hüsler [252], Hüsler et al. [260].

Let us discuss this approximation in detail. The derivation of the variance of $Y_n(t)$ is straightforward. For $i/n \leq t \leq (i+1)/n$ and $s = nt - i$ with $0 \leq s \leq 1$, we have

$$\sigma_n^2(s) = 2 \left((1 - r(\frac{s}{n}))(1 - s) + (1 - r(\frac{1-s}{n}))s - (1 - r(\frac{1}{n}))(1 - s)s \right)$$

which is 0 for $s = 0$ and $s = 1$. The variance is periodic, since it does not depend on i, and tends to 0 as $n \to \infty$. In case that $1 - r(t) \sim ct^\alpha$, we note that

$$\sigma_n^2(s) = \frac{2C}{n^\alpha}[s^\alpha(1 - s) + (1 - s)^\alpha s - (1 - s)s] + o(n^{-\alpha}).$$

The expression in the square brackets does not depend on n and has one or two maxima in $(0, 1)$ which are separated by some positive term. Denote the maximal variance of $Y_n(\cdot)$ by $\sigma_n^2 = \max_{[0,1]} EY_n^2(s)$. The maximal approximation error is of interest, hence

$$P\left(\max_{t \in [0, T/n]} |Y_n(t)| > u\right) = P\left(\max_{s \in [0, T]} |\tilde{Y}_n(s)| > u/\sigma_n\right)$$

as $n \to \infty$, where we transform the process $Y_n(t)$ to $\tilde{Y}_n(s) = Y_n(s/n)/\sigma_n$ with $s \in [0, T]$, having the maximal variance 1. Typically, $T = T(n) = n \to \infty$. The probability is dominated by the possible crossings in the neighborhood of the time points where $\tilde{Y}_n(\cdot)$ has variance 1. It implies that only the local behavior of the process in the corresponding intervals has to be restricted by some conditions. It means also that we may consider more general processes and more general approximation schemes than the particular examples above.

Hence, let us state our result for more general approximation processes $Y_n(\cdot)$ or $\tilde{Y}_n(\cdot)$. The behavior of $\sigma_n(t)$ can be approximated, as shown in the example of linear approximation, by $\sigma_n\sigma(t)$ as $n \to \infty$, assuming that $\sigma(t)$ satisfies the following conditions.

Let $\sigma(t), t \geq 0$, be a continuous function with $0 \leq \sigma(t) \leq 1$ and assume that

(i) $\{t : \sigma(t) = 1\} = \{t_k, \text{ with } t_{k+1} - t_k > h_0, k \geq 1\}$ for some $h_0 > 0$;

(ii) $T \leq Km_T$, for some $K > 0$, where m_T denotes the number of points t_k of maximal variance in $[0, T]$;

(iii) $\sigma(t)$ can be expanded in a neighborhood of t_k:

$$\sigma(t) = 1 - (a + \gamma_k(t))|t - t_k|^\beta$$

for some $a, \beta > 0$ where the functions $\gamma_k(t), k \geq 1$, are such that for any $\epsilon > 0$ there exists $\delta_0 > 0$ with

$$\max\{|\gamma_k(t)|, t \in J_k^{\delta_0}\} < \epsilon,$$

where $J_k^\delta = \{t : |t - t_k| < \delta\}$ for $\delta > 0$.

If these three conditions hold, we say that $\sigma(t)$ satisfies assumption (A).

Now the convergence of the variance $\sigma_n^2(s)$ and the correlation $r_n(t, s)$ of $Y_n(s)$ is restricted by the following four conditions:

(i) There exists $\delta_0 > 0$ such that $\sigma_n(t) = \sigma_n \sigma(t)(1 + \epsilon_n(t))$ for some $\sigma_n > 0$ with $\sup_{t \in [0,T]} |\epsilon_n(t)| \to 0$ and $\epsilon_n u^2 / \sigma_n^2 \to 0$, $(n \to \infty)$, where $u = u(n)$ and $\epsilon_n = \sup\{|\epsilon_n(t)|, t \in J^{\delta_0}\}$ with

$$J^\delta = \bigcup_{k \le m_T} J_k^\delta.$$

(ii) For any $\epsilon > 0$ there exists $\delta_0 > 0$ such that

$$r_n(t,s) = 1 - (b_n + \gamma_n(t,s))|t-s|^\alpha,$$

where $b_n \to b > 0$ as $n \to \infty$, $0 < \alpha \le \min(2, \beta)$ and the function $\gamma_n(t,s)$ is continuous at all points (t_k, t_k), $k \ge 1$, and $\sup\{|\gamma_n(t,s)|, t, s \in J^{\delta_0}, n \ge 1\} < \epsilon$.

(iii) There exist $\alpha_1, C > 0$ such that for any $t, s \le T$, and large n,

$$E\left((Y_n(t) - Y_n(s))^2\right) \le C|t-s|^{\alpha_1}.$$

(iv) For any $v > 0$ there exists $\delta > 0$ such that

$$\delta(v) = \sup\{|r_n(t,s)|, v \le |t-s|, t, s \le T, n \ge 1\} < \delta < 1.$$

If these four conditions hold, we say that the processes Y_n satisfy condition (B).

Under these assumptions the maximum of $|Y(t)|, t \le T$, can be characterized as both T and u tend to ∞.

Theorem 10.3.22. *Let $T = T(n), u = u(n)$ be such that $\min(T, u/\sigma_n) \to \infty$ as $n \to \infty$. Assume that conditions (A) and (B) hold for a sequence of Gaussian processes Y_n, with mean 0. If in addition (10.3) holds and*

$$m_T H_\alpha^{a/b} \phi(u/\sigma_n) \sigma_n / u \to \tau \ge 0$$

as $n \to \infty$ with some $H_\alpha^{a/b} > 0$, then

$$P(M_n(T) > u) \to 1 - e^{-\tau} \text{ as } n \to \infty.$$

In case T is fixed, the derivation of the result shows also how the maximum $M_n(T)$ tends to 0. One may also consider moments of $M_n(T)$ or define the process of exceedances, showing that this point process converges to a Poisson process, see [252].

A generalization of this approximation result is considered for $t \in \mathbb{R}$ in Hüsler et al. [260], and for $t \in \mathbb{R}^d$ in Seleznev [407]. We mention the moment result of the extension to random fields. Consider a sequence of mean zero Gaussian random fields $Y_n(t), t \in \mathbb{R}^d$, with (a.s.) continuous sample paths. Assume that the Gaussian random fields satisfy the Hölder condition

$$E(Y_n(t) - Y_n(s))^2 \leq C|t - s|^\alpha$$

for some positive constants C, α and all large n, where $|t| = (\sum_{i=1}^d t_i^2)^{1/2}$. Let $M_n(T_n) = \max_{t \in T_n} |Y_n(t)|$ be the uniform norm of the Gaussian random field on a set $T_n \subset \mathbb{R}^d$. Seleznjev [407] considered the moments of $M_n(T_n)$, where $T_n = [0, t_n]^d$ is the d-cube with $t_n \to \infty$.

Theorem 10.3.23. *Let $Y_n(t), t \in \mathbb{R}$, $n \geq 1$, be a sequence of Gaussian random fields with zero mean and variance functions $\sigma_n^2(t)$ with $\sigma_n^2(t) \leq 1$. Assume that the uniform norm $M_n(T_n)$ is such that, with suitable normalizations a_n, b_n,*

$$P(M_n(T_n) \leq b_n + xa_n) \to G(x) \qquad as\ n \to \infty$$

holds, where G is a non-degenerate distribution function, $T_n = [0, t_n]^d$, $t_n \to \infty$ and $a_n \sim b_n \sim \sqrt{2\log(n)}$. If the Hölder condition and $|T_n| = t_n^d \leq cn$ hold, for some $c > 0$, then for any positive p,

$$(EM_n^p(T_n))^{1/p} \sim \sqrt{2\log(n)} \qquad as\ n \to \infty.$$

The results hold also for $M_n(t)$ defined as maximum $M_n(T_n) = \max_{t \in T_n} Y_n(t)$. Typically, $G(x)$ is the Gumbel distribution. Applications of the first-order asymptotic result are given in Seleznjev [407]. For example, he discussed the approximation of a continuous Gaussian random field by an approximating sequence of Gaussian random fields which is based on a discrete time sampling scheme.

RUIN PROBABILITY AND GAUSSIAN PROCESSES WITH DRIFT

In risk theory of finance one considers the ruin probability which is often modelled with respect to a fractional Brownian motion $X(t)$ with parameter $\alpha = 2H < 2$ and mean 0. Then the ruin probability is defined as

$$P\left(\sup_{t \geq 0}(X(t) - ct^\beta) > u\right)$$

as $u \to \infty$, which is the distribution of the supremum of the Gaussian process $X(t) - ct^\beta$, having the trend $-ct^\beta$, where $\beta > H$ and $c > 0$. In this case we can interpret ct^β as sum of the premiums payed up to t, and $X(t)$ as sum of the claims up to t. Here, it is common to consider the fractional Brownian motion $X(t)$ with variance $|t|^\alpha$. Instead of this trend function we might consider some other function as long as the probability is well defined. Often we can put $\beta = 1$ by transformation. The problem occurs also in queuing theory and in telecommunications modelling see e.g. Norros [354], [355], Narayan [348], Hüsler and Piterbarg [257], [259], [258],

Choe and Shroff [68] and Dębicki [95], [96], [97] and Dębicki et al [99]. Note that again by standardizing the process, the event to be considered is

$$\{\exists t : X(t) > ct^\beta + u\} = \{\exists t : X(t)/\sigma(t) > u(t) = (ct^\beta + u)/t^H\}$$

since the variance of $X(t)$ is equal to t^{2H}. Again $u(t)$ has a minimum value at a unique point $t_0(u) = (uH/(c(\beta - H)))^{1/\beta}$ tending to ∞ with u. Thus it is convenient to transform the time $(t = su^{1/\beta})$ also such that the minimum of the boundary is taken uniquely in $s_0 = (H/c(\beta - H))^{1/\beta}$ and remains fixed, not depending on u. Hence the above mentioned approach where only the neighborhood of the unique point of minimal boundary plays the important role, can be applied in this case also.

We consider more generally a Gaussian process with mean 0 and variance $V^2(t)$, which is a regularly varying function at infinity with index $2H$, $0 < H < 1$. Assume that the paths are a.s. continuous and $X(0) = 0$ a.s.. Then we can derive not only the asymptotic behavior of the probability $P(\sup_{t \geq 0}(X(t) - ct^\beta) > u)$ as $u \to \infty$, but also the limiting distribution of the first ruin time τ_1, if it happens that $\tau_1 < \infty$. We define

$$\tau_1 = \tau_1(u) = \inf\{t : u + ct^\beta - X(t) \leq 0\} \leq \infty$$

with $\beta > H$ and $c > 0$. Applying the transformation in time and space, the processes

$$X^{(u)}(s) = \frac{X(su^{1/\beta})}{V(u^{1/\beta})(1 + cs^\beta)}, \quad s > 0,$$

are investigated. With the time change $t = su^{1/\beta}$, we have

$$P\left(\sup_{t \geq 0}(X(t) - ct^\beta) > u\right) = P\left(\exists s \geq 0 : X(su^{1/\beta}) > u(1 + cs^\beta)\right)$$

$$= P\left(\sup_{s \geq 0} \frac{X(su^{1/\beta})}{1 + cs^\beta} > u\right)$$

$$= P\left(\sup_{s \geq 0} X^{(u)}(s) > \frac{u}{V(u^{1/\beta})}\right),$$

and $\tau_1 = u^{1/\beta}\tilde{\tau}_1$, where

$$\tilde{\tau}_1 := \inf\left\{s \geq 0 : \frac{u}{V(u^{1/\beta})} - X^{(u)}(s) \leq 0\right\} \tag{10.27}$$

denotes the first ruin time in the changed time scale. The process $X^{(u)}(s)$ with mean zero is not standardized, its variance equals $v_u^{-2}(s)$, where

$$v_u(s) = \frac{s^H V(u^{1/\beta})}{V(su^{1/\beta})} v(s) \to v(s) \quad \text{as } u \to \infty, \quad \text{with } v(s) = s^{-H} + cs^{\beta - H}.$$

The function $v(s)$ has a unique minimum point

$$s_0 = \left(\frac{H}{c(\beta - H)}\right)^{1/\beta}$$

and the function $v(s)$ is locally a quadratic function in s_0:

$$v(s) = A + \frac{1}{2}B(s - s_0)^2 + o((s - s_0)^2), \quad \text{for } s \to s_0,$$

where

$$A := v(s_0) = \left(\frac{H}{c(\beta - H)}\right)^{-H/\beta} \frac{\beta}{\beta - H}$$

and

$$B := v''(s_0) = \left(\frac{H}{c(\beta - H)}\right)^{-(H+2)/\beta} H\beta.$$

The assumptions imply $v_u(s) \to v(s)$ and also that the (smallest) minimum point $s_0(u)$ of $v_u(s)$ (for fixed u) tends to s_0, as $u \to \infty$. If $v_u(s)$ can be well approximated by the function $v(s)$, assuming

$$\frac{v_u(s) - A(u)}{(s - s_0(u))^2} \to \frac{1}{2}B,$$

as $u \to \infty$ uniformly for s in a neighborhood of s_0, and if the standardized Gaussian process $X^{(u)}(s)v_u(s)$ is locally stationary in the vicinity of s_0 and satisfies a Hölder condition, then Hüsler and Piterbarg [261] showed that

$$P((\tau_1(u) - s_0(u)u^{1/\beta})/\sigma(u) < x \mid \tau_1(u) < \infty) \to \Phi(x) \qquad (10.28)$$

as $u \to \infty$, for all x, where $\sigma(u) := (AB)^{-1/2}u^{-1+1/\beta}V(u^{1/\beta})$.

The result follows by the main investigation of the ruin event in the neighborhood of s_0:

$$\sup_{s \in S_u(x)} X^{(u)}(s) > u/V(u^{1/\beta}),$$

where

$$S_u(x) = \left[s_0(u) - \delta(u), s_0(u) + x\tilde{u}^{-1}/\sqrt{A(u)B}\right].$$

It is shown that, for all x,

$$P\left(\sup_{s \in S_u(x)} X^{(u)}(s) > \tilde{u}\right) \sim \frac{D^{1/\alpha} A^{2/\alpha - 3/2} H_\alpha 2^{-1/\alpha} e^{-\frac{1}{2}A^2(u)\tilde{u}^2} \Phi(x)}{\sqrt{B} K^{-1}(\tilde{u}^{-1})\tilde{u}^2}$$

as $u \to \infty$, with $\tilde{u} = u/V(u^{1/\beta})$. By conditioning on $\tau_1(u) < \infty$, the limit distribution (10.28) of τ_u follows.

Also the last ruin time $\tau_2(u) = \sup\{t > 0 : X(t) - ct^\beta > u\}$ can be investigated. With the same methods, Hüsler and Zhang [267] showed that under the same conditions as in [261],

$$P\left((\tau_2(u) - s_0(u)u^{1/\beta})/\sigma(u) < x \mid \tau_1^{(u)} < \infty\right) \to \Phi(x)$$

as $u \to \infty$, for all x with the same $\sigma(u) := (AB)^{-1/2}u^{-1+1/\beta}V(u^{1/\beta})$. Furthermore, they derived the joint distribution of the first and last ruin times which is related to the normal distribution:

$$P\left((\tau_1(u) - s_0(u)u^{1/\beta})/\sigma(u) \geq x_1, (\tau_2(u) - s_0(u)u^{1/\beta})/\sigma(u) < x_2 \mid \tau_1(u) < \infty\right)$$
$$\to \Phi(x_2) - \Phi(x_1)$$

for any $x_1 < x_2$.

Transformations in time and space are rather convenient for dealing with such a problem to reduce it to a related problem with a known solution or a problem where a known method can be applied. In such investigations other processes are considered, in particular, e.g. scaled Brownian motions, filtered Brownian motions, self-similar Gaussian processes, integrated Gaussian processes, see the above given references or Dieker [112] where many of the above results are extended and combined, assuming a more general drift function $a(t)$ instead of the particular drift function ct^β.

Also the mean loss in case of ruin is of interest and is investigated by Boulogne et al. [52] for a particular ruin process $X(t)$. They considered the process $X(t) = u + ct - B_H(t)$ with drift ct, where $B_H(t)$ denotes the fractional Brownian motion. The conditioned average loss is defined as $E\left(-\inf_{[0,T]} X(t) \mid \inf_{[0,T]} X(t) < 0\right)$ as $u \to \infty$. Here T can be finite for the finite horizon investigations, or tending to ∞ as $u \to \infty$.

EXTREMES OF STORAGE MODELS

Related to the ruin probability analysis is the investigation of extremes of storage processes. The storage process is defined by

$$Y(t) = \sup_{s \geq t}(X(s) - X(t) - c(s - t)^\beta)$$

where $c > 0$ and $\beta > 0$. Transforming the time, we may also set $\beta = 1$, but one has to consider $Y(t) = \sup_{v \geq 0}(X(v^{1/\beta} + t) - X(t) - cv)$. The underlying input process $X(\cdot)$ is modelled in different ways, but often $X(t)$ is assumed to be a fractional Brownian motion with Hurst parameter H and variance t^{2H}. For proper definition of $Y(t)$, such that $Y(t)$ has finite values, we have to assume $H < \beta$. This model is considered in queueing applications to model teletraffic. Here $\beta = 1$ for linear service, see Norros [353], [354]. The process $Y(\cdot)$ is also applied in financial models, see Dacorogna et al. [88].

The particular probability $P(Y(0) > u) = P(\sup_{t \geq 0}(X(t) - ct) > u)$ with $\beta = 1$ is discussed above in relation to ruin probabilities, as $u \to \infty$. The extremes of the storage model $Y(t)$ on a finite interval $[0, T]$ have been investigated by Piterbarg [376], also as $u \to \infty$. He applied the relation

$$P\left(\sup_{t \in [0,T]} Y(t) \leq u\right) = P\left(\sup_{s \in [0,T/u],\, \tau \geq 0} Z(s, \tau) \leq u^{1-H}\right)$$

where $Z(s, \tau)$ is the Gaussian random field

$$Z(s, \tau) = [X(u(s + \tau)) - X(us)]/[u^H(1 + c\tau)].$$

$Z(s, \tau)$ is stationary in s for fixed τ and has a maximal variance

$$\sigma = \left(\frac{H}{c(1 - H)}\right)^H (1 - H)$$

at $\tau = \tau_0 = H/(c(1 - H))$. The supremum of $Z(s, \tau)$ occurs in a small neighborhood of τ_0, more precisely in a strip $\{0 \leq s \leq [T/u], |\tau - \tau_0| < \epsilon\}$, for a suitable $\epsilon = \epsilon(u)$. Hence it can be shown that

$$P\left(\sup_{t \in [0,T]} Y(t) > u\right) \sim \sqrt{2\pi} C_1 H_{2H}^2 T u^{\frac{2}{H}(1-H)+H-2} \bar{\Phi}(u^{1-H}/\sigma)$$

where $C_1 = \tau_0^{-3}(H(1 - H))^{-1/2} 2^{-2/H}$. This property holds even for T depending on u as long as T is not tending too fast to ∞ or to 0 (see Hüsler and Piterbarg [259]). Albin and Samorodnitsky [3] generalize the result of Piterbarg [376] for self-similar and infinitely divisible input processes.

The maximum of the storage process on a growing interval $[0, T]$ increases as $T \to \infty$, and has to be normalized if it grows fast at a certain rate in relation to the boundary u. If this maximum $M(T) = \max_{[0,T]} Y(t)$ is normalized suitably, then its distribution converges to a Gumbel distribution; for $\beta = 1$ (hence assuming $H < 1$)

$$P(M_T \leq b(T) + x\, a(T)) \to \exp(-e^{-x})$$

as $T \to \infty$ where

$$b(T) = (2\sigma^2 \log(T))^{1/(2(1-H))} + \left[\frac{h(2\sigma^2)^{1/(2(1-H))} \log(2\sigma^2 \log(T))}{4(1 - H)^2}\right.$$

$$\left. + \frac{(2\sigma^2)^{1/(2(1-H))} \log(\tilde{c})}{2(1 - H)}\right] (\log(T))^{-(1-2H)/(2(1-H))}$$

and

$$a(T) = \frac{(2\sigma^2)^{1/(2(1-H))}}{2(1 - H)} (\log(T))^{-(1-2H)/(2(1-H))}$$

with $h = 2(1 - H)^2/H - 1$, $\tilde{c} = 2^{-2/H} H_{2H}^2 \tau_0^{2H-5} H^{-1/2}(1 - H)^{3/2-2/H}$ and σ, τ_0 as above. This is shown in Hüsler and Piterbarg [259].

CHANGE-POINT REGRESSION AND BOUNDARY CROSSING

In statistical analysis of quality control we often have to investigate whether the production process is under control or whether it is getting worse, i.e., that there occurred a change point during the production where the production deviated from the standards. In the latter case the production process has to be fixed. Thus observing the process we have to decide by some suitable procedure whether a change point has occurred. Such a process is usually modelled as a regression model $X(t) = f(t) + Y(t)$ where f is the trend (regression) function, $Y(t)$ a process with mean μ and variance σ^2, and $X(t)$ the measurements of the production process. The process is under control when $f(t) \equiv \mu = 0$, say. The process is usually observed in a fixed interval, set to be $[0, 1]$. Hence the statistical hypotheses to be tested are H_0: $f \equiv 0$ against the alternative $H_1 : f(0) = 0, f(-1) = \lim_{t \uparrow 1} f(t) > 0$ and f is non-decreasing and right-continuous.

The partial sum process is the statistical process to be used for testing the hypotheses. A finite number of measurements are taken, often at equidistant points $t_{ni} = i/n, i = 1, \ldots, n$, say. In the limit ($n \to \infty$) the normalized partial sum process is under the null hypothesis a Brownian motion. The trend function is also replaced by the partial sums of $f(i/n)$. The class of alternative hypotheses can be written in the limit with a non-decreasing function g with $g(1) = 1$, say, as general trend and a parameter γ, setting $h = h_f = \gamma g$. A suitable test statistic for the given test situation is the Kolmogorov type statistic, i.e., the test function is $\sup_{t \in [0,1]} B(t)$. To control the α-level of the test we apply the critical value u_α such that $P(\sup_{[0,1]} B(t) > u_\alpha) = \alpha$. Now for comparisons with other tests or for a given application with a fixed trend we are interested in the power of the test. Thus we should determine

$$P(\sup_{[0,1]}(\gamma g(t) + B(t)) > u_\alpha)$$

for $\gamma > 0$. But this is equal to

$$1 - P(B(t)) \le u_\alpha - \gamma g(t), \text{ for all } t \in [0, 1]\}$$

a boundary crossing problem. Such problems are investigated for certain classes of trend functions g. E.g. if $g(t) = 0$ for $t \le \tau$ and linear after some point τ, which corresponds to a change point $\tau \in (0, 1)$ and level change of f, i.e., for some $a > 0$: $f(t) = a1(\tau \le t)$, then the power can be derived accurately (Wang and Pötzelberger [453], Janssen and Kunz [271] and Bischoff et al. [49]). Note that for $\gamma \to \infty$, the probability of not crossing the boundary $u_\alpha - \gamma g(t)$ by the Brownian motion tends to 0. Several authors dealt with such problems with certain classes of trend functions g, but accurate approximations are not always possible. In general by large deviations principles, a rough approximation is possible. For instance one can show for a certain class of g with $g(t) = \int_{[0,t]} f(s)ds, t \in [0, 1], f \in L_2([0, 1], \lambda)$,

λ Lebesgue measure, that

$$P\left(\sup_{t\in[0,1]} (\gamma g(t) + B(t)) > u_\alpha\right) = 1 - \exp\left(-\frac{1}{2}\gamma^2||\tilde{g}||^2 + O(\gamma)\right)$$

for $\gamma \to \infty$, where $\tilde{g}$ is the smallest concave non-decreasing majorant of g and $||g||^2 = \int_{[0,1]} f^2(s)\,ds$. Other related results with respect to the Brownian bridge are discussed e.g. in Bischoff et al [47], [48] and [50]. Some of these results can be applied for power calculation even for moderate γ with a good accuracy (Bischoff et al [47]).

10.4   Application: Empirical Characteristic Functions

Statistical procedures have been introduced which are based on the behavior of the real part of the empirical characteristic function

$$U_n(t) = \frac{1}{n}\sum_{j\leq n}\cos(tY_j),$$

where Y_j, $j \geq 1$, are iid rv with distribution F_Y. Let us denote the real part of the theoretical characteristic function by

$$u(t) = E(\cos(tY_1)).$$

THE FIRST ZERO

The statistical procedures based on the empirical characteristic functions and their efficiencies depend on a so-called 'working' interval $I_n = (-t_n, t_n)$ such that $U_n(t) \neq 0$ for all $t \in I_n$ (cf. Heathcote and Welsh [214], Welsh [460], Welsh and Nicholls [461], Csörgő and Heathcote [83]). This gives rise to the question of the maximal size of I_n determined by the empirical first zero R_n of $U_n(\cdot)$:

$$R_n = \inf\{t > 0 : U_n(t) = 0\}.$$

R_n depends on the first zero of $u(\cdot)$ denoted by

$$r_0 = \inf\{t > 0 : u(t) = 0\} \leq \infty.$$

As expected $R_n \to r_0$ (a.s.) as $n \to \infty$ under certain conditions (see Welsh [460]). We are interested in the limiting distribution of R_n in the particular case $r_0 = \infty$. (For $r_0 < \infty$, see Heathcote and Hüsler [213]).

    If $r_0 = \infty$ we can make use of the results derived in Section 10.3. The behavior of R_n as $n \to \infty$ depends on the behavior of $u(t)$ as $t \to \infty$ and as $t \to 0$. We discuss here only the special case where $u(t)$ decreases at the rate of a power function. We suppose that

$$E(|Y|^2) < \infty \tag{10.29}$$

and that, for some $p > 0$,

$$u(t) = at^{-p}(1 + \epsilon(t)/\log(t)), \quad \text{where} \quad \lim_{t \to \infty} \epsilon(t) = 0, \tag{10.30}$$

such that $r_0 = \infty$, $u''(\cdot)$ exists, $u'(t) \to 0$ and $u''(t) \to 0$ as $t \to \infty$. These conditions hold if for instance Y has an exponential or more general a gamma distribution.

(10.29) and (10.30) imply that the process $\{n^{1/2}(U_n(t) - u(t)), \ t > 0\}$ converges in distribution to a continuous Gaussian process $\{X(t), \ t > 0\}$, with mean 0 and variance $\sigma^2(t)$ and that the standardized process $\{X(\cdot)/\sigma(\cdot), t > \delta\}$ for some $\delta > 0$ is a locally stationary Gaussian process with $\alpha = 2$ and

$$C(t) = \left(-u''(0) + u''(2t) - 2u'^2(t) - (u'(2t) - 2u(t)u'(t))^2/(2\sigma^2(t)) \right) \Big/ \left(4\sigma^2(t) \right).$$

Here $\sigma^2(t) = (1 + u(2t) - 2u^2(t))/2$ with $\sigma^2(0) = 0$ and $\sigma^2(t) \to \sigma^2(\infty) = 1/2$ as $t \to \infty$. If the assumptions (10.29) and (10.30) are satisfied, then $C(t) \to C(\infty) = -u''(0)/2$ as $t \to \infty$.

CONVERGENCE RESULT

For some normalization t_n,

$$\begin{aligned}
P(R_n > t_n) &= P(\sqrt{n}(U_n(t) - u(t)) > -\sqrt{n}u(t), \ t \le t_n) \\
&\sim P(X(t) > -\sqrt{n}u(t), \ t \le t_n) \\
&= P(X(t) < \sqrt{n}u(t), \ t \le t_n) \\
&\sim P(X(t) \le \sqrt{n}u(t), \ \delta \le t \le t_n),
\end{aligned}$$

for any $\delta > 0$, since the probability of an exceedance of $\sqrt{n}u(\cdot)$ by $X(\cdot)$ in $[0, \delta)$ is asymptotically 0.

For $t_n \to \infty$ the approximation of $\{n^{1/2}(U_n(t) - u(t)), \ 0 < t \le t_n\}$ by $\{X(t), \ 0 < t \le t_n\}$ follows from a strong approximation theorem (given in Csörgő [82]). The probability

$$P(X(t) \le \sqrt{n}u(t), \ \delta \le t \le t_n)$$

is approximated by using Theorem 10.3.5, if all the conditions can be verified. We prove the following result using the normalization

$$t_n = (2pa^2n/\log(n))^{1/(2p)} \left(1 + \left(1 + \frac{1}{2p} \right) \frac{\log\log(n)}{\log(n)} + \frac{x - A}{\log(n)} \right) \tag{10.31}$$

where

$$A = \frac{1}{2}\log\left(C(\infty)/(2\pi^2)\right) + \frac{1}{2p}\log(2a^2p).$$

Theorem 10.4.1. *Assume that* (10.29) *and* (10.30) *hold for the iid rv* Y_j. *Then for* t_n *given in* (10.31) *and any* $\delta > 0$,

$$P(X(t) \le \sqrt{n}u(t),\ \delta \le t \le t_n) \to \exp(-e^x)$$

for all $x \in \mathbb{R}$, *as* $n \to \infty$ *and, thus,*

$$P(R_n \le t_n) \to 1 - \exp(-e^x).$$

Proof. We sketch only the verification of some conditions of Theorem 10.3.5. Note that $C(t) = C(\infty)(1 + o(1))$ and $\sigma^2(t) = 1/2 + O(t^{-p})$ for $t \to \infty$. Thus

$$nu^2(t)/\sigma^2(t) = 2na^2t^{-2p}(1 + \epsilon^*(t)/\log(t))$$

for $\epsilon^*(t) \to 0$. Hence $\min_{t \le t_n} u_n(t) := \sqrt{n}\min_{t \le t_n}(u(t)/\sigma(t)) \to \infty$. By Theorem 10.3.5 we have to calculate now the integral $J(t_n)$ for some $\delta > 0$:

$$J(t_n) = \frac{1}{\sqrt{\pi}}C^{1/2}(\infty)\int_\delta^{t_n} \varphi(u_n(t))\,dt + o(1).$$

This can be approximated by splitting up the interval $[\delta, t_n]$ into two parts with intermediate point $y_n = (2pa^2n/\log(n))^{(1-\theta)/2p}$ where $0 < \theta < 1$. The first part of the integral is asymptotically negligible since the boundary is tending fast to ∞, whereas the second part leads to $J(t_n) \to \exp(x)$ as $t_n \to \infty$. The verification of the smoothness condition (10.11) is somewhat tedious. Berman's condition can also be verified. For details see Heathcote and Hüsler [213]. $\qquad\square$

The limit distribution of R_n is similarly derived for cases where $u(\cdot)$ has a different rate of decrease or a different local behavior at $t = 0$ than assumed in Theorem 10.4.1, e.g., if Y_j has a normal or a stable law (cf. Heathcote and Hüsler [213], Hüsler [244] or Bräker and Hüsler [54]).

10.5   Extensions: Maxima of Gaussian Fields

Denote by $X_i(\mathbf{w}), i \ge 1$, independent copies of a Gaussian process $X(\mathbf{w})$ with $\mathbf{w} \in W \subset \mathbb{R}^d$, mean 0, variance 1 and continuous sample paths. Let W be an open set which contains $\mathbf{0}$. Assume that the covariance function $r(\mathbf{w}, \mathbf{w}')$ of $X(\mathbf{w})$ satisfies for some $0 < \alpha \le 2$ and $c_\alpha > 0$,

$$r(s\mathbf{w}, s\mathbf{w}') = 1 - \gamma(|\mathbf{w} - \mathbf{w}'|)L(s)s^\alpha + o(s^\alpha) \quad \text{for } s \to 0 \tag{10.32}$$

where the o-term is uniform in $\mathbf{w}$ and $\mathbf{w}'$, L a slowly varying function at 0, and $\gamma(\mathbf{w})$ a non-negative continuous function satisfying $\gamma(c\mathbf{w}) = c^\alpha\gamma(\mathbf{w})$ for any $\mathbf{w} \in D$

and $c \geq 0$. Here, $|\cdot|$ denotes the Euclidean norm in $\mathbb{R}^d$. In Theorem 7.4.3 we mentioned the limiting process if the Gaussian process is a Brownian motion or, by transformation, an Ornstein-Uhlenbeck process with $d = 1$ and $\alpha = 1$, a result of Brown and Resnick [55]. Kabluchko et al. [280] considered the more general case of the maximum process $M_n(\mathbf{w})$ of $X_i(\mathbf{w})$, $i \leq n$, with the usual normalization a_n and b_n of Gaussian rv, by introducing the suitable normalization c_n for the space parameter $\mathbf{w}$:

$$c_n = \inf\{s > 0 : L(s)s^\alpha = 1/b_n^2\}.$$

They showed in [280] that the linearly normalized maximum process $(M_n(c_n\mathbf{w}) - b_n)/a_n$ converges to a process η_α, called a Brown-Resnick process where $a_n = (2\log n)^{-1/2}$ and $b_n = \sqrt{2\log n} - \left(\frac{1}{2}\log\log n + \frac{1}{2}\log(2\pi)\right)/\sqrt{2\log n}$. The space of continuous real functions with a compact support K are denoted by $C(K)$.

Theorem 10.5.1. *Under the assumptions* (10.32),

$$(M_n(c_n\mathbf{w}) - b_n)/a_n \to \eta_\alpha(\mathbf{w})$$

weakly on $C(K)$ for every compact $K \subset D$.

The limiting process η_α can be characterized for general α. Kabluchko et al. [280] showed that the process $\eta_\alpha(\mathbf{w})$ is defined as

$$\eta_\alpha(\mathbf{w}) = \max_{i \geq 1}(U_i + W_i(\mathbf{w}))$$

where U_i are the points of a standard Poisson process, and $W_i(\mathbf{w})$ are iid fractional Brownian processes with Hurst parameter $\alpha/2$ and with drift

$$E(W_i(\mathbf{w})) = -|\mathbf{w}|^\alpha$$

and covariance function

$$\text{Cov}(W_i(\mathbf{w}), W_i(\mathbf{w}'))) = |\mathbf{w}|^\alpha + |\mathbf{w}'|^\alpha - |\mathbf{w} - \mathbf{w}'|^\alpha.$$

The process η_α is stationary with Gumbel marginal distributions and has continuous paths (see [280]). If the slowly varying function $L(s)$ tends to $c > 0$, then $c_n = (2c\log n)^{-1/\alpha}$.

This approach is extended by Kabluchko [281] for time-spatial Gaussian processes $X(t, \mathbf{w})$ with a time parameter $t > 0$ and a space parameter $\mathbf{w} \in \mathbb{R}^d$. One assumes again that $X(t, \mathbf{w})$ has mean 0 and variance 1, and defines the maximum

$$M_n(\mathbf{w}) = \max_{0 \leq t \leq n} X(t, \mathbf{w})$$

for the site $\mathbf{w}$. We need to scale the space parameter to derive a non-trivial limiting process, otherwise the maxima of $M_n(\mathbf{w})$ and $M_n(\mathbf{w}')$ for $\mathbf{w} \neq \mathbf{w}'$ are asymptotically independent, assuming that the autocorrelations between the processes

$X(t, \mathbf{w})$ and $X(s, \mathbf{w}')$ are not equal to 1 for some $t \neq s$. Here we have a continuous time parameter t instead of the index i, and we have not independent processes or copies $X_i(\mathbf{w})$ with respect to the discrete 'time' parameter i, as before. It is supposed that $X(s, \mathbf{w})$ is stationary with respect to s: it means that the covariance

$$E(X(s, \mathbf{w})X(s + t, \mathbf{w}')) =: r_t(\mathbf{w}, \mathbf{w}')$$

does not depend on s. Under adapted conditions, Kabluchko [281] showed that the same limiting process η_α holds for the normalized maximum process, showing that

$$\frac{M_n(c_n \mathbf{w}) - b_n}{a_n} \to \eta_\alpha, \quad \text{as } n \to \infty,$$

in the sense that the finite-dimensional distributions converge. He assumed the following conditions on the covariance function:

$$r_t(\mathbf{w}, \mathbf{w}') < 1$$

for $t \neq 0$, $\mathbf{w}, \mathbf{w}' \in D$ and the local behavior

$$r_{h^{1/\beta}t}(h^{1/\alpha}\mathbf{w}, h^{1/\alpha}\mathbf{w}') = 1 - (c_\alpha|\mathbf{w} - \mathbf{w}'|^\alpha + c_\beta|t|^\beta)h + o(h) \quad \text{for } h \to 0$$

for some $\alpha, \beta \in (0, 2]$ and positive constants c_α, c_β, where the o-term is uniform for $\mathbf{w}, \mathbf{w}' \in D$ and $t \in I$ with bounded I. Since the time-spatial process $X(t, \mathbf{w})$ does not consist of independent spatial processes $X_i(\mathbf{w})$ as in [280], one has to restrict the correlation for large gaps $t \to \infty$. As usual for the Gaussian processes, Berman's condition is assumed: $r_t(\mathbf{w}, \mathbf{w}') \log t \to 0$ as $t \to \infty$, for $\mathbf{w}, \mathbf{w}' \in D$.

Also the norming sequences have to be adapted, using the norming sequences of continuous time Gaussian processes. We set $a_n = (2 \log n)^{-1/2}$,

$$b_n = \sqrt{2 \log n} + \frac{1}{\sqrt{2 \log n}} \left(\frac{2 - \beta}{2\beta} \log \log n + \log \left(\frac{c_\beta^{1/\beta} H_\beta 2^{(2-\beta)/(2\beta)}}{\sqrt{2\pi}} \right) \right)$$

with Pickands constant H_β, and $c_n = (2c_\alpha \log n)^{-1/\alpha}$.

Chapter 11

Extensions for Rare Events

In the following sections we discuss some extensions which were mentioned in the previous chapters. Of main interest is the extension to general rare events in relation to a random sequence applying the same method as used for dealing with exceedances. In addition we treat now also the point process of all exceedances if clustering occurs. These results are applied to the processes of peaks over a threshold and of rare events. Finally, in the same way general rare events are considered without relation to a random sequence. Here triangular arrays of rare events will be analyzed by the same approach. This extension unifies easily the different local dependence conditions. Its application to multivariate extremes is then straightforward. As a particular case, triangular arrays of rare events in relation with exceedances of Gaussian sequences are considered since they are basic for maxima of a continuous Gaussian process. This analysis reveals also a new definition of Pickands constants.

11.1 Rare Events of Random Sequences

We mentioned at the end of Section 9.1 that the generalization of the theory for exceedances to a theory for rare events is rather straightforward. This generalization is the subject of this short section. Instead of the event $\{X_i > u_{ni}\}$ at time i or i/n, consider the event $\{X_i \in A_{ni}\}$ with A_{ni} a Borel set. Generalizing the uan condition, we assume

$$p_{n,max} = \sup_{i \le n} P(X_i \in A_{ni}) \to 0 \qquad \text{as} \quad n \to \infty \tag{11.1}$$

and replace condition (9.1) by

$$\limsup_{n \to \infty} \sum_{i \le n} P(X_i \in A_{ni}) < \infty, \tag{11.2}$$

M. Falk et al., *Laws of Small Numbers: Extremes and Rare Events*, 3rd ed.,
DOI 10.1007/978-3-0348-0009-9_11, © Springer Basel AG 2011

which means that $\{X_i \in A_{ni}\}$ are rare events. Using the notation introduced in Section 9.1, we will analyze the point processes of the rare events. Thus in analogy to the point process of exceedances, the point process of upcrossings and the point process of cluster positions, we define the point process of rare events, the point process of entering a rare set and the point process of cluster positions of rare events

$$N_n(B) = \sum_{i \in nB} 1\{X_i \in A_{ni}\},$$

$$\tilde{N}_n(B) = \sum_{i \in nB} 1\{X_{i-1} \notin A_{n,i-1}, \ X_i \in A_{ni}\}$$

and

$$N_n^*(B) = \sum_{j \le k_n} 1\{N_n(B_j \cap B) \ne 0\},$$

for some k_n. The intervals B_j are chosen in the following way: $B_1 = (0, i_1/n]$ with

$$\sum_{i \le i_1} P(X_i \in A_{ni}) \le \sum_{i \le n} P(X_i \in A_{ni})/k_n,$$

where i_1 is maximally chosen (compare Section 9.1), and in the same way

$$B_2 = (i_1/n, i_2/n], \ldots, B_{k_n} = (i_{k_n-1}/n, i_{k_n}/n]$$

are constructed.

Obviously, the conditions D and D^* have to be redefined. Let $\alpha_{n,m}$ be such that for fixed n and for any integers $1 \le j_1 < \cdots < j_p < j_1' < \cdots < j_q' \le n$ for which $j_1' - j_p \ge m$,

$$\left| P(X_{j_l} \notin A_{nj_l}, \ l \le p, \ X_{j_h'} \notin A_{nj_h'}, \ h \le q) \right.$$
$$\left. - P(X_{j_l} \notin A_{nj_l}, \ l \le p) \times P(X_{j_h'} \notin A_{nj_h'}, \ h \le q) \right| \le \alpha_{n,m}.$$

Definition 11.1.1. We say that condition $D(A_{ni})$ or just D holds if

$$\alpha_{n,m_n} \to 0 \qquad \text{as} \quad n \to \infty \tag{11.3}$$

for a sequence $\{m_n\}$ satisfying $m_n p_{n,max} \to 0$ as $n \to \infty$.

Again, we choose $\{k_n\}$ such that, as $n \to \infty$,

$$k_n(\alpha_{n,m_n} + m_n p_{n,max}) \to 0. \tag{11.4}$$

Let the sequence α_n^* be such that

$$\sum_{i<j,j+1\in I} P(X_i \in A_{ni},\ X_j \notin A_{nj}, X_{j+1} \in A_{n,j+1}) \leq \alpha_n^*,$$

for any $I = \{l_1 \leq i \leq l_2 \leq n\}$ with

$$\sum_{i\in I} P(X_i \in A_{ni}) \leq \sum_{i\leq n} P(X_i \in A_{ni})/k_n.$$

Definition 11.1.2. Condition $D^*(A_{ni})$ or just D^* holds if $k_n\alpha_n^* \to 0$ as $n \to \infty$, where $\{k_n\}$ satisfies (11.4).

CONVERGENCE OF THE POINT PROCESSES

Modifying the proofs of Section 9.1, we obtain the following results. We again assume that $\lambda(\cdot)$ and $\mu(\cdot)$ are continuous functions such that the limiting point processes are simple.

Theorem 11.1.3. *Suppose that conditions $D(A_{ni})$ and $D^*(A_{ni})$ hold for a random sequence $\{X_i,\ i \geq 1\}$ and Borel sets $\{A_{ni},\ i \leq n,\ n \geq 1\}$, satisfying (11.1) and (11.2). Then*

(i) *$\tilde{N}_n \to_D \tilde{N}$, a Poisson process with intensity measure $\mu(\cdot)$, with $\mu((0,1]) < \infty$ is equivalent to*

$$\sum_{i\leq nt} P(X_i \notin A_{ni},\ X_{i+1} \in A_{n,i+1}) \to \mu(t), \qquad t \leq 1, \qquad (11.5)$$

where $\mu(\cdot)$ is a continuous function with $\mu((0,t]) = \mu(t)$ and $\mu(1) < \infty$.

(ii) *If in addition to (11.5)*

$$\sum_{i\leq nt} P(X_i \in A_{ni}) \to \lambda(t), \qquad t \leq 1 \qquad (11.6)$$

with $\lambda(1) < \infty$, then $\lambda(1) = \mu(1)$ is equivalent to $E(\|N_n - \tilde{N}_n\|) \to 0$ and it follows that $N_n \to_D \tilde{N}$.

(iii) *$N_n^* \to_D \tilde{N}$, is equivalent to (11.5), and it follows that*

$$E(\|N_n^* - \tilde{N}_n\|) \to 0$$

as $n \to \infty$.

The condition $D^*(A_{ni})$ restricts the pattern of the occurrence of rare events in the same way as that of the exceedances. Instead of considering the events $E_{ni} = \{X_i \in A_{ni}\}$, we may treat any triangular array of general events E_{ni} not necessarily being represented in terms of a random sequence and Borel sets. The same approach can be used for general rare events $\{E_{ni}\}$. This is discussed briefly in Section 11.5.

Example 11.1.4. Consider again the stationary max-autoregressive sequence of Example 9.2.3:

$$X_i = c\max(X_{i-1},\, Y_i) \qquad \text{for} \quad i \ge 2,$$

where $0 < c < 1$, $\{Y_i\}$ is an iid sequence with distribution $H(\cdot)$, X_1 is a rv which is independent of $\{Y_i\}$. $\{X_i\}$ is stationary if $F(x) = F(x/c)H(x/c)$ is satisfied with $F(\cdot)$ denoting the distribution of X_i. We only deal with the case $F \in D(G_{1,\alpha})$, the case where clustering occurs. As rare sets we consider the intervals

$$A_n = (\beta a_n, a_n]$$

where $0 < \beta < 1$, and $a_n = a_n(x)$ denotes the norming sequence such that $n\bar{F}(a_n) \to x^{-\alpha} = \tau^*$, with some $x > 0, \alpha > 0$. Note that also $H \in D(G_{1,\alpha})$ and $n\bar{H}(a_n) \to \tau^*(c^{-\alpha} - 1)$.

It is easily seen that (11.1) and (11.2) hold, since

$$nP(X_i \in A_n) = n(\bar{F}(\beta a_n) - \bar{F}(a_n)) \to (\beta^{-\alpha} - 1)\tau^*.$$

Because of the stationarity we get immediately that $\lambda(t) = \lambda t$ and $\mu(t) = \mu t$ with

$$\lambda = (\beta^{-\alpha} - 1)\tau^*$$

and if $\beta \ge c$,

$$
\begin{aligned}
\mu &= \lim_{n\to\infty} nP(X_1 \notin A_n,\, X_2 \in A_n) \\
&= \lim_{n\to\infty} n\big(P(X_1 \le \beta a_n,\, X_2 \in A_n) + P(X_1 > a_n,\, X_2 \in A_n)\big) \\
&= \lim_{n\to\infty} n\big(P(X_1 \le \beta a_n,\, cY_2 \in A_n) + P(\beta a_n/c < X_1 \le a_n/c)\big) \\
&= \lim_{n\to\infty} n\big[\bar{H}(\beta a_n/c) - \bar{H}(a_n/c) + \bar{F}(\beta a_n/c) - \bar{F}(a_n/c)\big] \\
&= \tau^*\big[(\beta^{-\alpha} - 1)(1 - c^\alpha) + c^\alpha(\beta^{-\alpha} - 1)\big] \\
&= \tau^*(\beta^{-\alpha} - 1).
\end{aligned}
$$

If $\beta < c$ a similar derivation gives

$$
\begin{aligned}
\mu &= \tau^*\big[(\beta^{-\alpha} - 1)(1 - c^\alpha) + (1 - c^\alpha)\big] \\
&= \tau^*\beta^{-\alpha}(1 - c^\alpha).
\end{aligned}
$$

This implies that the clustering index of the rare events (cf. extremal index) can be defined as $\theta = \mu/\lambda$, hence

$$
\theta = \begin{cases} 1 & \text{if} \quad \beta \ge c, \\ (1 - c^\alpha)/(1 - \beta^\alpha) & \text{if} \quad \beta < c. \end{cases}
$$

The verification of the mixing conditions D and D^* is tedious but possible (cf. Alpuim et al. [10]). By Theorem 11.1.3 we get that the point process $\tilde{N}_n$ of the entrances into the sets A_n is asymptotically a Poisson process $\tilde{N}$ with intensity μ. $\tilde{N}$ is also the limiting point process of N_n^*, the point process of cluster positions. Moreover, if $\beta \geq c$ there is no clustering, hence $N_n \to_D \tilde{N}$. If $\beta < c$ clustering occurs. This is plausible, since if at some time point i we have $X_i \in A_n$, then usually $X_{i+1} = cX_i$. The probability of the event $\{X_i \in A_n, cX_i < X_{i+1} \in A_n\}$ is asymptotically negligible. But $cX_i \in A_n$ together with $X_i \in A_n$ is possible only if $\beta < c$ (otherwise $cX_i \leq \beta a_n$). In this case we are also interested in the cluster size distribution which is analyzed in Example 11.4.3.

If we compare the clustering index of the rare events with the extremal index, we see that the clustering index of the rare events depends on the rare sets A_{ni} through β, whereas the extremal index only depends on c, where the rare events $A_{n,i} = [a_n, \infty)$ consider exceedances of the boundary $a_n = a_n(x)$.

The relation of the two indexes are as follows. Instead of A_n we might use $A_n^* = (a_n, \beta a_n]$, $\beta > 1$. By a simple transformation of a_n we get for $\beta > 1/c$ from the above derivations that $\theta = \theta(\beta) = (1-c^\alpha)/(1-\beta^{-\alpha})$ and that $\theta(\beta) \to 1-c^\alpha = \theta(\infty)$, as $\beta \to \infty$. Note, $\theta(\infty)$ is the extremal index.

11.2 The Point Process of Exceedances

If exceedances do cluster, then the point process of exceedances N_n does not converge to a simple Poisson process in general, as was mentioned earlier. In this section we discuss this particular case by considering exceedances of constant boundaries.

We deal with (Radon) random measures η on $[0, 1]$ and denote their Laplace transforms by

$$L_\eta(f) = E\left(\exp\left(-\int_{[0,1]} f \, d\eta\right)\right),$$

where f is any non-negative measurable function on $[0, 1]$.

For a random sequence $\{X_i, \ i \geq 1\}$, we denote by N_n the point process of the exceedances

$$N_n = \sum_{i \leq n} \delta_{i/n} 1(X_i > u_n).$$

If $\{X(t), t \geq 0\}$ is a stochastic process in continuous time with continuous paths, then the exceedances of u_T by $X(t)$ define a random measure $\tilde{\zeta}_T$:

$$\tilde{\zeta}_T(B) = \int_{TB} 1(X(t) > u_T) \, dt,$$

where B is any Borel set in $[0, 1]$. In order to get non-trivial limits one has to normalize $\tilde{\zeta}_T$ usually by some constants a_T. Thus let

$$\zeta_T(B) = a_T \tilde{\zeta}_T(B).$$

In the following we present an approach which is applicable in both the discrete and the continuous time case. This approach extends the ideas and methods from the preceding sections on the point processes of exceedances. The stationary random measures were treated by Hsing et al. [231] and Leadbetter and Hsing [302] and extensions to non-stationary ones by Nandagopalan et al. [347]. For a detailed account of random measures see the monographs by Kallenberg [282], Resnick [393] and Reiss [387].

LONG RANGE DEPENDENCE

Instead of the long range mixing condition D we need a more restrictive mixing condition. For any $T > 0$ and $[a, b] \subset [0, 1]$, let

$$\mathcal{B}_T[a, b] = \sigma\{\zeta_T(B) : B \text{ is a Borel set in } [a, b]\}$$

be the σ-field generated by $\zeta_T(\cdot)$ and

$$\alpha_{T,l} := \sup \Big\{ |P(A \cap B) - P(A)P(B)| : A \in \mathcal{B}_T[0, s], B \in \mathcal{B}_T[s + l, 1],$$
$$0 \le s < s + l \le 1 \Big\}$$

with $0 < l < 1$.

Definition 11.2.1. The family of random measures $\{\zeta_T\}$ is called *strongly mixing*, if $\alpha_{T,l_T} \to 0$ as $T \to \infty$ for some $l_T \to 0$.

We assumed in Chapter 8 that a single exceedance of u_n by some rv X_i is negligible (uan condition). Here we assume that exceedances in small intervals of length less than l_T are negligible. Thus suppose that

$$\gamma_T := \sup\{1 - E(\exp(-\zeta_T(I))) : I \subset [0, 1] \text{ any interval with } m(I) \le l_T\}$$
$$\to 0, \tag{11.7}$$

as $T \to \infty$ where $m(\cdot)$ denotes the Lebesgue measure.

In the following let $\{k_T\}$ be such that

$$k_T (\alpha_{T,l_T} + \gamma_T) \to 0 \qquad \text{as} \quad T \to \infty. \tag{11.8}$$

Note that (11.8) holds for any bounded $\{k_T\}$, but we are interested in $k_T \to \infty$. This is possible by choosing for instance $k_T = \min(\alpha_{T,l_T}, \gamma_T)^{-1/2}$.

We assume that for each $T > 0$ there exists an interval partition $\{B_j, \ j \le k_T\}$ of $[0, 1]$ such that, for each $\epsilon > 0$,

$$\max_{1 \le j \le k_T} P(\zeta_T(B_j) > \epsilon) \to 0 \qquad \text{as} \quad T \to \infty. \tag{11.9}$$

These B_j's play the same role as the B_j's in Section 9.1.

EXTENDED COMPOUND POISSON PROCESS

The properties (11.7) and (11.9) imply for strongly mixing families of random measures ζ_T that if $\zeta_T \to_D \zeta$, then the limit random measure ζ is infinitely divisible and has independent increments. It is possible to derive the Laplace transform of ζ. Under the additional assumption

$$\limsup_{T \to \infty} P(\zeta_T([0,1]) = 0) > 0,$$

we get

$$L_\zeta(f) = \exp\left(-\int_{x \in [0,1]} \int_{y>0} \left[1 - \exp(-yf(x))\right] d\pi_x(y) d\nu(x)\right), \qquad (11.10)$$

where $\nu(\cdot)$ is a finite measure on $[0,1]$ and $\pi_x(\cdot)$ is a probability measure on $\mathbb{R}_+ \setminus \{0\}$ (ν-a.e. x). The limit random measure can be interpreted as an *extended* compound Poisson process in the sense that $\nu(\cdot)$ indicates the intensity of the Poisson events and $\pi_x(\cdot)$ the distribution of the atom size at x, given that there is an event at time point x. In the context of the point process of exceedances, $\nu(\cdot)$ is the intensity of occurrences of clusters and $\pi_x(\cdot)$ is the distribution of the cluster size at x, given that there is a cluster at time point x. In the context of sojourns, $\pi_x(\cdot)$ is the distribution of the length of a sojourn if it occurs at x. If $\pi_x(\cdot)$ does not depend on x, then ζ is a (usual) compound Poisson process.

For any finite T we have to define the analogous measures $\pi_{T,x}(\cdot)$ and $\nu_T(\cdot)$. Let, for any $y > 0$ and $1 \le j \le k_T$,

$$\pi_{T,j}(y) = P(\zeta_T(B_j) \le y \,|\, \zeta_T(B_j) \ne 0)$$

and write

$$\pi_{T,x} = \pi_{T,j} \quad \text{if} \quad x \in B_j.$$

Define $\nu_T(\cdot)$ by $\nu_T(B) = \sum_{j \le k_T} P(\zeta_T(B_j) \ne 0) m(B \cap B_j)/m(B_j)$.

In order for ζ to be an extended compound Poisson process, we suppose further conditions on ζ_T.

We replace (11.9) by a more restrictive condition. We assume that there exists an interval partition $\{B_j, \ 1 \le j \le k_T\}$ of $[0,1]$ such that (11.8) holds for k_T and

$$\max_{1 \le j \le k_T} P(\zeta_T(B_j) \ne 0) \to 0 \quad \text{as} \quad T \to \infty. \qquad (11.11)$$

Moreover, we need a smoothness condition: there exists an open non-empty subset G of $\mathbb{R}_+ \setminus \{0\}$, such that for every $a \in G$ the families $\{g_T(\cdot, a), \ T > 0\}$ with

$$g_T(x,a) = \int_{y>0} \left(1 - \exp(-ay)\right) d\pi_{T,x}(y)$$

are equicontinuous with respect to x, i.e., for every $x \in [0,1]$ and $\epsilon > 0$ there exist $T(x) > 0$ and $\delta(x) > 0$ such that

$$|g_T(x,a) - g_T(x',a)| < \epsilon \quad \text{for all} \quad T > T(x) \quad \text{and} \quad |x - x'| < \delta(x). \qquad (11.12)$$

In order for $\nu(\cdot)$ to be a finite measure, suppose that

$$\limsup_{T\to\infty} \nu_T([0,1]) < \infty. \tag{11.13}$$

Under these conditions, convergence of the normalized random measure ζ_T can be analyzed by means of the convergence of the measures $\nu_T(\cdot)$ and $\pi_{T,x}(\cdot)$ (Theorem 4.3 of Nandagopalan et al. [347]).

Theorem 11.2.2. *Let $\{\zeta_T\}$ be a family of strongly mixing random measures such that (11.7), (11.11), (11.12) and (11.13) hold. If $\nu_{T'} \to_w \nu'$ and $\pi_{T',x} \to_w \pi'_x$ (ν'-a.e. x) along a subsequence $T' \to \infty$, then $\zeta_{T'} \to_D \zeta$ (as $T' \to \infty$), where ζ has the Laplace transform given by (11.10), with π_x and ν satisfying*

$$d\nu(x) = (1 - \alpha(x))d\nu'(x)$$

and

$$\pi'_x = \alpha(x)\delta_0 + (1 - \alpha(x))\pi_x, \qquad \nu'\text{-a.e. } x.$$

$\alpha(x)$ is completely determined by $\pi'_x(\cdot)$ since $\alpha(x) = \pi'_x(\{0\})$ and $\pi_x(\cdot)$ is a probability measure on $\mathbb{R}_+ \setminus \{0\}$. $\alpha(x)$ accounts for the intervals of exceedances of asymptotical length 0. For a fixed x, $\alpha(x)$ can be positive, since $\{\pi_{T',x}\}$ are in general not tight at zero. The converse of Theorem 11.2.2 is also true (cf. Nandagopalan et al. [347]).

The generalization to d-dimensional random measures and point processes of multivariate extreme values is rather straightforward (cf. Nandagopalan et al. [347]).

In the following we discuss two special cases: stationary random measures and non-stationary point processes.

STATIONARY RANDOM MEASURES

Definition 11.2.3. A random measure ζ is called *stationary*, if

$$\zeta(B) =_D \zeta(B + s) \qquad \text{for any} \quad B \subset [0,1],\ s < 1,\ \text{with } B + s \subset [0,1].$$

In this case the interval partition is chosen to be $B_j = ((j-1)/k_T,\ j/k_T]$, $j \le k_T$ with $k_T \to \infty$ such that $k_T(\alpha_{T,l_T} + l_T) \to 0$ as $T \to \infty$. Then some of the assumptions and definitions are simplified. We have

$$\nu_T(B) = k_T P(\zeta_T(B_1) \ne 0)m(B) =: \tau_T m(B)$$

and

$$\pi_T(y) = P(\zeta_T(B_1) \le y \mid \zeta_T(B_1) \ne 0), \qquad y > 0.$$

Since $\pi_{T,x} = \pi_T$, (11.12) holds. (11.13) turns out to be the condition

$$\sup_T \tau_T < \infty. \tag{11.14}$$

Because of the stationarity the asymptotic negligibility assumptions (11.7) and (11.11) hold using (11.14).

Corollary 11.2.4. *Let* $\{\zeta_T\}$ *be a family of strongly mixing stationary random measures on* $[0,1]$ *such that* (11.14) *holds with* k_T *satisfying* $k_T(\alpha_{T,l_T}+l_T) \to 0$. *If* $\tau_{T'} \to \tau'$ *and* $\pi_{T'} \to_w \pi'$ *along a subsequence* $T' \to \infty$, *then* $\zeta_{T'} \to_D \zeta$, *as* $T' \to \infty$, *where* ζ *is a compound Poisson process with Laplace transform*

$$L_\zeta(f) = \exp\left(-\tau \int_{x \in [0,1]} \int_{y>0} \left(1 - \exp(-yf(x))\right) d\pi(y)\, dx\right) \qquad (11.15)$$

with

$$\tau = (1-\alpha)\tau'$$

and

$$\pi' = \alpha\delta_0 + (1-\alpha)\pi.$$

This result is due to Leadbetter and Hsing [302]. They applied it to obtain the limiting random measure for the sojourn times (above a given boundary) of a particular stationary Gaussian process (with $\alpha = 2$, cf. Chapter 10):

$$\zeta_T(B) = a_T \int_{TB} 1(X(t) > u_T)\, dt,$$

where $a_T = \sqrt{2\log(T)}$. Further results on sojourns for Gaussian processes are given in Berman [39]. The extension of Corollary 11.2.4 to multidimensional stationary random measures is given in Nandagopalan et al. [347], (see also Zhang ([472]).

POINT PROCESSES

Another special case arises when the ζ_T's are point processes, denoted by N_n, which are not necessarily stationary. In this case the probability measures $\pi_{n,x}$ are *tight* at 0:

$$\pi_{n,x}(y) = P(N_n(B_j) \le y \,|\, N_n(B_j) \neq 0) = 0$$

for $y \in [0,1)$. Thus $\alpha(x) = 0$ holds for all x implying $\nu' = \nu$ and $\pi'_x = \pi_x$.

Corollary 11.2.5. *Let* $\{N_n\}$ *be a family of strongly mixing point processes on* $[0,1]$ *such that* (11.7), (11.11), (11.12) *and* (11.13) *hold. If* $\nu_n \to_w \nu$ *and* $\pi_{n,x} \to_w \pi_x$, ν-*a.e.* x *as* $n \to \infty$, *then* $N_n \to_D N$, *where* N *has again the Laplace transform given by* (11.10).

If $\{N_n\}$ are stationary point processes, then the limiting point process N is again a compound Poisson process. With obvious notational changes we have the following result.

Corollary 11.2.6. *Let $\{N_n\}$ be a family of strongly mixing stationary point pro-cesses on $[0,1]$ such that (11.14) holds with $k_n \to \infty$ satisfying $k_n(\alpha_{n,m_n}+m_n) \to 0$ as $n \to \infty$. If $\tau_n \to \tau$ and $\pi_n \to_w \pi$ as $n \to \infty$, then $N_n \to_D N$, as $n \to \infty$, where N is a compound Poisson process with Laplace transform*

$$L_N(f) = \exp\left(-\tau \int_{x \in [0,1]} \int_{y>0} \left(1 - \exp(-yf(x))\right) d\pi(y)\, dx\right).$$

11.3 Application to Peaks-over-Threshold

We now analyze the point process of the excesses of X_i above the boundary u_{ni}. The excess $(X_i - u_{ni})_+$ is in general not integer-valued and the point process will be a marked one. From the discussion in Chapter 1, we know that the excesses follow asymptotically a generalized Pareto distribution (cf. Section 1.3). Hence, in the stationary case the limiting random measure can be a compound Poisson process and in the non-stationary case an extended compound Poisson process. In the following we will only deal with the stationary case assuming that the boundaries are constant. The following results are due to Leadbetter [301].

Let $\{X_i,\ i \geq 1\}$, be a stationary random sequence and u_n the boundary such that

$$n(1 - F(u_n)) \to \tau \in (0,\infty). \tag{11.16}$$

The interval partition of $[0,1]$ will again be given by $B_j = ((j-1)r_n^*/n, jr_n^*/n]$ with $r_n^* = [n/k_n]$, $j \leq k_n$, and $B_{k_n+1} = (k_n r_n^*/n, 1]$.

Exceedances may cluster because of the local dependence among the X_i. If there is an exceedance in B_j, we consider j/k_n as the cluster position. We may associate with j/k_n the corresponding maximum excess $M_n(B_j) - u_n$ and analyze then the resulting marked point process. In some applications (insurance, heavy rainfall) it would be more informative to consider the sum of all excesses within a block instead of only the largest one. However, considering only the largest excess simplifies the problem. Thus, define

$$\zeta_n^*(\cdot) = \sum_{j \leq k_n+1} \delta_{j/k_n}(\cdot)\, (M_n(B_j) - u_n)_+\,.$$

Note that the cluster positions are chosen as j/k_n being the right endpoint of the block B_j. Other choices for cluster positions are also possible and do not affect the asymptotic result. By stationarity, $(M_n(B_j) - u_n)_+$ are identically distributed. It can be shown that if a certain mixing condition is satisfied, then the maximum excesses of blocks are asymptotically independent. Moreover, it can be shown that they follow asymptotically a generalized Pareto distribution (GPD). In general, we have to normalize the largest excesses by a factor a_n^* ($a_n^* > 0$) in order to obtain a compound Poisson process as limiting random measure. Therefore, define

$$\zeta_n(\cdot) = \zeta_n^*(\cdot)/a_n^* = \sum_{j \leq k_n+1} \delta_{j/k_n}(\cdot)\, ((M_n(B_j) - u_n)/a_n^*)_+\,.$$

To guarantee that $\{\zeta_n\}$ is strongly mixing (Definition 11.2.1), we redefine the mixing coefficient $\alpha_{T,l}$ in this context, denoted by $\tilde{\alpha}_{n,m}$. (We use a different notation for this mixing coefficient since we define rather different σ-fields in this application). Let

$$\mathcal{B}_n[k,l] = \sigma\{(X_i - u_n)_+ : k \leq i \leq l\}$$

for $1 \leq k \leq l \leq n$ and the mixing coefficients

$$\tilde{\alpha}_{n,m} = \sup \Big\{|P(A \cap B) - P(A)P(B)| :$$
$$A \in \mathcal{B}_n[1,k], \ B \in \mathcal{B}_n[k+m,n], \ 1 \leq k \leq n-m\Big\}.$$

Then the strongly mixing condition holds if

$$\tilde{\alpha}_{n,m_n} \to 0 \qquad \text{for some sequence} \quad m_n = o(n). \tag{11.17}$$

This condition (11.17) is slightly stronger than $\Delta(u_n)$ used in Section 11.4 where the σ-fields $\mathcal{B}_n[1,k]$ are defined with respect to the events $\{X_i > u_n\}$.

To prove the convergence to a compound Poisson process, we use Corollary 11.2.4. Note again that one can always construct a sequence $\{k_n\}$ with $k_n \to \infty$ such that $k_n(\tilde{\alpha}_{n,m_n} + m_n) \to 0$ as $n \to \infty$.

The first two conditions (11.7) and (11.11) follow immediately since, for any interval $I = I(n)$ with $m(I) \to 0$ as $n \to \infty$,

$$1 - E\big(\exp(-\zeta_n(I))\big) \leq P(\zeta_n(I) \neq 0)$$
$$\leq \sum_{i \in nI} (1 - F(u_n))$$
$$= (\tau + o(1))m(I) \to 0.$$

(11.14) holds since by assumption $\tau_n := n(1 - F(u_n)) \to \tau$ as $n \to \infty$, hence $\limsup_n \nu_n((0,1]) \leq \limsup_n \tau_n = \tau < \infty$.

It remains to show that $\nu_n \to_w \nu$ and $\pi_{n,x} \equiv \pi_n \to_w \pi$, where

$$\nu_n(B) = \sum_{j \leq k_n+1} P(\zeta_n(B_j) \neq 0)m(B \cap B_j)/m(B_j)$$
$$= k_n P(\zeta_n(B_1) \neq 0)(m(B) + o(1))$$

and by stationarity

$$\pi_{n,x}(y) = \pi_n(y) = P(\zeta_n(B_1) \leq y \,|\, \zeta_n(B_1) \neq 0)$$

for all $x \in [0,1]$. We write again $\bar{F}(x) = 1 - F(x)$ and $\bar{G}(x) = 1 - G(x)$.

Lemma 11.3.1. *Let $\{X_i, \ i \geq 1\}$ be a stationary random sequence with marginal distribution F satisfying*

$$\lim_{u \to \omega(F)} \bar{F}(u + yg(u))/\bar{F}(u) = \bar{G}(y), \tag{11.18}$$

for all $0 \leq y < \omega(G)$ with some strictly positive function g. Assume that $\{u_n\}$ is such that (11.16) and (11.17) hold and that the extremal index θ of the random sequence exists. Let $r_n^* = [n/k_n]$ with k_n satisfying $k_n(\tilde{\alpha}_{n,m_n} + m_n) \to 0$. Then with $a_n^* = g(u_n)$, we have

(i) *for $0 < y < \omega(G)$,*

$$P(M_{r_n^*} - u_n \geq a_n^* y) = [\theta \tau \bar{G}(y) + o(1)]/k_n,$$

(ii) $P(M_{r_n^*} - u_n \leq a_n^* y \mid M_{r_n^*} - u_n > 0) \to G(y)$ *as $n \to \infty$.*

Remark 11.3.2. Note that G in (11.18) is a GPD and (11.18) holds for any $F \in D(G_i)$, G_i an extreme value distribution (EVD) $(i = 1, 2, 3)$.

Proof. We use (11.16), hence $\bar{F}(u_n)/\tau \sim 1/n$, to derive

$$n\bar{F}(u_n + a_n^* y) = (1 + o(1)) \frac{\bar{F}(u_n + yg(u_n))}{\bar{F}(u_n)/\tau} \to \tau \bar{G}(y)$$

by (11.18). This implies that

$$P(M_n \leq u_n + yg(u_n)) \to \exp(-\theta \tau \bar{G}(y))$$

since θ is the extremal index of the random sequence. It can be shown that (11.17) with respect to $\{u_n\}$ implies (11.17) with respect to $\{v_n\}$ if $v_n \geq u_n$ for all n. Since $v_n = u_n + yg(u_n) \geq u_n$, we conclude that (11.17) holds with respect to this particular sequence $\{v_n\}$. By Lemma 9.1.4,

$$P(M_n \leq v_n) - P^{k_n}(M_{r_n^*} \leq v_n) \to 0$$

and thus

$$P^{k_n}(M_{r_n^*} \leq v_n) \to \exp(-\theta \tau \bar{G}(y)).$$

This implies the first statement and also the second one, since as $n \to \infty$,

$$P(M_{r_n^*} \leq u_n + yg(u_n) \mid M_{r_n^*} > u_n)$$
$$= 1 - P(M_{r_n^*} > u_n + yg(u_n))/P(M_{r_n^*} > u_n)$$
$$= 1 - (1 + o(1)) \frac{\theta \tau \bar{G}(y)/k_n}{\theta \tau /k_n}$$
$$\to 1 - \bar{G}(y) = G(y). \qquad \square$$

This lemma implies now the convergence of ν_n and π_n:

$$\nu_n(B) = (m(B) + o(1))k_n P(\zeta_n(B_1) \neq 0)$$
$$= (m(B) + o(1))k_n P(M_{r_n^*} > u_n)$$

$$= (m(B) + o(1))\theta\tau$$
$$\to m(B)\theta\tau$$

and

$$\pi_n(y) = P(M_{r_n^*} \le u_n + yg(u_n) | M_{r_n^*} > u_n)$$
$$\to G(y) = \pi(y)$$

for all $0 < y < \omega(G)$.

Combining these statements, we obtain the following result of Leadbetter [301].

Theorem 11.3.3. *Let $\{X_i, \ i \ge 1\}$ be a stationary random sequence with extremal index θ and marginal distribution F satisfying (11.18). Assume that $\{u_n\}$ is such that (11.16) and (11.17) hold. Then*

$$\zeta_n \to_D \zeta,$$

a compound Poisson process with intensity $\theta\tau$ and compounding distribution G.

11.4   Application to Rare Events

In Section 11.1 we dealt with rare events $\{X_i \in A_{ni}\}$ instead of the exceedances. The results of Section 11.2 are now applied to the point process defined by these rare events.

Again, let $\{X_i, \ i \ge 1\}$ be a random sequence and $\{A_{ni}, \ i \le n, \ n \ge 1\}$ be a triangular array of Borel sets in $\mathbb{R}$ such that (11.1) and (11.2) hold. We assume that the point process N_n of the rare events defined by

$$N_n(\cdot) = \sum_{i \le n} \delta_{i/n}(\cdot) 1(X_i \in A_{ni})$$

is strongly mixing. This assumption is implied by the condition $\Delta(A_{ni})$ which is defined in the same way as $\Delta(u_n)$ in Section 11.3, where we use now obviously the σ-field

$$\mathcal{B}_n[k,l] = \sigma\{\{X_i \in A_{ni}\}, \ k \le i \le l\}$$

and as always $\alpha_{n,m}$ with respect to these σ-fields.

Definition 11.4.1. Condition $\Delta(A_{ni})$ is said to hold if $\alpha_{n,m_n} \to 0$ for some sequence $\{m_n\}$ satisfying $m_n p_{n,max} \to 0$ as $n \to \infty$.

This condition is more restrictive than $D(A_{ni})$ used in Section 11.1. Then Corollary 11.2.5 is used to derive the following general result. To define $\pi_{n,x}$ and ν_n, use the partition B_j, $j \le k_n$, of $[0,1]$ introduced in Section 11.1 with $k_n \to \infty$ as in (11.4). Note that in this case $\nu_n \to_w \nu = \mu$ where μ is given in (11.5), assuming $D^*(A_{ni})$. By the construction of these B_j's, condition (11.11) holds using (11.2).

Condition (11.7) is restated in terms of $N_n(I)$, i.e.,

$$\gamma_n = \sup\{1 - E(\exp(-N_n(I))) : I \subset [0,1] \text{ any interval with } m(I) \le m_n\}$$
$$\to 0 \quad \text{as} \quad n \to \infty. \tag{11.19}$$

Corollary 11.4.2. *Let $\{X_i, i \ge 1\}$ be a random sequence and $\{A_{ni}, i \le n, n \ge 1\}$ a triangular array of Borel sets such that $\Delta(A_{ni})$, $D^*(A_{ni})$, (11.1) and (11.2) hold. Suppose that also (11.5), (11.12) and (11.19) hold. If $\pi_{n,x} \to_w \pi_x$ (μ-a.e. x) as $n \to \infty$, then $N_n \to_D N$, where N has again the Laplace transform given by (11.10).*

This result is simplified further if we assume a stationary random sequence $\{X_i\}$ and a 'constant' array $A_{ni} = A_n$ for all $i \le n$, as in the following example.

Example 11.4.3. We apply this result to the stationary max-AR(1)-random sequence $\{X_i\}$ discussed in Section 11.1. We consider $A_n = (\beta a_n, a_n]$ with $0 < \beta < c < 1$. Obviously, (11.1), (11.2), (11.12) and (11.19) are satisfied. We mentioned in Section 11.1 that $D^*(A_n)$ and (11.5) hold. Note that (11.5) implies the weak convergence of ν_n. Also $\Delta(A_n)$ can be verified with some tedious calculations. This follows in a similar way as in Alpuim et al. [10].

Thus it remains to verify that $\pi_n \to_w \pi$. Let $J = \max\{j : c^j \ge \beta\}$. Since $A_n = (\beta a_n, a_n]$, the cluster size is bounded from above by $J + 1$. Note that a cluster starts at $i + 1$ if, for some $j \ge 1$,

$$X_i \notin A_n, X_{i+l} \in A_n, \quad \text{for all } 1 \le l \le j.$$

The cluster size is then at least j. We get for $j \le J + 1$,

$$P(X_i \notin A_n, X_{i+l} \in A_n, l \le j)$$
$$= P(X_i \notin A_n, c^l Y_{i+1} \in A_n, l \le j)$$
$$\quad + P(X_i \notin A_n, c^l X_i \in A_n, l \le j) + o(1/n)$$
$$= P(Y_i \in (\beta a_n/c^j, a_n/c]) + P(X_i \in (a_n, a_n/c])1(j \le J)$$
$$\quad + P(X_i \in (\beta a_n/c^{J+1}, a_n/c])1(j = J+1) + o(1/n)$$
$$= x^{-\alpha}\Big((1 - c^\alpha)[(\beta^{-\alpha}c^{\alpha(j-1)} - 1) + 1(j \le J)]$$
$$\quad + (\beta^{-\alpha}c^{\alpha J} - 1)c^\alpha 1(j = J+1)\Big)/n + o(1/n) .$$

Then with $B_1 = (0, 1/k_n]$ and $D^*(A_n)$ we get

$$P(N_n(B_1) \ge j) = \sum_{i \le n/k_n} P(X_i \notin A_n, X_{i+l} \in A_n, l \le j) + o(1/k_n)$$
$$= \begin{cases} x^{-\alpha}(1 - c^\alpha)\beta^{-\alpha}c^{\alpha(j-1)}/k_n + o(1/k_n) & \text{for } j \le J, \\ x^{-\alpha}(\beta^{-\alpha}c^{\alpha J} - 1)/k_n + o(1/k_n) & \text{for } j = J+1 . \end{cases}$$

This implies

$$P(N_n(B_1) \neq 0) = x^{-\alpha}(1 - c^\alpha)\beta^{-\alpha}/k_n + o(1/k_n)$$

and, thus,

$$1 - \pi_n(j) = \frac{P(N_n(B_1) > j)}{P(N_n(B_1) \neq 0)}$$

$$\rightarrow 1 - \pi(j)$$

$$= \begin{cases} c^{\alpha j} & \text{for } 1 \leq j < J, \\ (c^{\alpha J} - \beta^\alpha)/(1 - c^\alpha) & \text{for } j = J, \\ 0 & \text{for } j = J + 1. \end{cases}$$

Corollary 11.4.2 implies thus the convergence of the point process N_n of rare events $\{X_i \in A_n\}$ to a compound Poisson process which has Laplace transform given in (11.15), where $\pi(\cdot)$ is given above and $\tau = x^{-\alpha}\beta^{-\alpha}(1 - c^\alpha)$.

We can verify easily that the mean cluster size, which is the reciprocal of the clustering index of the rare events, is equal to $(1 - \beta^\alpha)/(1 - c^\alpha)$. Hence θ indicates the reciprocal value of the mean sojourn number of the random sequence in the rare set A_n.

Example 11.4.4. If the stationary sequence $X_i, i \geq 1$, is sampled only at certain time points, e.g. periodically, one may deal with the cluster behavior of the (sub)-sampled sequence. If the original sequence shows a clustering of exceedances, we may expect a clustering of such events also for the sampled version. But the clustering depends also on the sampling pattern. This is investigated e.g. by Robinson and Tawn [397], Scotto et al. [405], Hall and Scotto [200] and Martins and Ferreira [324] with further references. Define the sampling scheme by the function $g(i)$:

$$g(i) = 1 + (i - 1) \,(\text{mod}\, I) + T[\frac{i - 1}{I}]$$

for integers n, I and T, where $1 \leq I \leq T$, and $[x]$ denotes as usual the largest integer smaller than or equal to x. The sampled sequence consists of $Y_i = X_{g(i)}, i \geq 1$. If $I = T$, then all integers are selected, i.e., $Y_i = X_i$ for each $i \geq 1$. If $I = 1$, then the sampling is periodical, i.e., $Y_i = X_{1+(i-1)T}$. For the other values I, the sampling uses the first I consecutive observations of a block of T observations.

Since the sampling scheme is non-random, the events of exceedances can be imbedded in the general scheme of rare events $\{X_i \in A_{ni}\}$. This follows by setting $A_{ni}^{\complement} = \mathbb{R}$ for each index i which is not sampled, i.e., $i \neq g(k)$ for some k, and otherwise at sampled points $A_{ni} = [u_n, \infty)$ where u_n is the usual threshold of the original sequence X_i. Obviously, the cluster sizes of exceedances of the sequence $\{Y_i\}$ are smaller than the cluster sizes of the original sequence $\{X_i\}$, hence $\theta_Y \geq \theta_X$. An upper bound of the extremal index is derived by Robinson and Tawn [397] in the case of regular sampling with $I = 1$. Under a certain condition the extremal index can be derived precisely for the general sampling scheme defined above with respect to a function $g(\cdot)$, see e.g. Martins and Ferreira [324].

Random Thinning of a Stationary Sequence or Process

Instead of a deterministic sampling one might discuss a random sampling. This corresponds to the random thinning of the occurring exceedances. The situation occurs in random processes which model failures of measuring devices where either observations are lost or replaced by observations of independent sequences, see Hall and Hüsler [198]. It can be adequately investigated by the point process approach.

The problem is motivated from environmental applications and communication systems where the occurrence of large values (e.g. pollution levels, service or downloading times) may significantly affect health or quality of life and where not all data are available. Missing values may occur according to some random pattern.

The missing scheme for general processes has been considered by Weissman and Cohen [459] for the case of constant failure probability and independent failures by dealing with the related point processes of exceedances. A general sampling of non-stationary Gaussian sequences is investigated in Kudrov and Piterbarg [298]. They derived the joint distribution of the maxima M_n of all data up to time n and of the partial maxima M_{n,G_n} of the data with time points in a set $G_n \subset \{1, \ldots, n\}$, growing with n, by assuming certain restrictions on the correlation function of the stationary Gaussian sequence, the trend and the sampling scheme. The joint limiting distribution of the maxima M_n of all data up to time n and the randomly sampled maxima M_{n,G_n} is investigated for stationary sequences by Mladenović and Piterbarg [338] under similar dependence conditions as mentioned in Chapter 9. Peng et al. [361] considered this question in the particular case of stationary Gaussian sequences, where the correlation $r(k)$ does not always satisfy Berman's condition. In such a case, obviously the sample mean has to be subtracted from the maximum, following the results of McCormick [325].

For applications, it is necessary to introduce other random failure patterns and investigate the extremal properties of such incomplete sequences. One should consider not only failure patterns which are based on independent failures, but also where the random failure patterns satisfy a weak dependence structure. For instance, if a measurement device is failing because of technical problems at some random time, then it takes a while until it is working again. Hence, the independence of missing values is not realistic in such applications.

If a missing data value occurs, then several strategies are usually applied. Either the missing value is (i) replaced by a predefined value x_0 which can be sometimes 999 (if one is interested in small values and no such large values occur) or -1 (if one is interested in large values and no negative values occur), (ii) the data is completely lost and the time series is sub-sampled with a smaller sample size, or (iii) in some cases another automatic measurement device can be used for replacing the missing data by a proxy value.

In the following we consider only the simple case where the missing data is replaced by 0. The other cases are dealt with in Hall and Hüsler [198]. Instead of the time series or the sequence $X_k, k \geq 1$, we observe the time series $Y_k = U_k X_k$, where the U_k's are Bernoulli rv. If $U_k = 1$, then the value of X_k is not missing; otherwise $Y_k = 0$, for $U_k = 0$. Hence, the missing value is replaced by 0. Since we are interested in the maxima of the sequence Y_k, which is assumed to be positive with probability 1 (for simplicity), it does not matter that the missing value is set to 0. We may also consider the point process of exceedances of the sequence Y_k which is the point process of exceedances of the sequence X_k, thinned by the process U_k.

We assume that the random sequence U_k is independent of the random sequence X_k. However, we do not assume that the random sequence U_k consists of iid rv as in Weissman and Cohen [459]. We want to restrict the process U_k only by a weak dependence condition. As usual we assume that the sequence X_k satisfies weak dependence conditions, that either condition $\Delta(u_n)$ or both the conditions $D(u_n)$ and $D''(u_n)$ hold. See Chapters 9 and 11 for these conditions. We note that condition $D''(u_n)$ is condition (11.27) with $k = 2$.

Typically, one assumes that the boundary u_n satisfies the condition (9.2), i.e., $n(1 - F(u_n)) \to \lambda < \infty$ with the marginal distribution F of the sequence X_k. Condition (9.2) is usually convenient, but not adapted for a discrete distribution where its support consists of all sufficiently large integers. Hall and Hüsler [198] considered also distributions F, which satisfy the restriction used in Anderson [14]: $\lim_{n\to\infty} \frac{1-F(n)}{1-F(n+1)} = e^\alpha$ for some $\alpha > 0$ which implies that

$$e^{-\tau(x-1)} \leq \liminf_n F^n(x + b_n) \leq \limsup_n F^n(x + b_n) \leq e^{-\tau(x)}$$

where $\tau(x) = e^{-\alpha x}$. This condition holds, for example, for the negative binomial distribution, which is not in the max-domain of attraction of any extreme value distribution.

Hall [197] showed that

$$e^{-\theta\tau(x-1)} \leq \liminf_n P(M_n \leq u_n) \leq \limsup_n P(M_n \leq u_n) \leq e^{-\theta\tau(x)} \qquad (11.20)$$

iff $P(X_2 \leq u_n \mid X_1 > u_n) \to \theta$, assuming that the stationary sequence X_k satisfies the conditions $D(u_n)$ and $D''(u_n)$ with $u_n = x + b_n$, for any x, together with the Anderson condition. In this case $\tau(x)$ is of Gumbel type, i.e., $\tau(x) = e^{-x}$, for all $x \in \mathbb{R}$, and we say that the sequence X_k has the extremal index θ.

We mention here the result on the maxima of the randomly sub-sampled sequence Y_k, denoted by $M_{n,Y}$, assuming the weak distributional mixing conditions $D(u_n)$ and $D''(u_n)$. By assuming the stronger condition $\Delta(u_n)$, we can derive the result of the thinned point process of exceedances (see [198]).

Theorem 11.4.5. *Let $U_k, k \geq 1$, be a Bernoulli strongly mixing stationary sequence with $P(U_k = 1) = \beta$, being independent of the stationary sequence $X_k, k \geq 1$, for which conditions $D(u_n)$ and $D''(u_n)$ hold, having extremal index $\theta > 0$ and cluster size distribution π. Define $\nabla(i) = P(U_1 = 0, \ldots, U_i = 0)$ and*

$$\theta^* = \theta \left(1 - \sum_{j=1}^{\infty} \pi(j) \nabla(j) \right) / \beta.$$

(i) *If $P(M_n \leq u_n(\tau)) \to e^{-\theta\tau}$, $\tau > 0$, as $n \to \infty$, then as $n \to \infty$*

$$P(M_{n,Y} \leq u_n(\tau)) \to e^{-\theta^* \beta \tau}.$$

(ii) *If $\{X_n\}$ is such that (11.20) holds, then*

$$\limsup_{n\to\infty} P(M_{n,Y} \leq u_n) \leq e^{-\theta^* \beta \tau(x)},$$
$$\liminf_{n\to\infty} P(M_{n,Y} \leq u_n) \geq e^{-\theta^* \beta \tau(x-1)}.$$

The thinned sequence Y_k has now the extremal index θ^*. The non-thinning probability β has a simple and obvious impact on the limiting distribution. We can show that the condition $D(u_n)$ holds for the sequence Y_k if this condition holds for the sequence X_k, since we assume the sequence U_k to be strongly mixing, see [198].

If the sequence U_k consists of iid Bernoulli(β), then simply we have $\nabla(j) = (1 - \beta)^j$ and $\theta^* = \theta[1 - \sum \pi(j)(1 - \beta)^j]/\beta$. Let us consider more generally the example that U_n is a Markov chain. We assume that

$$P(U_n = 1 \mid U_{n-1} = 1) = P(U_n = 1 \mid U_{n-1} = 1, U_{n-i} = u_{n-i}, 2 \leq i \leq k) = \eta,$$
$$P(U_n = 1 \mid U_{n-1} = 0) = P(U_n = 1 \mid U_{n-1} = 0, U_{n-i} = u_{n-i}, 2 \leq i \leq k) = \mu$$

for all $k \geq 2$, $u_{n-2}, \ldots, u_{n-k} \in \{0, 1\}, n > k$. Given any values of $\eta, \mu \in [0, 1]$, it is easy to obtain

$$\beta = \frac{\mu}{1 - \eta + \mu}. \tag{11.21}$$

Given a value for $\beta \in [0, 1]$, the parameters η and μ are not entirely arbitrary due to (11.21). If for example $\beta > 0.5$, then η is restricted to be in $[2 - \frac{1}{\beta}, 1]$.

One can show that $\{U_n\}$ is regenerative with finite mean duration of renewal epochs and hence, that it is strongly mixing. Furthermore, we obtain

$$\nabla(j) = \frac{(1 - \mu)^{j-1}(1 - \eta)}{1 - \eta + \mu},$$

$$\bar{\nabla}(j) = P(U_1 = 1, \ldots, U_j = 1) = \frac{\mu \eta^{j-1}}{1 - \eta + \mu}.$$

If the sequence X_k satisfies the conditions of Theorem 11.4.5, it implies that the asymptotic results holds for the maximum $M_{n,Y}$ of the sequence Y_k with

$$\theta^* = \frac{(1 - e^{\alpha c})\,\mu\,(1 - e^{\alpha c}(\eta - \mu))}{(1 - \eta + \mu)\,(1 - e^{\alpha c}(1 - \mu))\,\beta}\,.$$

The failure of devices implying that data are missing can be extended to the multivariate situation, where $\mathbf{X}_n$ and $\mathbf{U}_n$ are multivariate sequences in $\mathbb{R}^d$. Again, one considers the componentwise maxima $M_{nj,Y}$ of the new process $\mathbf{Y}_n = (Y_{n1}, \ldots, Y_{nd})$ with $Y_{nj} = X_{nj} U_{nj}$ and $M_{nj,Y} = \max_{i \le n} Y_{ij}$, for $1 \le j \le d$. Each component defines on $[0, 1]$ also a point process N_{nj} of exceedances $Y_{ij} > u_{nj}$ (with $i \le n$) above a boundary u_{nj}. The components N_{nj} can be combined to a multivariate point process $\mathbf{N}_n = (N_{n1}, \ldots, N_{nd})$ or to a marked point process on $[0, 1]$ with marks in $\mathbb{R}^d$.

The multivariate or marked point processes can be dealt with in the same way as in Hall and Hüsler [198]. The convergence of multivariate point processes is considered in Nandagopalan [346]). The details of this generalization, the convergence of the point process $\mathbf{N}_n$ or the multivariate maximum $\mathbf{M}_n$, are worked out in Zhang [472] considering several models of failure of multivariate devices or the partial or complete missing of the multivariate data. The results follow from conditions which are multivariate versions of the univariate conditions used in Hall and Hüsler [198], as the multivariate versions $\Delta(\mathbf{u}_n), D(\mathbf{u}_n), D'(\mathbf{u}_n), D''(\mathbf{u}_n)$, which are special cases of the mixing conditions in Chapter 11.6.

RANDOM SEARCH FOR AN OPTIMAL VALUE

Rare events occur also in the optimization problem. Here the optimum, say the minimum of an objective function $g : \mathbb{R}^d \to \mathbb{R}$ $(d \ge 1)$, is investigated by a search procedure. Typically, the function g is too complicated to find analytically the minimum. By a suitable algorithm, one tries to find, or to approximate the minimum value. Quite often stochastic optimizers are applied, usually repeatedly because the computation is rather fast. The number of internal steps of the optimizers is denoted by n and the number of repetitions by k. The outcome of one repetition gives a random approximation of the optimum value, denoted by m_n, being the minimum of n steps.

A very simple optimizer is using the random search algorithm. Let $\mathbf{X}_j$, $j = 1, \ldots, n$, be iid rv in $\mathbb{R}^d$ with $\mathbf{X}_j \sim F$ and density f. These vectors will denote the points generated within an optimization run, the 'steps' of the run. Then $m_n = \min_{i \le n} g(\mathbf{X}_i)$.

Assume that the objective function g has a global minimum $g_{\min}$. If the set of global minimum points is countable, we denote the global minimum points by $\mathbf{x}_l, 1 \le l \le L$, where L is finite or infinite.

If an optimizer produces accurate solutions, the repeated independent outcomes $m_{nj}, j \le k$, should cluster near the optimum $g_{\min}$. Thus the limiting distribution of m_n (if n is large) is expected to be of Weibull type $(\gamma < 0)$, under certain conditions. Moreover, the greater the clustering effect, the smaller the absolute value of the shape parameter $\alpha = 1/\gamma$ and the better the solutions.

Define the domain $A_u = \{\mathbf{x} \in \mathbb{R}^d : 0 \le g(\mathbf{x}) - g_{\min} \le u\}$ for u small. The domain A_u can be bounded or unbounded depending on the function g. If the domain is bounded, we consider its d-dimensional volume $|A_u|$. If A_u is concentrated in a lower dimensional subspace $\mathbb{R}^r$, then $|A_u| = 0$ and one has to restrict the following derivations on this subspace $\mathbb{R}^r$. Hence, let us assume that $|A_u| > 0$. We consider the limiting behavior of the distribution of the outcomes m_n, as $n \to \infty$.

We get for the distribution of m_n, by the independence of the $\mathbf{X}_j$,

$$P(m_n > g_{\min} + u) = P\left(\min_{j \le n} g(\mathbf{X}_j) > g_{\min} + u\right)$$
$$= P\left(g(\mathbf{X}_j) > g_{\min} + u, \; j \le n\right)$$
$$= (1 - P(g(\mathbf{X}_1) \le g_{\min} + u))^n$$
$$= (1 - P(\mathbf{X}_1 \in A_u))^n = (1 - p(u))^n$$

where $p(u) = P(\mathbf{X}_1 \in A_u)$. We note that the asymptotic behavior of the minimum $\min_j g(\mathbf{X}_j)$ depends on the domain A_u or more precisely, on the probability $p(u)$ that $\mathbf{X}_j$ hits A_u. This probability $p(u)$ tends usually to 0 as $u \to 0$. If we can find a normalization sequence u_n such that $n\,p(u_n) \to \tau$, then we get immediately that $(1 - p(u_n))^n \to \exp(-\tau)$ as $n \to \infty$.

Theorem 11.4.6. *Assume that g has a global minimum value $g_{\min}$. Assume that A_u and the iid rv $\mathbf{X}_j$ $(j \ge 1)$ are such that $p(u) = P(\mathbf{X}_j \in A_u) \to 0$ as $u \to 0$. If there exists a normalization $u = u_n(x) \to 0$ (as $n \to \infty$) such that*

$$np(u_n(x)) \to h(x), \quad \text{for some } x \in \mathbb{R},$$

then as $n \to \infty$,

$$P\left(\min_{j \le n} g(\mathbf{X}_j) \le g_{\min} + u_n(x)\right) \to 1 - \exp(-h(x)).$$

If the function g has some isolated global minimum points $\mathbf{x}_l, l \le L$, we can derive a more explicit statement. Assume that the set A_u can be split into disjoint sets $A_u(\mathbf{x}_l) = \{\mathbf{x} : g(\mathbf{x}) - g_{\min} \le u \text{ and } |\mathbf{x} - \mathbf{x}_l| < \epsilon\}$ for some $\epsilon > 0$ and all sufficiently small u. The choice of ϵ has no impact on the result. It is only necessary that the sets $A_u(\mathbf{x}_l)$ are disjoint for all u sufficiently small. Such cases are discussed in the examples and simulations of Hüsler et al. [255] and Hüsler [254].

Theorem 11.4.7. *Assume that g has a countable number of isolated global minimum points $\mathbf{x}_l, 1 \le l \le L \le \infty$. Assume that each of the disjoint sets $A_u(\mathbf{x}_l)$ is bounded and concentrated in $\mathbb{R}^r$ with $r \le d$, for all small u and every $l \le L$, and that the iid rv $\mathbf{X}_j$ have a positive, uniformly continuous (marginal) density f_r at* *the global minimum points* $\mathbf{x}_l, l \leq L$, *where the marginal density* f_r *corresponds to the space of* A_u. *If* u_n *is such that for* $l \leq L$ *uniformly,*

$$n|A_{u_n}(\mathbf{x}_l)|_r \to \tau_l < \infty \qquad with \qquad \sum_{1 \leq l \leq L} f_r(\mathbf{x}_l)\tau_l < \infty,$$

then as $n \to \infty$,

$$P\left(\min_{j \leq n} g(\mathbf{X}_j) \leq g_{\min} + u_n\right) \to 1 - \exp\left(-\sum_{1 \leq l \leq L} f_r(\mathbf{x}_l)\tau_l\right).$$

If the density $f_r(x_l) = 0$, for some l, then one has to replace such a term in the sum by $\lim_n n \int_{x \in A_{u_n}(x_l)} f_r(x)\,dx$.

The examples in [255] consider the objective functions

(i) $g(x,y) = ax^2 + by^2$ for some $a, b > 0$ with minimum at the origin $\mathbf{0} = (0,0)$ and bounded domains $A_u = \{(x,y) : ax^2 + by^2 \leq u\}$;

(ii) the two-dimensional *sinc*-function $g(x,y) = \sin(x^2+y^2)/(x^2+y^2)$ with minimum $g_{\min} = \cos(r_0^2)$ at points (x,y) with $x^2 + y^2 = r_0^2$, where r_0 is the smallest positive solution of $r_0^2 \cos(r_0^2) = \sin(r_0^2)$, i.e., $r_0 = 2.1198$, and the domain A_u is a ring with center $\mathbf{0}$ and radii $r_0 \pm \sqrt{2u/c}$ for some constant $c = \tilde{g}''(r_0) + o(1)$, where $\tilde{g}(r) = \sin(r^2)/r^2$.

If the function g is rather regular, i.e., twice continuously differentiable at a unique global minimum point, another general result can be derived (see [255]).

11.5 Triangular Arrays of Rare Events

In the first section of this chapter we discussed the extension of the theory for extreme values to a theory for rare events. We restricted ourselves to the case where $\lambda(t)$ or $\mu(t)$ are continuous functions in t. This is a convenient assumption simplifying the proofs, since the asymptotic point process is still a simple Poisson process. If the limiting Poisson process is not simple, which is the case if for instance $\lambda(t)$ has some discontinuity points, the method for proving the analogous limiting result is more complicated. Basically we have to investigate the behavior of the Laplace transform of the point processes.

We can easily extend the theory further to more general rare events E_{ni} which are not necessarily of the form $E_{ni} = \{X_i \in A_{ni}\}$ as in Section 11.1.

Arrays of Rare Events

Let $\{E_{ni}, \ i \leq n, \ n \geq 1\}$ be any triangular array of events E_{ni} such that

$$p_{n,max} = \sup_{i \leq n} P(E_{ni}) \to 0 \qquad as \quad n \to \infty \qquad (11.22)$$

and

$$\lim_{n\to\infty} \sum_{i\leq n} P(E_{ni}) < \infty.$$

(11.23)

The point process N_n of the rare events is defined by

$$N_n(\cdot) = \sum_{i\leq n} \delta_{i/n}(\cdot) 1(E_{ni}).$$

We will apply the results of Section 11.2 in order to prove the convergence of N_n and related point processes. We also extend the method and some results of Section 11.1 and begin by reformulating the assumptions given there.

As mixing condition we are assuming either the condition $\Delta(E_{ni})$ or the weaker condition $D(E_{ni})$ which are generalizations of $\Delta(A_{ni})$ and $D(A_{ni})$, respectively. In both cases we denote the mixing coefficients by $\alpha_{n,m}$.

Definition 11.5.1. For every n and $m < n$, let $\alpha_{n,m}$ be such that

$$\left| P(E^{\complement}_{nj_l}, \; l \leq p, \; E^{\complement}_{nj'_h}, \; h \leq q) - P(E^{\complement}_{nj_l}, \; l \leq p) \times P(E^{\complement}_{nj'_h}, \; h \leq q) \right| \leq \alpha_{n,m}$$

for any choice of subsets $\{j_l, \; l \leq q\}$ and $\{j'_h, \; h \leq p\}$ of $\{1,\ldots,n\}$ which are separated by at least m. We say that condition $D(E_{ni})$ holds if

$$\alpha_{n,m_n} \to 0 \qquad \text{as} \quad n \to \infty$$

for some sequence $\{m_n\}$ satisfying $m_n p_{n,max} \to 0$ as $n \to \infty$.

Definition 11.5.2. For every n and $m < n$ let $\alpha_{n,m}$ be such that

$$\alpha_{n,m} \geq \sup \Big\{ |P(A \cap B) - P(A)P(B)| :$$
$$A \in \mathcal{B}_n[0,k], \; B \in \mathcal{B}_n[k+m,n], \; 0 < k \leq n-m \Big\},$$

where $\mathcal{B}_n[k,l] = \sigma\{E_{ni}, \; k \leq i \leq l\}$. We say that condition $\Delta(E_{ni})$ holds if

$$\alpha_{n,m_n} \to 0 \qquad \text{as} \quad n \to \infty.$$

for some sequence $\{m_n\}$ satisfying $m_n p_{n,max} \to 0$ as $n \to \infty$.

In both cases, let k_n again be a sequence such that as $n \to \infty$

$$k_n(\alpha_{n,m_n} + m_n p_{n,max}) \to 0.$$

(11.24)

Later we also will redefine $D^*(E_{ni})$ with respect to the rare events E_{ni}. We first reformulate $D'(E_{ni})$, the most restrictive local dependence condition. Let $C < \infty$ be some positive constant. For every n assume that there exists an interval partition $\{B_l(n), \; l \leq k_n\}$ of $(0,1]$ such that, for all $l \leq k_n$,

$$\sum_{i \in nB_l(n)} P(E_{ni}) \leq C/k_n \tag{11.25}$$

and let

$$\alpha'_n = \sup_{l \leq k_n} \sum_{i < j \in nB_l(n)} P(E_{ni} \cap E_{nj}).$$

Definition 11.5.3. We say that condition $D'(E_{ni})$ holds if a sequence of interval partitions $\{B_l(n), l \leq k_n\}$ of $(0,1]$ exists such that (11.25) holds and

$$k_n \alpha'_n \to 0 \qquad as \quad n \to \infty.$$

Denote now by $\lambda_n(\cdot)$ the mean number of occurrences of rare events:

$$\lambda_n(t) = \sum_{i \leq nt} P(E_{ni})$$

for $t \leq 1$. Let $\lambda(t) = \limsup_{n \to \infty} \lambda_n(t)$ for $t \leq 1$. Note that $\lambda_n(\cdot)$ and $\lambda(\cdot)$ are non-negative monotonically increasing functions. $\lambda_n(\cdot)$ and $\lambda(\cdot)$ define measures on $\mathcal{B}((0,1])$. We use again the same notation for functions and measures. Similar to Lemma 9.1.6 we have

Proposition 11.5.4. *Assume that the conditions $D(E_{ni})$ and $D'(E_{ni})$ hold for a triangular array of rare events. For any $t \leq 1$,*

$$P\left(\bigcap_{i \leq nt} E_{ni}^{\mathsf{C}} \right) \to \exp(-\lambda(t)) > 0$$

iff

$$\lambda_n(t) \to \lambda(t) < \infty. \tag{11.26}$$

POINT PROCESSES OF RARE EVENTS

As in previous sections the following can be stated on the point process N_n.

Theorem 11.5.5. *Assume that (11.22) and (11.23) hold for the triangular array of rare events $\{E_{ni}, \; i \leq n, \; n \geq 1\}$. If the conditions $\Delta(E_{ni})$ and $D'(E_{ni})$ hold, then $\lambda_n \to_w \lambda$ on $(0,1]$ is equivalent to*

$$N_n \to_D N$$

where N is a Poisson process with intensity measure $\lambda(\cdot)$.

The proof consists of verifying the assumptions of Corollary 11.2.5. Condition $D'(E_{ni})$ implies that $\pi_x(\cdot) = \delta_1(\cdot)$ for all $x \in (0,1]$. For details see Hüsler and Schmidt [265].

Weaker Restrictions on Local Dependence

As already mentioned in Section 9.1, Chernick et al. [67] introduced for the stationary case with a constant boundary u_n the weaker local conditions $D^{(k)}(u_n)$, $k \geq 1$, assuming

$$nP\Big(X_1 > u_n,\, X_2 \leq u_n, \ldots, X_k \leq u_n,\, \max_{k+1 \leq j \leq r_n^*} X_j > u_n\Big) \to 0$$

as $n \to \infty$ with $r_n^* = [n/k_n]$, where k_n satisfies (11.24). For applications the following simpler, but slightly more restrictive condition is reasonable:

$$n \sum_{k < j \leq r_n^*} P(X_1 > u_n,\, X_2 \leq u_n, \ldots, X_k \leq u_n,\, X_j > u_n) \to 0$$

as $n \to \infty$, or similarly

$$n \sum_{k < j \leq r_n^*} P(X_1 > u_n,\, X_{j-k+1} \leq u_n, \ldots, X_{j-1} \leq u_n,\, X_j > u_n) \to 0. \qquad (11.27)$$

In the case $k = 1$, (11.27) is equivalent to the condition $D'(u_n)$ (see Section 9.1) since the events simplify to $\{X_1 > u_n,\, X_j > u_n\}$. If $k = 2$, (11.27) is equivalent to the condition $D^*(u_n)$. As mentioned in Section 9.1, $D^{(k)}(u_n)$ implies $D^{(k+1)}(u_n)$ for $k \geq 1$, and similarly if (11.27) holds for some $k \geq 1$, then it also holds for $k+1$.

Such weaker conditions can be formulated for triangular arrays of the rare events. For every n, assume that there exists a sequence of interval partitions $\{B_l(n),\, l \leq k_n\},\, n \geq 1$, of $(0,1]$ such that (11.25) holds with E_{ni} replaced by $(\cap_{1 \leq l < k} E_{n,i-l}^{\complement}) \cap E_{ni}$. Condition $D^{(k)}(E_{ni})$ is said to hold if

$$k_n \sup_{l \leq k_n} \sum_{i \in nB_l(n)} P\Big(E_{ni} \cap \bigcup_{i < j,\, j \in nB_l(n)} \Big(\bigcap_{1 \leq h < k} E_{n,j-h}^{\complement} \cap E_{nj} \Big)\Big) \to 0$$

as $n \to \infty$. The following slightly stronger condition (11.28) generalizes (11.27) for the case of rare events and is easier to verify than $D^{(k)}(E_{ni})$. $D^{(k)}(E_{ni})$ is implied by the condition

$$k_n \alpha_n^{(k)} \to 0, \qquad (11.28)$$

where

$$\alpha_n^{(k)} = \sup_{l \leq k_n} \sum_{i < j \in nB_l(n)} P\Big(E_{ni} \cap \Big(\bigcap_{1 \leq h < k} E_{n,j-h}^{\complement} \Big) \cap E_{nj}\Big) \qquad (11.29)$$

with $B_l(n)$ as above. For $k = 1$, (11.28) is equivalent to $D'(E_{ni})$ since $\alpha_n' = \alpha_n^{(1)}$. Note also that for $k = 2$, the condition (11.28) redefines condition $D^* = D^*(E_{ni})$ for general rare events.

Another possibility is to use the following simple relationships. For a given triangular array E_{ni} and for any fixed $k \geq 1$ define the rare events

$$U_{ni}^{(k)} = \bigcap_{1 \leq h < k} E_{n,i-h}^{\mathsf{C}} \cap E_{ni}$$

with $1 \leq i \leq n$, letting $E_{ni} = \emptyset$ for $i < 1$ and $n \geq 1$, indicating the occurrence of a new rare event after at least $k - 1$ non-occurrences. Let us call such an event a *k-step rare event*. Note that $U_{ni}^{(1)} \equiv E_{ni}$.

If the triangular array E_{ni} satisfies (11.22), then (11.22) holds for the triangular array $U_{ni}^{(k)}$, with $k \geq 1$. Note that (11.28) is slightly more restrictive than $D'(U_{ni}^{(k)})$. The situation is similar for the long range mixing condition: $\Delta(E_{ni})$ implies $\Delta(U_{ni}^{(k)})$ for any fixed $k \geq 1$. Although it is not so easy to compare the conditions $D(E_{ni})$ and $D(U_{ni}^{(k)})$, the statement of Lemma 9.1.4 in terms of rare events $U_{ni}^{(k)}$ follows by assuming either $D(E_{ni})$ or $D(U_{ni}^{(k)})$, i.e.,

$$P\left(\bigcap_{j \leq k_n} \bigcap_{i \in B_j} [U_{ni}^{(k)}]^{\mathsf{C}} \right) - \prod_{j \leq k_n} P\left(\bigcap_{i \in B_j} [U_{ni}^{(k)}]^{\mathsf{C}} \right) \to 0.$$

In general, it is easier to verify $D(E_{ni})$ than $D(U_{ni}^{(k)})$.

λ_n and λ in terms of $U_{ni}^{(k)}$ are denoted by $\mu_n^{(k)}$ and $\mu^{(k)}$, respectively. For $t \in (0,1]$, let $\mu_n^{(k)}(t) = \sum_{i \leq nt} P(U_{ni}^{(k)})$ and $\mu^{(k)}(t) = \limsup_{n \to \infty} \mu_n^{(k)}(t)$. We have $\mu^{(1)}(\cdot) = \lambda(\cdot)$.

Reformulating N_n in terms of $U_{ni}^{(k)}$ and denoting this point process by $N_n^{(k)}$, we have the following implications of Theorem 11.5.5.

Corollary 11.5.6. *Assume that (11.22) and (11.23) hold for the triangular array of rare events $\{U_{ni}^{(k)}, \ i \leq n, \ n \geq 1\}$. If the conditions $\Delta(U_{ni}^{(k)})$ and $D'(U_{ni}^{(k)})$ are satisfied, for some $k \geq 1$, then $\mu_n^{(k)} \to_w \mu^{(k)}$ (as $n \to \infty$) on $(0,1]$ is equivalent to*

$$N_n^{(k)} \to_D N^{(k)},$$

where $N^{(k)}$ is a Poisson process with intensity measure $\mu^{(k)}(\cdot)$.

Proposition 11.5.4 can also be reformulated in terms of k-step rare events. We note that with the above definitions

$$\bigcap_{i \leq nt} E_{ni}^{\mathsf{C}} = \bigcap_{i \leq nt} (U_{ni}^{(k)})^{\mathsf{C}}$$

for any $k \geq 1$ and $t \in (0,1]$. Thus we get immediately

Corollary 11.5.7. *Assume that condition $D(E_{ni})$ and $D'(U_{ni}^{(k)})$ hold for a triangular array of rare events, for some $k \geq 1$. For any $t \leq 1$,*

$$P\left(\bigcap_{i \leq nt} E_{ni}^{\mathsf{C}} \right) \to \exp(-\mu^{(k)}(t)) > 0$$

iff

$$\mu_n^{(k)}(t) \to \mu^{(k)}(t) < \infty \qquad as \quad n \to \infty. \tag{11.30}$$

Condition $D'(U_{ni}^{(k)})$ implies $D'(U_{ni}^{(k+1)})$ for any $k \geq 1$. Hence from Corollary 11.5.7, if $\mu_n^{(k)}(t) \to \mu^{(k)}(t)$, then $\mu_n^{(k+1)}(t) \to \mu^{(k+1)}(t)$ and $\mu^{(k)}(t) = \mu^{k+1}(t)$. The point process $N_n^{(k)}$ of the k-step rare events is obtained by deleting the points of N_n of rare events E_{ni} which are not k-step rare events. This special thinning is such that the limit of $N_n^{(k)}$ is a non-homogeneous Poisson process instead of the non-homogeneous compound Poisson process N. If the conditions of Corollary 11.5.7 and (11.30) hold for some $k_0 \geq 1$, then such further thinning with $k \geq k_0$ has asymptotically no effect. We find that $N^{(k)}$ is the Poisson process underlying the compound Poisson process N or $N^{(j)}$ for $1 \leq j < k$. Note that the point process $N_n^{(2)}$ of 2-step rare events corresponds to the process $\tilde{N}_n$ of Section 11.1 for the special rare events $\{X_i \in A_{ni}\}$, as well as $\mu^{(2)}$ to μ.

Corollary 11.5.8. *Assume that (11.22) and (11.23) hold for the triangular array of rare events $\{E_{ni}, \; i \leq n, \; n \geq 1\}$ and that the conditions $\Delta(E_{ni})$ and $D'(U_{ni}^{(2)})$ are satisfied. If $\mu_n^{(2)} \to_w \mu^{(2)}$ $(n \to \infty)$, then the following statements are equivalent:*

(i) $\lambda(1) = \mu^{(2)}(1)$, *implying* $\lambda(\cdot) \equiv \mu^{(2)}(\cdot)$,

(ii) $E\left(\|N_n - N_n^{(2)}\| \right) \to 0$,

(iii) $N_n \to_D N^{(2)}$,

where $N^{(2)}$ is a Poisson process with intensity measure $\mu^{(2)}(\cdot)$.

This corollary compares the point processes $N_n^{(1)} = N_n$ and $N_n^{(2)}$. A similar result holds for the comparison of $N_n^{(k)}$ and $N_n^{(l)}$ for any $1 \leq k < l$ together with $\mu^{(k)}$ and $\mu^{(l)}$ instead of $\lambda(\cdot)$ and $\mu^{(2)}(\cdot)$.

Finally, we now consider the point process N_n^* of clusters of rare events which is defined by

$$N_n^*(\cdot) = \sum_{l \leq k_n} \delta_{t_l}(\cdot) 1(\cup_{i \in nB_l(n)} E_{ni})$$

with t_l the upper endpoint of $B_l(n)$.

Corollary 11.5.9. *Assume that* (11.22) *and* (11.23) *hold for the triangular array of rare events* $\{E_{ni}, \ i \leq n, \ n \geq 1\}$. *If the condition* $D'(U_{ni}^{(2)})$ *is satisfied, then*

$$\lim_{n \to \infty} E\left(\|N_n^* - N_n^{(2)}\|\right) = 0.$$

Moreover, if $\mu^{(2)} \to_w \mu^{(2)}$ *and* $\Delta(E_{ni})$ *holds, then*

$$N_n^* \to_D N^{(2)},$$

where $N^{(2)}$ *is a Poisson process with intensity measure* $\mu^{(2)}(\cdot)$.

Instead of $U_{ni}^{(k)}$ we might as well define the rare events

$$V_{ni}^{(k)} = E_{ni} \cap \bigcap_{1 \leq l < k} E_{n,i+l}^{\complement}$$

for $1 \leq i \leq n$ using $E_{ni} = \emptyset$ for $i > n$ and $n \geq 1$. These rare events correspond to the downcrossings in the case of exceedances.

Applications

(i) The exceedances of a multivariate random sequence are defined by

$$E_{ni} = \{\boldsymbol{X}_i \not\leq \boldsymbol{u}_n(\boldsymbol{x})\},$$

where $\{\boldsymbol{X}_i, \ i \geq 1\}$ denotes the multivariate random sequence, and $\boldsymbol{u}_n(\boldsymbol{x})$ the vector of boundaries. By applying the above results to this special triangular array of rare events we get the limit results for multivariate extremes. Details are discussed in the next section.

(ii) Davis [91], [92] derived limit results for the joint distribution of minima and maxima of a stationary random sequence. The above statements extend his result under the condition $D'(U_{ni}^{(k)})$ with respect to the rare events

$$E_{ni} = \{X_i > a_n x + b_n\} \cup \{X_i < c_n x + d_n\}.$$

and setting $k \geq 1$. Moreover, a similar analysis can be performed under these weaker dependence conditions on non-stationary random sequences by letting

$$E_{ni} = \{X_i > u_{ni}\} \cup \{X_i < v_{ni}\},$$

where $\{u_{ni}, \ i \leq n, \ n \geq 1\}$ and $\{v_{ni}, \ i \leq n, \ n \geq 1\}$ are two general boundaries.

(iii) For applications in the stationary case to moving average sequences see Chernick et al. [67].

Triangular Array of Rare Events Based on Gaussian Sequences

Let us consider a triangular array of standardized Gaussian sequences $X_{n,i}, i \leq n, n \geq 1$, being stationary for fixed n. Assume that the correlation function $r_{n,j} = E(X_{n,i}X_{n,i+j})$ depends on n also, not only on the lag j. If Berman's condition holds and if $r_{n,j}$ do not tend to 1 for fixed j, then the row-wise maxima $M_n = \max\{X_{n,i}, i \leq n\}$ behaves as the maxima of iid normal rv. Hence the normalized M_n has a Gumbel limit distribution G_3, discussed in Example 9.2.2. But if now $r_{n,j}$ tends to 1 for a fixed j, the speed of this convergence is important. Crucial is the condition

$$(1 - r_{n,j})\log(n) \to \delta_j \in (0, \infty]$$

for all $j \geq 1$. Obviously, $\delta_0 = 0$. If $\delta_j < \infty$ for some j, then the dependence is so strong that exceedances may cluster. Hence the maxima M_n shows a different behavior as if all $\delta_j = \infty, (j \geq 1)$.

Hsing et al [232] showed that the limit distribution of the normalized M_n is still the Gumbel distribution G_3, but the clustering is occurring with a cluster index θ. They showed under the long range mixing condition (a condition related to $D(E_{ni})$ with $E_{ni} = \{X_{n,i} > u_{ni}\}$) which is slightly more general than Berman's condition, that

$$P(M_n \leq a_n x + b_n) \to \exp(-\theta e^{-x}) = G_3(x - \log(\theta))$$

as $n \to \infty$, where

$$\theta = P(\eta/2 + \sqrt{\delta_k}W_k \leq \delta_k \text{ for all } k \geq 1 \text{ such that } \delta_k < \infty),$$

($\theta = 1$ if $\delta = \infty$ for all $k \geq 1$) with η denoting a standard exponential rv, independent of the W_k, and $\{W_k : \delta_k < \infty, k \geq 1\}$ being jointly normal with means 0 and

$$E(W_i W_j) = \frac{\delta_i + \delta_j - \delta_{|i-j|}}{2(\delta_i \delta_j)^{1/2}} \ .$$

The cluster index θ is related to Pickands constants H_α observed in the investigations of maxima of continuous Gaussian processes, dealt with in Chapter 10. Note that the approximation of $X(t), 0 \leq t \leq T$, is based on the $X(iq)$'s with grid mesh q such that $qu^{2/\alpha} = \gamma \to 0$. Fix for a moment $\gamma > 0$. Assuming condition (10.1) for the local behavior of the correlation function $r(t)$ of $X(t)$, i.e., $1 - r(t) \sim C|t|^\alpha$, for $t \to 0$ with $0 < \alpha \leq 2$ and $C > 0$, we get $(1 - r_{n,j})\log(n) = (1 - r(jq))\log(n) \sim C|jq|^\alpha \log(n) = C(\gamma j)^\alpha/2 = \delta_j$ using $n = [T/q]$ and $\log(n) \sim \log(T)$. This shows that the cluster index $\theta = \theta(\gamma) < 1$. If now $\gamma \to 0$ we expect that θ is related to Pickands constants H_α. This relation is discovered by Hüsler [251]. He shows that $H_\alpha = C^{-1/\alpha} \lim_{\gamma \to 0} \theta(\gamma)/\gamma$. This gives a new definition of H_α, mentioned in Chapter 10.2.

However, H_α is independent of C which can be shown by transforming the rv. Considering the limit w.r.t. $C^{-1/\alpha}\gamma$ instead of γ, we get

$$
\begin{aligned}
H_\alpha &= \lim_{\gamma \to 0} C^{-1/\alpha} \theta(\gamma)/\gamma = \lim_{\gamma \to 0} C^{-1/\alpha} \theta(C^{-1/\alpha}\gamma)/(C^{-1/\alpha}\gamma) \\
&= \lim_{\gamma \to 0} P(\eta + \max_{k \geq 1} V_k \leq 0)/\gamma \\
&= \lim_{\gamma \to 0} \int_0^\infty e^{-x} P(\max_{k \geq 1} V_k \leq -x)\, dx/\gamma\,.
\end{aligned}
$$

where V_k are normal variables with

$$
E(V_k) = -(\gamma k)^\alpha, \quad \mathrm{Cov}(V_k, V_j) = \gamma^\alpha (k^\alpha + j^\alpha - |k - j|^\alpha)
$$

(correcting a misprint in the paper). This shows that H_α can be interpreted indeed as a cluster index in the continuous case of Gaussian processes.

11.6 Multivariate Extremes of Non-IID Sequences

The theory of multivariate extremes of iid rv $\boldsymbol{X}_i \in \mathbb{R}^d$, $i \geq 1$, can be extended to cover extremes of non-iid rv applying similar techniques as in the univariate case. Again, these extensions lead to the study of the extremes of stationary and non-stationary sequences. To study the dependent sequences we will introduce the multivariate versions of the conditions on the local and long range dependence. An important additional aspect in the multivariate analysis is the dependence structure among the components of the rv. It is known that the multivariate EVD are positively dependent. The dependence structure may thus range from independence of the components to their complete dependence.

In the following we discuss briefly the limit results for the three classes of non-iid random sequences and the dependence structure of the limiting distributions.

We consider the partial maxima $\boldsymbol{M}_n = (M_{n1}, \ldots, M_{nd})$ up to 'time' n, defined componentwise as in Chapter 4 by

$$
M_{nj} = \max_{i \leq n} X_{ij},
$$

$j \leq d$, and its convergence in distribution. Let $\boldsymbol{a}_n = (a_{n1}, \ldots, a_{nd})$ and $\boldsymbol{b}_n = (b_{n1}, \ldots, b_{nd})$ be normalizing sequences such that

$$
P(\boldsymbol{M}_n \leq \boldsymbol{a}_n \boldsymbol{x} + \boldsymbol{b}_n) \to_w G(\boldsymbol{x}) \qquad \text{as} \quad n \to \infty.
$$

We denote the j-th marginal distribution of $G(\cdot)$ by $G_j(\cdot)$, $j \leq d$. (In general, for $j = 3$, the marginal distribution $G_3(\cdot)$ does not denote here the Gumbel distribution.

For fixed $\boldsymbol{x}$ let us define the rare events

$$
E_{ni} = \{\boldsymbol{X}_i \not\leq \boldsymbol{a}_n \boldsymbol{x} + \boldsymbol{b}_n\}
$$

for $i \leq n$, $n \geq 1$, and $U_{ni}^{(k)}$ as in the previous section to apply the results derived for rare events. In Section 11.5 we stated the convergence results for the normalized maxima and the point processes N_n, $\tilde{N}_n$, N_n^* or $N_n^{(k)}$ related to the exceedances, assuming certain conditions. These conditions can be restated in terms of multivariate extremes. For instance, (11.22) and (11.23) can be restated in terms of $U_{ni}^{(k)}$:

$$p_{n,max} = \sup_{i \leq n} P(\boldsymbol{X}_{i-l} \leq \boldsymbol{a}_n \boldsymbol{x} + \boldsymbol{b}_n, 1 \leq l < \min(i,k), \boldsymbol{X}_i \not\leq \boldsymbol{a}_n \boldsymbol{x} + \boldsymbol{b}_n) \to 0 \quad (11.31)$$

as $n \to \infty$, and

$$\lim_{n \to \infty} \sum_{i \leq n} P(\boldsymbol{X}_{i-l} \leq \boldsymbol{a}_n \boldsymbol{x} + \boldsymbol{b}_n, 1 \leq l < \min(i,k), \boldsymbol{X}_i \not\leq \boldsymbol{a}_n \boldsymbol{x} + \boldsymbol{b}_n) < \infty. \quad (11.32)$$

If $k = 1$ the events in (11.32) simplify to $\{\boldsymbol{X}_i \not\leq \boldsymbol{a}_n \boldsymbol{x} + \boldsymbol{b}_n\} = E_{ni}$.

The mixing conditions $\Delta(E_{ni})$, $D(E_{ni})$ and $D'(U_{ni}^{(k)})$, $k \geq 1$ fixed, can be reformulated similarly.

Theorem 11.5.5 and Corollary 11.5.6 imply the convergence of the point processes N_n and $N_n^{(k)}$, now defined in terms of k-step upcrossings $U_{ni}^{(k)}$, respectively. Again, $N_n^{(k)}$ is a specially thinned version of $N_n = N_n^{(1)}$, the point process counting all exceedances. $\mu_n^{(k)}$ and $\mu^{(k)}$ depend also on $\boldsymbol{x}$:

$$\mu_n^{(k)}(t, \boldsymbol{x}) = \sum_{1 \leq i \leq nt} P(\boldsymbol{X}_{i-l} \leq \boldsymbol{a}_n \boldsymbol{x} + \boldsymbol{b}_n, 1 \leq l < \min(i,k), \boldsymbol{X}_i \not\leq \boldsymbol{a}_n \boldsymbol{x} + \boldsymbol{b}_n).$$

Again, $\mu^{(1)}(\cdot, \boldsymbol{x})$ is the limit of the mean number of exceedances $\{\boldsymbol{X}_i \not\leq \boldsymbol{a}_n \boldsymbol{x} + \boldsymbol{b}_n\}$.

Theorem 11.6.1. *Let $\{\boldsymbol{X}_i, i \geq 1\}$ be a sequence of rv. Let $\boldsymbol{x} \in \mathbb{R}^d$ and $\boldsymbol{a}_n$, $\boldsymbol{b}_n$ be normalizing sequences such that (11.31), (11.32), $\Delta(E_{ni})$ and $D'(U_{ni}^{(k)})$ hold for some $k \geq 1$. Then*

$$\mu_n^{(k)}(\cdot, \boldsymbol{x}) \to_w \mu^{(k)}(\cdot, \boldsymbol{x}) \qquad on \quad (0,1] \qquad\qquad (11.33)$$

is equivalent to

$$N_n^{(k)} \to_D N^{(k)},$$

where $N^{(k)}$ is a Poisson process on $(0,1]$ with intensity measure $\mu^{(k)}(\cdot)$.

The Poisson process $N^{(k)}$ is in general non-homogeneous having fixed atoms at the possible discontinuity points of $\mu^{(k)}(t, \boldsymbol{x})$ with $\boldsymbol{x}$ fixed.

Assuming that the stated conditions hold for all $\boldsymbol{x}$, Theorem 11.6.1 immediately implies the convergence in distribution of $\boldsymbol{M}_n$, since by definition of $U_{ni}^{(k)}$, $P(\boldsymbol{M}_n \leq \boldsymbol{a}_n \boldsymbol{x} + \boldsymbol{b}_n) = P(N_n^{(k)}((0,1]) = 0)$. Moreover, if we only consider events $\{N_n^{(k)}((0,1]) = 0\}$, we can replace $\Delta(E_{ni})$ by $D(E_{ni})$ or $D(U_{ni}^{(k)})$ by Corollary 11.5.7.

However, the stronger conditions in the theorem are needed for the convergence of the point process $N_n^{(k)}$ which is a more informative result than just the convergence of the normalized maxima. A couple of results can be easily derived from this general convergence statement as mentioned also in earlier chapters on univariate maxima, as e.g. Chapter 9.2. For the convergence of $\boldsymbol{M}_n$ it is sufficient to assume (11.33) for $t = 1$ only:

$$\mu_n^{(k)}(1, \boldsymbol{x}) \to \mu^{(k)}(1, \boldsymbol{x}). \tag{11.34}$$

Theorem 11.6.2. *Let* $\{\boldsymbol{X}_i, \ i \geq 1\}$ *be a sequence of rv. If for some normalizing sequences* $\boldsymbol{a}_n$ *and* $\boldsymbol{b}_n$, (11.31), $D(E_{ni})$ *and* $D'(U_{ni}^{(k)})$, *for some* $k \geq 1$, *hold for all* $\boldsymbol{x}$ *where* $\mu^{(k)}(1, \boldsymbol{x}) < \infty$, *then* $P(\boldsymbol{M}_n \leq \boldsymbol{a}_n \boldsymbol{x} + \boldsymbol{b}_n) \to_w G(\boldsymbol{x}) = \exp(-\mu^{(k)}(1, \boldsymbol{x}))$ *iff*

$$\mu_n^{(k)}(1, \boldsymbol{x}) \to_v \mu^{(k)}(1, \boldsymbol{x}) \qquad on \ \mathbb{R}^d. \tag{11.35}$$

STATIONARY MULTIVARIATE SEQUENCES

Assume that the sequence of rv $\{\boldsymbol{X}_i, \ i \geq 1\}$ is stationary. The assumptions (11.31) and (11.32) are both implied by

$$nP(\boldsymbol{X}_l \leq \boldsymbol{a}_n \boldsymbol{x} + \boldsymbol{b}_n, 1 \leq l < k, \boldsymbol{X}_k \not\leq \boldsymbol{a}_n \boldsymbol{x} + \boldsymbol{b}_n) \to \mu^{(k)}(1, \boldsymbol{x}) < \infty \tag{11.36}$$

for some $k \geq 1$, with $P(\boldsymbol{X}_1 \not\leq \boldsymbol{a}_n \boldsymbol{x} + \boldsymbol{b}_n) \to 0$. This replaces condition (11.34) in the stationary case. We have $\mu^{(k)}(t, \boldsymbol{x}) = t\mu^{(k)}(1, \boldsymbol{x})$, being continuous in t. The term α_n' can be modified:

$$\alpha_n' = \sum_{k+1 \leq j \leq r_n^*} P(\boldsymbol{X}_1 \not\leq \boldsymbol{a}_n \boldsymbol{x} + \boldsymbol{b}_n, \boldsymbol{X}_{j-l} \leq \boldsymbol{a}_n \boldsymbol{x} + \boldsymbol{b}_n, 1 \leq l < k, \boldsymbol{X}_j \not\leq \boldsymbol{a}_n \boldsymbol{x} + \boldsymbol{b}_n),$$

where $r_n^* = [n/k_n]$ with k_n such that (11.24) holds (i.e., $k_n(\alpha_{n,m_n} + m_n p_{n,max}) \to 0$ as $n \to \infty$). Under these conditions, the convergence in distribution for $\boldsymbol{M}_n$ follows by Theorem 11.6.2.

If $k = 1$, the limit distribution $G(\cdot)$ is a multivariate EVD and hence max-stable (cf. Hsing [226] and Hüsler [245]). This implies that the components of the rv $\boldsymbol{Y}$ with distribution G are associated (cf. Resnick [393]).

The same holds also for $k > 1$. The proof follows by Theorem 11.6.2 and similar arguments as in Hüsler [245].

Corollary 11.6.3. *Let* $\{\boldsymbol{X}_i, \ i \geq 1\}$ *be a stationary sequence of rv. If for some normalizing sequences* $\boldsymbol{a}_n$ *and* $\boldsymbol{b}_n$, (11.31), $D(E_{ni})$ *and* $D'(U_{ni}^{(k)})$, *for some* $k \geq 1$, *hold for all* $\boldsymbol{x}$ *where* $\mu^{(k)}(1, \boldsymbol{x}) < \infty$, *then*

$$P(\boldsymbol{M}_n \leq \boldsymbol{a}_n \boldsymbol{x} + \boldsymbol{b}_n) \to_w G(\boldsymbol{x}) = \exp\left(-\mu^{(k)}(1, \boldsymbol{x})\right)$$

iff the convergence in (11.36) holds vaguely on $\mathbb{R}^d$. *Moreover,* $G(\cdot)$ *is max-stable.*

Let $k = 1$. If, in addition to the conditions of Corollary 11.6.3, for all $1 \leq j < j' \leq d$,

$$nP(X_{1j} > a_{nj}x_j + b_{nj}, X_{1j'} > a_{nj'}x_{j'} + b_{nj'}) \to 0 \qquad \text{as} \quad n \to \infty \qquad (11.37)$$

for some x with $G(x) \in (0,1)$, then the components of Y are independent: i.e., $G(x) = \prod_{j \leq d} G_j(x_j)$ (cf. Galambos [167], Takahashi [438], see Theorem 4.3.2). If the components of Y are completely dependent then $G(x) = \min_{j \leq d} G_j(x_j)$. Necessary and sufficient conditions can be formulated in a similar way. The dependence structure is reflected by the so-called dependence or copula function $D_G(\cdot)$ defined by $G(x) = D_G(G_1(x_1), \ldots, G_d(x_d))$. If $G(\cdot)$ is a multivariate EVD, then the max-stability of $G(\cdot)$ can be translated into an equivalent property of the dependence function $D_G(\cdot)$. For further discussion see for instance Galambos [167], Resnick [393], Hsing [226] and Section 4.2.

If $k > 1$, criteria for the complete dependence and the independence among the components of Y can also be derived. The independence among the components of Y does not follow from (11.37). We need in addition

$$nP(X_{1j} > a_{nj}x_j + b_{nj}, X_{ij'} > a_{nj'}x_{j'} + b_{nj'}) \to 0 \qquad \text{as} \quad n \to \infty \qquad (11.38)$$

to hold for all $1 \leq j \neq j' \leq d$ and $1 \leq i \leq k$. This excludes the possibility of an exceedance both in component j at time 1 and j' at time $i (\leq k)$. Based on a result of Takahashi [438] it is sufficient to show that (11.38) holds for some x with $\mu^{(k)}(1, x_j) \in (0, \infty)$, $j \leq d$, where $x_j = (\infty, \ldots, x_j, \infty, \ldots, \infty)$.

Corollary 11.6.4. *Under the assumptions of Corollary 11.6.3 we have*

(i) $G(x) = \prod_{j \leq d} G_j(x_j)$ *for some x with $G_j(x_j) \in (0,1)$, $j \leq d$, iff*

$$G(x) = \exp(-\mu^{(k)}(1, x)) = \prod_{j \leq d} G_j(x_j) = \prod_{j \leq d} \exp(-\mu^{(k)}(1, x_j)) \qquad (11.39)$$

   for all x.

(ii) *If (11.38) holds for some x where $\mu^{(k)}(1, x_j) \in (0, \infty)$, $j \leq d$, then $G(\cdot)$ has the representation (11.39) for all x.*

Statement (i) is the result of Takahashi [438], see Theorem 4.3.2, which is used to prove (ii) (cf. Hüsler [245]).

Example 11.6.5. Consider a stationary sequence of Gaussian random vectors $\{X_i, i \geq 1\}$. If the autocorrelations $r_{j,j'}(i, i') = Corr(X_{ij}, X_{i'j'})$, $j, j' \leq d$, $i, i' \geq 1$, satisfy Berman's condition

$$\sup_{|i-i'| \geq n} |r_{j,j'}(i, i')| \log(n) \to 0 \qquad \text{as} \quad n \to \infty \qquad (11.40)$$

and $\sup_{|i-i'| \geq 1} |r_{j,j'}(i, i')| < 1$ for all $1 \leq j \leq j' \leq d$, then the conditions

$D(E_{ni})$ and $D'(E_{ni})$ hold; (11.38) holds if $\sup_{|i-i'|\geq 0} |r_{j,j'}(i,i')| < 1$ for $j \neq j'$. Thus $P(\boldsymbol{M}_n \leq \boldsymbol{a}_n\boldsymbol{x} + \boldsymbol{b}_n) \to_w G(\boldsymbol{x}) = \prod_{j\leq d} G_j(x_j) = \exp\left(-\sum_{j\leq d} e^{-x_j}\right)$ (cf. Amram [12], Hsing [225], Hüsler and Schüpbach [266]).

We now consider the complete dependence. For $k = 1$ it can be shown that if for some $\boldsymbol{x}$ with $\max_{j\leq d} \mu^{(1)}(1, \boldsymbol{x}_j) = \mu^{(1)}(1, \boldsymbol{x}_{j'}) \in (0, \infty)$,

$$n\bigg(P(X_{1j} > a_{nj}x_j + b_{nj}, X_{1j'} > a_{nj'}x_{j'} + b_{nj'}) - P(X_{1j} > a_{nj}x_j + b_{nj}) \bigg) \to 0$$

as $n \to \infty$, for all $j \neq j' \leq d$, then the components of $\boldsymbol{Y} \sim G$ are completely dependent: $G(\boldsymbol{x}) = \min_{j\leq d} G_j(x_j)$. This condition is equivalent to

$$nP(X_{1j} > a_{nj}x_j + b_{nj}, \; X_{1j'} \leq a_{nj'}x_{j'} + b_{nj'}) \to 0. \tag{11.41}$$

For the case $k > 1$ this condition (11.41) has to be extended. For some $\boldsymbol{x}$ with $\max_{j\leq d} \mu^{(k)}(1, \boldsymbol{x}_j) = \mu^{(k)}(1, \boldsymbol{x}_{j'}) \in (0, \infty)$, let

$$nP(X_{lj} \leq a_{nj}x_j + b_{nj}, \; 1 \leq l < k, \; X_{kj} > a_{nj}x_j + b_{nj},$$
$$X_{ij'} \leq a_{nj'}x_{j'} + b_{nj'}, \; 1 \leq i \leq k) \to 0 \tag{11.42}$$

as $n \to \infty$, for all $j \neq j' \leq d$.

Corollary 11.6.6. *Under the assumptions of Corollary* 11.6.3 *we have*

(i) $G(\boldsymbol{x}) = \min_{j\leq d} G_j(x_j)$, *for all* $\boldsymbol{x}$ *iff* $G(\boldsymbol{x}) = \min_{j\leq d} G_j(x_j)$, *for some* $\boldsymbol{x}$ *with* $G(\boldsymbol{x}) \in (0, 1)$.

(ii) *If* (11.42) *holds for some* $\boldsymbol{x}$ *where* $\max_{j\leq d} \mu^{(k)}(1, \boldsymbol{x}_j) = \mu^{(k)}(1, \boldsymbol{x}_{j'}) \in (0, \infty)$, *then*

$$G(\boldsymbol{x}) = \min_{j\leq d} G_j(x_j)$$

for all $\boldsymbol{x}$, *i.e., the components of* $\boldsymbol{Y} \sim G(\cdot)$ *are completely dependent.*

Statement (i) is the result of Takahashi [438], see Theorem 4.3.2. Statement (ii) follows by using similar arguments as in Hüsler [245]. Note that if the components of $\boldsymbol{Y}$ with distribution G are pairwise completely dependent, then they are jointly completely dependent.

If, for instance, in Example 11.6.5 $r_{j,j'}(1, 1) = 1$ for all $j \neq j'$ and Berman's condition (11.40) holds, the conditions of Corollary 11.6.6 can be verified for $k = 1$ and therefore $G(\boldsymbol{x}) = \min_{j\leq d} G_j(x_j) = \exp(-\exp(-\min_{j\leq d} x_j))$.

INDEPENDENT MULTIVARIATE SEQUENCES

If (11.31) and (11.33) hold, then the class of non-degenerate limit distributions $G(\cdot)$ consists of distributions with the following property:

$$G_t^*(\boldsymbol{x}) = G(\boldsymbol{x})/G(\boldsymbol{A}(t)\boldsymbol{x} + \boldsymbol{B}(t)) \tag{11.43}$$

is a distribution in $\overline{I\!R}^d$ for all $t \leq 1$ where $\boldsymbol{A}(\cdot) : (0,1] \rightarrow \mathbb{R}_+^d$ and $\boldsymbol{B}(\cdot) : (0,1] \rightarrow \mathbb{R}^d$ are suitable functions with $\boldsymbol{A}(s)\boldsymbol{A}(t) = \boldsymbol{A}(st)$ and $\boldsymbol{B}(st) = \boldsymbol{A}(s)\boldsymbol{B}(t) + \boldsymbol{B}(s)$ for any $s,t \in (0,1]$ (see Hüsler [242]). $G_t^*(\boldsymbol{x})$ and $G(\boldsymbol{A}(t)\boldsymbol{x} + \boldsymbol{B}(t))$ are the limit distributions of $P(\boldsymbol{X}_i \leq \boldsymbol{a}_n\boldsymbol{x} + \boldsymbol{b}_n, nt \leq i \leq n)$ and $P(\boldsymbol{X}_i \leq \boldsymbol{a}_n\boldsymbol{x} + \boldsymbol{b}_n, 1 \leq i \leq nt)$, respectively. The distributions $G(\cdot)$ satisfying (11.43) are continuous in the interior of their support and form a proper subclass of the max-id distributions. They are all positively dependent and we have, for every $\boldsymbol{x}$,

$$\min_{j \leq d} G_j(x_j) \geq G(\boldsymbol{x}) \geq \prod_{j \leq d} G_j(x_j).$$

Obviously, the max-stable distributions belong to this class.

The dependence structure of the limit distribution is again determined by the asymptotic dependence structure among the components of $\boldsymbol{X}_i, i \leq n$. Assuming (11.31) for all $\boldsymbol{x}$ with $\mu^{(1)}(\cdot, \boldsymbol{x}) \in (0, \infty)$ and (11.35) for $k = 1$, a necessary and sufficient condition that $\boldsymbol{Y} \sim G(\cdot)$ has independent components, is given by

$$\sum_{i \leq n} P(X_{ij} > a_{nj}x_j + b_{nj}, \ X_{ij'} > a_{nj'}x_{j'} + b_{nj'}) \rightarrow 0 \qquad (11.44)$$

as $n \rightarrow \infty$ for any $1 \leq j < j' \leq d$, and all $\boldsymbol{x}$ such that $G(\boldsymbol{x}) > 0$ (i.e., $\mu^{(1)}(1, \boldsymbol{x}) < \infty$). This is shown in Hüsler [242] Theorem 3. Obviously, (11.44) reduces to (11.37) in the stationary case. Noting that the limit distributions are not max-stable in general, we cannot make use of Takahashi's result to derive criteria for independence or complete dependence.

Complete dependence is treated in a similar way as in the stationary case. We introduce a version of condition (11.41): Let for $\boldsymbol{x}$ with $\max_{j \leq d} \mu^{(1)}(1, \boldsymbol{x}_j) = \mu^{(1)}(1, \boldsymbol{x}_{j'}) \in (0, \infty)$,

$$\sum_{i \leq n} P(X_{ij} > a_{nj}x_j + b_{nj}, \ X_{ij'} \leq a_{nj'}x_{j'} + b_{nj'}) \rightarrow 0 \qquad (11.45)$$

as $n \rightarrow \infty$, for all $j \neq j'$. Condition (11.45 is assumed to hold for all $\boldsymbol{x}$ with $G(\boldsymbol{x}) \in (0,1)$. If (11.31) for all $\boldsymbol{x}$ with $\mu^{(1)}(\cdot, \boldsymbol{x}) \in (0, \infty)$ and (11.35) hold (with $k = 1$), then (11.45) is necessary and sufficient for the asymptotic complete dependence among the components M_{nj}.

GENERAL MULTIVARIATE SEQUENCES

Assume that the conditions $D(E_{ni})$ and $D^{(k)}(E_{ni})$, for some $k \geq 1$, hold for all $\boldsymbol{x}$ where $G(\boldsymbol{x}) \in (0,1)$. Combining the results for the independent sequences with Theorem 11.6.2, we derive statements about the class of limit distributions and criteria for independence and complete dependence. For $k = 1$ the class of limit distributions is identical to the one given by (11.43) (see Hüsler [243]). For $k > 1$

the same result can be derived. In general, the independence and complete dependence do not follow by (11.44) and (11.45), respectively. Therefore we introduce the following versions of (11.38) and (11.42):

$$\sum_{1\leq i\leq n-l+1} P(X_{ij} > a_{nj}x_j + b_{nj},\, X_{i+l-1,j'} > a_{nj'}x_{j'} + b_{nj'}) \to 0 \qquad (11.46)$$

as $n \to \infty$ for all $1 \leq j \neq j' \leq d$ and $1 \leq l \leq k$ for all $\boldsymbol{x}$ with $G(\boldsymbol{x}) \in (0,1)$; and for all $\boldsymbol{x}$ with $\max_{j\leq d} \mu^{(k)}(1,\boldsymbol{x}_j) = \mu^{(k)}(1,\boldsymbol{x}_{j'}) \in (0,\infty)$, let for all $j \neq j'$

$$\sum_{0\leq i\leq n-k} P(X_{i+l,j} \leq a_{nj}x_j + b_{nj},\, X_{i+k,j} > a_{nj}x_j + b_{nj},\, 1 \leq l < k,$$
$$X_{i+h,j'} \leq a_{nj'}x_{j'} + b_{nj'},\, 1 \leq h \leq k) \to 0 \qquad (11.47)$$

as $n \to \infty$.

Corollary 11.6.7. *Assume that the conditions of Theorem 11.6.2 and (11.35) hold. Then $G(\cdot)$ exists and*

$$\min_{j\leq d} G_j(x_j) \geq G(\boldsymbol{x}) \geq \prod_{j\leq d} G_j(x_j).$$

(i) *If (11.33) holds, the class of limit distributions is characterized by (11.43).*

(ii) *Independence: If (11.46) holds, then $G(\boldsymbol{x}) = \prod_{j\leq d} G_j(x_j)$ for all $\boldsymbol{x}$.*

(iii) *Complete Dependence: If (11.47) holds, then $G(\boldsymbol{x}) = \min_{j\leq d} G_j(x_j)$ for all $\boldsymbol{x}$.*

This presents an application of the general results of Section 11.5 for extremes and rare events of random sequences. However, arrays of random sequences are also of particular interest. We close this section by discussing such an example to indicate that the results of this section can easily be extended further for arrays of rv.

Example 11.6.8. Let $\{\boldsymbol{X}_i(n),\, i \leq n,\, n \geq 1\}$ be an array of standard Gaussian rv, where $\{\boldsymbol{X}_i(n),\, i \leq n\}$ are iid for every fixed n. For simplicity let $d = 2$. Denote by $\rho(n) = r_{1,2}(1,1) = E(X_{i1}(n)X_{i2}(n))$. Let $((1 - \rho(n)) \log(n))^{1/2} \to \lambda \in [0,\infty]$. Then $M_n = \max_{i\leq n} \boldsymbol{X}_i(n)$, suitably normalized, converges in distribution to $\boldsymbol{Y} \sim G$ with Gumbel marginal distributions. As mentioned in Example 4.1.4, if $\lambda = 0$, then the components of $\boldsymbol{Y}$ are completely dependent and if $\lambda = \infty$, then they are independent, as in Example 11.6.5. To prove these dependence structures, we might use the condition (11.38) and (11.41), respectively, adapted for the array of rv; for instance, instead of (11.38) we assume

$$nP(X_{1j}(n) > a_{nj}x_j + b_{nj},\, X_{1j'}(n) > a_{nj'}x_{j'} + b_{nj'}) \to 0$$

which holds in this example if $\lambda = \infty$. Thus we simply replace the rv $\boldsymbol{X}_i$ by $\boldsymbol{X}_i(n)$ in these conditions. As mentioned, if $\lambda \in (0, \infty)$, then the limit distribution G is max-stable, with a certain dependence among the components of $\boldsymbol{Y}$ (cf. Example 4.1.2). It is straightforward to extend these results for arrays of non-iid Gaussian sequences by assuming the conditions of Proposition 11.5.4, Theorem 11.6.2 or Berman's condition as in Example 11.6.5.

Chapter 12

Statistics of Extremes

We use in the following the theory developed in the preceding chapters to discuss a few nonstandard applications. Of interest are here the statistical estimation of the cluster distribution and of the extremal index in a stationary situation. In the last section we treat a frost data problem which is related to an extreme value problem of a nonstationary sequence.

12.1 Introduction

Let $\{X_i,\ i \geq 1\}$ be a stationary random sequence such that $D(u_n)$ holds. Under additional conditions it was shown in Chapter 9 that

$$P(M_n \leq a_n x + b_n) \to_w G^\theta(x) \tag{12.1}$$

as $n \to \infty$ with $0 < \theta \leq 1$. θ is called the extremal index (Leadbetter [300]). If $\theta < 1$, then the exceedances do cluster. Since G is an EVD and therefore max-stable, the limiting distribution G^θ of the maxima is still of the same type as G. Hence the statistical procedures developed for the iid case can also be applied to the case where exceedances do cluster. We approximate $P(M_n \leq y) \approx G^\theta((y - b_n)/a_n) = G((y-\beta)/\alpha)$ for some constants $\alpha > 0$ and β. Hence from m independent sequences of n observations each, we get m iid maxima $M_{n,i}, i \leq m$ and we can for instance fit an EVD G to the empirical distribution of $\{M_{n,i}, i \leq m\}$ and estimate α, β using a well-known statistical procedure, e.g. using the best linear unbiased estimators (see Gumbel [182], Lawless [299], Castillo [60]. Recent books on extreme values discussing statistical procedures and applications are by Embrechts et al. [122], Coles [71], Nadarajah and Kotz [293], Finkenstädt and Rootzén [158], Beirlant et al. [32] and de Haan and Ferreira [190].

Smith [420] presented another approach to apply the extreme value theory. He used a semiparametric approach to discuss the large values of ozone data; see also Davison and Smith [94] and for multivariate extensions Coles and Tawn [73].

M. Falk et al., *Laws of Small Numbers: Extremes and Rare Events*, 3rd ed.,
DOI 10.1007/978-3-0348-0009-9_12, © Springer Basel AG 2011

Another approach is used for extremes of nonstationary sequences in Hüsler [239] to discuss low temperatures in spring. This application is presented in Section 12.4.

Because of the clustering of the exceedances the asymptotic distributions of the extreme order statistics depend on the cluster size distribution. For stationary sequences it can be proved that the r-th largest order statistic has asymptotically the following distribution $M_n^{(r)}$, $r \geq 1$ fixed,

$$P(M_n^{(r)} \leq a_n x + b_n) \rightarrow_w G^\theta(x) \sum_{0 \leq s \leq r-1} \sum_{s \leq j \leq r-1} (-\log(G^\theta(x)))^s \pi_{s,j}/s! \quad (12.2)$$

where $\pi_{s,j} = (\pi * \pi * \cdots * \pi)(\{j\})$ denotes the value at j of the s-th convolution of the cluster size distribution π. In order to apply this result to concrete statistical problems, we have to estimate θ and π.

12.2 Estimation of θ and $\pi(\cdot)$

Under rather general conditions θ^{-1} is equal to the limit of the mean cluster size,

$$\theta^{-1} = \sum_{j \geq 1} j\pi(\{j\}) < \infty,$$

if $\theta > 0$. Therefore it is reasonable to propose

$$\hat{\theta}_n = Z/\sum_{i \leq Z} Y_i$$

as an estimator for θ, where Z denotes the number of clusters of exceedances of a given boundary u_n by $\{X_i, \ i \leq n\}$ and Y_i denotes the size of the i-th cluster. Denoting by N the total number of exceedances:

$$\sum_{i \leq Z} Y_i = N,$$

we can rewrite

$$\hat{\theta}_n = Z/N.$$

The estimator $\hat{\theta}_n$ depends on the definition of the cluster of exceedances. Since there are more than one definition of a cluster, we have to select the suitable one for applications. If the condition $D^*(u_n)$ holds, we observe asymptotically only 'runs' of exceedances. This means that the random sequence $\{X_i\}$ crosses the boundary u_n at some time point, remains above the boundary for a few time points and crosses u_n downwards to remain below u_n for a long time interval before eventually crossing u_n upwards again (Figure 9.2.1). It seems therefore reasonable to use the run definition in applications.

In the previous chapter we used another definition of a cluster in deriving the results. There all exceedances in a subinterval were considered to form one cluster. Such a cluster is called a *'block'* of exceedances. This definition does not consider the fact that a block may contain more than one upcrossing. This definition is more convenient for theoretical purposes.

Leadbetter and Nandagopalan [304] showed that, if $D^*(u_n)$ holds and the length of the subintervals tends to ∞, both definitions are asymptotically equivalent, resulting in the same cluster size distribution.

Since in applications the sample size is always finite, a run of exceedances is split up into two clusters if it overlaps two successive subintervals and runs of exceedances within a subinterval are joined to give a block of exceedances. Therefore it is obvious that the definition of cluster will influence the estimation of θ.

CHOICE OF THRESHOLD

The estimation of θ is also influenced by the choice of u_n. We know from the theory that the boundary u_n is chosen such that $n(1 - F(u_n)) \to \tau \in \mathbb{R}_+$. Since the number of exceedances has asymptotically a Poisson distribution, there is a positive probability that no exceedance above the boundary u_n occurs in which case θ cannot be estimated.

For some boundaries u_n we have $N = 1$ and therefore $Z = 1$ which gives $\hat{\theta}_n = 1$. If u_n is such that $N = 2$, then $\hat{\theta}_n = 1$ or $= 1/2$ depending on Z; similarly for $N > 2$. This implies that the estimator $\hat{\theta}_n$ cannot be consistent for every $\theta \in (0, 1]$, if u_n is chosen in such a way that N is fixed.

Therefore it was proposed to use $v_n < u_n$ such that N tends to ∞, i.e., $n(1 - F(v_n)) \to \infty$ (slowly). Hsing [227] and Nandagopalan [346] proved, that with this choice the estimator $\hat{\theta}_n$ of θ is consistent, whether one uses the run or the block definition for clusters.

Leadbetter et al. [307] proposed to use v_n in applications such that $\sharp\{i \leq n : X_i > v_n\}/n = N/n$ is approximately 5%, 7.5% or 10%. Obviously N gets large with increasing n.

Hsing [227] showed using the block definition for clusters and the above choice of v_n that under certain additional conditions the estimator $\hat{\theta}_n$ has asymptotically a normal distribution. Moreover, the asymptotic variance of the estimator depends on θ.

In applications the following version of Hsing's result ([227], Corollary 4.5) is used:

$$\left(\frac{\gamma_n}{\hat{\theta}_n(\hat{\theta}_n^2 \hat{\sigma}_n^2 - 1)}\right)^{1/2} (\hat{\theta}_n - \theta) \to_D N(0, 1),$$

with

$$\hat{\sigma}_n^2 = \sum_{i \leq Z} Y_i^2 / Z$$

and

$$\gamma_n = nP(X_1 > v_n),$$

where γ_n is substituted by N. In the following S_n denotes the estimate of the standard deviation of $\hat{\theta}_n$:

$$S_n^2 = \hat{\theta}_n(\hat{\theta}_n^2 \hat{\sigma}_n^2 - 1)/N.$$

The value $\pi(\{j\})$ of the cluster size distribution for $j \geq 1$ is estimated by the relative frequency of clusters with size $Y_i = j$, $i \leq Z$, with respect to the selected boundary v_n: $\hat{\pi}(\{j\}) = \sum_{i \leq Z} 1(Y_i = j)/Z$.

12.3 Application to Ecological Data

We present an example by Gentleman et al. discussed in Leadbetter et al. [307] on acid rain. The sulfate concentration was measured during $n = 504$ rainfalls in Pennsylvania (USA). Only the values larger than 93μmol/l were recorded.

TABLE 12.3.1. Sulfate values of rainfall i, $i \leq 504 = n$.

rain i	sulf.	rain i	sulf.	rain i	sulf.	rain i	sulf.	rain i	sulf.
55	140	102	130	228	280	353	100	415	150
60	95	129	98	229	160	374	110	439	100
64	150	150	110	237	96	375	110	452	100
65	110	160	95	242	95	376	95	453	100
73	99	168	110	247	98	377	99	455	190
74	130	176	130	253	150	397	340	469	130
75	120	177	130	315	94	398	99	470	130
77	110	184	150	317	110	402	140	476	105
83	110	187	110	324	200	404	95	480	110
85	120	188	110	334	330	405	100	485	150

In order to obtain approximately 10%, 7.5% and 5% for N/n, v_n should be 93, 99 and 110 μmol/l, respectively. Thus $N = 50, 38$ and 21, respectively. To analyze the behavior of the estimator with respect to v_n, we continuously change the value of v_n from 93 up to the maximal possible value 340.

CLUSTERS OF RUNS

In our analysis we first use the run definition for the clusters and later the block definition. Obviously, $\hat{\theta}_n$ is a piecewise constant function of v_n. One might think that the estimator is monotonically increasing. Although this is not true, one can observe that $\hat{\theta}_n$ has the tendency to increase in v_n (Figure 12.3.1).

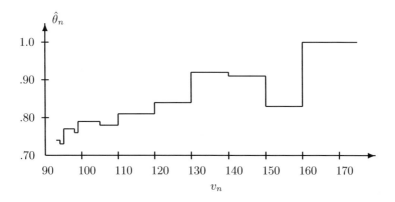

FIGURE 12.3.1. Estimation of θ by $\hat{\theta}_n$ using the run definition for clusters.

TABLE 12.3.2. Estimation of θ and the standard error of the estimator based on v_n and the run definition for clusters.

v_n	N	Cluster Size				Z	$\hat{\theta}_n$	S_n
		1	2	3	4			
93	50	27	8	1	1	37	.74	.060
94	49	26	8	1	1	36	.73	.061
95	44	25	8	1		34	.77	.053
96	43	24	8	1		33	.77	.053
98	41	22	8	1		31	.76	.055
99	38	22	8			30	.79	.050
100	33	19	7			26	.79	.054
105	32	18	7			25	.78	.055
110	21	13	4			17	.81	.067
120	19	13	3			16	.84	.069
130	13	11	1			12	.92	.068
140	11	9	1			10	.91	.078
150	6	4	1			5	.83	.124
160	5	5				5	1.00	

Note that the entries for $v_n = 110$ in the tables slightly differ from the values given in Leadbetter et al. [307], probably because of a misprint in their paper.

From Table 12.3.2 it is not clear which value of v_n gives an appropriate estimate of θ. This fact can also be deduced from Figure 12.3.1: There is no large enough interval of v_n where $\hat{\theta}_n$ is stable, which could be used as criterion.

CLUSTERS GIVEN BY BLOCKS

This problem also arises if the block definition for the clusters is used. To indicate the dependence of $\hat{\theta}_n$ on the length of the subintervals we carry out the analysis of the same data using subintervals of length 5 and 10 (as in Leadbetter et al. ([307]), again for all levels v_n larger than 93.

We deduce from this example, that the estimates of θ are equal for the three cluster definitions if $v_n \geq 110$. The same is true for the cluster size distributions. This gives a good reason for selecting $v_n=110$ as an appropriate level for the estimation. $\hat{\theta}_n = .81$ is selected as the estimate for θ.

Leadbetter et al. [307] proposed to use S_n as an additional criterion for the choice of v_n. We observe from Figure 12.3.3 that S_n remains stable for $110 \leq v_n \leq 140$. Hence a value v_n of this interval should be selected. Whether the proposed criterion leads to a satisfactory and efficient estimation of θ and the cluster size distribution is still an open question.

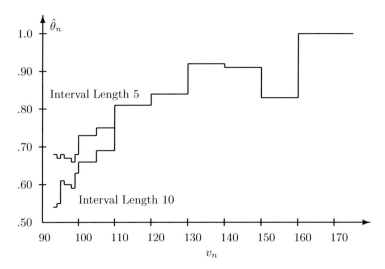

FIGURE 12.3.2. Estimation of θ by $\hat{\theta}_n$ by using the block definition for clusters and intervals of length 5 and 10.

TABLE 12.3.3. Estimation of θ and the standard error S_n of the estimator based on v_n and the block definition for clusters.

v_n	N	Interval Length 5			Z	$\hat{\theta}_n$	S_n	Interval Length 10				Z	$\hat{\theta}_n$	S_n
		Cluster Size						Cluster Size						
		1	2	3				1	2	3	4			
93	50	21	10	3	34	.68	.052	11	11	3	2	27	.54	.050
94	49	20	10	3	33	.67	.052	12	10	3	2	27	.55	.053
95	44	18	10	2	30	.68	.052	15	8	3	1	27	.61	.060
96	43	17	10	2	29	.67	.053	14	8	3	1	26	.60	.059
98	41	15	10	2	27	.66	.053	12	8	3	1	24	.59	.059
99	38	15	10	1	26	.68	.052	13	8	3		24	.63	.057
100	33	15	9		24	.73	.052	13	7	2		22	.66	.062
105	32	16	8		24	.75	.054	14	6	2		22	.69	.066
110	21	13	4		17	.81	.067	13	4			17	.81	.067
120	19	13	3		16	.84	.069	13	3			16	.84	.069
130	13	11	1		12	.92	.068	11	1			12	.92	.068
140	11	9	1		10	.91	.078	9	1			10	.91	.078
150	6	4	1		5	.83	.124	4	1			5	.83	.124
160	5	5			5	1.00		5				5	1.00	

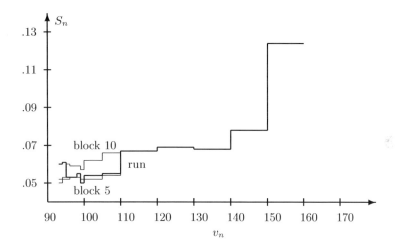

FIGURE 12.3.3: Estimated standard error S_n for interval length 5 and 10 and the run definition for clusters.

OTHER ESTIMATES

Hsing [229] proposed another estimator for the cluster size distribution and showed
that it has better properties than the one used so far by analyzing an AR(1)
random sequence with Cauchy noise. An adaptive procedure for the estimation of
the extremal index is given by Hsing [228]. He shows that this procedure is weakly
consistent and discusses the optimal choice of the level v_n.

Finally we mention a procedure to estimate θ proposed by Gomes [178]. She
used the following result of Leadbetter and Nandagopalan [304]:

$$\lim_{n\to\infty} \frac{P(X_1 > u_n, X_2 \le u_n)}{P(X_1 > u_n)} = \theta \qquad (12.3)$$

if $D^*(u_n)$ holds. Gomes proposed to estimate θ by replacing the probabilities in
(12.3) with the empirical relative frequencies. This estimator is up to a factor
$n/(n-1)$ equal to the estimator $\hat{\theta}_n$ based on the run definition. If n is large, the
two estimators are almost identical.

For the choice of v_n, Gomes proposed the following iterative adaptive pro-
cedure: In a first step $\hat{\theta}_0$ is estimated based on a rather low level $v_n = X_{[4n/5];n}$,
where $X_{i;n}$ denotes the i-th smallest order statistic of the rv $X_i, i \le n$. This implies
that $N \approx n/5$. In a further step $\hat{\theta}_n$ is reestimated by selecting the adapted level
$v_n = X_{n_1;n}$ where

$$n_1 = \begin{cases} n - \min\left(\dfrac{n}{2}, \dfrac{n(0.075 - 0.00005n)}{\hat{\theta}_0}\right), & \text{if} \quad n \le 1000, \\[3mm] n - \min\left(\dfrac{n}{2}, \dfrac{25}{\hat{\theta}_0}\right), & \text{if} \quad n > 1000. \end{cases}$$

(n_1 is selected as small as $n/2$ only if $\hat{\theta}_0$ is rather small, i.e., $\hat{\theta}_0 < 50/n \le 0.05$
for $n > 1000$). Gomes found out by simulating some special random sequences
that this adaptive procedure gives an estimation of θ with a small variability. In
our example the adapted level $v_n \approx 100$ or 110, as in the preceding discussion. In
statistical applications one often argues that if different methods lead to the same
estimate then the resulting value is considered to be a good estimate. Thus let
us accept $\hat{\theta}=0.81$. By using the asymptotic normality of $\hat{\theta}_n$ and the value for S_n,
we get the 95% confidence interval $(.81 \pm .067 \cdot 1.960) = (.68, .94)$. Since 1 is not
included in this confidence interval, we conclude that sulfate values larger than
$110\mu\text{mol/l}$ do cluster and the mean cluster size is $1.23 = 1/\hat{\theta}$.

Similar applications are e.g. given by Hsing [230] and Dietrich and Hüsler
[113] and in the book edited by Finkenstädt and Rootzén [158].

12.4   Frost Data: An Application

The following problem arises in agronomy: during the springtime temperatures
below a certain threshold temperature τ damage the plants. The lowest temperature X_i of the day i occurs in the early morning. This temperature is recorded

each day throughout the year. Of interest is the last day T in the spring with the lowest temperature below τ (last frost). Usually τ is chosen to be between $0°$ and $-4°C$. We want to derive the distribution of T and to estimate high quantiles of t_p (e.g. for $p = 0.90$ or 0.95) of this distribution. The last frost day is defined by

$$ T = \begin{cases} \max\{i : X_i < \tau\}, & \text{if } \exists i : X_i < \tau \\ 0, & \text{else.} \end{cases} $$

Figure 12.4.1 shows a realization of the random sequence $\{X_i, \ i \leq 365\}$ during the critical period. Here $T = 127$.

It is obvious that

$$ P(T \leq t) = P(X_i \geq \tau, \ \text{for all } i > t) $$

(a similar relation was discussed already in Example 8.3.3). To derive this distribution we need some information on the random sequence of the lowest temperatures $\{X_i\}$ or a random sample of T. We first analyzed data collected at a certain site in Switzerland during 18 years. As 18 realizations of T are not sufficient for estimating high quantiles of the distribution of T by a nonparametric confidence interval, we used a stochastic model for $\{X_i\}$.

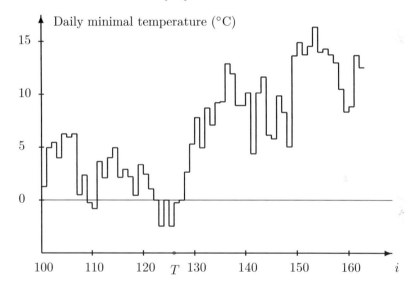

FIGURE 12.4.1. Daily minimal temperatures during the critical period and T $(\tau = 0°C)$.

It is rather obvious that $\{X_i\}$ is not a stationary sequence, because springtime temperatures have an increasing trend. The analysis of the data showed that the variability of X_i is not constant in the time period of interest, March to May, $i = 60$ to 160. The analysis is based on about 1800 data points. By detrending and standardizing the sequence $\{X_i\}$ we get

$$Y_i = (X_i - \mu_i)/\sigma_i,$$

where the mean μ_i and the standard deviation σ_i for day i are estimated from the data by $\bar{x}_i$ and s_i. The analysis of the random sequence $\{Y_i\}$ showed that it can be modelled as a Gaussian AR(1) random sequence. (For a detailed account on the analysis see Hüsler [238]).

Thus, let us assume in the following that

$$Y_i = \rho Y_{i-1} + Z_i$$

for all i, where Z_i are iid Gaussian rv with mean 0 and variance $1 - \rho^2$, each Z_i being independent of Y_{i-1}. From the data ρ is estimated to be 0.73. Note that Berman's condition (cf. Example 9.2.2) holds for this simple random sequence, since $E(Y_{i+1}Y_i) = \rho^i$ for $i \geq 1$ and $\rho^n \log(n) \to 0$ as $n \to \infty$.

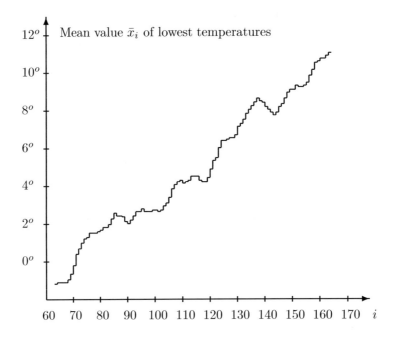

FIGURE 12.4.2. Mean values $\bar{x}_i$ of the lowest temperatures of each day during the critical period.

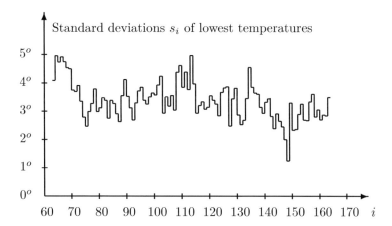

FIGURE 12.4.3. Standard deviations s_i of the lowest temperatures of each day during the critical period.

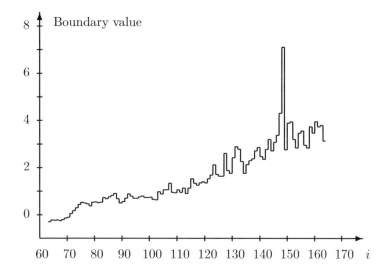

FIGURE 12.4.4. Standardized boundary values $(\bar{x}_i - \tau)/s_i$ of the lowest temperatures of day i, $\tau = 0°\mathrm{C}$.

Asymptotic Approximation

Using this model we have to calculate the following probabilities:

$$P(T \leq t) = P(Y_i \geq (\tau - \mu_i)/\sigma_i, \ i > t)$$
$$= P(Y_i \leq (\mu_i - \tau)/\sigma_i, \ i > t), \qquad (12.4)$$

by the symmetry of the Gaussian distribution. For large i the boundary values $(\mu_i - \tau)/\sigma_i$ are large compared with the upper tail of the Gaussian distribution (Figure 12.4.4). Since we are only interested in the tail behavior of the distribution of T, we apply the asymptotic theory for nonstationary sequences and nonconstant boundaries presented in Chapter 9. Using Theorem 9.1.7 or Lemma 9.1.6 (cf. Example 9.4.1) we approximate (12.4)

$$P\{Y_i \leq (\mu_i - \tau)/\sigma_i, \ i > t\} \approx \prod_{t < i \leq t_0} P(Y_i \leq (\mu_i - \tau)/\sigma_i)$$
$$= \prod_{t < i \leq t_0} \Phi((\mu_i - \tau)/\sigma_i)$$
$$=: \tilde{P}(t), \qquad (12.5)$$

where $\Phi(\cdot)$ denotes the standard normal df and t_0 ($= 163$) a day at the beginning of summer, after which there is no frost with probability 1.

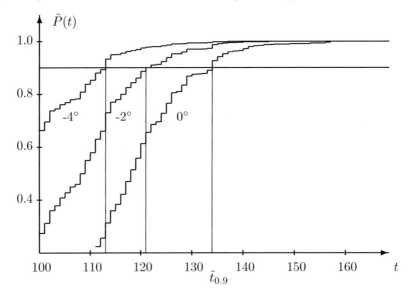

FIGURE 12.4.5. $\tilde{P}(t)$ for the threshold temperatures $\tau = -4°, -2°$ and $0°C$.

Approximation (12.5) is rather accurate for high boundary values. One can easily calculate $\tilde{P}(t)$ and $\tilde{t}_p = \min\{t : \tilde{P}(t) \geq p\}$ to estimate the quantiles t_p. Figure 12.4.5 shows the values of $\tilde{P}(t)$ for the three threshold temperatures $\tau = 0°, -2°, -4°$C. For example for $\tau = 0°$C we get $\tilde{t}_{0.9} = 134$.

This is an asymptotic approximation and its accuracy is not known. However, for applications an estimation of the approximation error is certainly needed. Therefore, we analyze in the following the goodness of this approximation. To find upper and lower bounds of the probability in (12.4), we use the Bonferroni inequalities, the Slepian inequality and the simulation technique.

BONFERRONI INEQUALITIES

Using the Bonferroni inequalities we get

$$
P(Y_i \leq (\mu_i - \tau)/\sigma_i, \quad t < i \leq t_0)
$$
$$
= 1 - P(Y_i > (\mu_i - \tau)/\sigma_i \quad \text{for some} \ i \in (t, t_0])
$$
$$
\geq 1 - \sum_{t < i \leq t_0} P(Y_i > (\mu_i - \tau)/\sigma_i)
$$
$$
= 1 - \sum_{t < i \leq t_0} (1 - \Phi((\mu_i - \tau)/\sigma_i))
$$
$$
=: \hat{P}(t)
$$

and

$$
P(Y_i \leq (\mu_i - \tau)/\sigma_i, \quad t < i \leq t_0)
$$
$$
= 1 - P(Y_i > (\mu_i - \tau)/\sigma_i \quad \text{for some} \ i \in (t, t_0])
$$
$$
\leq 1 - \sum_{t < i \leq t_0} (1 - \Phi((\mu_i - \tau)/\sigma_i))
$$
$$
+ \sum_{t < i < j \leq t_0} P(Y_i > (\mu_i - \tau)/\sigma_i, Y_j > (\mu_j - \tau)/\sigma_j)
$$
$$
=: \check{P}(t).
$$

The terms of the double sum can be evaluated using the bivariate normal distribution with correlation ρ^{j-i}. Figure 12.4.6 shows the goodness of these approximations. It is obvious that the upper bound can be larger than 1 and the lower bound can be negative. Nevertheless, the approximation turns out to be quite accurate for $p \geq 0.90$ in our analysis. Using $\hat{P}(t)$ and $\check{P}(t)$ upper and lower bounds for t_p are derived: $\hat{t}_p$ and $\check{t}_p$, respectively. For example for $\tau = 0°$C we get $\hat{t}_{0.90} = 134$ and $\check{t}_{0.90} = 128$.

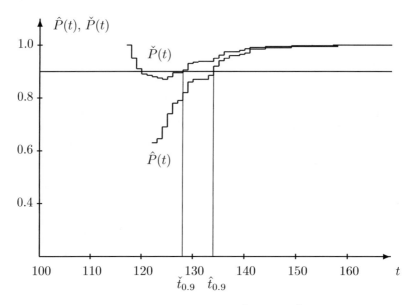

FIGURE 12.4.6. Bonferroni bounds $\hat{P}(t)$ and $\check{P}(t)$, $\tau = 0°$C.

SLEPIAN INEQUALITY

Other bounds for the distribution of T are obtained by using Slepian inequality for Gaussian sequences (Theorem 10.3.2). We use the fact that all autocorrelations of $\{Y_i\}$ are nonnegative and smaller than $\rho = .73$.

Denote by W_i an equally correlated stationary Gaussian random sequence with mean 0, variance 1 and autocorrelation $E(W_1 W_i) = \rho$, for all $i > 1$. W_i can be written as

$$W_i = \sqrt{\rho}\, U + \sqrt{1 - \rho}\, Z_i,$$

where $U, Z_1, Z_2, \ldots$ are iid standard normal rv. Since $\rho \geq E(Y_i Y_j)$ for all $i \neq j$, we get by Slepian inequality (Theorem 10.3.2)

$$P(Y_i \leq (\mu_i - \tau)/\sigma_i, \quad t < i \leq t_0)$$

$$\leq P(W_i \leq (\mu_i - \tau)/\sigma_i, \quad t < i \leq t_0)$$

$$= \int_{-\infty}^{\infty} \prod_{t < i \leq t_0} P\left(Z_i \leq [(\mu_i - \tau)/\sigma_i - \sqrt{\rho}\, u]/[\sqrt{1 - \rho}]\right) d\Phi(u)$$

$$= \int_{-\infty}^{\infty} \prod_{t < i \leq t_0} \Phi\left([(\mu_i - \tau)/\sigma_i - \sqrt{\rho}\, u]/[\sqrt{1 - \rho}]\right) d\Phi(u)$$

$$=: P_W(t), \tag{12.6}$$

and since $E(Y_i Y_j) \geq 0$ for all $i \neq j$

$$P(Y_i \leq (\mu_i - \tau)/\sigma_i, \quad t < i \leq t_0)$$
$$\geq P(Z_i \leq (\mu_i - \tau)/\sigma_i, \quad t < i \leq t_0)$$
$$= \prod_{t < i \leq t_0} P(Z_i \leq (\mu_i - \tau)/\sigma_i)$$
$$= \prod_{t < i \leq t_0} \Phi((\mu_i - \tau)/\sigma_i)$$
$$= \tilde{P}(t).$$

This shows that the above asymptotic approximation $\tilde{P}(t)$ in (12.5) is a lower bound for $P(t)$, therefore the quantiles $\tilde{t}_p$ of $\tilde{P}(t)$ are upper bounds for the exact quantiles t_p, i.e., $\tilde{t}_p$ is a *conservative* estimator for t_p. $\tilde{t}_p$ is a simple estimator for t_p but it is not very accurate for all p. However, a conservative estimator like $\tilde{t}_p$ is preferred in most applications. The integral in (12.6) can be calculated numerically. It gives a lower bound $t_p(W)$ for t_p, i.e., a nonconservative estimator. For example, for $\tau = 0°C$ we get $t_{0.90}(W) = 126$, which is close to $\check{t} = 128$. Figure 12.4.7 shows the behavior of $P_W(t)$ for the three threshold temperatures.

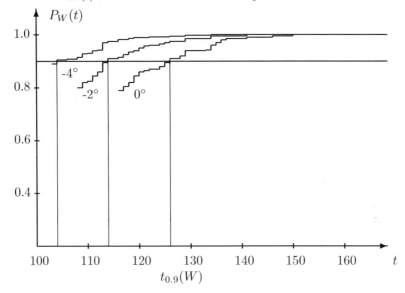

FIGURE 12.4.7. $P_W(t)$ for $\tau = -4°, -2°$ and $0°C$.

SIMULATION

If no theoretical tools are available for discussing the goodness of the approximation, we can simulate the proposed model. If the size of the simulated sample is sufficiently large, the simulation results are accurate. We simulated the Gaussian AR(1) random sequence $\{Y_i\}$ with $\rho = 0.73$ and obtained 5000 data points for T. An additional analysis of 100'000 simulated data points resulted in a difference of ± 1 day for the estimate of $t_{0.90}$. Figure 12.4.8 shows the behavior of the simulated distribution $P_S(t)$ of T for three threshold temperatures.

From $P_S(t)$ we can derive the simulated quantiles $t_p(S)$ for each threshold temperature. For example, for $\tau = 0°C$ we get $t_{0.90}(S) = 129$.

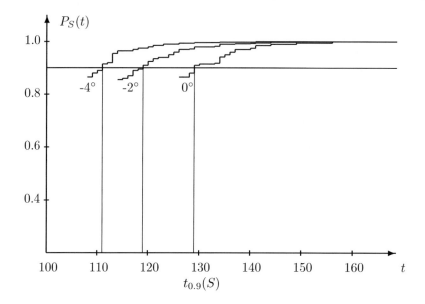

FIGURE 12.4.8. $P_S(t)$ for $\tau = -4°, -2°$ and $0°C$.

COMPARISON

The results obtained using different approaches for discussing the accuracy of (12.5) are summarized in Table 12.4.1.

Note the rather small differences between the estimates. This is certainly a consequence of the choice of the boundaries which are high, compared to the upper tail of the Gaussian distribution and of the fact that the autocorrelation $\rho = 0.73$, is not too large. The average of corresponding upper and lower bounds would give a very accurate estimate for the true quantile. Changes in the assumptions of the model, like smoothing the boundaries u_{ni} or choosing another value for ρ do not affect the estimates very much.

Further details about the analysis are given in Hüsler [238] and [239]. Similar applications of this method on data collected at other sites are given in Volz and Filliger [451]. They also compared this procedure with a nonparametric confidence interval estimation using data collected at a specific site for over 80 years and confirmed the rather good behavior of our estimators.

TABLE 12.4.1. Comparison of different estimators for $t_{0.90}$.

τ	$\tilde{t}_{0.90}$	$\check{t}_{0.90}$	$\hat{t}_{0.90}$	$t_{0.90}(W)$	$t_{0.90}(S)$
$-0°C$	134	128	134	126	129
$-1°C$	126	122	128	120	125
$-2°C$	121	114	122	114	119
$-3°C$	116	113	117	111	113
$-4°C$	113	102	113	104	111

$\tilde{t}_{0.90}$:   asymptotic approximation, independent X_i
$\hat{t}_{0.90}$:   Bonferroni inequality, upper bound for t_p
$\check{t}_{0.90}$:   Bonferroni inequality, lower bound for t_p
$t_{0.90}(W)$:   Slepian inequality, W_i a special Gaussian sequence
$t_{0.90}(S)$:   simulation of a Gaussian AR(1) sequence.

Author Index

M. Falk et al., *Laws of Small Numbers: Extremes and Rare Events*, 3rd ed.,
DOI 10.1007/978-3-0348-0009-9, © Springer Basel AG 2011

476 *Index*

Subject Index

Bibliography

[1] Abdous, B., Ghoudi, K., and Khoudraji, A. (1999). Non-parametric estimation of the limit dependence function of multivariate extremes. Extremes **2**, 245-268.

[2] Adler, R.J. (1990). An introduction to continuity, extrema and related topics for general Gaussian processes. Hayward, California, Inst. Math. Statistics.

[3] Albin, J.M.P., and Samorodnitsky, G. (2004). On overload in storage model with self-similar and infinitely divisible input. Ann. Appl. Probab. **14**, 820-844.

[4] Aldous, D. (1989). *Probability Approximations via the Poisson Clumping Heuristic*. Springer, New York.

[5] Al-Hussaini, A., and Elliott, R.J. (1984). Convergence of the empirical distribution to the Poisson process. Stochastics **13**, 299-308.

[6] Alpuim, M.T. (1988). High level exceedances in stationary sequences with extremal index. Stoch. Proc. Appl. **30**, 1-16.

[7] Alpuim, M.T. (1989) An extremal Markovian sequence. J. Appl. Probab. **26**, 219-232.

[8] Alpuim, M.T., and Athayde, E. (1990). On the stationary distribution of some extremal Markovian sequence. J. Appl. Probab. **27**, 291-302.

[9] Alpuim, M.T., Catkan, N., and Hüsler, J. (1993). Nonstationary max-autoregressive random sequences. Preprint.

[10] Alpuim, M.T., Catkan, N., and Hüsler, J. (1995). Extremes and clustering of nonstationary max-AR(1) sequences. Stoch. Process. Appl. **56**, 171-184.

[11] Alsina, C., Nelsen, R.B., and Schweizer, B. (1993). On the characterization of a class of binary operations on distribution functions. Statist. Probab. Letters **17**, 85-89.

[12] Amram, F. (1985). Multivariate extreme value distributions for stationary Gaussian sequences. J. Mult. Analysis **16**, 237-240.

[13] Andersen, P.K., Borgan, Ø., Gill, R.D., and Keiding, N. (1993). *Statistical Models based on Counting Processes*. Springer Series in Statistics, Springer, New York.

[14] Anderson, C.W. (1970). Extreme value theory for a class of discrete distributions, with applications to some stochastic processes. J. Appl. Prob. **7**, 99-113.

[15] Anderson, C.W. (1980). Local limit theorems for the maxima of discrete random variables. Math. Proc. Camb. Phil. Soc. **88**, 161-165.

[16] Anderson, C.W., Coles, S.G., and Hüsler, J. (1997). Maxima of Poisson-like variables and related triangular arrays. Annals Appl. Probab. **7**, 953-971.

[17] Arratia, R., Goldstein, L., and Gordon, L. (1990). Poisson approximation and the Chen-Stein method (with discussion). Statistical Science **5**, 403-434.

[18] Aulbach, S., Bayer, V., and Falk, M. (2009). A multivariate piecing-together approach with an application to operational loss data. Technical report, University of Würzburg.

[19] Azaïs, J.-M., Bardet, J.-M., and Wschebor, M. (2002). On the tails of the distribution of the maximum of a smooth stationary Gaussian process. ESAIM: Probab. Statist. **6**, 177-184.

[20] Azaïs, J.-M., and Wschebor, M. (2009). *Level sets and extreme of Random processes and fields*. Wiley, Hoboken, N.J., USA.

[21] Balkema, A.A., and Haan, L. de (1972). On R. von Mises' condition for the domain of attraction of $\exp(-e^{-x})$. Ann. Math. Statist. **43**, 1352-1354.

[22] Balkema, A.A., and Haan, L. de (1974). Residual life time at great age. Ann. Probab. **2**, 792-804.

[23] Balkema, A.A., Haan, L. de, and Karandikar, R.L. (1993). Asymptotic distribution of the maximum of n independent stochastic processes. J. Appl. Prob. **30**, 66-81.

[24] Balkema, A.A., and Resnick, S.I. (1977). Max-infinite divisibility. J. Appl. Prob. **14**, 309-319.

[25] Ballerini, R., and McCormick, W.P. (1989). Extreme value theory for processes with periodic variances. Comm. Statist.-Stoch. Models **5**, 45-61.

[26] Barbe, P., and McCormick, W.P. (2004). Second-order expansion for the maximum of a some stationary Gaussian sequences. Stoch. Proc. Appl. **110**, 315-342.

[27] Barbour, A.D., and Hall, P. (1984). On the rate of Poisson convergence. Math. Proc. Cambridge Philos. Soc. **95**, 473-480.

[28] Barbour, A.D., and Holst, L. (1989). Some applications of the Stein-Chen method for proving Poisson convergence. Adv. Appl. Probab. **21**, 74-90.

[29] Barbour, A.D., Holst, L., and Janson, S. (1992). *Poisson Approximation*. Oxford Studies in Probability, Oxford University Press.

[30] Barndorff-Nielsen, O., and Sobel, M. (1966). On the distribution of the number of admissible points in a random vector sample. Theor. Probab. Appl. **11**, 283-305.

[31] Barnett, V. (1976). The ordering of multivariate data. J. R. Statist. Soc. Ser. A **139**, 318-344.

[32] Beirlant, J., Goegebeur, Y., Segers, J., and Teugels, J. (2004). *Statistics of Extremes. Theory and Applications.* Wiley Series in Probability and Statistics, Wiley, Chichester.

[33] Berezovskiy, B.A., and Gnedin, A. (1984). *The Best Choice Problem.* Akad. Nauk, Moscow (in Russian).

[34] Beirlant, J. and Teugels, J. L. T. (1989). Asymptotic normality of Hill's estimator, In *Extreme Value Theory Proceedings* (Oberwolfach, 1987, Lecture Notes in Statistics), Vol.51, Springer, New York.

[35] Berman, S.M. (1961). Convergence to bivariate limiting extreme value distributions. Ann. Inst. Statist. Math. **13**, 217-223.

[36] Berman, S.M. (1964). Limit theorems for the maximum term in stationary sequences. Ann. Math. Statist. **33**, 502-516.

[37] Berman, S.M. (1971). Maxima and high level excursions of stationary Gaussian processes. Trans. Amer. Math. Soc. **160**, 65-85.

[38] Berman, S.M. (1974). Sojourns and extremes of Gaussian processes. Ann. Probab. **2**, 999-1026 (Correction (1980) **8**, 999).

[39] Berman, S.M. (1992). *Sojourns and Extremes of Stochastic Processes.* Wadsworth and Brooks, Pacific Grove.

[40] Bhattacharya, P.K. (1974). Convergence of sample paths of normalized sums of induced order statistics. Ann. Statist. **2**, 1034-1039.

[41] Bhattacharya, P.K., and Gangopadhyay, A.K. (1990). Kernel and nearest neighbor estimation of a conditional quantile. Ann. Statist. **18**, 1400-1415.

[42] Bhattacharya, P.K., and Mack, Y.P. (1987). Weak convergence of k-NN density and regression estimators with varying k and applications. Ann. Statist. **15**, 976-994.

[43] Bhattacharya, R.N., and Rao, R.R. (1976). *Normal Approximation and Asymptotic Expansion.* Wiley, New York.

[44] Bickel, P.J., Klaassen, C.A.J., Ritov, Y., and Wellner, J.A. (1993). *Efficient and Adaptive Estimation for Semiparametric Models.* Johns Hopkins University Press, Baltimore.

[45] Billingsley, P. (1971). *Weak Convergence of Measures: Applications in Probability.* SIAM Monograph No. 5, Philadelphia.

[46] Bingham, N.H., Goldie, C.M., and Teugels, J.L. (1987). *Regular Variation.* Cambridge University Press, Cambridge.

[47] Bischoff, W., Hashorva E., Hüsler, J., and Miller, F. (2003). Exact asymptotics for boundary crossings of the Brownian bridge with trend with application to the Kolmogorov test. Ann. Inst. Statist. Math. **55**, 849-864.

[48] Bischoff, W., Hashorva E., Hüsler, J., and Miller, F. (2004). On the power of the Kolmogorov test to detect the trend of a Brownian bridge with applications to a change-point problem in regression models. Statist. Probab. Letters **66**, 105-115.

[49] Bischoff, W., Hashorva E., Hüsler, J., and Miller, F. (2004). Analysis of a change-point regression problem in quality control by partial sums processes and Kolmogorov type tests. *Metrika* **62**, (2005) 85-98.

[50] Bischoff, W., Miller, F., Hashorva E., and Hüsler, J. (2003). Asymptotics of a boundary crossing probability of a Brownian bridge with general trend. Meth. Comp. Appl. Probab. **5**, 271-287.

[51] Bortkiewicz, L. von (1898). *Das Gesetz der kleinen Zahlen.* Teubner, Leipzig.

[52] Boulogne, P., Pierre-Loti-Viaud, D., and Piterbarg, V. (2008). On average losses in the ruin problem with fractional Brownian motion as input. Extremes **12**, 77-91.

[53] Bräker, H.U. (1995). High boundary excursions of locally stationary Gaussian processes. Proc. of the Conf. on Extreme Value Theory and Appl., Gaithersburg, (MA), 1993, Vol. 3, NIST special publ. **866**, 69-74.

[54] Bräker, H.U., and Hüsler, J. (1991). On the first zero of an empirical characteristic function. J. Appl. Probab. **28**, 593-601.

[55] Brown, B.M., and Resnick, S.I. (1977). Extreme values of independent stochastic processes. J. Appl. Prob. **14**, 732-739.

[56] Brozius, H. and Haan, L. de (1987). On limiting laws for the convex hull of a sample. J. Appl. Prob. **24**, 852-862.

[57] Bruss, F.T., and Rogers, L.C.G. (1991). Embedding optimal selection problems in a Poisson process. Stoch. Proc. Appl. **38**, 267-278.

[58] Buishand, T.A., Haan, L. de, and Zhou, C. (2006). On spatial extremes: with application to a rainfall problem. *Ann. Appl. Stat.* **2**, 624-642.

[59] Capéraà, P., Fougères, A.-L., and Genest, C. (1997). A nonparametric estimation procedure for bivariate extreme value copulas. Biometrika **84**, 567-577.

[60] Castillo, E. (1987). *Extreme Value Theory in Engineering.* Academic Press, New York.

[61] Castillo, E., Galambos, J., and Sarabia, J.M. (1989). The selection of the domain of attraction of an extreme value distribution from a set of data. In *Extreme Value Theory* (J. Hüsler and R.-D. Reiss, eds.), Lecture Notes in Statistics, Springer, New York, **51**, 181-190.

[62] Catkan, N. (1994). Aggregate excess measures and some properties of multivariate extremes. Ph.d. thesis, University of Bern.

[63] Chaudhuri, P. (1991). Nonparametric estimates of regression quantiles and their local Bahadur representation. Ann. Statist. **19**, 760-777.

[64] Chen, L.H.Y. (1975). Poisson approximation for dependent trials. Ann. Probab. **3**, 534-545.

[65] Cheng, K.F. (1984). Nonparametric estimation of regression function using linear combinations of sample quantile regression function. Sankhyā Ser. A, **46**, 287-302.

[66] Cheng, P.E. (1984). Strong consistency of nearest neighbor regression function estimators. J. Mult. Analysis **15**, 63-72.

[67] Chernick, M.R., Hsing, T., and McCormick, W.P. (1991). Calculating the extremal index for a class of stationary sequences. Adv. Appl. Probab. **23**, 835-850.

[68] Choe, J., and Shroff. N.B. (1999). On the supremum distribution of integrated stationary Gaussian proceeses with negative linear trend. Adv. Appl. Probab. **31**, 135-157.

[69] Christoph, G., and Falk, M. (1996). A note on domains of attraction of p-max stable laws. Statist. Probab. Letters **28**, 279-284.

[70] Coles, S.G. (1993). Regional modelling of extreme storms via max-stable processes. J. R. Statist. Soc. B **55**, 797-816.

[71] Coles, S.G. (2001). *An Introduction to Statistical Modeling of Extreme Values.* Springer Series in Statistics, Springer, New York.

[72] Coles, S.G., and Tawn, J.A. (1991). Modelling Extreme Multivariate Events. J. R. Statist. Soc. **53**, 377-392.

[73] Coles, S.G., and Tawn, J.A. (1994). Statistical methods for multivariate extremes: an application to structural design (with discussion). Appl. Statist. **43**, 1-48.

[74] Coles, S.G., Heffernan, J.E., and Tawn, J.A. (1999). Dependence measures for extreme value analyses. Extremes **2**, 339-365.

[75] Coles, S.G., and Pauli, F. (2001). Extremal limit laws for a class of bivariate Poisson vectors. Statist. Probab. Letters **54**, 373-379.

[76] Cormann, U., and Reiss, R.-D. (2009). Generalizing the Pareto to the log-Pareto model and statistical inference. Extremes **12**, 93-105.

[77] Cormann, U. (2010). Statistical models for exceedances with applications to finance and environmental statistics. PhD Thesis, Dept. Math., University of Siegen.

[78] Cover, T.M. (1968). Estimation by the nearest neighbor rule. IEEE Trans. Inform. Theory IT-**14**, 50-55.

[79] Cramér, H., and Leadbetter, M.R. (1967). *Stationary and Related Stochastic Processes.* Wiley, New York.

[80] Csiszár, L. (1963). Eine informationstheoretische Ungleichung und ihre Anwendung auf den Beweis der Ergodizität von Markoffschen Ketten. Publ. Math. Inst. Hungar. Acad. Sci. **8**, Ser.A, 85-198.

[81] Csörgő, M., and Horváth, L. (1988). Convergence of the empirical and quantile distributions to Poisson measures. Statist. Decisions **6**, 129-136.

[82] Csörgő, S. (1981). Limit behaviour of the empirical characteristic function. Ann. Probab. **9**, 130-144.

[83] Csörgő, S., and Heathcote, C.R. (1987). Testing for symmetry. Biometrika 74, 177-184.

[84] Csörgő, S., and Mason, D.M. (1985). Central limit theorems for sums of extreme values. *Mathematical Proceedings of the Cambridge Philosophical Society*, **98**, 547-558.

[85] Csörgő, S., and Mason, D.M. (1989). Simple estimators of the endpoint of a distribution. In *Extreme Value Theory* (J. Hüsler and R.-D. Reiss eds.), Lecture Notes in Statistics, Springer, New York, bf 51, 132-147.

[86] Cuculescu, I., and Theodorescu, R. (2001). Copulas: diagonals, tracks. Rev. Roumaine Math. Pures Appl. **46**, 731-742.

[87] Cuzick, J.M. (1981). Boundary crossing probabilities for stationary Gaussian processes and Brownian motion. Trans. Amer. Math. Soc. **263**, 469-492.

[88] Dacorogna, M.M., Gençay, R., Müller, U.A., and Pictet, O.V. (2001). Effective return, risk aversions and drawdowns. Physics A, **289**, 229-248.

[89] Daley, D.J., and Hall, P. (1984). Limit laws for the maximum of weighted and shifted iid random variables. Ann. Probab. **12**, 571-587.

[90] Daley, D.J., and Vere-Jones, D. (1988). *An Introduction to the Theory of Point Processes*. Springer Series in Statistics, Springer, New York.

[91] Davis, R.A. (1979 a). Maxima and minima of stationary sequences. Ann. Probab. **3**, 453-460.

[92] Davis, R.A. (1979 b). Limit laws for the maximum and minimum of stationary sequences. Z. Wahrsch. Verw. Geb. **61**, 34-42.

[93] Davis, R.A., and Resnick, S.I. (1984). Tail estimates motivated by extreme value theory. Ann. Statist. **12**, 1467-1487.

[94] Davison, A.C., and Smith, R.L. (1990). Models for exceedances over high thresholds (with discussions). J. Roy. Statist. Soc., Ser. B **52**, 393-442.

[95] Dębicki, K. (1999). A note on LDP for supremum of Gaussian processes over infinite horizon. Statist. Probab. Letters **44**, 211-219.

[96] Dębicki, K. (2001). Asymptotics of the supremum of scaled Brownian motion. Probab. Math. Statist. **21**, 199-212.

[97] Dębicki, K. (2002). Ruin probability for Gaussian integrated processes. Stoch. Proc. Appl. **98**, 151-174.

[98] Dębicki, K., and Kisowski, P. (2008). Asymptotics of supremum distribution of $\alpha(t)$-locally stationary Gaussian processes. Stoch. Proc. Appl. **118**, 2022-2037.

[99] Dębicki, K., Michna, Z., and Rolski, T. (1998). On the supremum from Gaussian processes over infinite horizon. Probab. Math. Statist. **18**, 83-100.

[100] Desgagné, A., and Angers, J.-F. (2005). Importance sampling with the generalized exponential power density. Statist. Comp. **15**, 189-195.

[101] Deheuvels, P. (1978). Caractèrisation complète des lois extrème multivariées et de la convergene des types extrèmes. *Publ. Inst. Statist. Univ. Paris* **23**, 1-36.

[102] Deheuvels, P. (1984). Probabilistic aspects of multiariate extremes. In *Statistical Extremes and Applications* (J. Tiago de Oliveira ed.), 117-130, D. Reidel Publishing Company.

[103] Deheuvels, P. (1991). On the limiting behavior of the Pickands estimator for bivariate extreme-value distributions. Statist. Probab. Letters **12**, 429-439.

[104] Deheuvels, P., and Pfeifer, D. (1988). Poisson approximation of multinomial distributions and point processes. J. Mult. Analysis **25**, 65-89.

[105] Deheuvels, P., and Martynov, G.V. (1996). Cramér-von Mises-type tests with applications to tests of independence for multivariate extreme-value distributions. Comm. Statist. - Theory Methods **25**, 871-908.

[106] Deheuvels, P., and Tiago de Oliveira, J. (1989). On the non-parametric estimation of the bivariate extreme-value distributions. Statist. Probab. Letters **8**, 315-323.

[107] Dekkers, A.L.M., Einmahl, J.H.J., and Haan, L. de (1989). A moment estimator for the index of an extreme-value distribution. Ann. Statist. **17**, 1833-1855.

[108] Dekkers, A.L.M., and Haan, L. de (1989). On the estimation of the extreme-value index and large quantile estimation. Ann. Statist. **17**, 1795-1832.

[109] Devroye, L. (1982). Necessary and sufficient conditions for the pairwise convergence of nearest neighbor regression function estimates. Z. Wahrsch. Verw. Geb. **61**, 467-481.

[110] Diebolt, J., El-Aroui, M.-A., Garrido, M., and Girard, S. (2003). Quasi-conjugate Bayes estimates for GPD parameters and application to heavy tails modelling. Technical Report.

[111] Diebolt, J., El-Aroui, M.-A., Garrido, M., and Girard, S. (2005). Quasi-conjugate Bayes estimates for GPD parameters and application to heavy tails modelling. Extremes **8**, 57-78.

[112] Dieker, A.B. (2004) Extremes of Gaussian processes over an infinite horizon. Stochastic Processes and their Applications **115**, 207-248.

[113] Dietrich, D., and Hüsler, J. (2001). Are low temperatures increasing with the global warming? In *Statistical Analysis of Extreme Values*, 2nd ed. R.-D. Reiss and M. Thomas, Birkhäuser, pp. 367-372.

[114] Drees, H. (1995). Refined Pickands estimator of the extreme value index. Ann. Statist. **23**, 2059-2080.

[115] Drees, H., and Huang, X. (1998). Best attainable rates of convergence for estimators of the stable tail dependence function. J. Mult. Analysis **64**, 25-47.

[116] Dupuis, D.J., and Tawn, J.A. (2001). Effects of mis-specification in bivariate extreme value problems. Extremes **4**, 315-330.

[117] Dwass, M. (1966). Extremal processes II. Illinois J. Math. **10**, 381-391.

[118] Eddy, W.F. (1985). Ordering of multivariate data. In *Computer Science and Statistics*. (Ed. L. Billard), pp. 25-30. North-Holland.

[119] Eddy, W.F., and Gale, J.D. (1981). The convex hull of a spherically symmetric sample. Adv. Appl. Prob. **13**, 751-763.

[120] Einmahl, J.H.J., Haan, L. de, and Huang, X. (1993). Estimating a multidimensional extreme-value distribution. J. Mult. Analysis **47**, 35-47.

[121] Einmahl, J.H.J., Haan, L. de, and Piterbarg, V. (2001). Nonparametric Estimation of the Spectral Measure of an Extreme Value Distribution. Ann. Stat **29**, 1401-1423.

[122] Embrechts, P., Klüppelberg, C., and Mikosch, T. (1997). *Modelling Extremal Events*. Applications of Mathematics, Springer, New York.

[123] Embrechts, P., McNeil, A., and Straumann, D. (2002). Correlation and dependency in risk management: properties and pitfalls. In *Risk Management: Value at Risk and Beyond* (M. Dempster ed.), Cambridge University Press, Cambridge, 176-223.

[124] Eubank, R.L. (1988). *Spline Smoothing and Nonparametric Regression*. Marcel Dekker, New York.

[125] Falk, M. (1985). Uniform convergence of extreme order statistics. Habilitationsschrift. Universität GH Siegen.

[126] Falk, M. (1986). Rates of uniform convergence of extreme order statistics. Ann. Inst. Statist. Math. **38**, 245-262.

[127] Falk, M. (1989). A note on uniform asymptotic normality of intermediate order statistics. Ann. Inst. Statist. Math. **41**, 19-29.

[128] Falk, M. (1989). Best attainable rate of joint convergence of extremes. In *Extreme Value Theory* (J. Hüsler and R.-D. Reiss eds.), Lecture Notes in Statistics, Springer, New York, 1-9.

[129] Falk, M. (1990). A note on generalized Pareto distributions and the k upper extremes. Probab. Th. Rel. Fields **85**, 499-503.

[130] Falk, M. (1993). Asymptotically optimal estimators of general regression functionals. J. Mult. Analysis **47**, 59-81.

[131] Falk, M. (1994 a). Extreme quantile estimation in δ-neighborhoods of generalized Pareto distributions. Statist. Probab. Letters, **20**, 9-21.

[132] Falk, M. (1994 b). Efficiency of convex combinations of Pickands estimator of the extreme value index. J. Nonparametric Statist. **4**, 133-147.

[133] Falk, M. (1995 a). LAN of extreme order statistics. Ann. Inst. Statist. Math. **47**, 693-717

[134] Falk, M. (1995 b). Some best parameter estimates for distributions with finite endpoint. Statistics **27**, 115-125.

[135] Falk, M. (1995 c). On testing the extreme valued index via the POT-method. Ann. Statist. **23**, 2013-2035.

[136] Falk, M. (1998). Local asymptotic normality of truncated empirical processes. Ann. Statist. **26**, 692-718.

[137] Falk, M. (2003). Domain of multivariate attraction via angular decomposition. Comm. Statist. - Theory Methods **32**, 2109-2116.

[138] Falk, M. (2008). It was 30 years ago today when Laurens de Haan went the multivariate way. Extremes **11**, 55-80.

[139] Falk, M., and Guillou, A. (2008). Peaks-over-threshold stability of multivariate generalized Pareto distributions. J. Multivar. Anal. **99**, 715-734.

[140] Falk, M., and Liese, F. (1998). LAN of thinned empirical processes with an application to fuzzy set density estimation. Extremes **1**, 323-349.

[141] Falk, M., and Marohn, F. (1992). Laws of small numbers: some applications to conditional curve estimation. In *Probability Theory and Applications* (J. Galambos and I. Kátai eds.). Kluwer Academic Publishers, Dordrecht, 257-278.

[142] Falk, M., and Marohn, F. (1993 a). Asymptotic optimal tests for conditional distributions. Ann. Statist. **21**, 45-60.

[143] Falk, M., and Marohn, F. (1993 b). Von Mises conditions revisited. Ann. Probab. **21**, 1310-1328.

[144] Falk, M., and Marohn, F. (2000). On the loss of information due to nonrandom truncation. J. Mult. Analysis **72**, 1-21.

[145] Falk, M., Marohn, F., and Tewes, B. (2002). *Foundations of Statistical Analyses and Applications with SAS*, Birkhäuser, Basel.

[146] Falk, M., and Michel, R. (2006). Testing for Tail Independence in Extreme Value Models. Ann. Inst. Statist. Math. **58**, 261-290.

[147] Falk, M., and Michel, R. (2009). Testing for a Multivariate Generalized Pareto Distribution. Extremes **12**, 33-51.

[148] Falk, M., and Reiss, R.-D. (1992a). Poisson approximation of empirical processes. Statist. Probab. Letters **14**, 39-48.

[149] Falk, M., and Reiss, R.-D. (1992b). Statistical inference of conditional curves: Poisson process approach. Ann. Statist. **20**, 779-796.

[150] Falk, M., and Reiss, R.-D. (1993). A Hellinger distance bound for the nearest neighbor approach in conditional curve estimation. Statist. Decisions, Supplement Issue No. 3, 55-68.

[151] Falk, M., and Reiss, R.-D. (2001). Efficiency of estimators of the dependence parameter in a class of bivariate Peaks-Over-Threshold models. Statist. Probab. Letters **52**, 233-242.

[152] Falk, M., and Reiss, R.-D. (2002). A characterization of the rate of convergence in bivariate extreme value models. Statist. Probab. Letters **59**, 341-351.

[153] Falk, M., and Reiss, R.-D. (2003 a). Efficient estimators and LAN in canonical bivariate POT models. J. Mult. Analysis **84**, 190-207;.

[154] Falk, M., and Reiss, R.-D. (2003 b). A characterization of the Marshall-Olkin dependence function and a goodness-of-fit test. Sankhyā, Series A, **65**, 807-820.

[155] Falk, M., and Reiss, R.-D. (2005). On Pickands coordinates in arbitrary dimensions. J. Mult. Analysis **92**, 426-453.

[156] Feller, W. (1971). *An Introduction to Probability Theory and Its Applications.* Vol II, 2nd. ed., John Wiley, New York.

[157] Fernique, X. (1964). Continuité des processus Gaussiens. Comptes Rendus **258**, 6058-6060.

[158] Finkenstädt, B., and Rootzén, H. (eds.) (2004). *Extreme Value in Finance, Telecommunications, and the Environment.* Chapman&Hall/CRC, Boca Raton.

[159] Fraga Alves, M.I., Haan, L. de, and Neves, C. (2009). Statistical Inference for heavy and super-heavy tailed distributions. J. Stat. Plan. Inf. **139**, 213-227.

[160] Frees, W.E., and Vadez, E.A. (1998). Understanding relationships using copulas. *North American Actuarial Journal* **2**, 1-25.

[161] Frick, M., Kaufmann, E., and Reiss, R.-D. (2007). Testing the tail-dependence based on the radial component. Extremes **10**, 109-128.

[162] Frick, M., and Reiss, R.-D. (2009). Expansions of multivariate Pickands densities and testing the tail dependence. J. Multivariate Anal. **100**, 1168-1181.

[163] Frick, M., and Reiss, R.-D. (2009). Limiting distributions of maxima under triangular schemes. Submitted.

[164] Fuller, W.A. (1976). *Introduction to Statistical Time Series.* John Wiley, New York.

[165] Gänssler, P., and Stute, W. (1987). *Seminar on Empirical Processes.* DMV Seminar Band 9. Birkhäuser, Basel.

[166] Galambos, J. (1978). *The Asymptotic Theory of Extreme Order Statistics.* 1st edition, Krieger, Malabar.

[167] Galambos, J. (1987). *The Asymptotic Theory of Extreme Order Statistics.* 2nd edition, Krieger, Malabar.

[168] Geffroy, J. (1958). Contribution à la théorie des valeurs extrêmes. Publ. l'Inst. Statist. l'Univ. Paris **7**, 37-121.

[169] Geffroy, J. (1959). Contribution à la théorie des valeurs extrêmes, II. Publ. l'Inst. Statist. l'Univ. Paris **8**, 3-65.

[170] Geluk, J. L. and de Haan, L. (1987). *Regular Variation, Extensions and Tauberian theorems*, CWI Tract 40, Center for Mathematics and Computer Science, Amsterdam, the Netherlands.

[171] Geluk, J. L. and de Haan, L. (2000). Stable probability distributions and their domains of attraction: a direct approach, *Probability and Mathematical Statistics*, 20, 169–188.

[172] Geluk, J.L., Haan, L. de, and de Vries, C.G. (2007). Weak and strong financial fragility. Tinbergen Institute Discussion Paper. TI 2007-023/2.

[173] Genest, C., Quesada-Molina, J.J., Rodríguez-Lallena, J.A., and Sempi, C. (1999). A characterization of quasi-copulas. J. Mult. Analysis **69**, 193-205.

[174] Gerber, H.U. (1984). Error bounds for the compound Poisson approximation. Insurance: Mathematics and Economics **3**, 191-194.

[175] Giné, E., Hahn, M.G., and Vatan, P. (1990). Max-infinitely divisible and max-stable sample continuous processes. Probab. Th. Rel. Fields **87**, 139-165.

[176] Gnedenko, B. (1943). Sur la distribution limite du terme maximum d'une série aléatoire. Ann. Math. **44**, 423-453.

[177] Goldstein, L., and Messer, K. (1992). Optimal plug-in estimators for non-parametric functional estimation. Ann. Statist. **20**, 1306-1328.

[178] Gomes, M.I. (1991). Statistical inference in an extremal Markovian Model, Preprint.

[179] Ghoudi, K., Khoudraji, A., and Rivest, L.P. (1998). Statistical properties of couples of bivariate extreme-value copulas. Canad. J. Statist. **26**, 187-197.

[180] Groeneboom, P. (1988). Limit laws for convex hulls. Probab. Th. Rel. Fields **79**, 327-368.

[181] Groeneboom, P., and Wellner, J.A. (1992). *Information Bounds and Non-parametric Maximum Likelihood Estimation.* DMV Seminar Band 19, Birkhäuser, Basel.

[182] Gumbel, E.J. (1958). *Statistics of Extremes.* Columbia University Press, New York.

[183] Gumbel, E.J. (1960). Bivariate exponential distributions. *J. Amer. Statist. Assoc.* **55** 698-707.

[184] Haan, L. de (1970). *On Regular Variation and its Application to the Weak Convergence of Sample Extremes.* Math. Centre Tracts 32, Amsterdam.

[185] Haan, L. de (1971). A form of regular variation and its application to the domain of attraction of the double exponential distriubtion. Z. Wahrsch. Verw. Geb. **17**, 241-258.

[186] Haan, L. de (1981). Estimation of the minimum of a function using order statistics. J. American Statist. Assoc. **76**, 467-469.

[187] Haan, L. de (1984). A spectral representation of max-stable processes. Ann. Probab. **12**, 1194-1204.

[188] Haan, L. de (1985). Extremes in higher dimensions: The model and some statistics. In *Proc. 45th Session ISI* (Amsterdam), 26.3.

[189] Haan, L. de (1988). Statistical work related to the height of sea-dikes. Lecture given at the 18th European Meeting of Statisticians, Berlin, GDR, August 22-26.

[190] Haan, L. de, and Ferreira, A. (2006). *Extreme Value Theory. (An Introduction).* Springer Series in Operations Research and Financial Engineering, Springer, New York.

[191] Haan, L. de, and Peng, L. (1997). Rates of convergence for bivariate extremes. J. Mult. Analysis **61**, 195-230.

[192] Haan, L. de, and Pickands, J. (1986). Stationary min-stable processes. Probab. Th. Rel. Fields **72**, 478-492.

[193] Haan, L. de, and Resnick, S.I. (1977). Limit theory for multivariate sample extremes. Z. Wahrsch. Verw. Gebiete **40**, 317-337.

[194] Haan, L. de, and Ronde, J. de (1998). Sea and wind: multivariate extremes at work. Extremes **1**, 7-45.

[195] Haight, F.A. (1967). *Handbook of the Poisson Distribution*. Wiley, New York.

[196] Hájek, J. (1970). A characterization of the limiting distributions of regular estimates. Z. Wahrsch. Verw. Geb. **12**, 21-55.

[197] Hall, A. (1996) Maximum term of a particular autoregressive sequence with discrete margins, Comm. Statist. - Theory and Methods **25**, 721-736.

[198] Hall, A., and Hüsler, J. (2006). Extremes of stationary sequences with failures. Stoch. Models **22**, 537-557.

[199] Hall, A., Scotto, M., and Ferreira, H. (2004) On the extremal behavior of generalised periodic sub-sampled moving average models with regularly varying tails. Extremes **7**, 149-160.

[200] Hall, A., and Scotto, M. (2003) Extremes of sub-sampled integer-valued moving average models with heavy-tailed innovations, Statistics and Probability Letters **63**, 97-105.

[201] Hall, P. (1982 a). On estimating the endpoint of a distribution. Ann. Statist. **10**, 556-568.

[202] Hall, P. (1982 b). On some simple estimates of an exponent of regular variation. J. Royal Statist. Soc. B **44**, 37-42.

[203] Hall, P., and Tajvidi, N. (2000). Distribution and dependence function estimation for bivariate extreme-value distributions. Bernoulli **6**, 835-844.

[204] Hall, P., and Welsh, A.H. (1985). Adaptive estimates of parameters of regular variation. Ann. Statist. **13**, 331-341.

[205] Härdle, W. (1990). *Applied Nonparametric Regression*. Cambridge University Press, Cambridge.

[206] Härdle, W., Janssen, P. and Serfling, R.J. (1988). Strong uniform consistency rates for estimators of conditional functionals. Ann. Statist. **16**, 1428-1449.

[207] Hashorva, E. (2002) Asymptotics of dominated Gaussian maxima. Extremes **5**, 353-368.

[208] Hashorva, E., and Hüsler, J. (1999). Extreme values in FGM random sequences. J. Mult. Analysis **68**, 212-225.

[209] Hashorva, E., and Hüsler, J. (2000). Extremes of Gaussian processes with maximal variance near the boundary points. Meth. Compt. Appl. Probab. **2**, 255-269.

[210] Hashorva, E., and Hüsler, J. (2002). Remarks on compound Poisson approximation of Gaussian random sequences. Statist. Probab. Letters **57**, 1-8.

[211] Hashorva, E., and Hüsler, J. (2003). On multivariate Gaussian tails. Ann. Inst. Statist. Math. **55**, 507-522.

[212] Häusler, E., and Teugels, J.L. (1985). On asymptotic normality of Hill's estimator for the exponent of regular variation. Ann. Statist. **13**, 743-756.

[213] Heathcote, C.R., and Hüsler, J. (1990). The first zero of an empirical characteristic function. Stoch. Processes Appl. 35, 347-360.

[214] Heathcote, C.R., and Welsh, A.H. (1983). The robust estimation of autoregressive processes by functional least squares. J. Appl. Probab. **20**, 737-753.

[215] Heffernan, J.E. (2000). A directory of coefficients of tail dependence. Extremes **3**, 279-290.

[216] Hendricks, W., and Koenker, R. (1992). Hierarchical spline models for conditional quantiles and the demand for electricity. J. Amer. Statist. Assoc. **87**, 58-68.

[217] Hill, B.M. (1975). A simple approach to inference about the tail of a distribution. Ann. Statist. **3**, 1163-1174.

[218] Holst, L., and Janson, S. (1990). Poisson approximation using the Stein-Chen method and coupling: number of exceedances of Gaussian random variables. Ann. Probab. **18**, 713-723.

[219] Höpfner, R. (1997). On tail parameter estimation in certain point process models. J. Statist. Plann. Inference **60**, 169-187.

[220] Höpfner, R., and Jacod, J. (1994). Some remarks on the joint estimation of the index and the scale parameter for stable processes. In *Asymptotic Statistics* (P. Mandl and M. Huskova eds.). Proceedings of the Fifth Prague Symposium 1993, Physica, Heidelberg, 273-284.

[221] Hofmann, D. (2006). *On the Representation of Multivariate Extreme Value Distributions via Norms.* Diploma-thesis, University of Würzburg (in German).

[222] Hofmann, D. (2009). *Characterization of the D-Norm Corresponding to a Multivariate Extreme Value Distribution.* PhD-thesis, University of Würzburg.

[223] Hogg, R.V. (1975). Estimates of percentile regression lines using salary data. J. Amer. Statist. Assoc. **70**, 56-59.

[224] Hosking, J.R.M. and Wallis, J.R. (1987). Parameter and quantile estimation for the generalized Pareto distribution. Technometrics, **29**, 339-349.

[225] Hsing, T. (1988). On extreme order statistics for a stationary sequence. Stoch. Proc. Appl. **29**, 155-169.

[226] Hsing, T. (1989). Extreme value theory for multivariate stationary sequences. J. Mult. Analysis **29**, 274-291.

[227] Hsing, T. (1991). Estimating the parameters of rare events. Stoch. Processes Appl. **31**, 117-139.

[228] Hsing, T. (1993 a). Extremal index estimation for a weakly dependent stationary sequence. Ann. Statist. **21**, 2043-2071.

[229] Hsing, T. (1993 b). On some estimates based on sample behaviour near high level excursions. Probab. Th. Rel. Fields **95**, 331-356.

[230] Hsing, T. (1997) A case study of ozone pollution with XTREMES. In Reiss, R.-D. and Thomas, M. *Statistical Analysis of Extreme Values*. Birkhäuser, Basel.

[231] Hsing, T., Hüsler, J., and Leadbetter, M.R. (1988). On the exceedance point process for a stationary sequence. Probab. Th. Rel. Fields **78**, 97-112.

[232] Hsing, T., Hüsler, J., and Reiss, R.D. (1996). On the extremes of a triangular array of Gaussian random variables. Ann. Appl. Probab. **6**, 671-686.

[233] Hsing, T., and Leadbetter, M.R. (1997). On multiple-level excursions by stationary processes with deterministic peaks. Stoch. Proc. Appl. **71**, 11-32.

[234] Huang, X. (1992). *Statistics of Bivariate Extreme Values*. PhD-thesis, Tinbergen Institute Research Series.

[235] Hüsler, J. (1979). The limiting behaviour of the last exit time for sequences of independent, identically distributed random variables. Z. Wahrsch. Verw. Geb. **50**, 159-164.

[236] Hüsler, J. (1980). Limit distribution of the last exit time for stationary random sequences. Z. Wahrsch. Verw. Geb. **52**, 301-308.

[237] Hüsler, J. (1983 a). Asymptotic approximations of crossing probabilities of random sequences. Z. Wahrsch. Verw. Geb. **63**, 257-270.

[238] Hüsler, J. (1983 b). Anwendung der Extremwerttheorie von nichtstationären Zufallsfolgen bei Frostdaten. Techn. Bericht 10, Inst. f. math. Statistik, Univ. Bern.

[239] Hüsler, J. (1983 c). Frost data: A case study on extreme values of nonstationary sequences. In *Statistical Extremes and Applications* (J. Tiago de Oliveira ed.), NATO ASI Series C, Vol. 131, Reidel Dordrecht, pp. 513-520.

[240] Hüsler, J. (1986 a). Extreme values of non-stationary random sequences. J. Appl. Probab. **23**, 937-950.

[241] Hüsler, J. (1986 b). Extreme values and rare events of non-stationary random sequences. In *Dependence in Probability and Statistics*, (E. Eberlein and M.S.Taqqu eds.). Birkhäuser, Boston.

[242] Hüsler, J. (1989 a). Limit properties for multivariate extreme values in sequences of independent, non-identically distributed random vectors. Stoch. Process. Appl. **31**, 105-116.

[243] Hüsler, J. (1989 b). Limit distributions of multivariate extreme values in nonstationary sequences of random vectors. In *Extreme Value Theory* (J. Hüsler and R.-D. Reiss eds.), Lecture Notes in Statistics **51**, Springer, New York.

[244] Hüsler, J. (1989 c). First zeros of empirical characteristic functions and extreme values of Gaussian processes. In *Statistical Data Analysis and Inference* (Y. Dodge, ed.) North Holland, Amsterdam, pp. 177-182.

[245] Hüsler, J. (1990 a). Multivariate extreme values in stationary random sequences. Stoch. Process. Appl. **35**, 99-108.

[246] Hüsler, J. (1990 b). Extreme values and high boundary crossings of locally stationary Gaussian processes. Ann. Probab. **18**, 1141-1158.

[247] Hüsler, J. (1993). A note on exceedances and rare events of non-stationary sequences. J. Appl. Probab. **30**, 877-888.

[248] Hüsler, J. (1994 a). Limit results for univariate and multivariate nonstationary sequences. In *Extreme Value Theory and Applications* (J. Galambos, J. Lechner and E. Simiu eds.), Vol. 1. Kluwer, Dordrecht, 283-304.

[249] Hüsler, J. (1995). A note on extreme values of locally stationary Gaussian Processes. J. Statist. Plann. Inference **45**, 203-213.

[250] Hüsler, J. (1997). Global warming in relation with clustering of extremes. In *Statistical analysis of extreme values*, R.-D. Reiss and M. Thomas, Birkhäuser, 257-264.

[251] Hüsler, J. (1999). Extremes of a Gaussian process and the constant H_α. Extremes **2**, 59-70.

[252] Hüsler, J. (1999). Extremes of Gaussian processes, on results of Piterbarg and Seleznjev. Statist. Probab. Letters **44**, 251-258.

[253] Hüsler, J. (2004). Dependence between extreme values of discrete and continuous time locally stationary Gaussian processes. Extremes **7**, 179-190.

[254] Hüsler, J. (2010). On Applications of Extreme Value Theory in Optimization. In *Experimental Methods for the Analysis of Optimization Algorithms.* (Th. Bartz-Beielstein, M. Ciarandini, L. Paquete and M. Preuss eds.), Springer, Berlin.

[255] Hüsler, J., Cruz, P., Hall, A., and de Fonseca, C. (2003). On Optimization and Extreme Value Theory. Meth. Comp. Appl. Probab. **5**, 183-195.

[256] Hüsler, J., and Kratz, M.F. (1995) Rate of Poisson approximation of the number of exceedances of nonstationary normal sequences. Stoch. Proc. Appl. **55**, 301-313.

[257] Hüsler, J., and Piterbarg, V. (1999). Extremes of a certain class of Gaussian processes. Stoch. Proc. Appl. **83**, 257-271.

[258] Hüsler, J., and Piterbarg, V. (2004). On the ruin probability for physical fractional Brownian motion. Stoch. Proc. Appl. **113**, 315-332.

[259] Hüsler, J., and Piterbarg, V. (2004). Limit theorem for maximum of the storage process with fractional Brownian motion as input. Techn. Report 2003-040, Eurandom. Stoch. Proc. Appl. **114**, 231-250.

[260] Hüsler, J., Piterbarg, V., and O. Seleznjev (2003). On convergence of the uniform norms for Gaussian processes and linear approximation problems. Ann. Appl. Probab. **13**, 1615-1653.

[261] Hüsler, J., and Piterbarg, V. (2008). A limit theorem for the time of ruin in a Gaussian ruin problem. Stoch. Proc. Appl. **118**, 2014-2022.

[262] Hüsler, J., Piterbarg, V., and Rumyantseva, E. (2010). Extremes of Gaussian processes with simple random variance. Submitted.

[263] Hüsler, J., Piterbarg, V., and Zhang, Y. (2010). Extremes of Gaussian processes with general random variance. In preparation.

[264] Hüsler, J., and Reiss, R.-D. (1989). Maxima of normal random vectors: between independence and complete dependence. Statist. Probab. Letters **7**, 283-286.

[265] Hüsler, J., and Schmidt, M. (1996). A note on point processes of rare events. J. Appl. Probab. **33**, 654-663.

[266] Hüsler, J., and Schüpbach, M. (1988). Limit results for maxima in non-stationary multivariate Gaussian sequences. Stoch. Processes Appl. **27**, 91-99.

[267] Hüsler, J., and Zhang, Y. (2008). On first and last ruin times of Gaussian processes. Stat. Probab. Letters **78**, 1230-1235.

[268] Ibragimov, I.A., and Has'minskii, R.Z. (1981). *Statistical Estimation.* Springer, New York.

[269] Jacod, J., and Shiryaev, A.N. (1987). *Limit Theorems for Stochastic Processes.* Springer, New York.

[270] Janssen, A., and Marohn, F. (1994). On statistical information of extreme order statistics, local extreme value alternatives, and Poisson point processes. J. Mult. Analysis **48**, 1-30.

[271] Janssen, A., and Kunz, F. (2004). Brownian type boundary crossing probabilities for piecewise linear boundary functions. Comm. Statist. - Theory Methods **33**, 1445 - 1464.

[272] Jiménez, J.R., Villa-Diharce, E., and Flores, M. (2001). Nonparametric estimation of the dependence function in bivariate extreme value distributions. J. Mult. Analysis **76**, 159-191.

[273] Joe, H. (1993). Parametric families of multivariate distributions with given marginals. J. Mult. Analysis **46**, 262-282.

[274] Joe, H. (1994). Multivariate extreme-value distributions with applications to environmental data. Canad. J. Statist. **22**, 47-64.

[275] Joe, H. (1997). *Multivariate Models and Dependence Concepts.* Chapman & Hall, London.

[276] Joe, H., Smith, R.L., and Weissman, I. (1992). Bivariate threshold method for extremes. J. R. Statist. Soc. B **54**, 171-183.

[277] Jones, M.C., and Hall, P. (1990). Mean squared error properties of kernel estimates of regression quantiles. Statist. Probab. Letters **10**, 283-289.

[278] Juncosa, M.L. (1949). The asymptotic behaviour of the minimum in a sequence of random variables. Duke Math. J. **16**, 609-618.

[279] Juri, A., and Wüthrich, M.V. (2004). Tail dependence from a distributional point of view. Extremes **6**, 213-246.

[280] Kabluchko, Z., Schlather, M., and Haan, L. de (2009). Stationary max-stable fields associated to negative definite functions, Ann. Probab. **37**, 2042-2065.

[281] Kabluchko, Z. (2009). Extremes of space-time Gaussian processes. Stoch. Proc. Appl. **119**, 3962-3980.

[282] Kallenberg, O. (1983). *Random Measures*, 2nd ed., Akademie-Verlag, Berlin.

[283] Kaufmann, E. (1992). *Contributions to Approximations in Extreme Value Theory*. PhD thesis, University of Siegen.

[284] Kaufmann, E., and Reiss, R.-D. (1992). On conditional distributions of nearest neighbors. J. Mult. Analysis **42**, 67-76.

[285] Kaufmann, E., and Reiss, R.-D. (1993). Strong convergence of multivariate point processes of exceedances. Ann. Inst. Statist. Math. **45**, 433-444.

[286] Kaufmann, E., and Reiss, R.-D. (1995). Approximation rates for multivariate exceedances. J. Statist. Plann. Inference **45**, 235-245.

[287] Kaufmann, E., and Reiss, R.-D. (2002). An upper bound on the binomial process approximation to the exceedance process. Extremes **5**, 253-269.

[288] Kimber, A. C. (1983). A note on Poisson maxima. Z. Wahrsch. Verw. Geb. **63**, 551-552.

[289] Koenker, R., and Bassett, G. (1978). Regression quantiles. Econometrica **46**, 33-50.

[290] Koenker, R., and D'Orey, V. (1987). Computing regression quantiles. J. Roy. Statist. Soc. Ser. C **36**, 383-393.

[291] Kolchin, V.F. (1969) The limiting behavior of extreme terms of a variational series in polynomial scheme. Theory Probab. Appl. **14**, 458-469.

[292] Konstant, G., and Piterbarg, V.I. (1993). Extreme values of cyclostationary Gaussian random processes. J. Appl. Probab. **30**, 82-97.

[293] Kotz, S., and Nadarajah, S. (2000). *Extreme Value Distributions. Theory and Applications*. Imperial College Press, London.

[294] Kratz, M.F., and León, J.R. (2006). On the second moment of the number of crossings by a stationary Gaussian process. Annals Probability **34**, 1601-1607.

[295] Kratz, M.F., and León, J.R. (2010). Level curves crossings and applications in Gaussian models. To be published in Extremes.

[296] Kratz, M.F., and Rootzén, H. (1997). On the rate of convergence for extremes of mean square differentiable stationary normal processes. J. Appl. Probab. **34**, 908-923.

[297] Kremer, E. (1986). Simple formulas for the premiums of the LCR and ECOMOR treaties under exponential claim sizes. Blätter der DGVM, Band **XVII**, 237-243.

[298] Kudrov, A.V., and Piterbarg, V.I. (2007). On maxima of partial samples in Gaussian sequences with pseudo-stationary trends. Lithuanian Math. J. **47**, 48-56.

[299] Lawless, J.F. (1982). *Statistical Models and Methods for Lifetime Data*. Wiley, New York.

[300] Leadbetter, M.R. (1983). Extremes and local dependence in stationary sequences. Z. Wahrsch. Verw. Geb. **65**, 291-306.

[301] Leadbetter, M.R. (1991). On a basis for 'Peaks over Threshold' modeling. Statist. Probab. Letters **12**, 357-362.

[302] Leadbetter, M.R., and Hsing, T. (1990). Limit theorems for strongly mixing stationary random measures. Stoch. Proc. Appl. **36**, 231-243.

[303] Leadbetter, M.R., Lindgren, G., and Rootzén, H. (1983). *Extremes and Related Properties of Random Sequences and Processes.* Springer Series in Statistics, Springer, New York.

[304] Leadbetter, M.R., and Nandagopalan, S. (1989). On exceedance point processes for stationary sequences under mild oscillation restrictions. In *Extreme Value Theory* (J. Hüsler and R.-D. Reiss eds.). Lecture Notes in Statistics **51**, Springer, New York, 69-80.

[305] Leadbetter, M.R., and Rootzén, H. (1988). Extremal theory for stochastic processes. Ann. Probab. **16**, 431-478.

[306] Leadbetter, M.R., and Rootzén, H. (1998). On extreme values in stationary random fields. In *Stochastic Processes and Related Topics*, Birkhäuser, Boston, MA, pp. 275-285.

[307] Leadbetter, M.R., Weissman, I., Haan, L. de, and Rootzén, H. (1989). On clustering of high values in statistically stationary series. In *Proc. 4th Int. Meet. Statistical Climatology* (J. Sanson ed), Wellington, New Zealand Meteorological Service, pp. 217-222.

[308] LeCam, L. (1986). *Asymptotic Methods in Statistical Decision Theory.* Springer Series in Statistics. Springer, New York.

[309] LeCam, L., and Yang, G.L. (1990). *Asymptotics in Statistics (Some Basic Concepts).* Springer Series in Statistics. Springer, New York.

[310] Ledford, A.W., and Tawn, J.A. (1996). Statistics for near independence in multivariate extreme values. Biometrika **83**, 169-187.

[311] Ledford, A.W., and Tawn, J.A. (1997). Modelling dependence within joint tail regions. J. R. Statist. Soc. **B 59**, 475-499.

[312] Ledford, A.W., and Tawn, J.A. (1998). Concomitant tail behaviour for extremes. Adv. Appl. Probab. **30**, 197-215.

[313] Liese, F., and Vajda, I. (1987). *Convex Statistical Distances.* Teubner-Texte zur Mathematik, Bd. 95. Teubner, Leipzig.

[314] Lindgren, G., Maré, J. de, and Rootzén, H. (1975). Weak convergence of high level crossings and maxima for one or more Gaussian processes. Ann. Probab. **3**, 961-978.

[315] Lu, J.-C., and Bhattacharyya, G.K. (1991). Inference procedures for bivariate exponential model of Gumbel. Statist. Probab. Letters **12**, 37-50.

[316] Mack, Y.P. (1981). Local properties of k-NN regression estimates. Siam J. Alg. Disc. Meth. **2**, 311-323.

[317] Manteiga, W.G. (1990). Asymptotic normality of generalized functional estimators dependent on covariables. J. Statist. Plann. Inference **24**, 377-390.

[318] Mardia, K.V. (1970). *Families of Bivariate Distributions.* Griffin, London.

[319] Mardia, K.V., Kent, J., and Bibby, J. (1979). *Multivariate Analysis.* Academic, London.

[320] Marohn, F. (1995). *Contributions to a Local Approach in Extreme Value Statistics.* Habilitation thesis, Katholische Universität Eichstätt.

[321] Marohn, F. (1999). Local asymptotic normality of truncation models. Statist. Decisions **17**, 237-253.

[322] Marron, J.S. (1988). Automatic smoothing parameter selection: A survey. Empirical Economics **13**, 187-208.

[323] Marshall, A.W., and Olkin, I. (1967). A multivariate exponential distribution. J. Amer. Statist. Assoc. **62**, 30-44.

[324] Martins, A.P., and Ferreira, H. (2004). The extremal index of sub-sampled processes. J. Statist. Plann. Infer. **124**, 145-152.

[325] McCormick, W.P. (1980). Weak convergence for the maxima of stationary Gaussian processes using random normalization. Annals of Probability **8**, 483–497.

[326] Meerschaert, M.M., and Scheffler, H.-P. (2006). Stochastic model for ultraslow diffusion. Stoch. Process. Appl. **116**, 1215-1235.

[327] Meijzler, D.G. (1950). On the limit distribution of the maximal term of a variational series. Dopovidi Akad. Nauk Ukrain. SSR **1**, 3-10.

[328] Mercadier, C. (2007). Numerical bounds for the distributions of the maxima of some one- and two-dimensional Gaussian processes. Adv. Appl. Probab. **38**, 149-170.

[329] Michel, R. (1987). An improved error bound for the compound Poisson approximation of a nearly homogeneous portfolio. Astin Bulletin **17**, 165-169.

[330] Michel, R. (2006). *Simulation and Estimation in Multivariate Generalized Pareto Models.* PhD-thesis, University of Würzburg. http://www.opus-bayern.de/uni-wuerzburg/volltexte/2006/1848/

[331] Michel, R. (2007). Simulation of certain multivariate generalized Pareto distributions. Extremes **10**, 83-107.

[332] Michel, R. (2008). Some notes on multivariate generalized Pareto distributions. J. Mult. Analysis **99**, 1288-1301.

[333] Michel, R. (2009). Parametric estimation procedures in multivariate generalized Pareto models. Scan. Jour. Stat. **36**, 60-75.

[334] Michel, R. (2009). Estimation of the angular density in bivariate generalized Pareto models. Statistics **43**, 187-202.

[335] Michna, Z. (1999). On tail probability and first passage time for fractional Brownian motion. Math. Oper. Research **49**, 335-354.

[336] Mises, R. von (1936). La distribution de la plus grande de n valeurs. Reprinted in Selected papers II. Amer. Math. Soc., Providence, R.I., 1954, 271-294.

[337] Mittal, Y., and Ylvisaker D. (1975). Limit distributions for the maxima of stationary Gaussian processes. Stoch. Proc. Appl. **3**, 1-18.

[338] Mladenović, P., and Piterbarg, V. (2006). On asymptotic distribution of maxima of complete and incomplete samples from stationary sequences. Stoch. Proc. Appl. **116**, 1977-1991.

[339] Mohan, N.R., and Ravi, S. (1993). Max domains of attraction of univariate and multivariate p-max stable laws. Theory of Probability and its Applications **37**, 632-643.

[340] Mohan, N.R. and Subramanya, U.R. (1991). Characterization of max domains of attraction of univariate p-max stable laws. *Proceedings of the Symposium on Distribution Theory*, Kochi, Kerala, India, 11-24.

[341] Molchanov, I. (2007). Convex geometry of max-stable distributions. http://www.citebase.org/abstract?id=oai:arXiv.org:math/0603423. Extremes **11**, (2008) 235-259.

[342] Müller, H.G. (1988). *Nonparametric Regression Analysis of Longitudinal Data.* Lecture Notes in Statistics, Springer, New York.

[343] Nadarajah, S. (2000). Approximation for bivariate extreme values. Extremes **3**, 87-98.

[344] Nadarajah, S., and Mitov, K. (2002). Asymptotics of maxima of discrete random variables. Extremes **5**, 287-294.

[345] Nadaraya, E.A. (1964). On estimating regression. Theor. Probab. Appl. **9**, 141-142.

[346] Nandagopalan, S. (1990). Multivariate extremes and estimation of the extremal index. PhD thesis, University of North Carolina, Chapel Hill.

[347] Nandagopalan, S., Leadbetter, M.R., and Hüsler, J. (1992). Limit theorems of multi-dimensional random measures. Techn. report.

[348] Narayan, O. (1998). Exact asymptotic queue length distribution for fractional Brownian traffic. Advances in Performance Analysis, **1**, 39-63.

[349] Nelsen, R.B. (1999). *An Introduction to Copulas.* Springer, New York.

[350] Nelsen, R.B. (2006). *An Introduction to Copulas.* 2nd. ed., Springer, New York.

[351] NERC (1975). *The Flood Studies Report.* The Natural Environmental Research Council, London.

[352] Neves, C., and Fraga Alves, M. I. (2008). The ratio of maximum to the sum for testing super heavy tails. In *Advances in Mathematical and Statistical Modeling* (B. Arnold, N. Balakrishnan, J. Sarabia and R. Mínguez eds). Birkhäuser, Boston.

[353] Norros, I. (1994). A storage model with self-similar input. Queueing Systems **16**, 387-396.

[354] Norros, I. (1997). Four approaches to the fractional Brownian storage. In J. Lévy Véhel, E. Lutton and C. Tricot, editors, *Fractals in Engineering*, Springer.

[355] Norros, I. (1999). Busy periods of fractional Brownian storage: a large deviations approach. Advances in Performance Analysis, **1**, 1-19.

[356] Nussbaum, M. (1985). Spline smoothing in regression models and asymptotic efficiency in L_2. Ann. Statist. **13**, 984-997.

[357] O'Brien, G. (1987). Extreme values for stationary and Markov sequences. Ann. Probab. **15**, 281-291.

[358] Omey, E., and Rachev, S.T. (1988). On the rate of convergence in extreme value theory. Theory Probab. Appl. **33**, 601-607.

[359] Omey, E., and Rachev, S.T. (1991). Rates of convergence in multivariate extreme value theory. J. Mult. Analysis **38**, 36-50.

[360] Pancheva, E.I. (1985). Limit theorems for extreme order statistics under nonlinear normalization. In *Stability Problems for Stochastic Models*, (V.V. Kalashnikov and V.M. Zolotarev eds.), Lecture Notes in Math. **1155**, 248-309, Springer, Berlin.

[361] Peng, Z., Wang, P., and Nadarajah, S. (2009). Limiting distributions and almost sure limit theorems for the normalized maxima of complete and incomplete samples from Gaussian sequence. Electronic Journal of Statistics **3**, 851-864.

[362] Perfekt, R. (1994). Extremal behaviour of stationary Markov chains with applications. Ann. Appl. Probab. **4**, 529-548.

[363] Perfekt, R. (1997). Extreme value theory for a class of Markov chains with values in $\mathbb{R}^d$. Adv. Appl. Probab. **29**, 138-164.

[364] Perreira, L., and Ferreira, H. (2006). Limiting crossing probabilites of random fields. J. Appl. Prob. **43**, 884-891.

[365] Pfanzagl, J. (1974). *Allgemeine Methodenlehre der Statistik II*. Walter de Gruyter, Berlin.

[366] Pfanzagl, J. (1990). *Estimation in Semiparametric Models (Some Recent Developments)*. Lecture Notes in Statistics **63**. Springer, Berlin-Heidelberg.

[367] Pfanzagl, J. (1994). *Parametric Statistical Theory*. De Gruyter, Berlin.

[368] Pfeifer, D. (1989). *Einführung in die Extremwertstatistik*. Teubner, Stuttgart.

[369] Pickands III., J. (1969). Upcrossing probabilities for stationary Gaussian processes. Trans. Amer. Math. Soc. **145**, 75-86.

[370] Pickands III, J. (1971). The two-dimensional Poisson process and extremal processes. J. Appl. Probab. **8**, 745-756.

[371] Pickands III., J. (1975). Statistical inference using extreme order statistics. Ann. Statist. **3**, 119-131.

[372] Pickands III, J. (1981). Multivariate extreme value distributions. Proc. 43th Session ISI (Buenos Aires), pp. 859-878.

[373] Pickands III., J. (1986). The continuous and differentiable domains of attractions of the extreme value distributions. Ann. Probab. **14**, 996-1004.

[374] Piterbarg, V.I. (1981) Comparison of distribution functions of maxima of Gaussian processes. Theory Probab. Appl. **26**, 687-705.

[375] Piterbarg, V. (1996). *Asymptotic Methods in the Theory of Gaussian Processes and Fields.* AMS, Providence.

[376] Piterbarg, V. (2001). Large deviations of a storage process with fractional Browinan motion as input. Extremes **4**, 147-164.

[377] Piterbarg, V. (2004). Discrete vs. continuous time for large extremes of Gaussian processes. Extremes **7**, 161-178.

[378] Piterbarg, V., and Romanova T.A. (2002). Numerical estimation of the Pickands' constant. Techn. Report, IMSV, Univ. Bern.

[379] Piterbarg V.I., and Weber M. (1997). Tail distribution for Gaussian Suprema-Standard methods. Technical Report No. 491, Center for Stochastic Processes, Chapel Hill, North Carolina.

[380] Qualls, C., and Watanabe, H. (1972). Asymptotic properties of Gaussian processes. Ann. Math. Statist. **43**, 580-596.

[381] Rachev, S.T. (1991). *Probability Metrics and the Stability of Stochastic Models.* Wiley, Chichester.

[382] Radtke, M. (1988). *Rates of Convergence and Asymptotic Expansions under von Mises Conditions.* PhD thesis, University of Siegen (in German).

[383] Reiss, R.-D. (1981). Uniform approximation to distributions of extreme order statistics. Adv. Appl. Probab. **13**, 533-547.

[384] Reiss, R.-D. (1987). Estimating the tail index of the claim size distribution. Blätter der DGVM, Band XVIII, 21-25.

[385] Reiss, R.-D. (1989). *Approximate Distributions of Order Statistics. (With Applications to Nonparametric Statistics).* Springer Series in Statistics, Springer, New York.

[386] Reiss, R.-D. (1990). Asymptotic independence of marginal point processes of exceedances. Statist. Decisions **8**, 153-165.

[387] Reiss, R.-D. (1993). *A Course on Point Processes.* Springer Series in Statistics, Springer, New York.

[388] Reiss, R.-D., and Thomas, M. (1997). *Statistical Analysis of Extreme Values* 1st ed., Birkhäuser, Basel.

[389] Reiss, R.-D. and Thomas, M. (2001). *Statistical Analysis of Extreme Values* 2nd ed., Birkhäuser, Basel.

[390] Reiss, R.-D., and Thomas, M. (2007). *Statistical Analysis of Extreme Values,* 3rd ed., Birkhäuser, Basel.

[391] Rényi, A., and Sulanke, R. (1963). Über die konvexe Hülle von n zufällig gewählten Punkten. Z. Wahrsch. Verw. Geb. **2**, 75-84.

[392] Resnick S.I. (1986). Point processes, regular variation and weak convergences. Adv. Appl. Probab. **18**, 66-138.

[393] Resnick, S.I. (1987). *Extreme Values, Regular Variation, and Point Processes.* Applied Prob. Vol. 4, Springer, New York.

[394] Rice, S.O. (1944). Mathematical analysis of random noise. Bell Syst. Tech. J. **23**, 282-332.

[395] Rice, S.O. (1945). Mathematical analysis of random noise. Bell Syst. Tech. J. **24**, 45-156.

[396] Roberts, A.W., and Varberg, D.E. (1973). *Convex Functions.* Academic Press, New York.

[397] Robinson, M.E., and Tawn, J.A. (2000) Extremal analysis of processes sampled at different frequencies. J. Roy. Statist. Soc. Ser. B **62**, 117-135.

[398] Rootzén, H. (1983). The rate of convergence of extremes of stationary random sequences. Adv. Appl. Probab. **15**, 54-80.

[399] Rootzén, H. (1986). Extreme value theory for moving average processes. Ann. Probab. **14**, 612-652.

[400] Rootzén, H. (1988). Maxima and exceedances of stationary Markov chains. Adv. Appl. Probab. **20**, 371-390.

[401] Rootzén, H., and Tajvidi, N. (2006). Multivariate generalized Pareto distributions. Bernoulli **12**, 917-930.

[402] Royall, R.M. (1966). A class of nonparametric estimators of a smooth regression function. PhD thesis, Dept. of Statistics, Stanford University.

[403] Rychlik, T. (1992). Weak limit theorems for stochastically largest order statistics. In *Order Statistics and Nonparametrics: Theory and Applications* (P.K. Sen and I.A. Salama eds.), North-Holland, Amsterdam, pp. 141-154.

[404] Samanta, M. (1989). Non-parametric estimation of conditional quantiles. Statist. Probab. Letters **7**, 407-412.

[405] Scotto, M.G., Turkman, K.F., and Anderson, C.W. (2003). Extremes of some sub-sampled time series. J. Time Ser. Anal. **24**, 579-590.

[406] Seleznjev, O. (1996) Large deviations in the piecewise linear approximation of Gaussian processes with stationary increments. Adv. Appl. Probab. **28**, 481-499.

[407] Seleznjev, O. (2006). Asymptotic behavior of mean uniform norms for sequences of Gaussian processes and fields. Extremes **8**, 161–169.

[408] Serfling, R.J. (1980). *Approximation Theorems of Mathematical Statistics.* Wiley, New York.

[409] Serfozo, R. (1980). High level exceedances of regenerative and semi-stationary processes. J. Appl. Probab. **17**, 432-431.

[410] Serfozo, R. (1988). Extreme values of birth and death processes and queues. Stoch. Proc. Appl. **27**, 291-306.

[411] Shao, Qi-M. (1996) Bounds and estimators of a basic constant in extreme value theory of Gaussian processes. Statistica Sinica **6**, 245-257.

[412] Sheskin, D.J. (2003). *Parametric and Nonparametric Statistical Procedures.* 3rd ed., Chapman&Hall/CRC, Boca Raton.

[413] Shi, D. (1995). Multivariate extreme value distribution and its Fisher information matrix. ACTA Math. Appl. Sinica **11**, 421-428.

[414] Shiryaev, A.N., and Spokoiny, V. (1999). *Statistical Experiments and Decisions: Asymptotic Theory.* World Scientific Publ., London, Singapore.

[415] Sibuya, M. (1960). Bivariate extreme statistics. Ann. Inst. Statist. Math. **11**, 195-210.

[416] Slepian, D. (1962). The one-sided barrier problem for Gaussian noise. Bell System Techn. J. **41**, 463-501.

[417] Smith, R.L. Statistics of extreme values. Proc. 45th Session I.S.I., Paper 26.1, Amsterdam.

[418] Smith, R.L. (1987). Estimating tails of probability distributions. Ann. Statist. **15**, 1174-1207.

[419] Smith, R.L. (1988). Extreme value theory for dependent sequences via the Stein-Chen method of Poisson aproximation. Stoch. Proc. Appl. **30**, 317-327.

[420] Smith, R.L. (1989). Extreme value analysis of environmental time series: an application to trend detection in ground-level ozone (with discussion). Statistical Science **4**, 367-393.

[421] Smith, R.L. (1990). Max-stable processes and spatial extremes. Preprint, Univ. North Carolina.

[422] Smith, R.L. (1992). The extremal index for a Markovian chain. J. Appl. Probab. **29**, 37-45.

[423] Smith, R.L., Tawn, J.A., and Yuen, H.K. (1990). Statistics of multivariate extremes. Int. Statist. Review **58**, 47-58.

[424] Stein, C. (1986). *Approximate Computation of Expectations.* Institute of Mathematical Statistics Lecture Notes - Monograph Series, vol. 7, Hayward, California.

[425] Stephenson, A. (2003). Simulating multivariate extreme value distributions of logistic type. Extremes **6**, 49-59.

[426] Stone, C.J. (1975). Nearest neighbor estimators of a nonlinear regression function. In Proc. Computer Sci. Statist. 8th Ann. Symp. Interface, pp. 413-418.

[427] Stone, C.J. (1977). Consistent nonparametric regression (with discussion). Ann. Statist. **5**, 549-645.

[428] Stone, C.J. (1980). Optimal rates of convergence for nonparametric estimators. Ann. Statist. **8**, 1348-1360.

[429] Stone, C.J. (1982). Optimal global rates of convergence for nonparametric regression. Ann. Statist. **10**, 1040-1053.

[430] Strasser, H. (1985). *Mathematical Theory of Statistics*. De Gruyter Studies in Math. **7**. De Gruyter, Berlin.

[431] Stute, W. (1982). Asymptotic normality of nearest neighbor regression function estimates. Ann. Statist. **10**, 917-926.

[432] Stute, W. (1986). On almost sure convergence of conditional empirical distribution functions. Ann. Probab. **14**, 891-901.

[433] Subramanya, U.R. (1994). On max domains of attraction of univariate p-max stable laws. Statist. Probab. Letters **19**, 271-279.

[434] Sweeting, T.J. (1985). On domains of uniform local attraction in extreme value theory. Ann. Probab. **13**, 196-205.

[435] Sweeting, T.J. (1989). Recent results on asymptotic expansions in extreme value theory. In *Extreme Value Theory* (J. Hüsler and R.-D. Reiss eds.), Lecture Notes in Statistics, Springer, New York, 10-20.

[436] Tajvidi, N. (1996). Characterization and some statistical aspects of univariate and multivariate generalised Pareto distributions. PhD Thesis, Dept. Math., University of Göteborg.

[437] Takahashi, R. (1987). Some properties of multivariate extreme value distributions and multivariate tail equivalence. Ann. Inst. Statist. Math. **39**, Part A, 637-647.

[438] Takahashi, R. (1988). Characterizations of a multivariate extreme value distribution. Adv. Appl. Prob. **20**, 235-236.

[439] Taleb, N.N. (2007). *The Black Swan: the Impact of the Highly Improbable*. Random House.

[440] Tawn, J. (1990). Modelling multivariate extreme value distributions. Biometrika **77**, 245-253.

[441] Teugels, J.L. (1981). Remarks on large claims. Bull. Inst. Internat. Statist. **49**, 1490-1500.

[442] Teugels, J.L. (1984). Extreme values in insurance mathematics. In *Statistical Extremes and Applications*. NATO ASI Series, Reidel, Dordrecht.

[443] Tiago de Oliveira, J. (1962/63). Structure theory of bivariate extremes: extensions. Estudos de Math. Estat. Econom. **7**, 165-195.

[444] Tiago de Oliveira, J. (1989 a). Intrinsic estimation of the dependence structure for bivariate extremes. Statist. Probab. Letters **8**, 213-218.

[445] Tiago de Oliveira, J. (1989 b). Statistical decisions for bivariate extremes. In *Extreme Value Theory*. (J. Hüsler and R.-D. Reiss eds.), Lect. Notes Statistics **51**, pp. 246-259, Springer, New York.

[446] Todorovic, P., and Zelenhasic, E. (1970). A stochastic model for flood analysis. Water Resources Research **6**, 1641-1648.

[447] Truong, Y.K. (1989). Asymptotic properties of kernel estimators based on local medians. Ann. Statist. **17**, 606-617.

[448] Turkman, K.F., and Oliveira, M.F. (1992). Limit laws for the maxima of chain dependent sequences with positive extremal index. J. Appl. Probab. **29**, 222-227.

[449] Van der Vaart, A.W. (1998). *Asymptotic Statistics*. Cambridge University Press.

[450] Villaseñor-Alva, J.A., González-Estrada, E., and Reiss, R.-D. (2006). A nonparametric goodness of fit test for the generalized Pareto distribution. Unpublished manuscript.

[451] Volz, R., and Filliger, P. (1982). Ein Wahrscheinlichkeitsmodell zur Bestimmung des letzten Spätfrosttermins. Geophysik, Beiheft zum Jahrbuch der SNG, 21-25.

[452] Vatan, P. (1985). Max-infinite divisibility and max-stability in infinite dimensions. In *Probability in Banach Spaces V*, Lect. Notes Mathematics 1153, 400-425, Springer, New York.

[453] Wang, L., and Pötzelberger, K. (1997). Boundary crossing probability for Brownian motion and general boundaries. J. Appl. Probab. **34**, 54-65.

[454] Watson, G.S. (1964). Smooth regression analysis. Sankhyā, Ser. A, **26**, 359-372.

[455] Wei, X. (1992). Asymptotically efficient estimation of the index of regular variation. PhD thesis, University of Michigan.

[456] Weinstein, S.B. (1973). Theory and applications of some classical and generalized asymptotic distributions of extreme values. IEEE Trans. Inf. Theory **19**, 148-154.

[457] Weiss, L. (1971). Asymptotic inference about a density function at an end of its range. Nav. Res. Logist. Quart. **18**, 111-114.

[458] Weissman, I. (1978). Estimation of parameters and large quantiles based on the k largest observations. J. Amer. Statist. Assoc. **73**, 812-815.

[459] Weissman, I., and Cohen, U. (1995). The extremal index and clustering of high values for derived stationary sequences. J. Appl. Prob. **32**, 972-981.

[460] Welsh, A.H. (1984). A note on scale estimates based on the empirical characteristic function and their application to test for normality. Statist. Probab. Letters **2**, 345-348.

[461] Welsh, A.H., and Nicholls, D.F. (1986). Robust estimation of regression models with dependent regressors: the functional least squares approach. Econom. Theory **2**, 132-150.

[462] Witte, H.-J. (1993). *Extremal Points, Extremal Processes, Greatest Convex Minorants, Martingales and Point Processes*. Habilitationsschrift, University of Oldenburg.

[463] Witting, H. (1985). *Mathematische Statistik I*. Teubner, Stuttgart.

[464] Worms, R. (1998). Propriété asymptotique des excès additifs et valeurs extrêmes: le cas de la loi de Gumbel. C. R. Acad. Sci. Paris, t. 327, Série **I**, 509-514.

[465] Wu, W.B. (2002). Central limit theorems for functionals of linear processes and their applications. Statist. Sinica **12**, 635–649.

[466] Wu, W.B. (2003). Additive functionals of infinite-variance moving averages. Statist. Sinica **13**, 1259–1267.

[467] Wüthrich, M.V. (2004). Bivariate extension of the Pickands-Balkema-de Haan theorem. Annales de L'Institut Henry Poincaré-Probabilités & Statistiques **40**, 33-41.

[468] Zadeh, L.A. (1965). *Fuzzy Sets. Information and Control, Vol. 8.* Academic Press, New York, 338-353.

[469] Zaliapin, I.V., Kagan, Y.Y., and Schoenberg, F.P. (2005). Approximating the Distribution of Pareto Sums. Pure Appl. Geophys., **162**, 1187-1228.

[470] Zeevi, A., and Glynn, P.W. (2004). Recurrence properties of autoregressive processes with super-heavy-tailed innovation. J. Appl. Probab. **41**, 639-653.

[471] Zhang, D., Wells, M.T., and Peng, L. (2008). Nonparametric estimation of the dependence function for a multivariate extreme value distribution. J. Mult. Analysis **99**, 577-588.

[472] Zhang, Y. (2009). Limit theorems for missing values and ruins. Ph.d. thesis, University of Bern.

[473] Zhao, Z., and Wu, W.B. (2007). Asymptotic theory for curve-crossing analysis. Stochastic Processes and their Applications **117**, 862–877.

[474] Zimmermann, H.-J. (1996). *Fuzzy Set Theory - And its Applications.* 3rd ed., Kluwer Academic Publishers, Dordrecht.

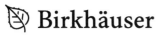

Foundations of Statistical Analyses and Applications with SAS

Falk, Michael, University of Würzburg, Germany / **Marohn, Frank,** Catholic University of Eichstätt-Ingolstadt, Germany / **Tewes, Bernward**, Catholic University of Eichstätt-Ingolstadt, Germany

2002. X, 402 p. Softcover
ISBN 978-3-7643-6893-7

Commonly there is no natural place in a traditional curriculum for mathematics or statistics, where a bridge between theory and practice fits into. On the other hand, the demand for an education designed to supplement theoretical training by practial experience has been rapidly increasing.

There exists, consequently, a bit of a dichotomy between theoretical and applied statistics, and this book tries to straddle that gap. It links up the theory of a selection of statistical procedures used in general practice with their application to real world data sets using the statistical software package SAS (Statistical Analysis System). These applications are intended to illustrate the theory and to provide, simultaneously, the ability to use the knowledge effectively and readily in execution.

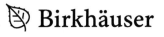

Birkhäuser

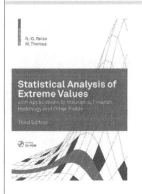

Statistical Analysis of Extreme Values

with Applications to Insurance, Finance, Hydrology and Other Fields

Reiss, Rolf-Dieter / Thomas, Michael
University of Siegen, Germany

3rd ed., 2007
XVII, 511 p.
With CD-ROM. Softcover
ISBN 978-3-7643-7230-9

The statistical analysis of extreme data is important for various disciplines, including hydrology, insurance, finance, engineering and environmental sciences. This book provides a self-contained introduction to the parametric modeling, exploratory analysis and statistical interference for extreme values.

The entire text of this third edition has been thoroughly updated and rearranged to meet the new requirements. Additional sections and chapters, elaborated on more than 100 pages, are particularly concerned with topics like dependencies, the conditional analysis and the multivariate modeling of extreme data. Parts I–III about the basic extreme value methodology remain unchanged to some larger extent, yet notable are, e.g., the new sections about "An Overview of Reduced-Bias Estimation" (co-authored by M.I. Gomes), "The Spectral Decomposition Methodology", and "About Tail Independence" (co-authored by M. Frick), and the new chapter about "Extreme Value Statistics of Dependent Random Variables" (co-authored by H. Drees). Other new topics, e.g., a chapter about "Environmental Sciences", (co-authored by R.W. Katz), are collected within Parts IV–VI.